THEMATIC CARTOGRAPHY AND VISUALIZATION

TERRY A. SLOCUM

Prentice Hall
Upper Saddle River, New Jersey 07458

Library of Congress Cataloging-in-Publication Data
Slocum, Terry A.
Thematic cartography and visualization/Terry A. Slocum.
p. cm.
Includes bibliographical references and index.
ISBN 0-13-209776-1
1. Cartography. 2. Visualization. I. Title.
GA108.7.S58 1999
526—dc21 98-42704
CIP

Acquisitions Editor: Daniel Kaveney
Executive Managing Editor: Kathleen Schiaparelli
Assistant Managing Editor: Lisa Kinne
Art Director: Jayne Conte
Cover Designer: Bruce Kenselaar
Manufacturing Manager: Trudy Pisciotti
Production Supervision/Composition: York Production Services
Map/Cover Art: Jennifer Smith-Kastens
All artwork is courtesy of Terry A. Slocum unless otherwise indicated.

Simon & Schuster/A Viacom Company
Upper Saddle River, New Jersey 07458

Appendices C–I can be found on the *Thematic Cartography and Visualization* Web page. Please point your browser to www.prenhall.com/slocum to access this information.

About the Cover
The front cover illustrates four possible color schemes for a map depicting the percentage of the population having Dutch ancestry in counties in Michigan. The upper left is an unordered hue-based scheme because it does not follow the progression of colors in the electromagnetic spectrum (red, orange, yellow, green, blue, indigo, and violet). The lower left is a lightness-based scheme with varying values of a blue hue. The upper right is a combined hue and lightness scheme in which the colors extend from yellow to green to blue. The lower right is a diverging scheme where varying values of orange and purple diverge from a neutral point in the middle. An important function of this book is to present various forms of symbology—color is one of several variables that is examined (Maps courtesy of Jennifer Smith-Kastens).

Printed in the United States of America
10 9 8 7 6 5 4 3 2 1

ISBN 0-13-209776-1

Prentice-Hall International (UK) Limited, *London*
Prentice-Hall of Australia Pty. Limited, *Sydney*
Prentice-Hall Canada Inc., *Toronto*
Prentice-Hall Hispanoamericana, S.A., *Mexico*
Prentice-Hall of India Private Limited, *New Delhi*
Prentice-Hall of Japan, Inc., *Tokyo*
Simon & Schuster Asia Pte. Ltd., *Singapore*
Editora Prentice-Hall do Brasil, Ltda., *Rio de Janeiro*

In memory of
George F. Jenks

Contents

Preface ix

Chapter 1 Introduction 1

1.1 What Is a Thematic Map? *2*
1.2 How Are Thematic Maps Used? *3*
1.3 Basic Steps for Communicating Map Information *3*
1.4 Consequences of Technological Change in Cartography *6*
1.5 Visualization and Cartography *11*
1.6 Cognitive Issues in Cartography *13*
1.7 Postmodernism in Cartography *14*
1.8 Scope of This Book *15*

Chapter 2 Symbolizing Spatial Data: Terminology and General Guidelines 18

2.1 Spatial Arrangement of Geographic Phenomena *19*
2.2 Levels of Measurement *20*
2.3 Visual Variables *22*
2.4 Comparison of Choropleth, Proportional Symbol, Isopleth, and Dot Mapping *25*
2.5 Selecting Visual Variables for Choropleth Maps *29*
2.6 Data Standardization *32*
2.7 Some Basic Map Design Elements *33*

Chapter 3 Statistical and Graphical Foundation 40

3.1 Population and Sample *41*
3.2 Descriptive versus Inferential Statistics *41*
3.3 General Methods for Analyzing Spatial Data, Ignoring Location *41*
3.4 Numerical Summaries in Which Location Is an Integral Component *52*

Chapter 4 Data Classification 60

4.1 Common Methods of Data Classification *61*
4.2 Classed versus Unclassed Mapping *75*
4.3 Using Spatial Context to Simplify Choropleth Maps *78*

Chapter 5 Principles of Color 83

5.1 How Color Is Processed by the Human Visual System *84*
5.2 Hardware Considerations in Producing Color Maps *88*
5.3 Models for Specifying Color *95*
5.4 Methods for Disseminating Maps to Users *101*

Chapter 6 Color Schemes For Univariate Choropleth and Isarithmic Maps 105

6.1 Map Use Tasks *106*
6.2 Kind of Data *106*
6.3 Type of Map *107*
6.4 Other Factors in Selecting Color Schemes *108*
6.5 Details of Color Specification *111*

Chapter 7 Proportional Symbol Mapping 118

7.1 Selecting Appropriate Data *119*
7.2 Kinds of Proportional Symbols *121*
7.3 Scaling Proportional Symbols *121*
7.4 Legend Design *130*
7.5 Handling Symbol Overlap *131*
7.6 Redundant Symbols *133*

Chapter 8 Interpolation Methods for Smooth Continuous Phenomena 136

8.1 Triangulation *137*
8.2 Inverse Distance *139*
8.3 Kriging *141*
8.4 Criteria for Selecting an Interpolation Method for True Point Data *145*
8.5 Limitations of Automated Interpolation Approaches *147*
8.6 Tobler's Pycnophylactic Approach: an Interpolation Method for Conceptual Point Data *148*

Chapter 9 Symbolizing Smooth Continuous Phenomena 153

9.1 Methods for Symbolizing Topography or Nontopographic Phenomena *154*
9.2 Methods for Symbolizing Topography *156*

Chapter 10 Dot and Dasymetric Mapping 168

10.1 Types of Ancillary Variables *169*
10.2 Manual versus Automated Production *169*
10.3 Creating a Dot Map *169*
10.4 Mapping Population Density: Langford and Unwin's Approach *173*

Chapter 11 Developments in Univariate Mapping Methods 177

11.1 Framed-Rectangle Symbols *178*
11.2 Modeling Geographic Phenomena and the Chorodot Map *179*
11.3 Dorling's Cartograms *181*
11.4 Novel Methods for Flow Mapping *184*
11.5 Mapping True 3-D Phenomena *187*

Chapter 12 Bivariate and Multivariate Mapping 193

12.1 Bivariate Mapping *195*
12.2 Multivariate Mapping *201*

Chapter 13 Data Exploration 210

13.1 Moellering's 3-D Mapping Software *211*
13.2 ExploreMap *211*
13.3 Project Argus *213*
13.4 MapTime *214*
13.5 Vis5D *215*
13.6 Aspens *217*
13.7 Transform *218*
13.8 ArcView *219*
13.9 Other Exploration Software *220*

Chapter 14 Map Animation 222

14.1 Graphic Scripts *222*
14.2 The Fundamental Work of DiBase and Colleagues *223*
14.3 Examples of Animations *225*

Chapter 15 Electronic Atlases and Tools for Developing Your Own Software 231

15.1 Defining Electronic Atlases *232*
15.2 Examples of Electronic Atlases *233*
15.3 Tools for Developing Your Own Software *237*

Chapter 16 Recent Developments 241

16.1 Depicting Data Quality *242*
16.2 Using Sound to Represent Spatial Data *248*
16.3 Recent Research on Color *249*
16.4 Virtual Reality and Visual Realism *251*
16.5 *Computers and Geosciences* Special Electronic Issue *253*
16.6 Keeping Pace with Recent Developments *253*

Appendix A CMYK Specifications for Color Choropleth Maps 257

Appendix B Using the CIE L*u*v* Uniform Color Space to Create Equally Spaced Colors 262

Glossary 263

References 274

Index 288

Preface

This text is a blend of the old and the new. By old, I refer to traditional univariate maps: choropleth, proportional symbol, isarithmic, and dot. Such maps have normally been discussed in the context of a communication model whose purpose is to impart spatial concepts or ideas to readers via a single "best" map. Although some have criticized the use of communication models, I feel they are useful when the intent is to design static stand-alone maps, as is often the case in printed newspapers, magazines, and academic journals. Moreover, communications models are useful for encouraging the mapmaker to select a symbology that will optimize the relationship between a real-world phenomenon and the mapped representation of that phenomenon.

By new, I refer to recently developed cartographic techniques that do not presume a single "best" map, and that are normally not discussed in association with communication models. Examples include data exploration, animation, and electronic atlases. It has become common to discuss such methods under the rubric of *visualization.* More generally, it has become common to consider any recently-developed novel cartographic technique as an approach for *visualizing* data, whether it be a static or dynamic map. In this context, this text considers a number of novel approaches for visualizing static maps.

In addition to considering visualization, there are several other important features of this text. One feature is that I contrast four common univariate methods (Chapter 2), pointing out that the choice of symbology depends not only on what is being mapped, but also on the message the mapmaker wishes to convey. At this point, I stress the issue of data standardization, which is often ignored by mapmakers.

A second feature is a focus on computer-based symbolization. Chapter 8 covers the use of digital methods for interpolating between point locations on isarithmic maps, while Chapter 10 stresses the role that the digital methods of geographic information systems (GIS) and remote sensing can play in creating dot and dasymetric maps. A consideration of digital methods is critical, as most maps are now produced by computer.

A third feature is a focus on the use of color in cartography. Principles of color related to visual processing, map production, specification, and dissemination are introduced in Chapter 5, while Chapter 6 addresses the use of color on univariate choropleth and isarithmic maps. Later chapters also mention color frequently, as indicated by the approximately 60 color illustrations included in this book.

A fourth feature is the inclusion of a separate chapter on bivariate and multivariate mapping. Traditionally, books on thematic cartography have focused on univariate mapping, but one of the corollaries of the trend toward visualization is a greater need to analyze multivariate data. Thus, students need to develop a basic foundation in bivariate and multivariate mapping.

I have included a chapter on graphical and statistical principles. Although statistical concepts are normally taught in a separate course, it is my experience that students forget many of the basics and often need a review. This chapter also covers graphical techniques not usually taught in a statistics course, such as the scatterplot matrix.

I have chosen not to include a chapter specifically entitled Choropleth Mapping. The fundamental concept of choropleth mapping is introduced in Chapter 2. Related concepts of data classification and selection of color schemes are covered in Chapters 4 and 6, respectively.

I mention particular software throughout the text. To assist readers in locating such software, Appendices C to I contain sources for thematic mapping and graphic design software, data exploration software, map animations, electronic atlases, tools for developing your own software, and geographic information systems. Appendices C to I can be found on the Internet at http://www.prenhall.com/slocum.

One concern I had in writing the text was deciding how to incorporate the field of geographic information systems. I considered developing a separate chapter dealing with cartographic issues most closely allied with GIS, but the issues did not mesh well with one another, nor did there seem to be any limit to the number of such issues if GIS was broadly defined. Therefore, I mention GIS where I think the linkage between the text and GIS is most obvious; notable examples include the use of GIS to assist in dot and dasymetric mapping (Chapter 10) and the importance of data exploration for GIS software (Chapter 13).

This book could be used in either a one- or a two-semester course on thematic cartography. In a two-semester course, Chapters 1 to 10 would be appropriate for the first semester (the focus would be on static univariate mapping), with the remaining chapters (11 to 16) used during the following semester (the focus would be on visualization and bivariate and multivariate mapping). For a one-semester course, I suggest the following chapters: 1, 2, 4, 6, 7, 10, 13, and 14. Wherever possible, I have tried to write the chapters as self-contained units, so the instructor should be able to select topics as desired.

Ideally, those reading this book would have had an introductory cartography course, but I have written it so that it can be understood by those who have had no prior training in cartography. Basic terms normally used in an introductory class are defined throughout the text, and a glossary is included at the end of the text.

ACKNOWLEDGMENTS

This book would not have been possible without the assistance of many people. First, I would like to thank Ray Henderson and Dan Kaveney, the acquisitions editors at Prentice Hall. Ray provided much encouragement when I was initially considering the project, and, when Ray moved on to other opportunities, Dan filled in admirably and patiently answered my many questions.

Cartographic Service at the University of Kansas assisted me in creating the more than 250 illustrations for the book. In her role as director of Cartographic Service, Barbara Shortridge provided sound design advice and much encouragement. More recently, Darin Grauberger took over as director of the lab and provided invaluable assistance in getting the illustrations to the final printing stage. Graphic designers in the lab who assisted included Keith Shaw, Jennifer Smith Kastens, and John Banning. To all of you, I say thanks for being so patient with all of the changes that I requested. The effort in creating illustrations was eased by the digital files provided by Ian Bishop, Cynthia Brewer, Keith Clarke, David DiBiase, Jason Dykes, Dynamic Graphics, William Hibbard, David Howard, Kansas Applied Remote Sensing Program, John Krygier, Mark Lindberg, Alan MacEachren, Mark Monmonier, Gail Thelin, Lloyd Treinish, and Stephen Yoder. Assistance also was provided by Greg Carbone, Dietrich Kastens, Fritz Kessler, Jon Kimerling, and Hal Moellering.

Numerous people assisted in editing the text and illustrations. At the University of Kansas, Barbara Shortridge graciously edited all of my initial drafts (sometimes a painstaking exercise!). An early draft of several chapters was reviewed by Sona Andrews, Douglas Banting, Nick Chrisman, Jeremy Crampton, and John Krygier. A more complete draft was reviewed by Nick Chrisman, Keith Clarke, Jeremy Crampton, Elisabeth Nelson, and Eugenie Rovai. I am especially thankful to Nick and Keith for the detailed editing they provided. Gretchen Miller and the staff at York Production Services did an admirable job of copyediting the manuscript.

I also would like to thank two professors from whom I learned much about cartography: Michael Dobson and George Jenks. Michael nurtured my interest in cartography at the bachelor's and master's levels and encouraged me to pursue further training. George Jenks taught me much about cartography but also about ephernistaphors, frog hairs, and other interesting things. To George, I dedicate this book.

Finally, I would like to thank those who provided moral and spiritual support. First, my wife, Arlene, and my children, Diane, Kevin, and Danny, who were patient with the evening and weekend writing and always had confidence that eventually I would finish. My Tae Kwon Do instructor, Master Ki-June Park, taught me a lot about the martial arts and provided great workouts that helped me repeatedly return to the book with a fresh mind.

Terry A. Slocum

1

Introduction

OVERVIEW

This book is about thematic mapping and the associated expanding area of visualization. A ***thematic map*** *(or* ***statistical map****) is used to display the spatial pattern of a particular theme or attribute. A familiar example is the temperature map shown in daily newspapers; the theme in this case is the anticipated high temperature for the day. The notion of a thematic map is described in detail in section 1.1, and contrasted with the* ***general-reference map,*** *which focuses on geographic location (for example, a topographic map may show the location of rivers). In section 1.2 the different uses for thematic maps are described: to provide* specific information *about particular locations, to provide* general information *about spatial patterns, and to compare patterns on two or more maps.*

An important function of this book is to assist mapmakers in selecting appropriate techniques for representing spatial data. For example, imagine that you wish to map the forest cleared for agriculture in each country during the preceding year, and you have been told that the number of acres of forest cleared is available by country on the Internet. You wonder whether such data will be sufficient and how the data should be symbolized. Cartographers have traditionally approached this problem by using a ***communication model,*** *a set of idealized steps for creating a map. These steps are introduced in section 1.3 and are as follows: (1) consider what the real-world distribution of the phenomenon might look like, (2) determine the purpose for making the map, (3) collect data appropriate for the map purpose, (4) construct the map, and (5) determine whether users find the map useful and informative. Although some have criticized the appropriateness of such steps, they are helpful in avoiding design blunders that can result from using the most readily available data and software.*

Like many disciplines, the field of cartography has undergone major technological changes in recent years. As recently as the 1970s, the bulk of maps were still produced by manual and photomechanical methods, but today this is no longer true, as students in cartography classes and employees of private firms and government agencies are using computer technology to produce the majority of their maps. Furthermore, it is apparent that in the near future virtually all maps will be computer-generated. Section 1.4 considers some of the consequences of this technological change, including (1) the ability of virtually anyone to create maps using personal computers, (2) new mapping methods, such as ***animation,*** *(3) the ability to explore geographic data in an interactive graphics environment, (4) the ability to link maps, text, pictures, video, and sound in* ***multimedia*** *presentations, (5) access to maps and related information via the Internet, and (6) the proliferation of color maps. Section 1.4 also considers the impact of technological change on* ***geographic information systems (GIS), remote sensing,*** *and* ***scientific visualization.***

The growth of scientific visualization has led to questions concerning the role of visualization in cartography. In this context, section 1.5 considers two definitions for visualization, a broad one that encompasses virtually all maps, and a narrower one that contrasts visualization with communication. In the narrower definition, which we will focus on, ***visualization*** *is considered a private activity in which unknowns are revealed in a highly interactive environment.* ***Communication*** *involves the opposite: a public activity in which knowns are presented in a noninteractive environment. When defined in this manner, we will see that the word* *"exploration" can be substituted for "visualization." Although not part of a formal definition, cartographers frequently have placed any recently developed, novel method for displaying data (such as animation and electronic atlases) under the rubric of visualization.*

While technological advances have had a major impact on cartography, the discipline has also experienced changes in its philosophical outlook. Section 1.6 deals with the increasing role that ***cognition*** *now plays in cartography. Traditionally, cartographers approached mapping with a behaviorist view, in which the human mind was treated like a black box. The trend today is toward a cognitive view, in which cartographers hope to find* why *symbols work effectively. Section 1.7 deals with the postmodernist notion that maps can be viewed from a variety of perspectives, with the potential for hidden agendas and meanings, some of which can be uncovered through map deconstruction.*

The chapter concludes with an overview of the remainder of the book. Roughly speaking, the book can be broken into two major parts. The first (Chapters 2 to 10) focuses on basic principles and traditional univariate thematic maps. The second (Chapters 11 to 16) covers recently developed methods (novel univariate maps, ***data exploration,*** *animation, and* ***electronic atlases****), and more complex techniques (****multivariate mapping****).*

1.1 WHAT IS A THEMATIC MAP?

Cartographers commonly distinguish between two types of maps: general reference and thematic. **General-reference maps** are used to emphasize the location of spatial phenomena. Topographic maps, such as those produced by the United States Geological Survey (USGS), are examples of general reference maps. On topographic maps readers can determine the location of streams, roads, houses, and many other natural and cultural features.

Thematic maps (or **statistical maps**) are used to emphasize the spatial distribution of one or more geographic attributes or variables. A familiar thematic map is the **choropleth map,** in which enumeration units (or data collection units) are shaded to represent different magnitudes of a variable (Color Plate 1.1). Although choropleth maps are frequently used, a variety of thematic maps are available, including proportional symbol, isarithmic, dot, and flow maps (Figure 1.1). A major purpose of this book is to introduce these and other types of thematic maps.

Although cartographers commonly distinguish between general reference and thematic maps, it should be recognized that they do so largely for the convenience of categorizing maps. The general reference map can be

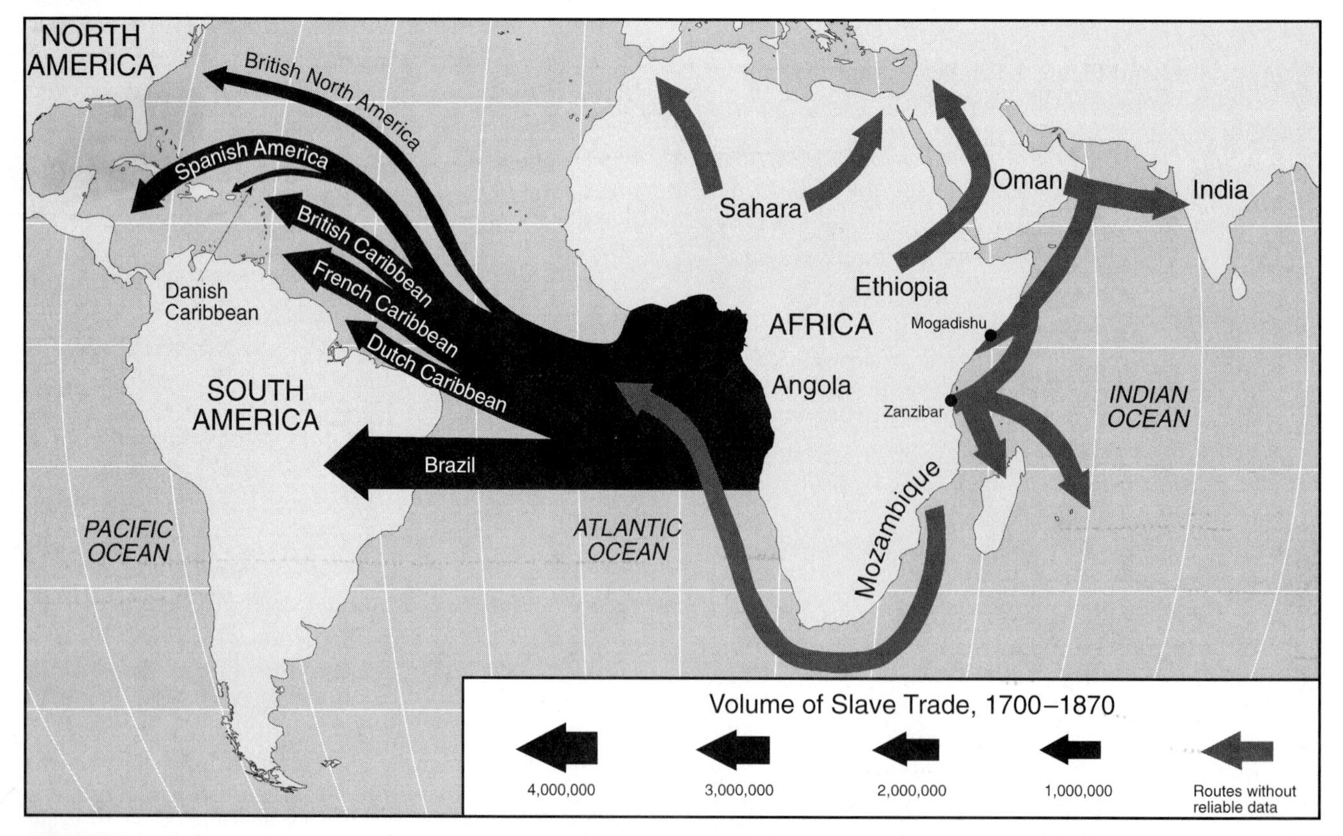

FIGURE 1.1 A flow map: an example of a thematic map. (HUMAN GEOGRAPHY: CULTURE, CONNECTIONS AND LANDSCAPES by Bergman, Edward, © 1995. Adapted by permission of Prentice-Hall, Inc., Upper Saddle River, NJ.)

viewed as a thematic map in which multiple variables are displayed simultaneously; thus it can be termed a multivariate thematic map. Furthermore, although the major emphasis of general-reference maps is on *location* of spatial phenomena, they can also portray the *distribution* of a particular variable (for example, the distribution of forest areas on a USGS topographic sheet).

1.2 HOW ARE THEMATIC MAPS USED?

Cartographers generally have recognized that thematic maps are used in three ways: to provide specific information about particular locations, to provide general information about spatial patterns, and to compare patterns on two or more maps. As an example of specific information, map A of Color Plate 1.1 indicates that between 8.8 and 12.0 percent of the people in Louisiana voted for Perot in the 1992 U.S. presidential election. As another example, Figure 1.1 indicates that approximately 2 million slaves were transported from Africa to Spanish America between 1700 and 1870.

In contrast to specific information, obtaining general information requires an overall analysis of the map. For example, map B of Color Plate 1.1 illustrates that a low percentage of people voted for Perot in the southeastern part of the United States, while a higher percentage voted for him in the central and northwestern states; and Figure 1.1 indicates that the bulk of the slave trade between 1700 and 1870 occurred outside North America.

A pitfall for naive mapmakers is that they often place inordinate emphasis on specific information. Map A of Color Plate 1.1 is illustrative of this problem. Here one can discriminate the classes based on strikingly different colors and thus determine which class each state belongs in (as we did for Louisiana), but it is difficult to acquire general information (such as which part of the country the Perot vote was concentrated in). In map B the reverse is the case: determining class membership is more difficult because the map is all brown, but the overall spatial pattern of voting is readily apparent.

As an illustration of pattern comparison, consider the **dot maps** of corn and wheat shown in Figure 1.2. Note that the patterns on these two maps are quite different. The distribution for corn is concentrated in the traditional corn belt region of the Midwest, while that for wheat is focused in the Great Plains, with the Palouse region of eastern Washington also having a notable concentration. Traditionally, the comparison of patterns such as these was limited by their fixed placement on pages of paper atlases. Fortunately, interactive graphics now provide the option of looking at selected distributions simultaneously, a topic that will be considered more fully in Chapter 15.

Two further issues are important when considering the ways in which thematic maps are used. First, one should distinguish between **information acquisition** and **memory for mapped information.*** The above discussion focused on information acquisition, or acquiring information while the map is being used. We could also consider memory for map information and how that memory is integrated with other spatial information (obtained through either maps or field work). For example, a cultural geographer might note that houses in a particular area are predominately built of limestone. Recalling a geologic map of bedrock, the geographer might mentally correlate the distribution of limestone in the bedrock with the pattern of limestone houses.

A second issue is that terms other than *specific* and *general* can be found in the literature. I have used these terms (developed by MacEachren 1982b) because they appear most frequently. Other researchers have developed a more complex set of terms. For example, Robertson (1991, 61) distinguished among three kinds of information: values at a point, local distributions characterized by "gradients and features," and the global distribution characterized by "trends and structure"; he argued that these levels corresponded closely with Bertin's (1981) elementary, intermediate, and superior levels. Additional terminology can be found in Olson (1976a), Board (1984), and Muehrcke and Muehrcke (1992).

1.3 BASIC STEPS FOR COMMUNICATING MAP INFORMATION

Traditionally, cartography has been taught within the framework of **map communication models** (for example, Dent 1996, 12–14; Robinson et al. 1984, 15–16), in which it is presumed that a cartographer wishes to communicate information to a group of map readers. Although communication models recently have received criticism (for example, MacEachren 1995, 3–11), their use can often lead to better-designed maps.

The map communication model that we will use is shown as a set of five idealized steps in Figure 1.3. Let's examine these steps by assuming that you wish to map the distribution of total population in the United States from the last decennial census. As a mapmaker, your tasks would be as follows:

Step 1. Consider what the real-world distribution of the phenomenon might look like. One way to implement this

* Technically, memory for mapped information would be equivalent to what psychologists term "long-term memory," but for simplicity the word "memory" is normally used.

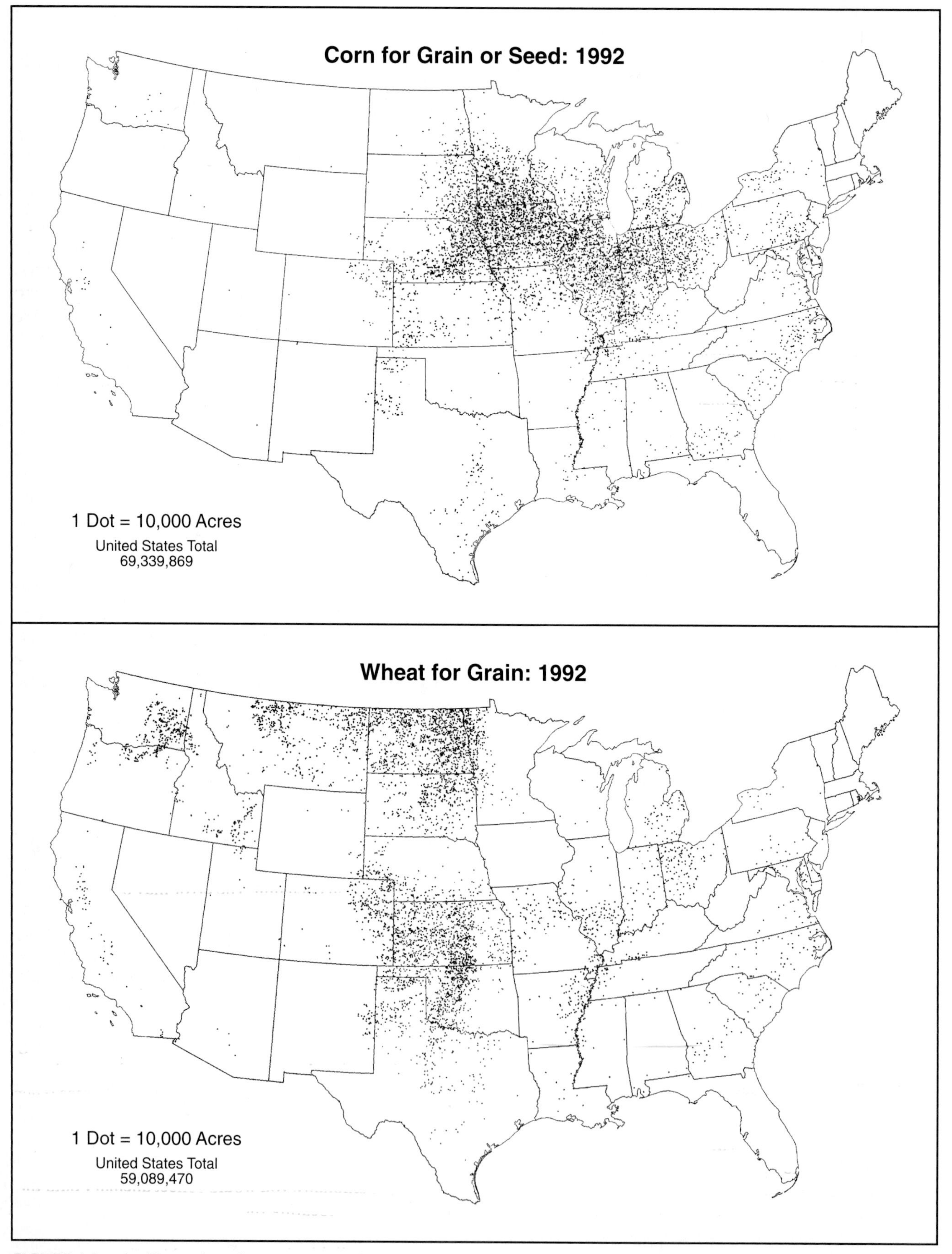

FIGURE 1.2 An illustration of pattern comparison, one of the fundamental ways in which thematic maps are used. (From U.S. Bureau of the Census 1995.)

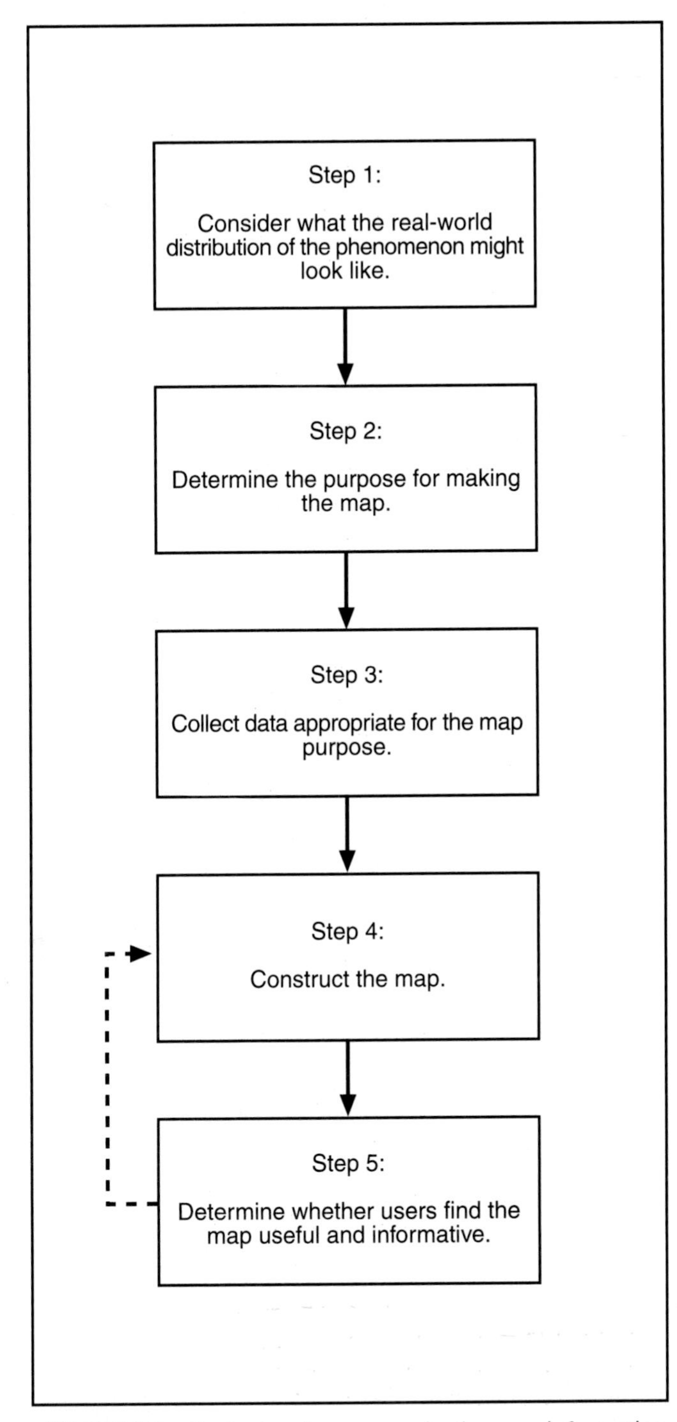

FIGURE 1.3 Basic steps for communicating map information to others. See text for discussion.

step is to ask yourself, "What would the distribution of the phenomenon look like if I were to view it while traveling across the landscape?" In the case of population, you might know (based on your travels or knowledge as a geographer) that a large percentage of people were concentrated in major cities and that such cities were much more densely populated than rural areas.

It is, of course, unrealistic to presume a single "objective" real world. In the case of population, "correct" population values are unknown for several reasons. The U.S. Bureau of the Census is never able to make an exact count of population; after every census, city officials dispute the census figures for their city. Also, census figures do not necessarily count the homeless or illegal aliens (the latter would account for a significant percentage of the population in some areas of the United States, such as California). Another problem is that population obviously varies locationally during the day and throughout the year; people commute to work, travel to the beach on weekends, and take vacations far away from home.

In spite of these problems, it is useful to think of a "real-world" distribution. Such an approach forces the mapmaker to think about the distribution at its most detailed level, and then decide what degree of complexity meets the purpose of the map. This first step is critical. Mapmakers too often display data at the level of a convenient political unit (county, state, or nation) because data are available for that level, rather than considering the purpose of the map.

Step 2. Determine the purpose for making the map. One purpose would be to attempt to match the real world as closely as possible (within the constraints of the map scale used). In the case of population, this would mean that you might want to distinguish clearly between urban and rural population.

Another purpose might be to map the distribution at the county level; such views are often sought by government officials for political reasons. From the viewpoint of the mapmaker, it is important to realize that this approach can introduce error into the resulting map because each enumeration unit (county in this case) is represented by a single value, and does not portray the variation within each unit. The error may be unimportant if the focus is on how one county in general compares to another county, but it can be a serious problem if readers infer more from the map than was intended; for example, readers might erroneously assume that the population density is uniform across a county on a choropleth map.

Step 3. Collect data appropriate for the map purpose. In general, spatial data can be collected from primary sources (such as field studies) or secondary sources (such as census data). For something close to the real-world view of population, you would likely consult the U.S. Census of Population for information on urban and rural population; additionally, you would collect ancillary data that could assist in locating the population data within rural areas. For a county-level view of population, the Census figures for individual counties would suffice.

Step 4. Construct the map. This step is a very

complex one that involves assessing the following criteria:

1. How the map will be used. Will it be used to portray specific or general information?
2. The spatial dimension of the data. Are the data point, linear, or areal?
3. The level at which the data are measured—(nominal, ordinal, interval, ratio)
4. Whether data standardization is necessary. If the data are in count form, do they need to be adjusted?
5. How many variables are to be mapped.
6. Whether there is a temporal component to the data.
7. Any technical limitations the cartographer has. For example, a journal may not be willing to reproduce maps in color.
8. The characteristics of the intended audience. Is the map intended for the general public or professional geographers?
9. Time and cost constraints. For example, creating a high-quality dot map will cost more than a choropleth map, regardless of the technical capabilities available.
10. Aesthetics. Some symbols are more visually appealing than others.

A full consideration of these criteria will occupy the rest of this book. The following, however, are two maps that could result from efforts to construct a population map of the United States: a combined proportional symbol–dot map (Color Plate 1.2) for the real-world view, and a choropleth map for the county-level view (map A in Figure 1.4). The combined proportional symbol–dot map is particularly illustrative of how one can attempt to match the real world. Note that the overall population is split into urban and rural, and that the urban population is further subdivided into "urbanized areas" and "places outside urbanized areas."

Step 5. Determine whether users find the map useful and informative. Possibly the most important point to keep in mind is that you are designing the map for others, not for yourself. You may find a particular color scheme pleasing, for example, but you should ask yourself whether users will also find it attractive. More importantly, will they understand the logic of it? Ideally, you should answer such questions by getting feedback about the map from potential users. Admittedly, time and cost constraints may make this task difficult, but it is desirable because you may discover not only whether the symbol scheme used was a logical one, but also the nature of concepts that users acquire from the map. Moreover, if you plan on employing the map to illustrate a *particular* concept (as for a class lecture), then you would want to know whether users acquired this *same* concept from using the map.

If the analysis in step 5 reveals that the map is not useful and informative, then the map may have to be redesigned. This possibility is shown as a dashed line in Figure 1.3. It is conceivable that you might also have to return to an earlier step, but it is more likely that you will have to modify some design aspect, such as choosing a different color scheme.

Unfortunately, naive mapmakers are unlikely to follow the five steps I have outlined. Instead, their decisions are frequently based on readily available data and mapping software. As an example, imagine that for a term paper a student wished to map the distribution of population we have been discussing. Rather than considering steps 1 and 2 of the model, the student might simply use state totals either because fewer numbers would have to be entered into the computer or because the data were readily available (as at an Internet site).

Furthermore, in step 4 the student might choose a choropleth map (map B of Figure 1.4) because software for creating such a map is readily available. Presuming that the student had collected data in raw count form, the choropleth map would be a poor choice because it requires standardized data (as we will see in Chapter 2). Cartographers would argue that a proportional symbol map (map C of Figure 1.4) would be a better choice if raw counts are to be mapped.

1.4 CONSEQUENCES OF TECHNOLOGICAL CHANGE IN CARTOGRAPHY

Over the last 35 years, the field of cartography has undergone major technological change, evolving from a discipline based on pen and ink to one based on computer technology. One consequence of technological change is that map production is no longer the sole province of trained cartographers; virtually anyone with access to a personal computer can create maps. Although this is desirable because it enables more people the opportunity to make maps, it is also problematic because there is no guarantee that the resulting maps will be well designed and accurate. The maps shown previously in Color Plate 1.1 are a good example of this. Map A uses an illogical set of unordered hues, while map B uses logically ordered shades from the same hue. Although map A may allow users to discriminate easily between individual states, it does not readily permit perception of the overall spatial pattern, which is one of the major reasons for creating a map.

Another error commonly committed by naive map-

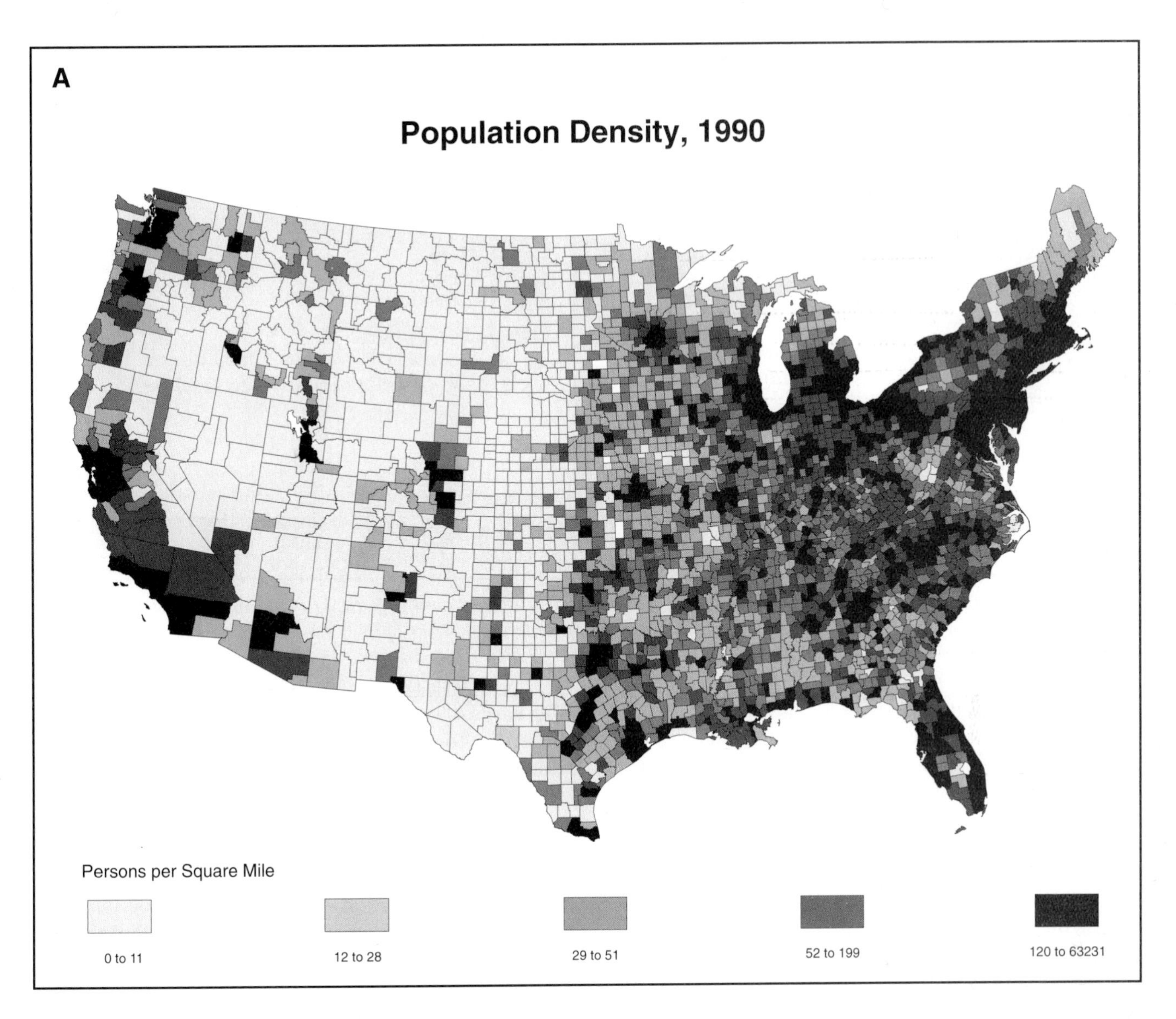

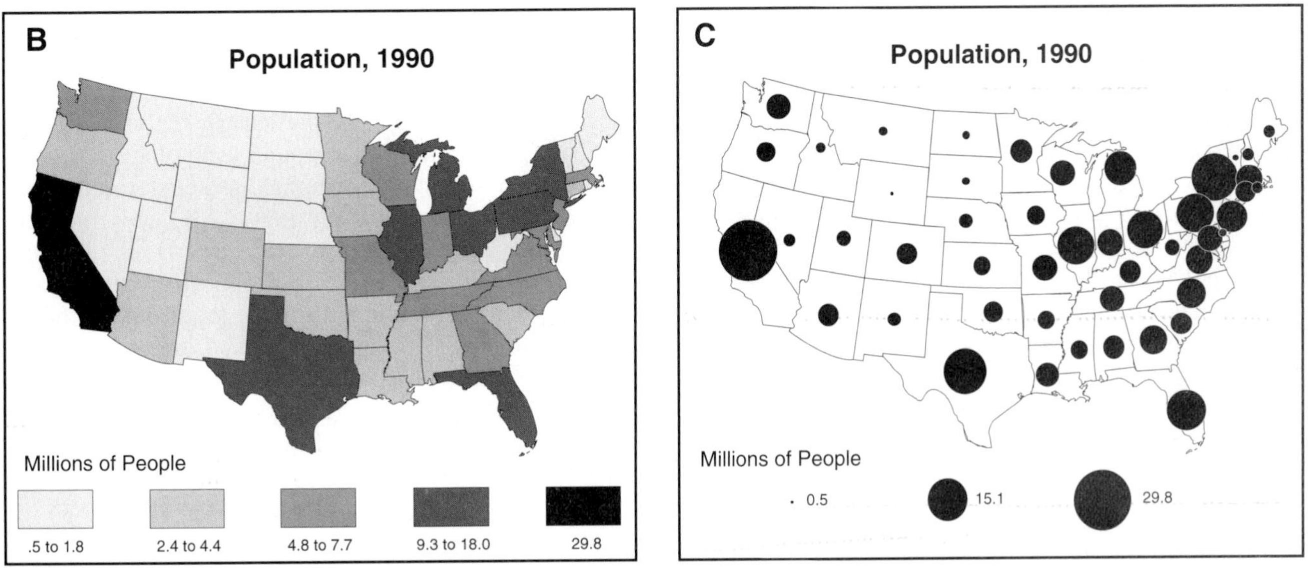

FIGURE 1.4 Potential maps of the population distribution in the United States in 1990. (A) a standardized choropleth map at the county level; (B) an unstandardized choropleth map at the state level; and (C) a state-level proportional symbol map. (Data Source: U.S. Bureau of the Census 1994a.)

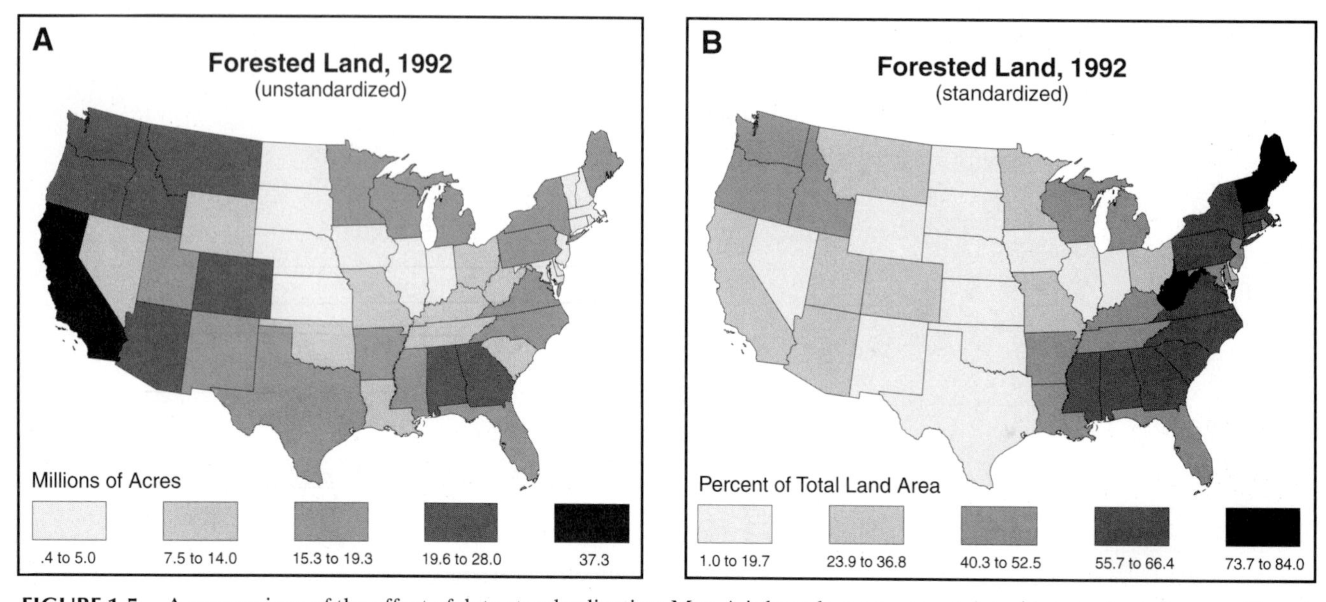

FIGURE 1.5 A comparison of the effect of data standardization. Map A is based on raw-count data (the number of acres of forested land), while map B is based on standardized data (the number of acres of forested land relative to the area of each state). Map A is misleading because states with large areas tend to have more forest. (Data Source: Powell et al. 1992.)

makers is illustrated in Figure 1.5. Map A suggests that forested land is more likely to be found in the western part of the United States, while map B suggests another pattern with an eastern dominance. Map B illustrates a more accurate distribution because it is based on **standardized data** (the number of acres of forested land relative to the area of each state); in contrast, map A is based on **raw-count data** (the number of acres of forested land). Choropleth maps of raw-count data tend to portray large areas as having high values of the mapped variable; in the case of map A, readers may incorrectly interpret the resulting dark shades as representing a high proportion of land in forests.

One purpose of this book is to teach you how to avoid these and other map symbolization problems. For example, Chapter 6 discusses how to select appropriate color schemes for choropleth and isarithmic maps, while Chapter 2 covers data standardization. One alternative to cartographic instruction is the development of **expert systems,** in which a computer automatically makes decisions on symbolization by using a knowledge base provided by experienced cartographers. Although prototypical expert systems have been developed (Buttenfield and Mark, 1991, provide an overview; see Zhan and Buttenfield, 1995, for more recent work), commonly used cartographic software has not implemented expert systems to date. In part, this is because cartographers have not been able to agree on the rules for symbolization (Wang and Ormeling 1996).

A second consequence of technological change is the ability to produce maps that would have been difficult or impossible to create by manual methods. One of the earliest examples was the unclassed map, introduced by Tobler (1973).* Figure 1.6 compares a traditional classed map (A) with its unclassed counterpart (B). On the **classed map,** data for enumeration units are grouped into classes of similar value and a progressively darker gray tone is assigned to each class, while on the **unclassed map,** gray tones are assigned to each enumeration unit in proportion to the data value associated with that unit. Some cartographers have argued that the unclassed map is preferable because it comes closer to reflecting the real-world distribution, while others have argued for the classed map on the grounds that it is easier to interpret. We will consider this issue in more detail in Chapter 4.

Animated maps (or maps characterized by continuous or dynamic change) are particularly representative of the capability of modern computer technology. Although the notion of animated mapping has been around since at least the 1930s (Thrower 1959, 1961), only recently have cartographers begun to recognize its full potential (Campbell and Egbert 1990). Probably the most common forms of animation are those representing changing cloud cover, precipitation, and fronts on daily television weather reports. Animations of spatial data are now also found in popular computerized encyclopedias, such as Grolier's (DiBiase 1994). Chapter 14 of this book will consider a wide variety of animations that geographers and other spatial scientists have developed; for

* Unclassed maps had been produced by manual methods, but they were difficult to create (see Robinson, 1982, for a review of the history of thematic mapping).

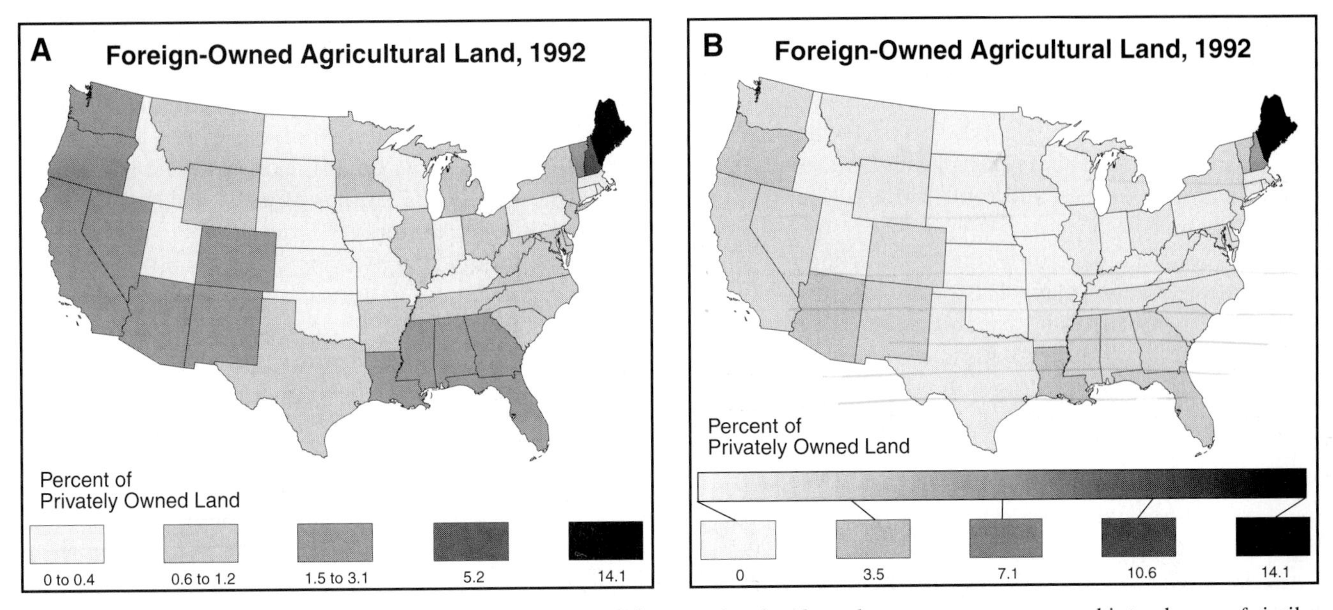

FIGURE 1.6 A comparison of classed (A) and unclassed (B) maps. On the classed map, states are grouped into classes of similar value and a gray tone is assigned to each class, while on the unclassed map, gray tones are selected proportional to the data value associated with each state. (Data Source: DeBraal 1992.)

example, we will consider an animation developed by Treinish (1992) for visualizing the formation of the ozone hole over Antarctica.

A third consequence of technological change is that it alters our fundamental way of using maps. With the communication model approach, cartographers generally created one "best" map for users. In contrast, interactive graphics now permits users to examine spatial data dynamically and thus develop several different representations of the data; this process is termed **data exploration.**

The software MapTime (Yoder 1996), which permits the display of temporal data associated with point locations, exemplifies the nature of data exploration. MapTime permits users to explore the data using three approaches: animation, **small multiples** (maps are shown for individual time periods), and **change maps** (a map is computed representing the difference between two points in time). Examination of population values for 196 U.S. cities from 1790 to 1990 (a data set distributed with MapTime) illustrates the different perspective provided by these approaches. For example, an animation reveals major growth in northeastern cities over most of the period, with an apparent drop for some of the largest cities from about 1950 to the present. In contrast, a change map, showing the percent population gains or losses between 1950 and 1990, reveals a distinctive pattern of population decrease throughout the northeast (Color Plate 1.3). One of the keys to data exploration is that displays such as these can be created in a matter of seconds. MapTime and other data exploration software will be covered in more detail in Chapter 13.

A fourth consequence of technological change is that it enables mapmakers to link maps, text, pictures, video, and sound in **multimedia** presentations. For example, in the Grolier encyclopedia already mentioned, animations include sound clips to assist in understanding. These animations are integrated, of course, with the rest of the encyclopedia, which includes text, pictures, and videos. In a more sophisticated form of multimedia known as **hypermedia,** the various forms of media can be linked transparently in ways not anticipated by system designers (Buttenfield and Weber 1994).

Closely associated with animation, data exploration, and multimedia is the **electronic atlas.** Initially, electronic atlases emulated the appearance of paper atlases (such as the *Electronic Atlas of Arkansas,* as described in Smith 1987). More recently, however, electronic atlases have begun to incorporate animation, data exploration, and multimedia capability. Examples of electronic atlases will be described in Chapter 15.

A fifth consequence of technological change is the ability to access maps and related information via the Internet (or information superhighway). The Internet has, of course, changed our daily lives as evidenced by the common listing of web addresses in newspaper and television advertisements. From the standpoint of cartography, the Internet can serve as a source of spatial data, attribute data, finished maps (both static and animated), software for creating static maps, software for exploring data, electronic atlases, tools for developing your own software, course materials for students, and course materials for instructors (Table 1.1). Covering the full breadth of the Internet relevant to cartography is beyond the scope of a single text. This text will focus on novel static maps, software for exploring data, animated maps, electronic atlases, and tools for developing your own software.*

A sixth consequence of technological change is the proliferation of color maps. This proliferation has been driven largely by increasing capabilities (and decreasing costs) of graphic display systems and color printers, and the ease of transferring information digitally via the Internet. Consider the color capability that can now be achieved in our own homes: for under $1500, personal computers are available that will display 256 colors at one time out of a palette of 16.8 million, and color printers of good graphic quality are now available for under $500. Admittedly, color reproduction in book or journal form is still expensive, but in spite of these high costs, we are seeing greater use of color, in both the academic literature and the popular press. In response to the proliferation of color maps, Chapters 5 and 6 of this text focus on color issues.

In addition to its impact on cartography, technological change also has affected other techniques commonly taught within geography departments; **geographic information systems (GIS)** is particularly noteworthy because of its close association with cartography. The notion of using GIS to analyze layers of spatial data has been around since the 1960s, but only in the last 15 years has major expansion taken place, largely because of advances in hardware and software.

Developments in GIS are important to cartographers for several reasons. One is that, by using a GIS, mapmakers can reduce the time it takes to create a detailed map. For example, the tedious process of manually compiling layers of information for dot mapping can now be handled automatically by using GIS. This notion will be considered in Chapter 10.

Developments in GIS have also accelerated some of the changes already mentioned. For example, ArcView® GIS Software provides the ability to explore geographic data during the course of a spatial analysis. We will consider a specific case of this in Chapter 13. As another example, Van Voris et al. (1993) have used animation to visualize forest growth as a function of global warming. When viewing a video distributed in association with their work, one can actually see individual trees grow and species composition change over time. By changing the parameters of a model for global warming (say, the level of carbon dioxide in the atmosphere), scientists can use this approach to see the impact of climate change on forest growth.

Remote sensing is another technique that has been dramatically affected by technological change. In 1972, remote sensing entered the digital age with the launching of Landsat, the "first satellite tailored specifically for broad-scale observation of the earth's land surface" (Campbell 1996, 158). Although it was clear to some that remote sensing would have great potential for mapping spatial phenomena, initially the technique was used relatively little by cartographers. One reason for this slow adoption is that the techniques of cartography, remote

* See Peterson (1997) for an introduction to Internet applications for cartography (also available at http://maps.unomaha.edu/NACIS/cp26).

TABLE 1.1 Internet uses for cartography

Function	*Example*	*Address (URL)*
Spatial data	Longitude and latitude of major U.S. cities	http://www.bcca.org/misc/qiblih/latlong_us.html
Attribute data	Population estimates for counties	http://www.census.gov/population/www/estimates/county.html
Finished maps		
Static	3D representation of pollution in New York harbor	http://www.rpi.edu/locker/69/000469/dx/harbor.www/harbor.html
Animated	Animation of urban sprawl in the San Francisco Bay area	http://edcwww2.cr.usgs.gov/umap/umap.html
Software for creating static maps	Software for creating choropleth maps at the state level	http://maps.esri.com/ESRI/mapobjects/tmap.htm
Software for exploring data	Project Argus	http://midas.ac.uk/argus
Electronic atlases	Massachusetts Electronic Atlas	http://icg.harvard.edu/~maps/maatlas.htm
Tools for developing your own software	Khoros	http://www.tnt.uni-hannover.de/soft/imgproc/khoros
Course materials for students	The Geographer's Craft	http://www.utexas.edu/depts/grg/gcraft/contents.html
Course materials for instructors	Cartography: The Virtual Department	http://www.utexas.edu/depts/grg/virtdept/contents.html

sensing, and GIS evolved separately. Although some professionals had knowledge of more than one technique, most had expertise in only one.

It also took time for experts to develop methods for interpreting remotely sensed digital data and for the technology to become sophisticated enough for detailed mapping tasks (such as a spatial resolution fine enough to easily detect individual agricultural fields). Today, we are beginning to see the adoption of remote-sensing technology by cartographers. A good example is Langford and Unwin's (1994) use of remote sensing to create population density surfaces, a technique that will be described in Chapter 10.

Improvements in technological capabilities have also led to what is essentially a new field of inquiry, that of **scientific data visualization,** or simply **scientific visualization,** a notion first presented by McCormick et al. (1987) in an issue of *Computer Graphics* dedicated solely to the topic. To McCormick et al., the objective of scientific visualization was "to leverage existing scientific methods by providing . . . insight through visual methods" (p. 3). Today scientific visualization generally involves using sophisticated workstations to explore large multivariate data sets (Color Plate 1.4). Work in scientific visualization extends far beyond the realm of spatial data, which geographers normally deal with, to include topics such as medical imaging and visualization of molecular structure and fluid flows (Keller and Keller, 1993, provide numerous examples).

1.5 VISUALIZATION AND CARTOGRAPHY

Although cartographers have been aware of developments taking place in scientific visualization, they have not been able to agree on how the term *visualization* should be used in the field of cartography. Thus far, two basic definitions for visualization have been developed. The first is a broad one that encompasses both paper and computer-displayed maps. According to MacEachren et al. (1992, 101):

> Geographic visualization will be defined here as the use of concrete visual representations—whether on paper or through computer displays or other media—to make spatial contexts and problems visible, so as to engage the most powerful human information-processing abilities, those associated with vision.

Using this definition, the word "visualization" could be applied to the visual analysis of a paper map created by nonautomated methods or to the visual analysis of maps created on sophisticated interactive graphic displays.

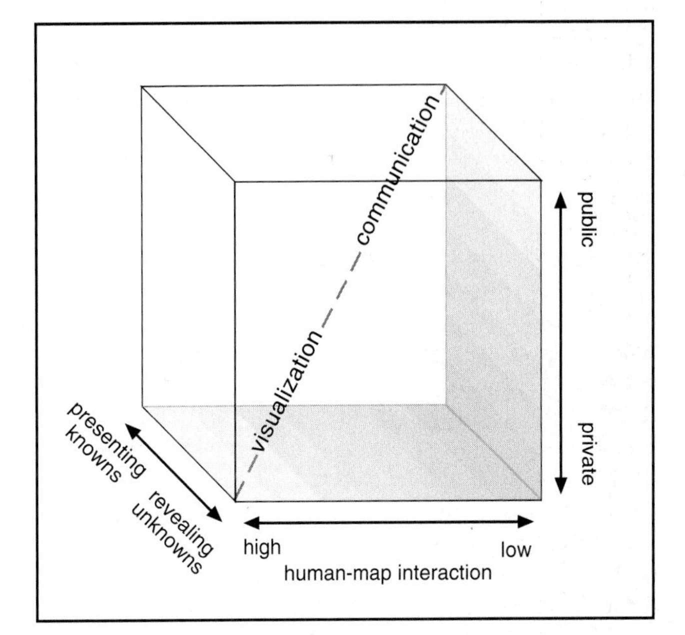

FIGURE 1.7 A graphical representation of how maps are used. Note that visualization is contrasted with communication along three dimensions: private versus public, revealing knowns versus presenting knowns, and the degree of human-map interaction. (After MacEachren 1994b, 7.)

The second, and narrower definition, is based on MacEachren's (1994b, 6) cartography-cubed representation of how maps are used, which is shown in Figure 1.7.* In this graphic, visualization is contrasted with communication along three dimensions: private versus public, revealing knowns versus presenting knowns, and the degree of human-map interaction. MacEachren argues that **visualization** is a private activity in which unknowns are revealed in a highly interactive environment, while **communication** involves the opposite: a public activity in which knowns are presented in a noninteractive environment.

As an example of visualization, MacEachren employs Ferreira and Wiggin's (1990) "density dial," in which class break points on choropleth maps are manipulated to identify and enhance spatial patterns. In contrast, MacEachren argues communication is exemplified by "you are here" maps used to locate oneself in a shopping mall. MacEachren stresses that certain map uses do not fit neatly into either category (and thus the need for the cartography-cubed representation). For example, Thelin and Pike's (1991) dramatic shaded relief map shown in Figure 1.8 (to be discussed in Chapter 9) fits into the communication realm because it is available to a wide readership (making it "public") and the user cannot interact with the paper version of it. The Thelin and Pike map,

* MacEachren used the term $(\text{Cartography})^3$; I use the word "cubed" to avoid confusion with a "superscript 3."

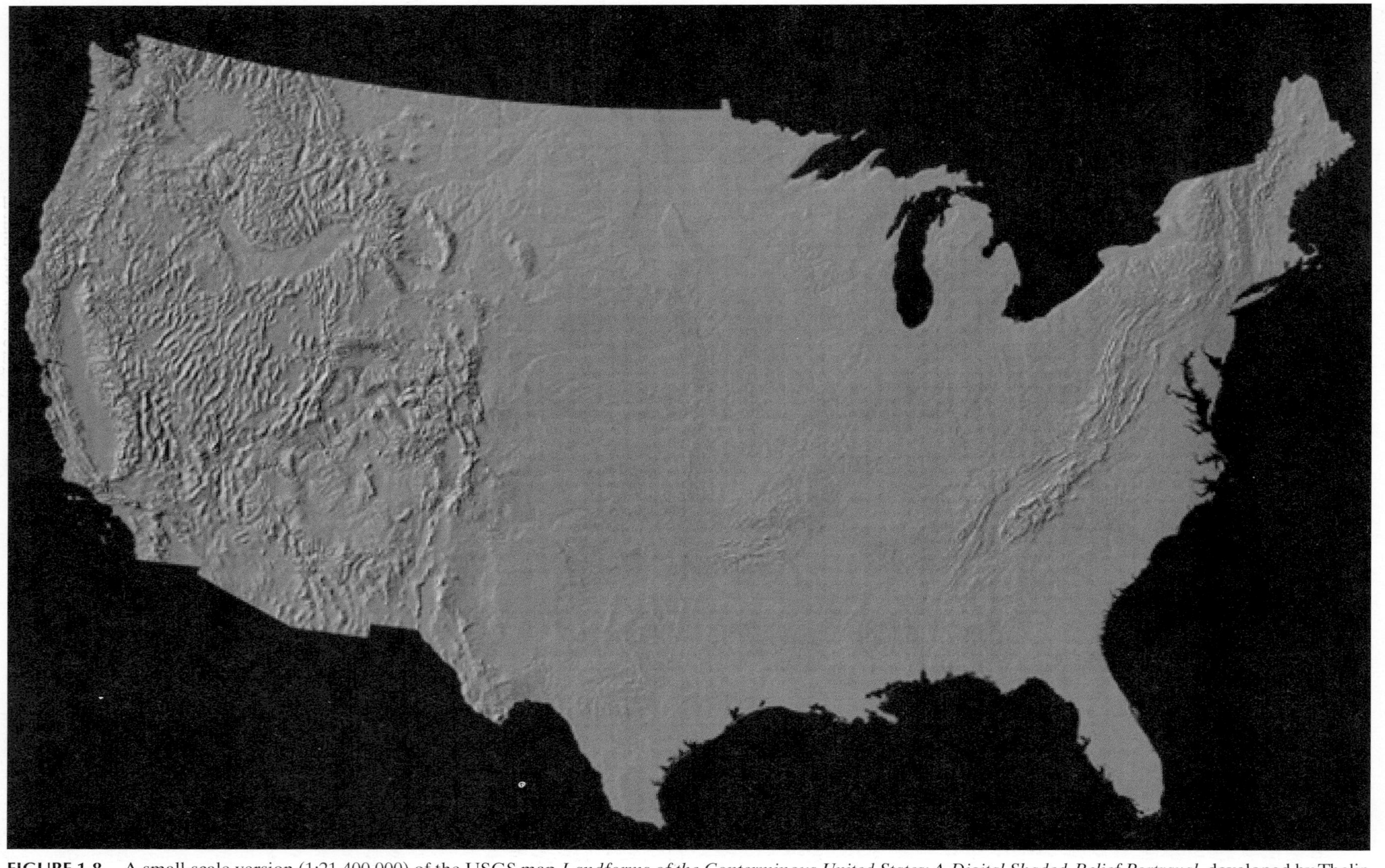

FIGURE 1.8 A small-scale version (1:21,400,000) of the USGS map *Landforms of the Conterminous United States: A Digital Shaded-Relief Portrayal,* developed by Thelin and Pike (1991). (Courtesy of Gail P. Thelin.)

however, fits visualization in the sense that it often reveals unknowns. The net result is that the Thelin and Pike map falls in the upper right corner of MacEachren's diagram.

It is apparent that the notions of map communication models and data exploration introduced previously can be associated with communication and visualization respectively in MacEachren's cartography-cubed representation. Thus, when using the five steps of the communication model presented in section 1.3, the intention generally is that the map is being made for the general public, there will be low human-map interaction, and the focus is on presenting knowns. In data exploration, the emphasis is on revealing unknowns via high human-map interaction in a private setting; in this sense the word "exploration" could just as easily be substituted for "visualization" in MacEachren's graphic.

On an informal basis, visualization has been used to describe any recently developed novel method for displaying data. Thus, cartographers have placed animation and electronic atlases under the rubric of "visualization." Additionally, novel methods that may result in static maps (such as Dorling's cartograms, see Chapter 11) are also placed under the heading of visualization.

1.6 COGNITIVE ISSUES IN CARTOGRAPHY

While technological change certainly has been important to cartography, there have also been major changes in philosophical thinking about maps. Sections 1.6 and 1.7 discuss two of these changes: the rise of cognition and postmodernism. Understanding the role that cognition plays in cartography requires contrasting it with perception. *Perception* deals with our initial reaction to map symbols (that a symbol is there, that one symbol is larger or smaller than another, or that symbols have different colors). In contrast, *cognition* deals not only with perception, but with our thought processes, prior experience, and memory.* For example, contour lines on a topographic map can be interpreted without looking at a legend because of one's past experience with such maps. Alternatively, one might correlate the pattern of a distribution of soils on a particular map with the distribution of vegetation seen on a previous map.

The principles of cognition are important to cartographers because they can provide an explanation for why certain map symbols work (that is, communicate information effectively). Traditionally, cartographers were not so concerned with why symbols worked, but rather with determining which worked best. This was known as a *behaviorist view,* in which the human mind was treated like a black box. The trend today is toward a *cognitive view,* in which cartographers hope to find why symbols work effectively. Such an approach should build a theoretical basis for map symbol processing that will, hopefully, not only assist in telling us why particular symbols being tested work, but provide a basis for evaluating other map symbols, even symbols that have not yet been developed.

To illustrate the difference between the behaviorist and cognitive views, consider an experiment in which a cartographer wishes to compare the effectiveness of two color sequences for numerical data: five hues from a yellow-to-red progression; and five hues from the electromagnetic spectrum (red, orange, yellow, green, blue). Let us presume that the results of such an experiment reveal that the yellow-red progression works best. The traditional behaviorist would report these results, but probably not provide any indication as to why one sequence works better than another. In contrast, the cognitivist would consider how color is processed by the eye-brain system. The cognitivist might theorize that the yellow-red progression works best because of opponent process theory (see Chapter 5). Effectiveness of the spectral hues might be argued for on the grounds that different hues will appear to be at slightly different distances from the eye, and thus form a logical progression (for example, red will appear nearer than, say, blue, as discussed in Chapter 6). Spectral hues might also be considered effective because of their common use on maps, and the likelihood that readers have experience in using them.

An important concept of cognition is the three types of memory: iconic memory, short-term visual store, and long-term visual store (Peterson 1995). **Iconic memory** deals with the initial perception of a object (in our case, a map or portion thereof) by the retina of the eye (see Chapter 5 for a detailed discussion of the retina). Calling this "memory" is somewhat of a misnomer, since it exists for only about one-half second and we have no control over it. Iconic memory, however, has relatively unlimited capacity and is unaffected by pattern complexity.

Visual information initially recorded in iconic memory is passed on to the **short-term visual store.** Only selected information is passed on at this stage; for example, the boundary of Texas shown in Figure 1.9 will likely be simplified to some extent in moving from iconic to short-term visual store. Keeping information in short-term visual store requires constant attention (or activation). This is accomplished by rehearsal of the items being memorized (for example, staring at the map of Texas and telling yourself to remember its shape).

After the information has been rehearsed in short-term visual store, it is ultimately passed on to **long-term visual memory** (Figure 1.9). Note that arrows are shown in both

* For a more in-depth discussion of perception and cognition, see Goldstein (1989) and Anderson (1990) respectively.

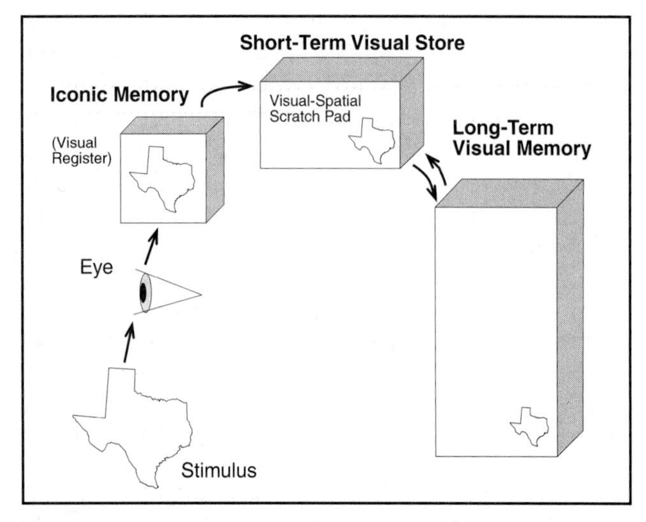

FIGURE 1.9 Three forms of memory used in cartography. A perceived map is initially stored in iconic memory (the retina of the eye). The image is then passed to short-term visual store in the brain (where it is rehearsed). Finally, information is stored for later use in the long-term visual memory in the brain. (INTERACTIVE & ANIMATED CARTOGRAPHY by Peterson, M. P., © 1995. Adapted by permission of Prentice-Hall, Inc., Upper Saddle River, NJ.)

directions between the short-term visual store and long-term visual memory. When something is initially memorized, information must be moved from short-term store to long-term memory; when something is retrieved from memory, the opposite is the case. As an example of the latter, imagine that you are shown a map of Texas and asked to indicate which state it is. To make your decision, you must retrieve the image of Texas from long-term memory, and compare it with the image in short-term store to make your decision.

I have covered here only some of the basic concepts of cognitive psychology that are necessary for understanding this text. For a broader overview, see Peterson (1995, Chapter 2) and MacEachren (1995).

1.7 POSTMODERNISM IN CARTOGRAPHY

Some geographers (and most scholars in the humanities and social sciences) hold that in this postmodern world of jarring juxtapositions of cultures and economies it is difficult to trust any single grand theory or explanation. These people believe that problems can best be approached from multiple perspectives or viewpoints; for example, in a study of an urban neighborhood, a postmodernist would want to acquire not only the perspective of those in positions of political and economic power, but as wide a sample of inhabitants as possible: men and women, children, the elderly, each of the social classes, ethnic and racial groups, or any others who might shed light on the dynamics of the neighborhood (Cloke et al. 1991, 170–201).

An important notion of postmodernism is that a text (or map) is not an objective form of knowledge, but rather has numerous hidden agendas or meanings. **Deconstruction** enables us to uncover these hidden agendas and meanings. Harley (1989, 3) states:

> Deconstruction urges us to read between the lines of the map . . . to discover the silences and contradictions that challenge the apparent honesty of the image. We begin to learn that cartographic facts are only facts within a specific cultural perspective.

To illustrate deconstruction, consider Bockenhauer's (1994) examination of the official state highway maps of Wisconsin over a seven-decade period. Although the major purpose of highway maps is presumably to assist a motorist in getting from one place to another, Bockenhauer (p. 17) argues that "three dominant 'cultures' could be identified as factors influencing the map product: a transportation/modernization culture, a culture of promotion, and a subtle, beneath-the-surface culture of dominion." As an example of the transportation-modernization culture, he recounts the removal of surfaced county roads from recent editions of the map, and thus the greater emphasis on "getting us from here to there by encouraging travelers onto the freeways" (p. 21).

An illustration of highway maps as promotional devices is their use by government officials; for example, in the 1989–1990 edition, Governor Thompson showcased himself next to a replica of a Duesenberg automobile. Finally, an example of the culture of dominion was the portrayal of women on the maps: "Among the most common and prominent images appearing on the . . . maps . . . are those of women in swim suits and fishermen. Nearly all of the photos of people enjoying Wisconsin fishing . . . are of white men. . . . [The] women seem to be part of the package of 'pleasure' offered to white men in Wisconsin" (p. 24). Although some would disagree with Bockenhauer's interpretations of these maps, his point is made clear that maps may convey information other than their supposed primary purpose.

Harley (1989) and Wood and Fels (1992) have argued that virtually any map, when deconstructed, can be interpreted from a variety of perspectives, and thus have hidden or unintended meanings. A seemingly innocent map of percent forest cover, for example, may mean different things to different people. To those wishing to extract lumber, it might be an indicator of areas of potential economic value, while for others it may signify safe havens for wildlife. Also note that the variable "percent forest cover" may not be useful for many applications; such a variable says nothing about the age of the trees, the species composition, or associated wildlife.

Although this text will not focus on postmodernism and map deconstruction, both mapmakers and map users need to recognize their importance. Mapmakers must realize that maps may communicate unintended messages, and that the data they have chosen to map or the method of symbolizing that data may be a function of the culture of which they are a part. Conversely, map users must recognize that a single map may depict only one representation of a spatial phenomenon (for example, a map of percent forest cover is one representation of vegetation).

Note that there is some overlap between the notions of data exploration and postmodernism, since both promote the notion of multiple representations of data. In the context of data exploration, "multiple representations" refers to the various methods of symbolizing the data (such as using MapTime to display population data as both an animation and change map). The postmodernist should be interested in this approach because it concurs with the notion that there is no single "correct" way of visualizing data. Additionally, however, the postmodernist would be interested in the multiple meanings and potentially hidden agendas found in a particular thematic map, as suggested above.

1.8 SCOPE OF THIS BOOK

Chapters 2 to 10 of this book focus on basic principles and traditional univariate thematic maps. Chapter 2 presents terminology and general guidelines for symbolizing spatial phenomena that, if followed, should avoid many common design blunders. A key element is an introduction to and comparison of various forms of thematic mapping (choropleth, isopleth, proportional symbol, and dot).

A statistical and graphical foundation for the remainder of the book is provided in Chapter 3. Those with at least an introductory course in quantitative methods in geography will find some of the material in this chapter to be a review, although much of the graphical material will be new.

Basic principles of data classification are covered in Chapter 4. The chapter considers not only many traditional classification methods (equal intervals, quantiles, and optimal), but also contrasts classed versus unclassed maps and introduces approaches for simplifying spatial patterns. Such material would normally be covered in a chapter on choropleth mapping, but I have chosen to present it as a separate chapter because much of it has wider applicability.

Chapters 5 and 6 provide a foundation on the use of color on thematic maps. Chapter 5 covers basic principles of color, including how color is processed by the human visual system (section 5.1), hardware considerations in producing maps in both soft-copy and hard-copy forms (section 5.2), models for specifying colors used on maps (section 5.3), and how maps are disseminated to potential users (section 5.4). Sections 5.2 and 5.4 might be placed in a separate chapter entitled "Production and Reproduction" issues, but are included with the color material because they are critical to understanding the use of color. Chapter 6 covers the utilization of color on two of the more common forms of thematic map: choropleth and isarithmic. Many of the factors for selecting color discussed here, however, will have wider applicability.

Chapter 7 describes **proportional symbol mapping,** a common technique for representing numerical data occurring at point locations (such as the volume of water released by a geyser). Unlike choropleth mapping, which is spread over several chapters, proportional symbol mapping is concentrated in this single chapter. Thus, a broad range of issues are discussed, including selecting appropriate data, kinds of symbols, scaling symbols, legend design, symbol overlap, and **redundant symbols** (representing the same data with different types of symbols).

When mapping **smooth continuous phenomena** (phenomena that change gradually, such as snowfall or the earth's topography), mapmakers commonly encounter two important issues. One is that data representing such phenomena generally are available only at a limited set of irregularly spaced **control points** (weather stations, in the case of snowfall), and thus it is necessary to interpolate data for locations between the control points. Chapter 8 provides an overview of common automated methods of interpolation, including triangulation, inverse distance, and kriging. Also considered in Chapter 8 are criteria for selecting an interpolation method, general limitations of automated contouring, and the **pycnophylactic approach,** a specialized interpolation method for handling data associated with enumeration units.

A second issue involved in mapping smooth continuous phenomena is deciding on an appropriate symbology for representing a phenomenon. Chapter 9 covers a range of symbolization methods. A familiar example is the **isarithmic** (or **contour**) map in which isarithms (or contours) are used to depict lines of equal value. The areas between lines may also be shaded using **hypsometric tints** (as on temperature maps found in daily newspapers). When symbolizing the earth's topography, we will see that a broader range of methods are possible, including hachures, shaded relief, and the use of color to depict aspect and slope.

Automated methods for dot and dasymetric mapping are covered in Chapter 10. **Dot maps** are constructed by letting one dot equal a certain number of a phenomenon (for example, a dot might equal 500 people), while **dasymetric maps,** like choropleth maps, use areal symbols to represent regions of uniformity. Traditionally, cartographic texts have presented these methods within separate chapters; I present them together because they both make use of ancillary information, or information that assists in locating the phenomenon of interest.

Chapters 11 to 16 of the book cover recently developed methods (novel univariate maps, data exploration, animation, and electronic atlases), and more complex techniques (multivariate mapping). Chapter 11 describes several recently developed methods for univariate thematic mapping. Topics covered include framed-rectangle symbols (a point symbol that is an alternative to the choropleth map), models for representing geographic phenomena (which can be used to assist in selecting an appropriate symbology), Dorling's novel method for creating **cartograms** (a cartogram is constructed by scaling enumeration units in proportion to associated data), software for depicting migration flows and continuous vector-based flows (such as wind speed and direction), and **true 3-D phenomena** (such as the level of carbon dioxide in the earth's atmosphere).

Chapter 12 extends the concepts of univariate mapping into the multivariate realm by considering both bivariate and multivariate mapping. **Bivariate mapping** involves displaying two variables simultaneously; for example, you might want to examine the relationship between education and income for counties within the United States. **Multivariate mapping** involves displaying three or more variables simultaneously; for instance, you might wish to visualize the relationship among temperature, salinity, and current speed within the Atlantic Ocean. A fundamental decision in both bivariate and multivariate mapping is whether variables are displayed on separate maps (maps are compared) or variables are displayed on the same map (maps are combined). Chapter 12 will cover issues related to map comparison and consider various methods for combining two or more variables on the same map.

Chapters 13 to 15 focus on several methods resulting from the technological change in cartography introduced in the present chapter: data exploration (Chapter 13), animation (Chapter 14), and electronic atlases (Chapter 15). The latter section of Chapter 15 also contains information on tools for those who wish to develop their own data exploration software, animations, or electronic atlases. In contrast to previous chapters, these chapters are heavily software-dependent. I focus on particular software because these methods have only recently evolved and thus basic principles underlying them are still in development.

Given the rapidly changing nature of cartography, Chapter 16 considers a number of more recent developments in thematic mapping and visualization. Topics covered include the depiction of data quality, the use of sound to represent spatial data, research on the use of color, virtual reality and visual realism, and a special electronic issue of *Computers & Geosciences* focusing on visualization. Chapter 16 also provides information on journals, conferences, and useful Internet sites that can assist in keeping pace with developments in this rapidly changing discipline.

Appendixes C to I provide extensive sources for thematic mapping and graphic design software, data exploration software, map animations, electronic atlases, tools for developing your own software and geographic information systems. Many of these sources are accessible via the Internet, while others are available in more traditional diskette, CD-ROM, or video form. The appendixes can be found on the Internet at http://www.prenhall.com/slocum.

FURTHER READING

Bertin J. (1981) *Graphics and Graphic Information-Processing.* Berlin: Walter de Gruyter. (Translated by W. J. Berg and P. Scott)

A text by the French cartographer Bertin, whose work is widely cited. We will have occasion to reference his work at several places in the present text. Also see Bertin (1983).

Board, C. (1984) "Higher-order map-using tasks: Geographical lessons in danger of being forgotten." *Cartographica* 21, no. 1:85–97.

Discusses some issues relevant to the kinds of information that can be acquired from maps.

Buttenfield, B. P., and Mark, D. M. (1991) "Expert systems in cartographic design." In *Geographic Information Systems: The Microcomputer and Modern Cartography,* ed. D. R. F. Taylor, pp. 129–150. Oxford: Pergamon Press.

Although a bit dated, this work provides a good summary of the potential for using expert systems in map design.

Campbell, C. S., and Egbert, S. L. (1990) "Animated cartography: Thirty years of scratching the surface." *Cartographica* 27, no. 2:24–46.

An overview of early work in animated cartography, along with some suggestions for the potential of animation.

Crampton, J. (1995) "Cartography resources on the World Wide Web." *Cartographic Perspectives* no. 22:3–11.

The World Wide Web is expanding very rapidly, so this article is already dated. But it is still useful for those who are beginning to look for cartographic resources on the Web.

DiBiase, D. (1990b) "Visualization in the earth sciences." *Earth and Mineral Sciences* 59, no. 2:13–18.

Discusses visualization and its role in geographic research.

Keller, P. R., and Keller, M. M. (1993) *Visual Cues: Practical Data Visualization.* Los Alamitos, Calif.: IEEE Computer Society Press.

Examples of visualization from a wide variety of disciplines.

Koláčný, A. (1969) "Cartographic information: A fundamental

concept and term in modern cartography." *The Cartographic Journal* 6, no. 1:47–49.

A cartographic communication model that MacEachren (1995) claims had "the greatest initial impact on cartography" (p. 4).

MacEachren, A. M. (1994b) "Visualization in modern cartography: Setting the agenda." In *Visualization in Modern Cartography,* ed. A. M. MacEachren and D. R. F. Taylor, pp. 1–12. Oxford: Pergamon Press.

Discusses definitions for visualization.

MacEachren, A. M. (1995) *How Maps Work: Representation, Visualization, and Design.* New York: Guilford Press.

An advanced treatment of cognitive issues in cartography.

McCormick, B. H., DeFanti, T. A., and Brown, M. D. (1987) "Visualization in scientific computing." *Computer Graphics* 21, no. 6:entire issue.

A classic early work on visualization.

Monmonier, M. S. (1985) *Technological Transition in Cartography.* Madison, Wisc.: University of Wisconsin Press.

A text dealing with technological change in cartography.

Peterson, M. P. (1997) "Cartography and the Internet: Introduction and research agenda." *Cartographic Perspectives* no. 26:3–12.

An overview of the Internet and its role in cartography. Other articles in this issue also deal with the Internet. The complete issue can be found at http://maps.unomaha.edu/NACIS/cp26.

Rundstrom, R. A. (Ed.) (1993) "Introducing cultural and social cartography." *Cartographica* 30, no. 1:entire issue.

A set of articles dealing with postmodern issues in cartography. Note that Rundstrom, the editor of this work, describes the study of postmodern issues as "cultural and social cartography."

Sheppard, E., and Poiker, T. (Eds.) (1995) "GIS and society." *Cartography and Geographic Information Systems* 22, no. 1:entire issue.

A set of articles covering postmodern issues in GIS.

Wood, D., and Fels, J. (1992) *The Power of Maps.* New York: Guilford Press.

An extensive essay on postmodern issues in cartography. Also see Wood and Fels (1986).

2

Symbolizing Spatial Data: Terminology and General Guidelines

OVERVIEW

The purpose of this chapter is to introduce terminology and provide general guidelines for symbolizing spatial data. One goal is to assist you in selecting among four common mapping techniques: choropleth, proportional symbol, isopleth, and dot. For example, imagine that you wish to map the spatial pattern of income in Washington, D.C., and that you have collected data on the annual income of all families in each census tract of the city. You might wonder which of the above four techniques would be appropriate. Reaching this goal will require that we first consider (1) the spatial arrangement of geographic phenomena, (2) the various levels at which we can measure geographic phenomena, and (3) the types of symbols that can be used to represent spatial data.

Section 2.1 discusses the spatial arrangement of geographic phenomena. One way to think about spatial arrangement is to consider ***spatial dimension*** *or extent—whether a phenomenon can be conceived of as* ***points, lines, areas,*** *or* ***volumes.*** *For example, water well sites in a rural area constitute a point phenomenon, while a city boundary is representative of a linear phenomenon. Another way of thinking about spatial arrangement is to contrast discrete and continuous phenomena.* ***Discrete phenomena*** *occur at isolated point locations, while* ***continuous phenomena*** *occur everywhere. For example, water towers in a city would be discrete, while the distribution of solar insolation during the month of January is continuous. Discrete and continuous phenomena may also be classified as smooth or abrupt. For instance, rainfall and sales tax rates for states are both continuous in nature, but the former is smooth, while the latter is abrupt (varying at state boundaries).*

Section 2.2 considers ***levels of measurement,*** *which refers to the various ways of measuring a phenomenon when a data set is created. For instance, we might specify the soil type of a region as an entisol, as opposed to a mollisol; such a categorization of soils would be termed a nominal level of measurement. In total, we will consider four levels of measurement:* ***nominal, ordinal, interval,*** *and* ***ratio.*** *The latter two levels are commonly combined into numerical data, which is the focus of this book.*

The term ***visual variables*** *is commonly used to describe the various perceived differences in map symbols that are used to represent spatial data. For example, the visual variable spacing involves varying the distance between evenly spaced objects (such as horizontal lines). Section 2.3 will cover a host of visual variables, including* ***spacing, size, perspective height, orientation, shape, arrangement, hue, lightness, saturation,*** *and* ***location.***

Section 2.4 introduces four common mapping techniques (choropleth, proportional symbol, isopleth, and dot) and considers how a mapmaker selects among them. We will see that the choice of a mapping technique is a function of both the nature of the underlying phenomenon and the purpose for making the map. Section 2.4 also introduces the notion of standardizing data to account for the area over which it is collected (such as enumeration units on a choropleth map); here we will consider the most direct form of standardization, which involves dividing a count value by the area of the enumeration unit (for example, dividing acres of wheat for each county by the area of each county).

Section 2.5 considers the issue of selecting an appropriate visual variable for choropleth mapping, which has traditionally been the most common thematic mapping method. Selecting an appropriate visual variable requires creating a logical match between the level of measurement of the data and the visual variable (for example, if data are numerical, the visual variable should appear to reflect the numerical character of the data).

Section 2.6 will extend the basic concept of standardization introduced in section 2.4 to include several other standardization methods. For example, areas of enumeration units can be accounted for indirectly by taking the ratio of two count variables that do not involve area, such as a ratio of males to females.

A map intended for presentation (as opposed to data exploration) consists of a pattern of symbols depicting the theme being mapped and map design elements that assist in interpreting the theme (such as the title and legend). Section 2.7 considers (1) which of these map design elements should be included; (2) what their content should be; (3) how they should be positioned; and (4) how font, style, and size of typography should be selected.

2.1 SPATIAL ARRANGEMENT OF GEOGRAPHIC PHENOMENA

2.1.1 Spatial Dimension

One way to think about the spatial arrangement of geographic phenomena is to consider their **spatial dimension,** or extent. For our purposes, we will consider five types of phenomena with respect to spatial dimension: point, linear, areal, 2½-D, and true 3-D.

Point phenomena are assumed to have no spatial extent and can thus be termed "zero-dimensional." Examples include locations of religious worship, oil wells, and locations of nesting sites for eagles. Locations for point phenomena can be specified in either two- or three-dimensional space; for example, places of religious worship can be specified by x and y coordinate pairs (longitude and latitude), while nesting sites for eagles could be specified by x, y, and z coordinates (the z coordinate would be the height above the earth's surface).

Linear phenomena are one-dimensional in spatial extent, having length, but essentially no width. Examples include a boundary between countries and the path of a stunt plane during an air show. Locations of linear phenomena are defined as an unclosed series of x and y coordinates (in two-dimensional space), or an unclosed series of x, y, and z coordinates (in three-dimensional space).

Areal phenomena are two-dimensional in spatial extent, having both length and width. An example would be a lake (assuming that we focus on its two-dimensional surface extent). Data associated with political units (such as counties) can also fit into this framework, since the location of each county can be specified as an enclosed region. In two-dimensional space, areal phenomena are defined by a series of x and y coordinates that completely enclose a region.

When we move into the realm of volumetric phenomena, it is convenient to consider two types: 2½-D and true 3-D. The first of these, **2½-D phenomena,** can be thought of as a surface, in which each point on the surface is defined by longitude, latitude, and value above a zero point (or alternatively a value below a zero point). Probably the easiest example to understand is elevation above sea level, because we can actually see the surface in the landscape. More abstract examples, such as the precipitation falling over a region, can also be considered as surfaces.

Another way of thinking about 2½-D surfaces is that they are single-valued in the sense that each longitude and latitude has a single value associated with it. In contrast, **true 3-D phenomena** are multivalued because each longitude and latitude may have multiple values associated with it. In particular, any point associated with a true 3-D phenomenon can be specified by four values: longitude, latitude, height above (or depth below) a zero point, and the value of the phenomenon. Consider mapping carbon dioxide (CO_2) concentrations in the atmosphere. At any point, it is possible to define longitude, latitude, height above sea level, and an associated level of CO_2. Color Plate 2.1 illustrates an example of a true 3-D phenomenon, layers of material beneath the earth's surface.

It is important to realize that map scale plays a major role in defining the spatial dimension of a phenomenon. For example, on a **small-scale map** (such as a page-size map of France) places of religious worship occur at points, but on a **large-scale map** (such as a map of a local neighborhood) individual buildings would likely be apparent, and thus the focus might be on the area covered by the place of worship. Similarly, a river could be considered a linear phenomenon on a small-scale map, but on a large-scale map, the emphasis could be on the area covered by the river.

2.1.2 Discrete versus Continuous and Smooth versus Abrupt Phenomena

Another way of thinking about spatial arrangement is to contrast discrete versus continuous and smooth versus abrupt phenomena. Here we consider basic definitions for these terms. A more complete discussion of these terms will be provided in section 11.2.

The terms "discrete" and "continuous" are often used in statistics courses to describe different types of data along a number line; here we consider their use by cartographers in a spatial context. **Discrete phenomena** are presumed to occur at distinct locations (with space in between). Individuals associated with a population would be an example of a discrete phenomenon; for an instant in time, a location can be specified for each person, with space between individuals. Of course, the average space between people may vary considerably, from miles in a rural setting to inches in a crowded elevator.

Continuous phenomena occur throughout a geographic region of interest. The examples presented previously for 2½-D phenomena would also be considered continuous phenomena. For instance, when considering elevation, every longitude and latitude position has a value above or below sea level.

Both discrete and continuous phenomena can also be described as either smooth or abrupt. **Smooth phenomena** change in a gradual fashion, while **abrupt phenomena** change suddenly. This concept is most easily understood for continuous phenomena. The distribution of total precipitation over the course of a year for a humid region would be a smooth continuous phenomenon because we would not expect such a distribution to exhibit abrupt discontinuities. In contrast, sales tax rates by state for the United States would be considered an abrupt continuous phenomenon because although each enumeration unit (a state) has a value, there would be abrupt changes at the boundaries between states.

2.1.3 Phenomena versus Data

When mapping geographic phenomena, it is important to distinguish between the actual phenomenon and the data collected to represent that phenomenon. For example, imagine that we wish to map the percentage of forest cover in South Carolina. If we try to visualize the phenomenon, we can conceive of it as smooth and continuous in some portions of the state where the percentage gradually increases or decreases. In other areas, we can conceive of relatively abrupt changes where the percentage shifts very rapidly (when, say, an urban area is bounded by a hilly forested region).

One form of data that we might use to represent percentage of forest cover would be individual values for counties, which can be found in the state statistical abstract for South Carolina (South Carolina State Budget and Control Board 1994, 45). We might consider mapping these data directly by creating the **prism map** shown in Figure 2.1B. Note that in this case there are abrupt changes at the boundaries of each county. Such a map might be appropriate if we wished to provide a typical value for each county, but it obviously hides the variation within each county and misleads the reader into thinking that changes take place only at county boundaries.

Potentially, a better approach would be the smooth, continuous map **(fishnet map)** shown in Figure 2.1A; this map indicates that the percentage of forest cover does not coincide with county boundaries, but rather changes in a gradual fashion. A still better map would be one that shows some of the abrupt changes that are likely to occur. Creating such a map would require more information about the location of forest within the state, as might be available from a remotely sensed image. Our purpose at this point in the text is not to create the most representative map of the phenomenon, but to stress that the mapmaker must carefully distinguish between data that have been collected and the phenomenon that is being mapped. Which type of map is used will be a function of both the nature of the underlying phenomenon and the purpose of the map. We will consider this issue in greater depth in section 2.4.

2.2 LEVELS OF MEASUREMENT

When a geographic phenomenon is measured to create a data set, we commonly speak of the **level of measurement** associated with the resulting data.* Four levels are commonly recognized—nominal, ordinal, interval, and ratio—with each subsequent level including all characteristics of the preceding levels. A **nominal** level of measurement involves a grouping (or categorization), but no ordering. The classic example is religion, in which individuals might be identified as Catholic, Protestant, Jewish, or other; here each religious group is different, but one is not more or less in value than another. Another example would be classes on a land use/land cover map; for example, grassland, forest, urban, water, and cropland differ from one another, but one is not more or less in value.

The second level of measurement, **ordinal,** includes categorization plus an ordering (or ranking) of the data. For example, a geologist asked to specify the likelihood of finding oil at each of 50 well sites might be unwilling to provide numerical data, but would feel comfortable specifying a low, moderate, or high potential at each site. Here three categories (low, moderate, and high) are provided, with a distinct ordering among them.

Another example of ordinal data would be rankings resulting from a map comparison experiment. Imagine that you constructed dot maps for 10 different phenomena and asked people to compare those maps with another dot map (say, of population) and to rank the maps from "most like" to "least like" the population map. The 10 maps ranked by each person would represent a distinct ordering, and thus represent ordinal data.

An **interval** level of measurement involves an ordering of the data plus an explicit indication of the numeri-

* See Chrisman (1997) for a more complete discussion of levels of measurement.

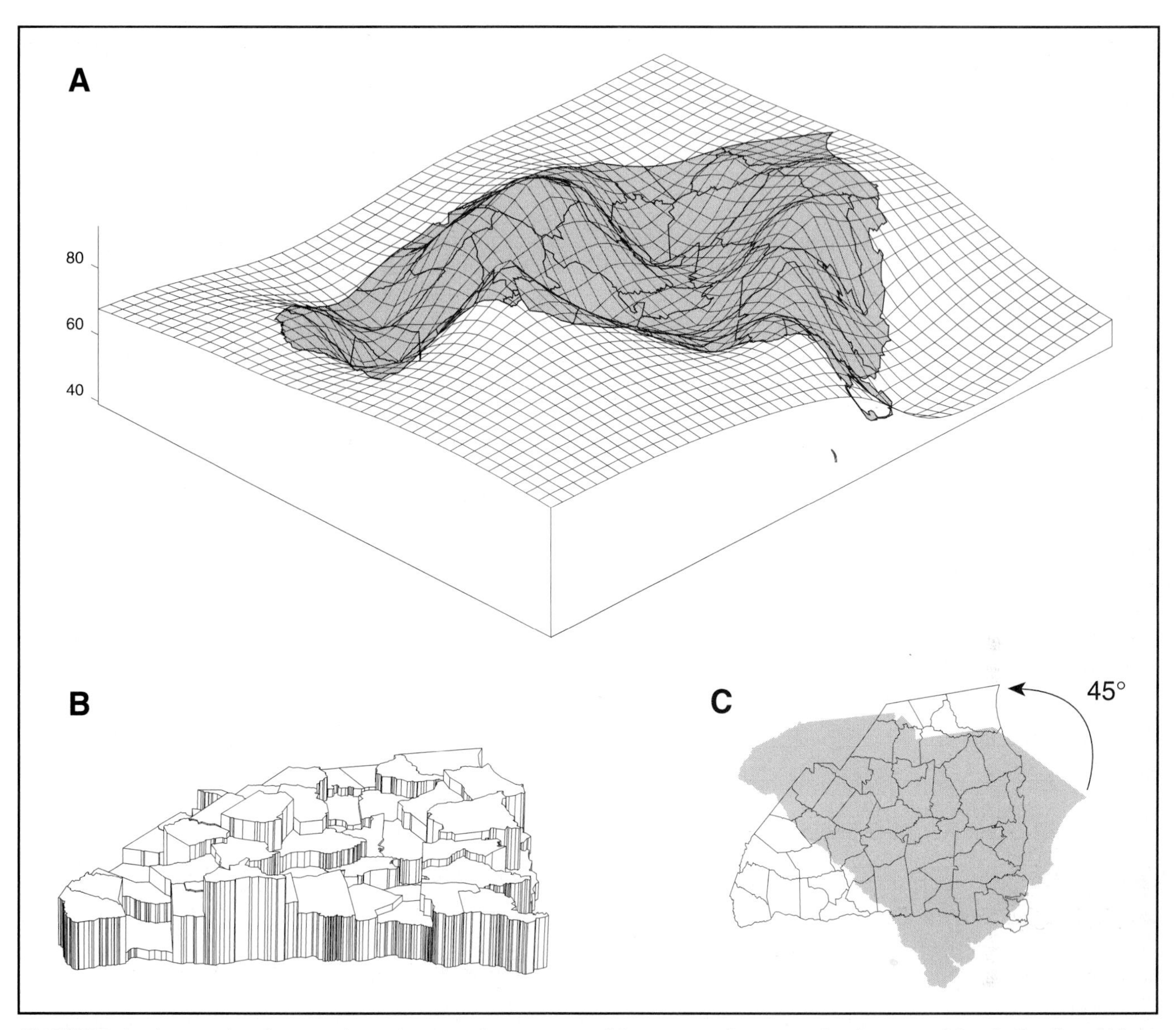

FIGURE 2.1 Approaches for mapping a data set of percentage of forest cover by county for the state of South Carolina: (A) the data are treated as coming from a smooth continuous phenomenon; (B) the data are treated as an areal phenomenon. Map C illustrates that maps A and B have been rotated 45 degrees from a traditional north-oriented map. For A, values outside the state are extrapolated, and thus must be treated with caution. (Data Source: South Carolina State Budget and Control Board 1994.)

cal difference between two categories. Classic examples are the Fahrenheit and Celsius temperature scales. Consider temperatures of 20°F and 40°F recorded in Fairbanks, Alaska, and Chattanooga, Tennessee, respectively. These two values are ordered, and they reveal the precise numerical difference between the two cities.

One problem with interval scales is the arbitrary nature of the zero point. In the case of the Celsius scale, 0 is the freezing point for pure water, while on the Fahrenheit scale, 0 is the lowest temperature obtained by mixing salt and ice. An *arbitrary zero point* means that ratios of two interval values cannot be interpreted correctly; for example, 40°F is numerically twice the value of 20°F, but it is not twice as warm (in terms of the kinetic energy of the molecules). An example of an interval scale familiar to academics is SAT scores, which range from a minimum of 200 to a maximum of 800. Note that it is not possible to say that an individual scoring 800 on an SAT exam did four times better than an individual scoring 200; all that can be said is that the individual scored 600 points better.

A **ratio** level of measurement has all the characteristics of an interval level, plus a *nonarbitrary zero point*. Continuing with the temperature example, the Kelvin scale is ratio in nature because at 0°K all molecular motion ceases; thus, a temperature of 40°K is twice as warm as 20°K (in terms of the kinetic energy of the molecules). Ratio data sets are more common than interval ones. For example, a perusal of maps shown in this text will reveal that most are based on ratio-level data.

Because many symbol forms can be used with both interval and ratio scales, these two levels of measurement are often grouped together and referred to as **numerical data.** Another way of grouping the levels of measurement is to combine ordinal, interval, and ratio to create **quantitative data,** which are contrasted with **qualitative data** (nominal level data).*

In addition to distinguishing among the levels of measurement, we will find it necessary to differentiate among three kinds of numerical data: bipolar, balanced, and unipolar (Eastman 1986). **Bipolar data** are characterized by either natural or meaningful dividing points. A *natural* dividing point is one that is part of the data and can be used intuitively to divide the data into two parts. An example would be a value of 0 for percentage of population change, which would divide the data into positive and negative percent changes. A *meaningful* dividing point does not occur inherently in the data, but can logically divide the data into two parts. An example would be the mean of the data, which would enable differentiating values above the mean from those below.

Balanced data are characterized by two phenomena that coexist in a complementary fashion. A straightforward example is the percentage of English and French spoken in Canadian provinces, as a high percentage of English-speaking people implies a low percentage of French-speaking people (the two are in "balance" with one another). **Unipolar data** have no natural dividing points and do not involve two complementary phenomena. Per capita income associated with the countries of Africa would be an example of unipolar data.

2.3 VISUAL VARIABLES

The term **visual variables** is commonly used to describe the various perceived differences in map symbols that are used to represent spatial phenomena. The notion of visual variables was developed by the French cartographer Bertin (1983) and subsequently modified by others, including McCleary (1983), Morrison (1984), DiBiase et al. (1991), and MacEachren (1994a). The approach used here is similar to MacEachren's, but differs primarily in the inclusion of 2½-D and 3-D phenomena and the use of the perspective-height visual variable. This chapter considers only visual variables for static maps. Additional visual variables for animated maps and for depicting data quality will be covered in sections 14.2.1 and 16.1.2, respectively.

* Statistics texts often include ordinal data within the qualitative data category.

The visual variables to be discussed are illustrated in Figure 2.2 and Color Plate 2.2. Note that the visual variables appear in the rows, while the columns represent the dimensions of spatial phenomena discussed in the preceding section. In discussing the visual variables, we will sometimes need to distinguish between the overall symbol and the marks making up a symbol. For example, note that the spacing visual variable shown for point phenomena consists of rectangular symbols, and that each rectangle is composed of parallel horizontal marks.

2.3.1 Spacing

The **spacing** visual variable involves changes in the distance between the marks making up the symbol (Figure 2.2). Cartographers traditionally used the term **texture** to describe these changes (Castner and Robinson 1969), but we will use the term spacing because texture has varied usages in the literature.

2.3.2 Size

Cartographers have used size as a visual variable in two different ways. One has been to change the size of the entire symbol, as is shown for the point and linear phenomena (Figure 2.2). Another is to change the size of individual marks making up the symbol, as for the areal, 2½-D, and true 3-D phenomena. This inconsistency may be a bit confusing, but the term *size* seems to reflect the visual differences that arise in each case. Note that for areal phenomena, the size of the entire areal unit could also be changed, as is done on cartograms, to be discussed in section 11.3.

2.3.3 Perspective Height

Perspective height refers to the perspective three-dimensional view of the phenomenon (Figure 2.2). It is interesting to consider some of the potential applications of this visual variable. In the case of point phenomena, oil production at well locations might be represented by raised sticks (or lollipops) above each well, with the stick height proportional to well production. For linear phenomena, total traffic flow between two cities over some time period could be represented by a fencelike structure above each roadway, with the height of the "fence" proportional to traffic flow. In the case of areal and 2½-D phenomena, we have already discussed examples for the forest cover data in South Carolina (Figure 2.1). Perspective height cannot be used for true 3-D phenomena because three dimensions are needed to locate the phenomenon being mapped.

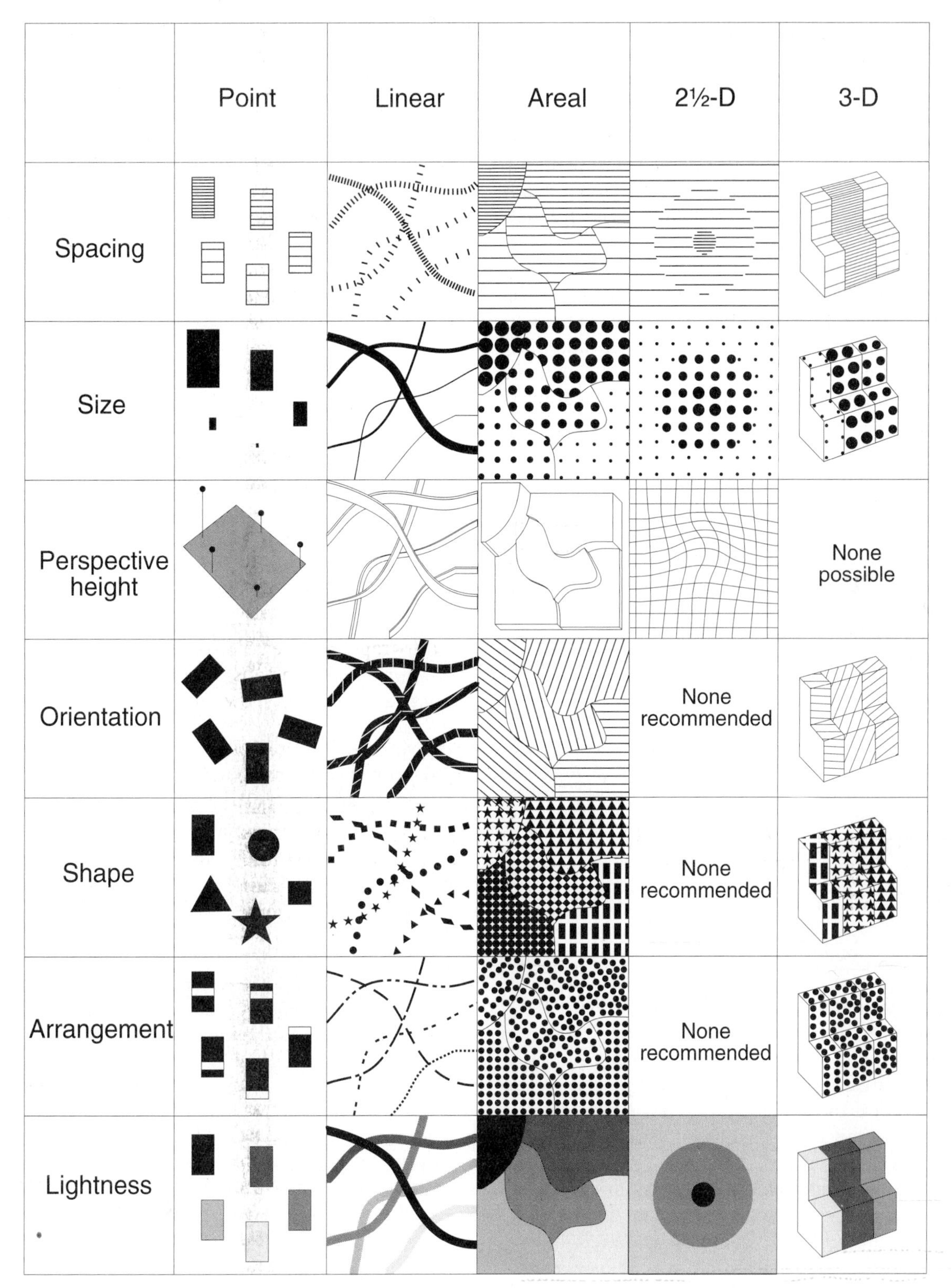

FIGURE 2.2 Visual variables for black-and-white maps. For visual variables for color maps, see Color Plate 2.2.

2.3.4 Orientation and Shape

As with the size visual variable, the **orientation** visual variable differs for the various types of spatial phenomena. For linear, areal, and true 3-D phenomena, orientation refers to the various directions of individual marks making up the symbol. In contrast, for point phenomena, orientation refers to the direction of the entire point symbol (Figure 2.2). (The logic is that the small size of symbols associated with point phenomena would make it difficult to determine the orientation of marks within a symbol.) Since orientation is most appropriate for representing qualitative data, orientation is not recommended for 2½-D phenomena, which are inherently numerical (we will discuss this issue more fully in section 2.5). Note that the **shape** visual variable is handled in a fashion similar to orientation.

2.3.5 Arrangement

Understanding the **arrangement** visual variable requires a careful examination of Figure 2.2. For areal and true 3-D phenomena, note that arrangement refers to how marks making up the symbol are distributed; marks for some areas are part of a square pattern, while marks for others appear to be randomly placed. For linear phenomena, arrangement refers to how lines are broken into a series of dots and dashes, as might be found on a map of political boundaries. Finally, for point phenomena, arrangement refers to changing the position of the white marker within the black symbol.

2.3.6 Hue, Lightness, and Saturation

The visual variables hue, lightness, and saturation are commonly recognized as basic components of color.* **Hue** is the dominant wavelength of light making up a color (the notion of wavelengths of light and the associated electromagnetic spectrum will be considered in detail in section 5.1.1). Another way of thinking about hue is that in everyday life it is the parameter of color most often used; for example, you might note that one person has on a red shirt and another a blue shirt. Color Plate 2.2 illustrates how red, green, and blue hues can be used to depict spatial phenomena.

Lightness (or **value**) refers to how dark or light a particular color is, while holding hue constant; for example, in Color Plate 2.2 different lightnesses of a red hue are shown. Lightness also can be shown as shades of gray (in the absence of what we commonly would call color), as in Figure 2.2.

Saturation (or **chroma**) can be thought of as a mixture of gray and a pure hue. This concept is illustrated in Color Plate 2.3, where the areal symbols shown for saturation in Color Plate 2.2 are arranged along a continuum from a desaturated red (grayish red) to a fully saturated red (while holding lightness constant).

* See Brewer (1994a) for a discussion of terminology associated with color.

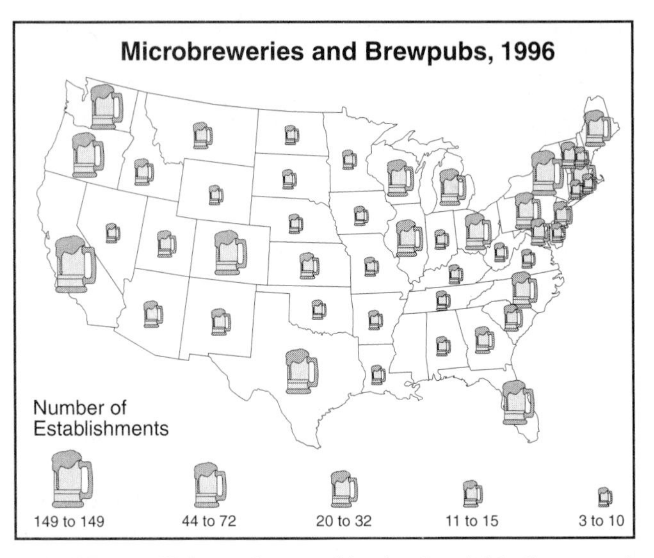

FIGURE 2.3 Using a pictographic visual variable (beermugs) to represent the number of microbreweries and brewpubs in each U.S. state. The construction of this map will be discussed in more detail in Chapter 7. (Data Source: http://www.beertown.org.)

2.3.7 Some Considerations in Working with Visual Variables

The reader should bear in mind that Figure 2.2 and Color Plate 2.2 depict only a fraction of the many symbols that could be used to depict the visual variables; for example, either circles or squares might be used with the size visual variable. A major group of symbols not shown in the figures are **pictographic symbols** (as opposed to **geometric symbols**), which are intended to look like the phenomenon being mapped. For instance, Figure 2.3 illustrates the use of beer mugs (based on the size visual variable) to represent the number of microbreweries and brewpubs in each U.S. state. Another interesting example of pictographic symbols can be found at http://www.geog.ucsb.edu/~kclarke/bigelectionmap.jpg, which depicts voting results by state in the 1996 U.S. presidential election using faces of the candidates.

Also keep in mind that the visual variables can serve as basic building blocks for creating more complex representations. For example, Figure 2.4 illustrates

FIGURE 2.4 Combining the visual variables spacing and size. (After MacEachren 1994a, p. 26.)

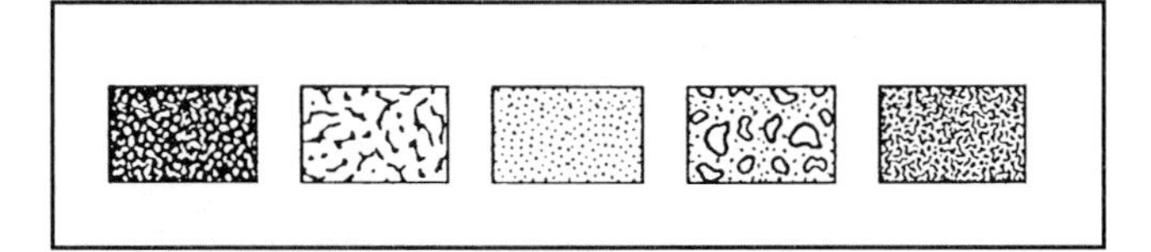

FIGURE 2.5 Different "patterns" or "textures" that can be created to portray nominal information. Note that these are not readily described in terms of the visual variables shown in Figure 2.2.

how the visual variables spacing and size might be combined. MacEachren (1994a, 27) called the resulting symbol a form of texture. In my opinion, such an approach produces a rather coarse-looking map, but it can highlight various aspects of a distribution; for example, MacEachren noted that it emphasized the northeast-southwest trend on the map (see Figure 2.19 of his work).

In Figure 2.2 and Color Plate 2.2, the term for each visual variable used appears to be a clear expression of the visual differences that we see; for example, in the case of the orientation visual variable for point phenomena, we see that one rectangle is at a different orientation than another. Moreover, if we wanted, we could compute a mathematical expression of this difference (that one rectangle is rotated 40 degrees from a vertical, while another is rotated 50 degrees). Sometimes, describing the visual difference between symbols is not so easy. For example, try describing the differences between the symbols shown in Figure 2.5. Such symbols are often referred to as differing in "pattern" or "texture" and are frequently used to symbolize nominal data.

It also should be noted that the visual variable location was not explicitly depicted in the illustrations. **Location** refers to the position of individual symbols. I chose not to illustrate this visual variable because it is an inherent part of mapping (for example, each symbol shown for point phenomena can be defined by the x and y coordinate values of its center). If location were illustrated, it would be represented by constant symbols (identical dots for point phenomena) that varied only in position.

2.4 COMPARISON OF CHOROPLETH, PROPORTIONAL SYMBOL, ISOPLETH, AND DOT MAPPING

This section defines and contrasts four common mapping techniques: choropleth, proportional symbol, isopleth, and dot. For illustrative purposes, we will contrast these techniques by mapping data for acres of wheat harvested in Kansas counties. We will see that which technique is appropriate is a function of both the nature of the underlying phenomenon and the purpose for making the map. Here we consider a basic introduction to these mapping techniques; more advanced concepts are covered in subsequent chapters (4 and 6 for choropleth, 7 for proportional symbol, 8 for isopleth, and 10 for dot).

2.4.1 Choropleth Mapping

Choropleth maps are created by shading enumeration units with an intensity proportional to the data values associated with those units. The choropleth map is clearly appropriate for areal phenomena, in which values change abruptly at enumeration unit boundaries. Choropleth maps may also be appropriate when the mapmaker wishes to focus on "typical" values for individual enumeration units, even though the underlying phenomenon does not change abruptly at enumeration unit boundaries. For example, politicians and government officials might use this approach when stressing how one county or state compares with another. Although choropleth maps are commonly used in this fashion, it is important to recognize two major limitations: (1) they do not portray the variation that may actually occur within enumeration units, and (2) the boundaries of enumeration units are arbitrary, and thus unlikely to be associated with major discontinuities in the actual phenomenon.*

An important consideration in constructing choropleth maps is *data standardization,* in which raw-count data are adjusted for differing sizes of enumeration units. To understand the need to standardize, consider map A of Figure 2.6, which portrays a hypothetical distribution consisting of three distinct regions: S, T, and U. Note that regions S and T have equal-sized enumeration units, each 16 acres in size. In contrast, region U has enumeration units four times the size of those in S and T, or 64 acres in size. Let's presume that the number of acres of wheat harvested from enumeration units in each region is as follows: 0 in S, 16 in T, and 64 in U (these numbers are shown within each enumeration unit in Figure 2.6A).

The acres of wheat harvested from each enumeration unit represent *counts* (or *totals*). Mapping these counts with the choropleth method produces the result shown in Figure 2.6B. A user examining this map would likely conclude that since region U is the darkest, it must have more wheat grown in it. Unfortunately, this conclusion would be inappropriate because no account has been made for the sizes of the enumeration units. One approach to

* See Langford and Unwin (1994) for a more detailed discussion of these limitations. We will consider their work in section 10.4.

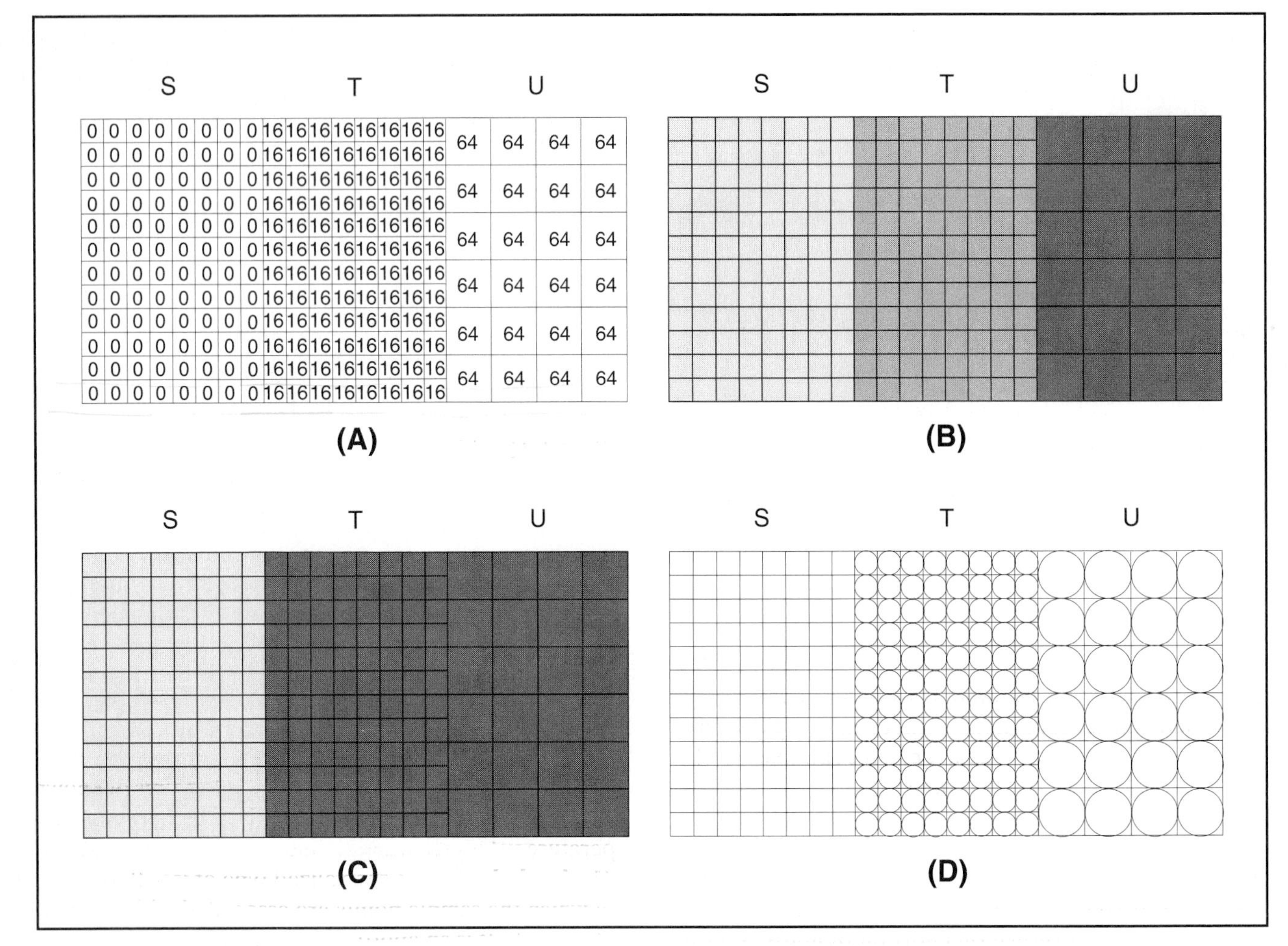

FIGURE 2.6 A hypothetical illustration of the effect of data standardization: (A) the raw data—number of acres of wheat harvested in each enumeration unit; (B) a choropleth map of the raw data; (C) a choropleth map of standardized data achieved by dividing the raw data by the area of the corresponding enumeration unit; and (D) a proportional circle map of the raw data.

adjust (or standardize) for the sizes of enumeration units is to divide each of the raw counts by the area of the associated enumeration unit; the resulting values are 0/16, or 0 for region S; 16/16, or 1 for region T; and 64/64, or 1 for region U. Mapping these values with the choropleth method results in Figure 2.6C; note that regions T and U are now identically shaded.

The Kansas wheat data can be standardized by dividing the acres harvested within each county by the area of the corresponding county. Figure 2.7 portrays maps of both unstandardized and standardized data. The results are not as dramatic as the hypothetical example because the areas of Kansas counties are similar in size, but some differences are still apparent. For example, note that the relatively large county in the northwestern part of the state (Thomas) shifts from the highest to the adjacent lower class, and that the irregularly shaped county in the southwestern part of the state (Finney) shifts from the highest to the middle class.

Standardization not only adjusts for the differing sizes of counties, but also provides a very useful variable, namely the proportion (or percentage) of each county from which wheat was harvested. Such a variable provides an indication of the probability that one might see wheat being cut at harvest time while driving through the county. The resulting map is shown in Figure 2.8A for comparison with the other methods of mapping discussed in this section. (Note that the *lightness* visual variable has been used for the choropleth map.)

2.4.2 Proportional Symbol Maps

A **proportional symbol map** is constructed by scaling symbols in proportion to the magnitude of data occurring at point locations, whether these locations are *true points*, such as an oil well, or *conceptual points*, such as the center of an enumeration unit; for the wheat data, the latter is the case since the data were collected over coun-

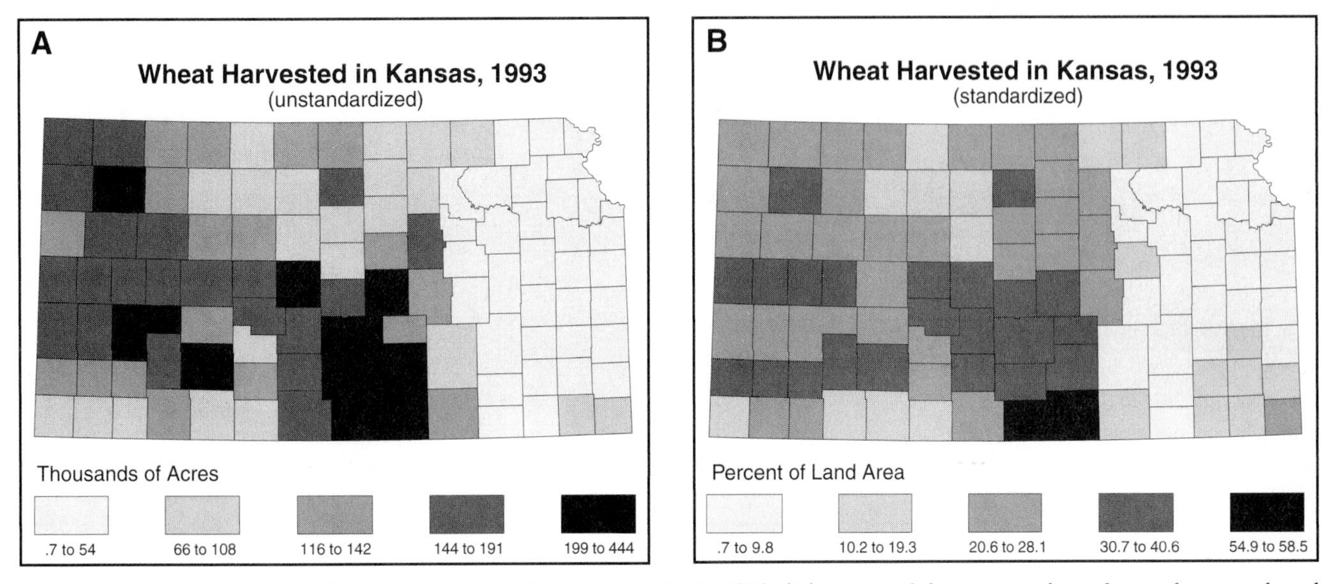

FIGURE 2.7 Standardizing wheat production in Kansas counties in 1993: (A) a map of the raw number of acres harvested; and (B) a standardized map resulting from dividing number of acres harvested by the area of each county. (Data Source: Kansas Agricultural Statistics 1994.)

ties. In contrast to the standardized data depicted on choropleth maps, proportional symbol maps are normally used to display raw-count data. Thus, the raw magnitudes for acres of wheat harvested are shown as proportional circles in Figure 2.8C. (Note that the visual variable used here is *size.*)

The raw magnitudes depicted on proportional symbol maps provide a necessary complement to the standardized data shown on choropleth maps. Magnitudes are important because a high proportion or rate may not be meaningful if there is not also a high magnitude. As an example, consider counties of the same size having populations of 100 and 100,000 in which 1 and 1,000 people, respectively, have some rare form of cancer. Dividing the cancer values by the population values yields the same proportion of people suffering from cancer (.01), but the rate for the less populous county would be of lesser interest to the epidemiologist.

Although the proportional symbol map may be a better choice than the choropleth map for count data, care should be taken in using it. To illustrate, consider map D of Figure 2.6, which displays the hypothetical wheat data using proportional circles. Note that all circles in region U are larger than those in region T. This may lead to the mistaken impression that counties in region U are more important in terms of wheat production than those in region T. Counties in region U may be more important to a politician in assigning tax dollars (more wheat harvested indicates a greater tax is appropriate), but in terms of the density of wheat harvested, regions T and U are identical.

2.4.3 Isopleth Map

An **isarithmic map** (or **contour map**) is created by interpolating a set of isolines between sample points of known values; for example, we might draw isolines between temperature values recorded for individual weather stations. The **isopleth map** is a specialized type of isarithmic map in which the sample points are associated with enumeration units. It is an appropriate alternative to the choropleth map when one can assume that the data collected for enumeration units are part of a smooth continuous (2½-D) phenomenon. For example, in the case of the wheat data, it might be argued that the proportion of land in wheat changes in a relatively gradual (smooth) fashion, as opposed to changing just at county boundaries (as on the choropleth map).

In a fashion similar to a choropleth map, an isopleth map also requires standardized data. Referring again to the hypothetical raw wheat harvested data shown in Figure 2.6A, imagine drawing contours through such data. High-valued contour lines would tend to occur in region U, where there are high values in the raw data; but as has already been shown for the choropleth case, region U is really no different from region T. Dividing the raw data by the area of each enumeration unit would result in standardized data that could be appropriately contoured.

The isopleth map resulting from contouring the standardized Kansas wheat data is shown in Figure 2.8B. (Again, note that the visual variable lightness has been used.) Although this map may be more representative of the general distribution of wheat harvested than the choropleth map, the assumption of continuity and the

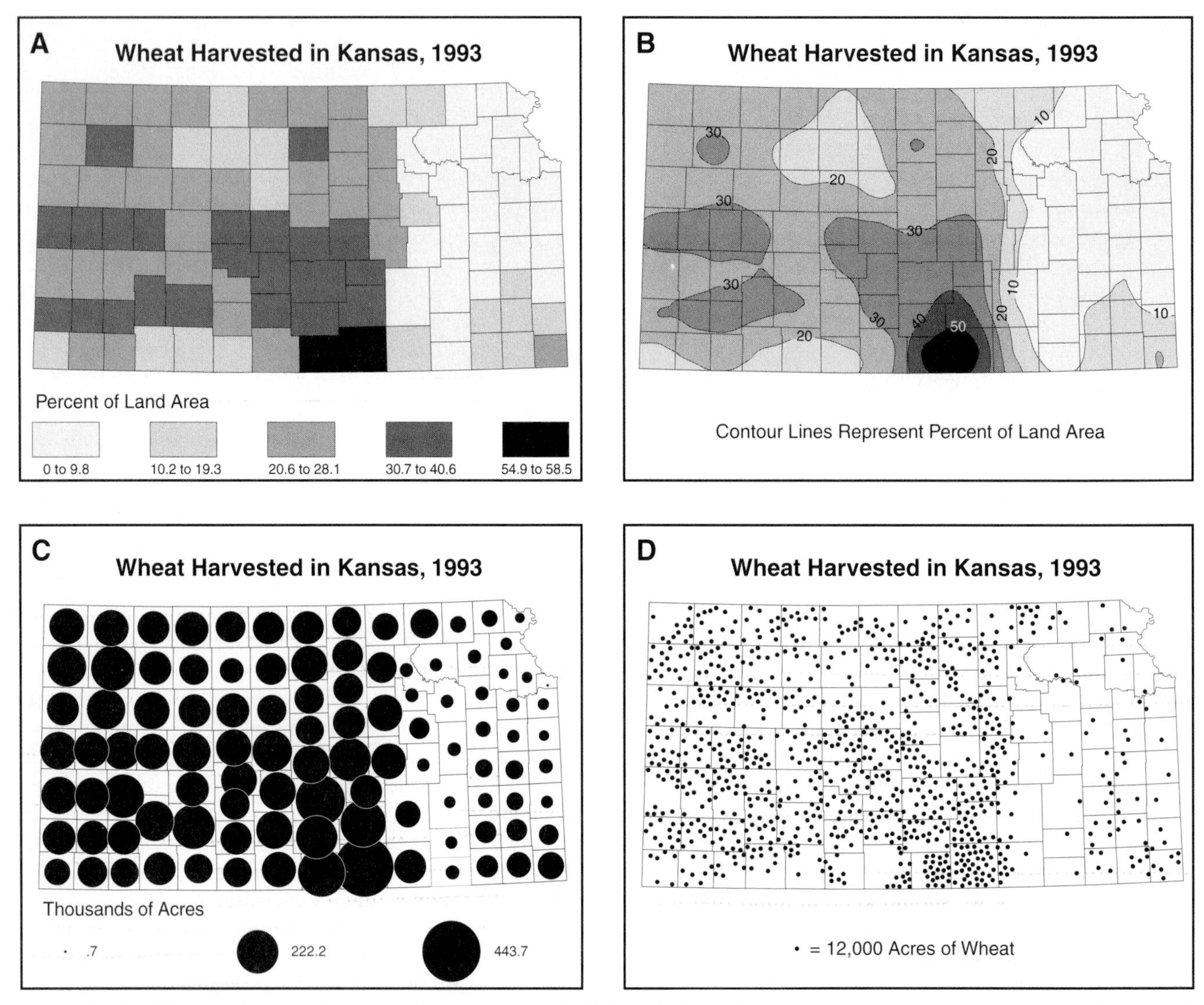

FIGURE 2.8 Maps of wheat production in Kansas counties in 1993: (A) choropleth map of the percentage of land area from which wheat was harvested; (B) isopleth map of the percentage of land area from which wheat was harvested; (C) proportional symbol map of acres of wheat harvested; and (D) dot map of acres of wheat harvested.

use of county-level data produce some questionable results. For example, note the island of higher value near the center of the extreme southeastern county (Cherokee). It seems unlikely that you would find a higher value here in reality; the high value is more likely a function of the fact that the centers of counties were used as a basis for contouring and Cherokee's value was higher than any of the surrounding counties. Note that a similar problem occurs within two northern counties (Figure 2.8B). The dot map may be a solution to this type of problem.

2.4.4 Dot Mapping

To create a **dot map,** one dot is set equal to a certain amount of a phenomenon, and dots are placed where that phenomenon is most likely to occur. The phenomenon may actually cover an area or areas (e.g, a field or fields of wheat), but for sake of mapping, the phenomenon is represented as located at points. Constructing an accurate dot map requires collecting ancillary information that indicates where the phenomenon of interest (wheat, in our case) is likely found. For the wheat data, this was accomplished using the cropland category of a land use/land cover map (the detailed procedures are described in Chapter 10). The resulting dot map is shown in Figure 2.8D. (In this case, the visual variable *location* is used.) Clearly, the dot map is able to represent the underlying phenomenon with much more accuracy than any of the other methods we have discussed. Also note that parts of the distribution exhibit sharp discontinuities that would be difficult to show with the isopleth method (which presumes smooth changes).

2.4.5 Discussion

An examination of Figure 2.8 reveals that each of the four maps provides a quite different picture of wheat production in the state of Kansas. It can be argued that the method used should depend on the purpose of the map. If the purpose is to focus on "typical" county-level information, then the choropleth and proportional symbol maps are appropriate. The choropleth map provides standardized information, while the proportional symbol map provides raw-count information. It must be emphasized that neither map depicts the detail of the underlying phenomenon, which is unlikely to follow enumeration unit boundaries.

When data are collected in the form of enumeration units, the dot and isopleth methods should be considered as two possible solutions for representing an underlying phenomenon that is not coincident with enumeration unit boundaries. In the case of the wheat data, the dot method is probably the more appropriate approach because it can capture some of the discontinuities in the phenomenon. The isopleth method, however, could probably be improved on with a finer grid of enumeration units (such as townships); of course, this would also be true of the choropleth and proportional symbol maps.*

It must be noted that we have only considered four of the more common methods of thematic mapping. Another alternative would be a **dasymetric map,** which, like the dot map, can show very detailed information, but uses standardized data. The dasymetric map will be covered in Chapter 10. Finally, we should keep in mind that if the maps were to be viewed in an interactive graphics environment, the mapmaker would have the option of showing several of them, and thus provide the user with various perspectives on the distribution of wheat harvested in Kansas.

2.5 SELECTING VISUAL VARIABLES FOR CHOROPLETH MAPS

In the preceding section, the visual variable lightness was utilized on both the choropleth and isopleth maps. An examination of Figure 2.2 and Color Plate 2.2 reveals that there are a number of other visual variables that might be used to represent a phenomenon that is treated as areal in nature. This section considers how we might select among these various visual variables. We will see that the basic solution is to select a visual variable that appears to "match" the level of measurement of the data. For illustrative purposes, we will again use the Kansas wheat data, presuming that a choropleth map is to be produced.

The specific visual variables we will discuss are illustrated in Figure 2.9 and Color Plate 2.4. In examining these figures, you will note that they depict classed maps using *maximum-contrast symbolization,* which means that symbols for classes have been selected so that they are maximally differentiated from one another. An alternative approach would be to create an unclassed map in which symbols are directly proportional to the value for each enumeration unit (as in Figure 1.6B). The maximum-contrast approach is used here because it is common and more easily constructed (particularly in the case of the size visual variable).

In addition to discussing Figure 2.9 and Color Plate 2.4, we will also consider Figure 2.10, which summarizes the use of visual variables for various levels of measurement. Note that the body of this figure is shaded and labeled to indicate various levels of acceptability: Poor (P), Marginally effective (M), and Good (G). MacEachren (1994a, 33) developed a similar figure, which he appeared to apply to all kinds of spatial phenomena. I intend the reader to use Figure 2.10 only for areal phenomena; as an exercise, students might consider developing such a figure for other kinds of phenomena.

We'll consider the perspective-height and size visual variables first because they have the greatest potential for logically representing the numerical data shown on choropleth maps. Use of perspective height produces what is commonly termed a *prism map* (Figure 2.9A). In Figure 2.10, note that perspective height is the only visual variable receiving a "good" rating for numerical data. The justification is that an unclassed map based on perspective height can portray ratios correctly (a data value twice as large as another will be represented by a prism twice as high), and that readers perceive the height of resulting prisms as ratios (Cuff and Bieri 1979).

There are two problems, however, that complicate the extraction of numerical information from prism maps. One is that tall prisms may sometimes hide smaller ones. One solution to this problem is to rotate the map so that blockage is minimized; for example, the map for wheat has been rotated so that it is viewed from the lower-valued northeast. A second solution is to manipulate the map in an interactive graphics environment. If a flexible program is available, it may even be possible to suppress selected portions of the distribution so that other portions may be seen. A third solution is to use the perspective height variable but also symbolize the distribution with another visual variable; for example, Figure 2.9D might be displayed in addition to Figure 2.9A.

Another problem with prism maps is that rotation may produce a view that is unfamiliar to readers who normally see maps with north at the top. This problem can be handled by showing a second map (as suggested above)

* Data at the township level are not released to the general public in order to protect the confidentiality of individual farm production.

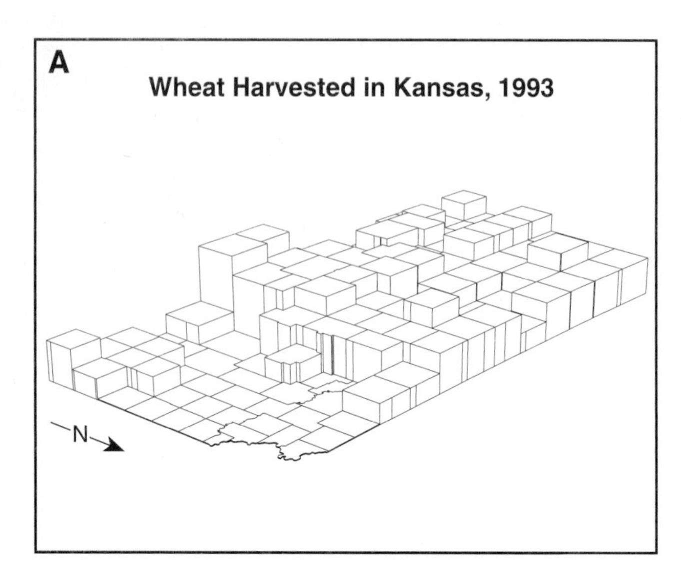

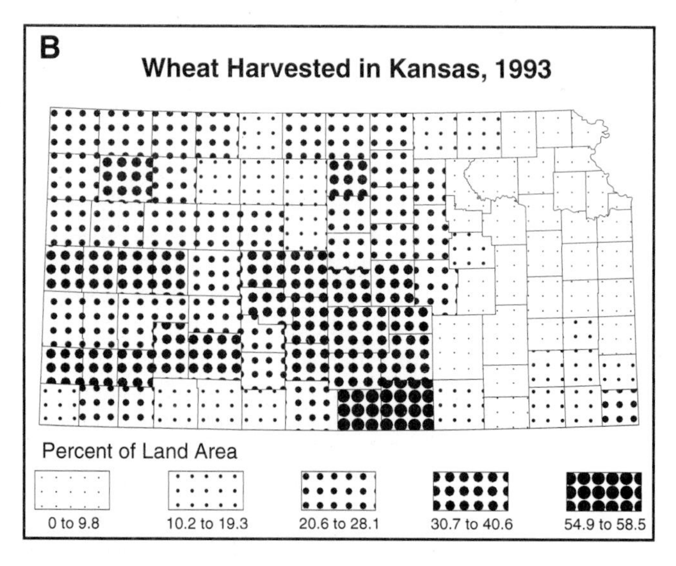

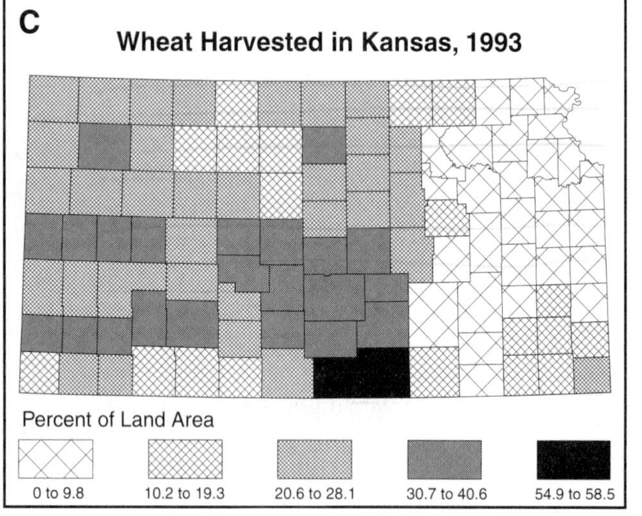

D

Wheat Harvested in Kansas, 1993

Percent of Land Area

0 to 9.8
10.2 to 19.3
20.6 to 28.1
30.7 to 40.6
54.9 to 58.5

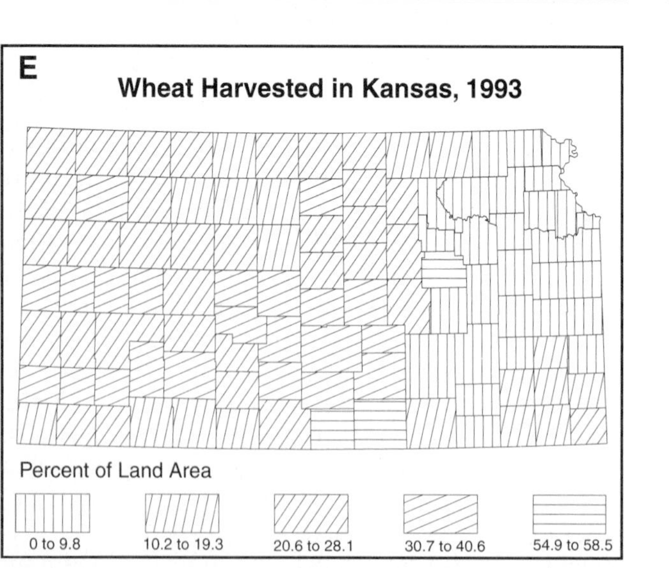

FIGURE 2.9 Representing the percentage of wheat harvested in Kansas counties using different visual variables: (A) perspective height, (B) size, (C) spacing, (D) lightness, and (E) orientation. For color visual variables, see Color Plate 2.4.

	Nominal	Ordinal	Numerical
Spacing	P	M[c]	M[c]
Size	P	M	M
Perspective height	P	M[a]	G[b]
Orientation	G	P	P
Shape	G	P	P
Arrangement	G	P	P
Lightness	P	G	M
Hue	G	G[d]	M[d]
Saturation	P	M	M

P = Poor **M** = Marginally Effective **G** = Good

[a] Since height differences are suggestive of numerical differences, use with caution for ordinal data.
[b] Hidden enumeration units and lack of a north orientation are problems.
[c] Not aesthetically pleasing.
[d] The particular hues selected must be carefully ordered, such as yellow, orange, red.

FIGURE 2.10 Effectiveness of visual variables for each level of measurement for areal phenomena. (After MacEachren 1994a, 33.)

or by using an overlay of the base to show the amount of rotation (as in Figure 2.1C).

The size visual variable is illustrated in Figure 2.9B; note that here the size of individual marks making up the areal symbol has been varied. Size might be argued to be appropriate for representing numerical relations because circles can be constructed in direct proportion to the data (a data value twice another can be represented by a circle twice as large in area). Furthermore, readers should see the circles in approximately the correct relations. (However, we will see in Chapter 7 that a correction factor may have to be implemented to account for underestimation of larger circles.)

Although some cartographers (most notably Bertin) have used the above sort of argument to promote the use of the visual variable size on choropleth maps, two problems are apparent. One is that it is questionable whether map users actually consider the sizes of circles when used as part of an areal symbol. Users might analyze circle size when trying to acquire specific information, but it seems unlikely that they would do so when analyzing overall map pattern. Rather, it is more likely that they would perceive areas of light and dark, in a fashion similar to the lightness visual variable. Another problem is that many cartographers (and presumably map users) find the coarseness of the resulting symbols unacceptable, and would prefer the finer shades typically used for hue, lightness, and saturation. The latter problem in particular caused me to give the size variable only a moderate rating for portraying numerical data (Figure 2.10).

Note also that both perspective height and size have been given only moderate ratings for portraying ordinal data. The logic is that if such variables are used to illustrate numerical relations, users may perceive such relations when only ordinal relations are intended. Obviously, both variables are inappropriate for nominal data because different heights and sizes suggest quantitative rather than qualitative information.

Although other visual variables can be manipulated mathematically to create proportional (ratio) relationships, individual symbols cannot be interpreted easily in a ratio fashion. For example, consider the lightness variable shown in Figure 2.9D. It is easy to see that one shade is darker or lighter than another, but it is difficult to establish ratio relations (that one shade is twice as dark as another).

Comments similar to that made for lightness can be applied to the visual variables spacing, saturation, and hue, with the following caveats. First, note that I have given spacing only a moderate rating for ordinal information because of the coarseness of symbols for low data values (see Figure 2.9C). Such coarseness is not aesthetically pleasing and seems to imply that low data values are qualitatively different from high data values. For saturation, only a moderate rating is given for ordinal information because it is my experience that people have a difficult time understanding what a "greater" saturation means.

I rated hue as "good" for both nominal and ordinal data because certain hues work well for nominal data, while others work better for ordinal data. For example, to display different soil types (alfisols, entisols, mollisols) within soil-mapping units, red, green, and blue hues might be deemed appropriate (one of these hues does not inherently represent more than another). For ordinal and higher-level data, logically ordered hues are necessary; for example, a yellow, orange, and red scheme (Color Plate 2.4B) is one possibility because orange is seen as a mixture of yellow and red (based on opponent process theory, a topic to be covered in Chapters 5 and 6).

The remaining visual variables (orientation, shape, and arrangement) are appropriate only for creating nominal differences. As an example, consider the orientation variable. The mapmaker might try to create a logical progression of symbols by starting with vertical lines for low values and then gradually change the angle of the lines so that the highest class is represented by horizontal lines, but an examination of Figure 2.9E reveals that this approach is not effective; the changing angle of the lines appears to create nominal differences, and the resulting map is "busy."

It should be noted that ratings provided in Figure 2.10 are based on my experience as a cartographer; individual readers of this text may develop slightly different ratings. For example, one of my students rated the orientation variable "poor" (even for nominal data) because he felt it lacked aesthetic quality and that it was difficult to discriminate among different orientations.

2.6 DATA STANDARDIZATION

The basic concept of data standardization was introduced in section 2.4. Data were standardized by dividing an area-based count variable (acres of wheat harvested) by the areas of enumeration units. The numerator and denominator were both in the same units of measurement (acres), so a proportion (or percentage) resulted. This section considers a number of additional approaches for standardizing data.

(1) One approach is to divide an area-based count variable by some other area-based count variable. For example, for acres of wheat harvested, we might divide by acres of wheat planted. The resulting ratio would provide a measure of success of the wheat crop (Figure 2.11A). As an

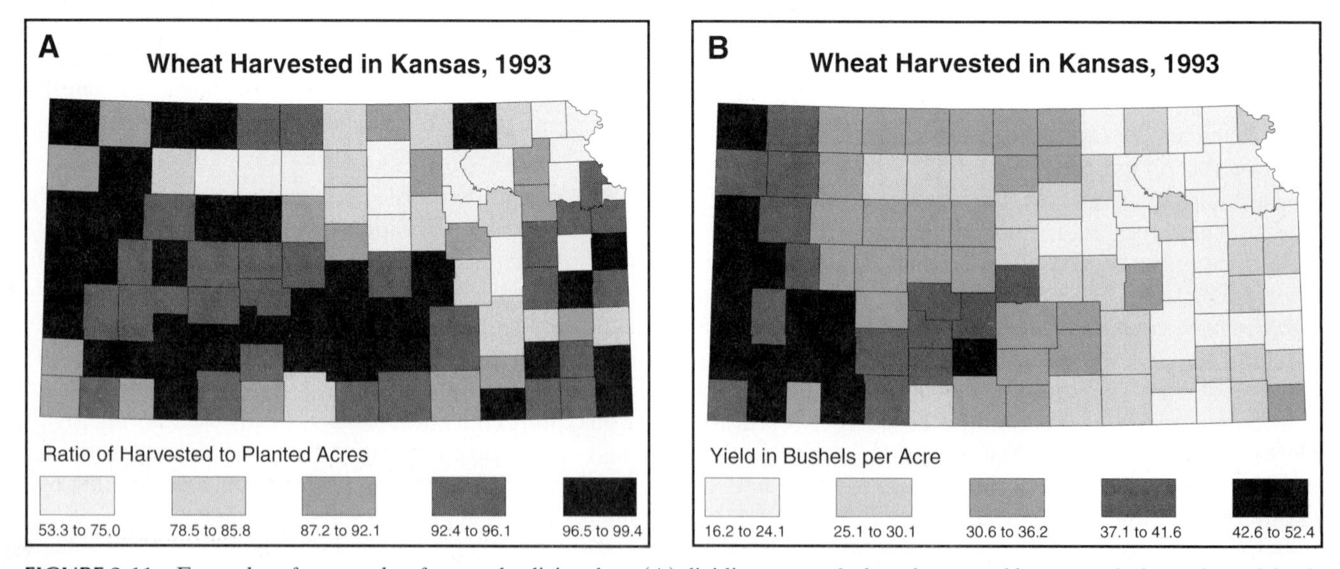

FIGURE 2.11 Examples of approaches for standardizing data: (A) dividing acres of wheat harvested by acres of wheat planted; both are area-based count variables and thus a proportion (or percentage) results. (B) Dividing bushels of wheat harvested by acres of wheat harvested; the result is a density measure (bushels per acre). (Data Source: Kansas Agricultural Statistics 1994.)

alternative, we might divide by acres harvested for all major crops, producing a map illustrating the relative importance of wheat to the agricultural economy of each county.

A second approach is to create a density measure by dividing a count variable not involving area by either the areas of enumeration units or some area-based count variable. For example, if bushels of wheat harvested are divided by the area of the county (in acres), then the result is bushels of wheat harvested per acre. This approach may not be meaningful, however, if the count data are known to occur only within a portion of the enumeration unit. For example, with the wheat data, it makes more sense to divide bushels harvested by acres of wheat harvested (Figure 2.11B).

A third approach is to compute the ratio of two count variables, neither of which involves area. For example, we might divide the value of the wheat harvested (in dollars) by the value of all crops harvested (in dollars). The resulting proportion would indicate the relative value of wheat in each county. Although area would not be included in the formula, the approach would indirectly standardize for area because larger areas would tend to have larger values for both variables.

When computing the ratio of two count variables not involving area, it is common to express the result as a *rate*. Although we are all familiar with rates, we often aren't aware of how they are computed. To illustrate, consider how cancer death rates are computed for counties in the United States. First, we establish a simple proportional relationship as follows:

$$\frac{\text{Cancer deaths for county}}{\text{Population for county}} = \frac{\text{number of cancer deaths}}{100{,}000 \text{ people}}$$

We then solve for the number of cancer deaths by rewriting the equation as follows:

$$\text{Number of cancer deaths} = \frac{\text{cancer deaths for county}}{\text{population for county}} \times 100{,}000$$

The resulting number of cancer deaths is termed the cancer death rate. More generally, the formula for rates is:

$$\text{Rate} = \frac{\text{magnitude for category of interest}}{\text{maximum possible magnitude}} \times \text{units of the rate}$$

A fourth standardization approach is to compute a summary numerical measure (such as the mean or standard deviation) for each enumeration unit. For example, we could compute the average size of farms in each county by dividing the acreage of all farms by the number of farms. Note that this approach accounts for the larger acreage in a larger county by dividing by a greater number of farms.

In general, note that all of the standardization approaches discussed thus far involve ratios. Thus, one might suggest that simply computing ratios is the key to standardization. A simple example illustrates that this is not the case. Imagine that you computed the mean number of acres of wheat harvested over a 10-year period for each county. Clearly, the values would be ratios (the numerator would be the sum of acreages over 10 years and the denominator would be 10), but the data would not be standardized because the denominator in each case would be 10. No adjustment would have been made to account for the fact that larger counties tend to have larger acres harvested in each year.

The major purpose of this section has been to illustrate various approaches for data standardization, and to stress that data standardization is necessary in order for a choropleth map to be meaningful. It is also important to realize that the various approaches lead to quite different maps of the basic phenomenon (wheat harvested) being investigated. For example, Figure 2.8A (representing the proportion of land area from which wheat was harvested) reveals a peak in the south-central part of the state, with a general tendency for higher values in the western two-thirds of the state. In contrast, Figure 2.11A (illustrating the percentage of planted wheat actually harvested) reveals a relatively indistinct pattern, with the exception of the lowest class extending across the northeastern and north-central sections of the state. The map of wheat yield (Figure 2.11B) produces yet another distinctive pattern, with progressively higher values as one moves westward across the state. Clearly, the method of standardization has a major impact on our view of wheat harvested in Kansas.

This discussion has focused on standardizing data for choropleth maps. Data standardization can (and often should) be implemented for other mapping techniques. For example, consider making a proportional symbol map of deaths resulting from AIDS for major U.S. cities. One could make a map of the raw death data, but this would most likely correlate with city population, as cities of larger population would tend to have a larger number of deaths. More interesting would be a map of death rates (for example, computing the number of AIDS deaths per 100,000 population). We will consider the issue of standardization for proportional symbol maps in more detail in Chapter 7.

2.7 SOME BASIC MAP DESIGN ELEMENTS

When a map is used for presentation (as opposed to data exploration), decisions must be made concerning several map design elements, including the title, legend, source, north arrow, scale, labeling of thematic information (for example, the data values or names associated with enumeration units), and **base information** (information that is not the major theme, but that provides a frame of reference for the theme). This section considers several issues related to these design elements, including (1) which of them are necessary, (2) what their content should be, (3) how they should be positioned, and (4) selection of appropriate font, style, and type size.

For illustrative purposes, we will map the percentage of religious or private secondary schools in Kenya in 1984. We will also consider the relationship of railroads (a form of base information) to the distribution of schools. (An Africanist in my department hypothesized that railroads would tend to be found in association with a high percentage of religious and private schools because such schools were developed by Europeans, who settled near railroads.)

2.7.1 Which Design Elements to Include?

Some cartographic texts have suggested that five basic design elements (title, legend, source, north arrow, and scale) should be included on any thematic map intended for presentation. The title, legend, and source are clearly critical to understanding the map, but I believe the north arrow and scale are optional, especially for areas that the reader is apt to be familiar with. Rather than incorporating these elements simply because they are commonly included by others, the map designer should ask whether they enhance map understanding. Incorporating such elements when they are not essential for interpretation only adds visual clutter and detracts from communicating the primary theme being mapped. With these thoughts in mind, a well-designed map of the secondary school data might include only a title, legend, and source (Figure 2.12).

An important question is whether individual data values for the theme being mapped should be labeled. Oftentimes, values are labeled because the naive map designer deems it important that the reader acquire specific information. Remember, however, that a major reason for making a map is to examine spatial pattern. If it is critical that individual values be acquired, they might better be placed in a table; if they are shown on the map, they should be small enough not to detract from the spatial pattern. Note that in the case of Kenya this would be difficult, as many of the enumeration units are small.

Traditionally, political boundaries have been the only base information included on thematic maps, but this has changed with the availability of inexpensive GIS software: With the click of a mouse, it is now easy to add a layer of base information (major roads, streams, or topography) to the map. Although highly desirable for data exploration, additional base information should be used with care when the map is intended for presentation. Rather than adding base information simply because the software permits it, the mapmaker should select such information based on its relevance to the theme being mapped, and the recognition that the added information may detract from communicating the major theme.

Figure 2.13 illustrates how the base information for railroads in Kenya can be combined with the secondary schools theme. Note that the white railroads contrast with the gray tones used to display the secondary school data. Also note that the location of the railroads appears to coincide with districts having the highest percentage of secondary schools.

Other elements that may be incorporated on thematic maps intended for presentation include graphical displays (such as a histogram), numerical summaries (such as the mean and standard deviation), and a map border (or frame). Graphical displays and numerical summaries should be used only for maps that are likely to be used by more statistically sophisticated readers. A map border is useful for visually separating the map from the text.

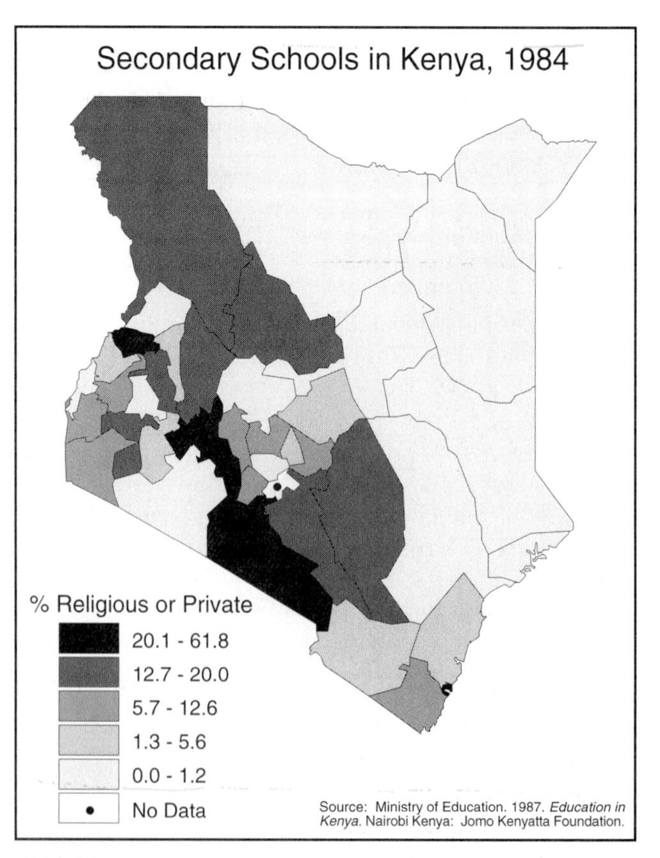

FIGURE 2.12 A well-designed map (compare with Figures 2.15 and 2.19).

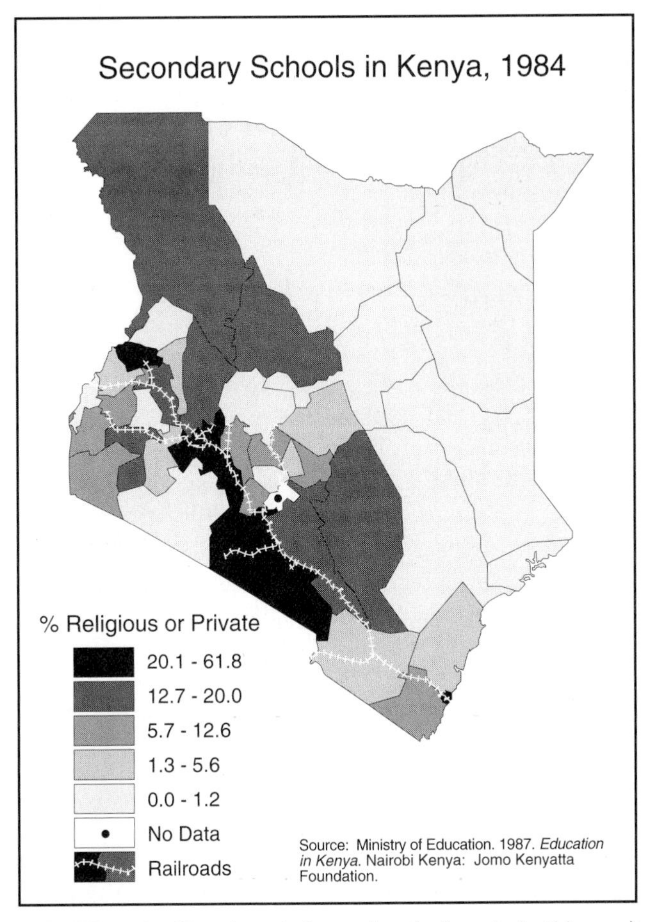

FIGURE 2.13 How base information (railroads, in this case) can be combined with a major theme. (Compare with Figure 2.12.)

2.7.2 Specifying the Contents of the Title, Legend, and Source

Generally, the title should indicate the major theme being mapped, the geographic region, and the date of the data. Thus, potential titles for the Kenyan example would be "Secondary Schools in Kenya, 1984" or "Secondary Schools in Kenya by District, 1984." A short, easily understood title is preferred over a long jargon-filled one such as "A Map of Concentrations of Secondary Schools in Enumeration Units of Kenya." It may be possible to eliminate either the region or date from the title if the framework within which the map is used makes such information readily apparent (as when all maps in an atlas are of the same geographic region for a particular time period).

The legend should include a clear explanation of what each symbol associated with the theme represents; base information should also be explained in the legend if it is not obvious. The legend title generally refines the topic presented in the overall map title ("% Religious or Private" in the Kenyan example). If no refinement of the map title seems essential, then no legend title need be shown; avoid using terms such as "Legend" and "Key" since the intent of the legend is obvious.

Text for each symbol element within the legend should be clearly associated with that element. For example, Figure 2.14 illustrates an incorrect approach for constructing a horizontally arranged legend (it may not be readily apparent which gray tone the values 5.7–12.6 are associated with). Since a common task is to locate a symbol on the map and determine the meaning of that symbol in the legend, text should be shown immediately to the right of each symbol. Furthermore, larger values should be shown at the top of the legend because "people associate 'up' with 'higher' and 'higher' with larger data values" (MacEachren, personal communication, 1996). Figures 2.12 and 2.15 illustrate correct and incorrect applications of these principles, respectively.

The source should include a clear description of where the data were acquired. For the Kenyan map, note that I used "Source: Ministry of Education. 1987. *Education in Kenya.* Nairobi Kenya: Jomo Kenyatta Foundation." For temporal data, the source should state the period over which the data were collected.

2.7.3 Positioning Major Design Elements

In positioning major design elements (title, legend, and source), mapmakers should avoid both large empty areas (to avoid wasting valuable graphic space) and maps that are either square or oblong. With respect to the latter, Robinson et al. (1995, 335) suggest a ratio of 3:5 for the area enclosed by the map border. When finished, the map should appear visually "balanced." Ideally, the title should be shown near the top of the map (the logic being that we normally read from top to bottom and thus would be inclined to examine the title first), and the source should

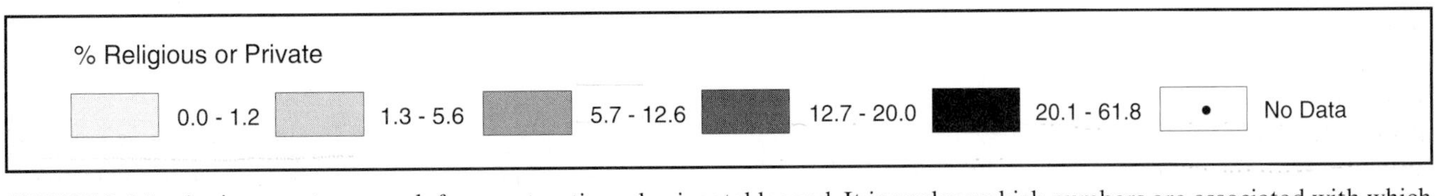

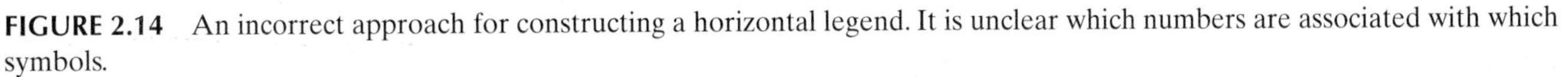

FIGURE 2.14 An incorrect approach for constructing a horizontal legend. It is unclear which numbers are associated with which symbols.

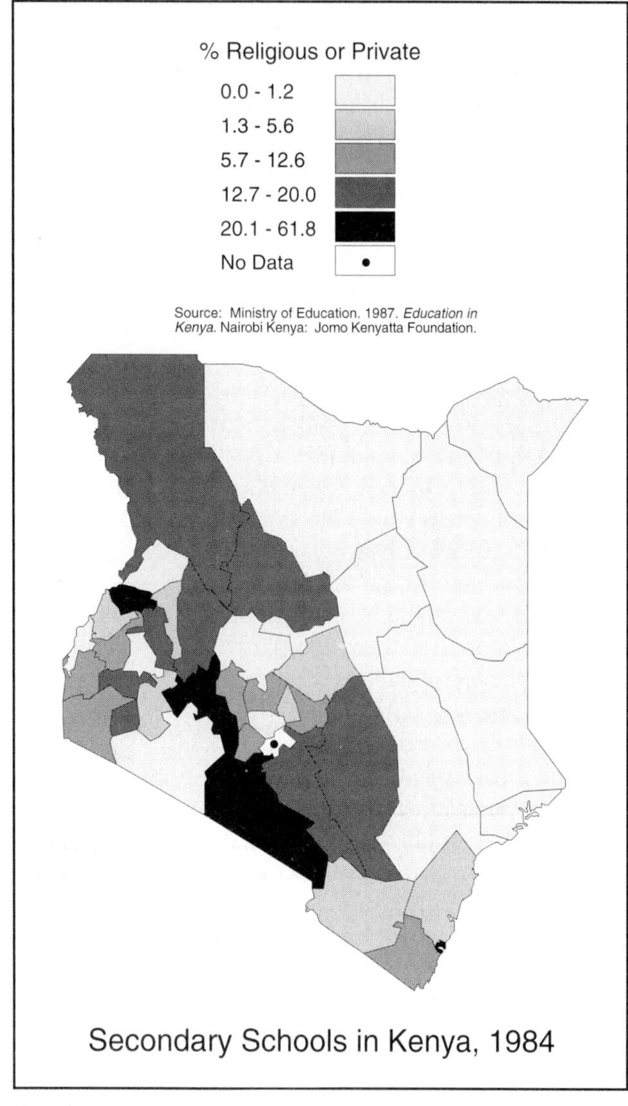

FIGURE 2.15 An example of poor legend construction and poor positioning of the title, legend, and source. (Compare with Figure 2.12.)

appear in the lower left-hand (or right-hand) corner. Figure 2.15 shows a poor example of positioning the title, legend, and source (compare with Figure 2.12).

2.7.4 Selecting Appropriate Typography

Mapping and graphic design software generally include three basic parameters for specifying type: font, style, and size.* **Font** refers to the basic design of the type. For example, Robinson et al. (1995, 407–408) distinguish between classic (or oldstyle), modern, and sans serif (Figure 2.16). Classic fonts tend to reproduce the look of fine freehand calligraphy, while modern fonts provide a greater exaggeration in the width of strokes within a letter (compare the "o"s for classic and modern in Figure 2.16).

Note that both classic and modern fonts have short extensions **(serifs)** at the ends of each major stroke of the letter (Figure 2.17); san serif typefaces lack such extensions. Serifed type is preferred for reading continuous text because of the cues it provides to tie subsequent letters together. In cartography, however, sans serif is preferred because of the need to scan for unfamiliar words.

Style refers to whether a particular font is normal, bold, italic, or a combination of bold and italic (Figure 2.18). **Size** deals with how large the type is, and is normally expressed in points, where a point is 1/72 of an inch (0.35 mm). Since the point size actually refers to the height of the metal block on which type was created prior to computer automation, point sizes must be considered an approximation to the size actually seen on a printed map.

* The reader will find that a different set of terms is used by cartographers; for example, Robinson et al. (pp. 407–414) use the term *style* instead of *font,* and *form* instead of *style.*

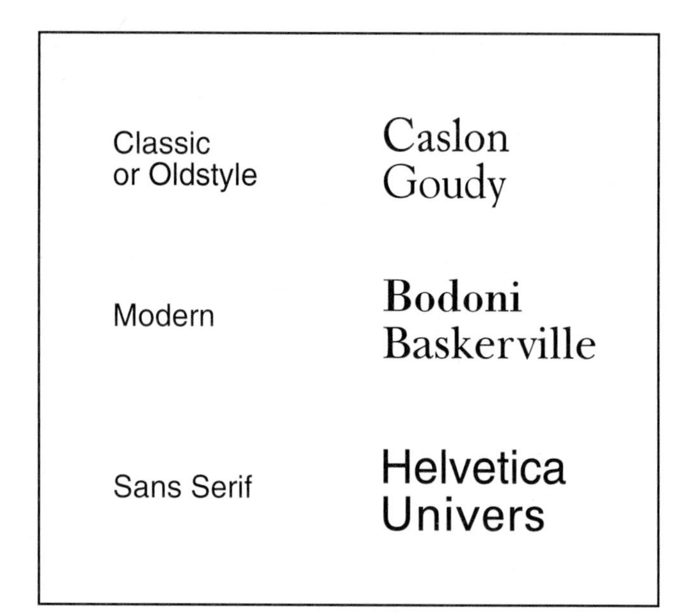

FIGURE 2.16 Examples of different types of fonts.

FIGURE 2.17 Serifs are short extensions at the end of major strokes of letters in classic and modern type.

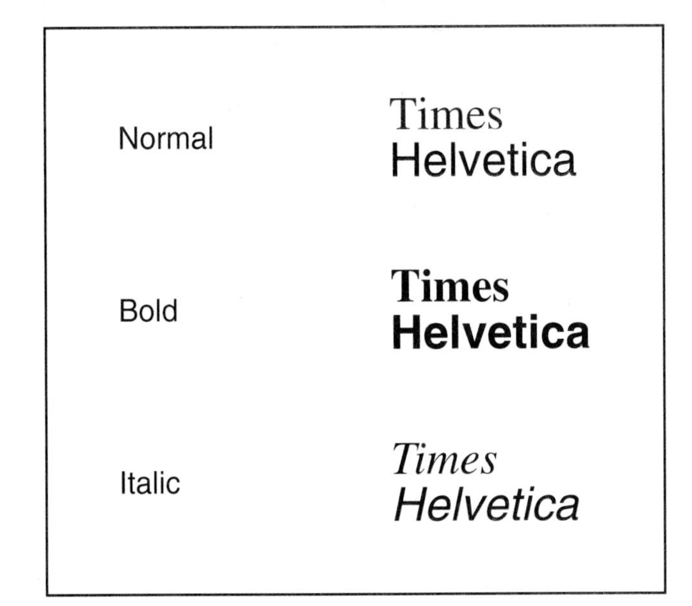

FIGURE 2.18 Style refers to whether a particular font is normal, bold, or italic.

The following are some basic guidelines for using typography associated with major design elements (title, legend, and source):

1. Use only a single font for the entire display, as multiple fonts tend to create unnecessary visual static.
2. Avoid selecting fonts based on their decorative appeal; legibility is more important.
3. Use size to distinguish the major elements.
4. A combination of uppercase and lowercase lettering is more readable than uppercase alone. Uppercase, however, may be used for short titles.
5. Avoid type smaller than 10 points for major elements such as the title.
6. Use bold lettering for important elements only, such as titles. Using bold for all lettering will draw attention away from the thematic material being mapped.

Figure 2.19 illustrates a map that violates a number of these guidelines (compare with Figure 2.12).

Typography is especially important on general-reference maps, as numerous hierarchical features (such as cities) must be labeled within the body of the map. Although general-reference maps are not the focus of this text, it is useful to consider some basic guidelines:

1. Use style (for example, serif versus sans serif) for nominal data and size for ordinal or higher-level data.
2. Avoid using multiple fonts, but if they must be combined use a sans serif with a classic or modern for greater contrast.

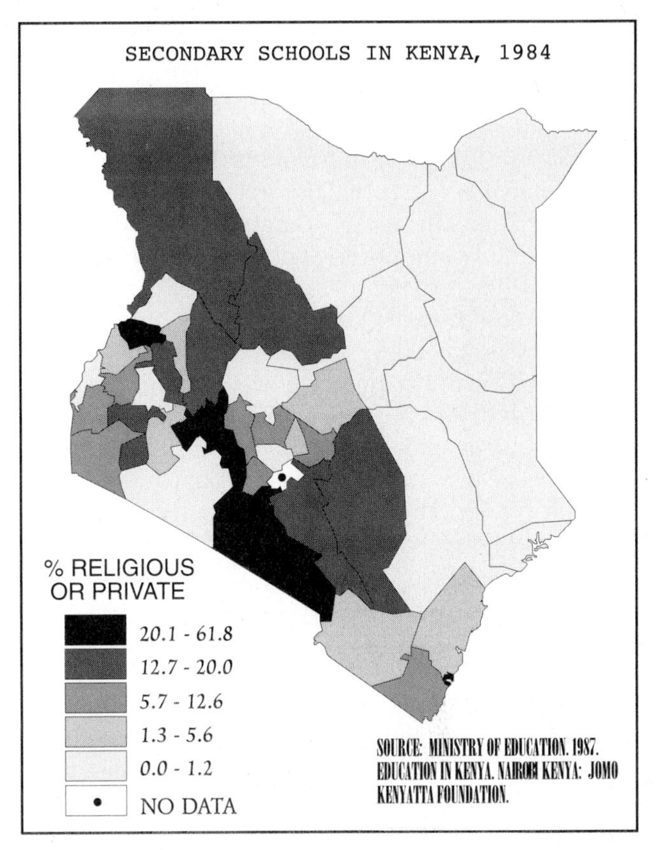

FIGURE 2.19 A map that violates a number of guidelines for typography associated with the title, legend, and source. (Compare with Figure 2.12.)

3. Use italic for hydrographic features, a long-standing convention within cartography.
4. Avoid type smaller than 6 points.
5. Use at least a 2-point difference between type sizes, since map users are not sensitive to slight differences in type size.
6. Curve the lettering along geographic features that curve (such as rivers).
7. The preferred location for labeling point locations is the upper right, with a secondary preference for the lower right.

For a more detailed discussion of typographic guidelines, see Robinson et al. (1995, 416–419), Dent (1996, 290–298), and Imhof (1975).

2.7.5 Automated Positioning of Type

Throughout this section, we have assumed that the mapmaker makes decisions regarding type placement. When using general-purpose graphic design software (such as Freehand), this is almost certainly the case. The reader must bear in mind, however, that specialized **labeling software** has been developed for automatically positioning type (van Roessel 1989; Ebinger and Goulette 1990; Doerschler and Freeman 1992; Sadahiro 1995). Labeling software focuses on the placement of type within the map body, as opposed to positioning major map elements (such as the title). One problem has restricted the widespread use of such software: The solution is computer-intensive when a considerable amount of type must be positioned. Mower (1993) has dealt with this problem by using **parallel computers** (computers that use multiple processors). Another problem is that the finished product normally requires some interactive editing to arrive at a solution that is visually acceptable. As more sophisticated algorithms are developed and the speed of computers increases, it appears inevitable that typography will become a more fully automated process.

SUMMARY

In this chapter, we have covered basic terminology and general guidelines for symbolizing spatial data. We have discovered that the spatial arrangement of the underlying phenomenon is an important consideration in selecting an appropriate symbology. For example, if the underlying phenomenon is **smooth** and **continuous** (such as yearly snowfall for Russia), then a **contour map** would be appropriate, but a **choropleth map** would not be.

Another important consideration in selecting symbology is the **level of measurement** of the data. Ideally, there should be a logical match between the level of measurement and the symbology (or **visual variable**) used to represent the data. For instance, if data are numerical (such as the magnitude of electrical generation at power plants in kilowatt hours), then the symbology should be capable of enabling a map reader to obtain numerical relations (for example, **proportional symbols** would be appropriate). Keep in mind, though, that we generally do not expect readers to acquire precise numerical information from maps; maps are primarily used to show spatial patterns.

Although the underlying phenomenon is an important consideration in selecting symbology, we have seen that map purpose can also play an important role. For example, if the mapmaker wishes to show "typical" values for enumeration units, then a choropleth map may be appropriate even when the underlying phenomenon is not coincident with enumeration units, as was the case with the distribution of wheat harvested in Kansas counties. The mapmaker should realize, however, that in this case a choropleth map may lead to incorrect perceptions of the underlying phenomenon. Therefore, when a choropleth map is deemed necessary, I would suggest also including a more accurate representation of the phenomenon (such as a **dot map**).

We have also learned about the importance of **data standardization,** that **raw counts** (or **totals**) need to be adjusted to account for the area over which the data have been collected (typically an enumeration unit). The simplest form of adjustment is to divide raw counts by the areas of enumeration units (for example, dividing the number of people in counties by the areas of counties to create a map of population density). Although such a map could be useful, we have seen that many other methods of standardization are possible (for example, creating a ratio of males to females). Ultimately, the purpose of the map should determine the appropriate method of standardization.

Finally, we have learned about some of the basic design parameters that can assist in creating a thematic map that is to be presented to a larger group of readers. These are some key recommendations: (1) the title should be prominent and easily understood; (2) the legend should clearly indicate what each symbol associated with the mapped theme represents; (3) variety in lettering fonts should be avoided; and (4) the overall map should appear visually balanced, with approximately a 3:5 ratio. From Chapter 1, also recall that it is desirable to get feedback from a group of potential readers before actually publishing the map.

FURTHER READING

Bertin, J. (1983) *Semiology of Graphics: Diagrams, Networks, Maps.* Madison, Wisc.: University of Wisconsin Press. (Translated by W. J. Berg.)

Chapter 2 of this widely cited text focuses on visual variables. The text is a bit difficult to read as it has been translated from French to English.

Brewer, C. A. (1994a) "Color use guidelines for mapping and visualization." In *Visualization in Modern Cartography,* ed. A. M. MacEachren and D. R. F. Taylor, pp. 123–147. Oxford: Pergamon Press.

Pages 124–126 cover terminology for using color. We will consider Brewer's work more fully in Chapter 6.

Chrisman, N. (1997) *Exploring Geographic Information Systems.* New York: John Wiley & Sons.

See Chapter 1 for a more complete discussion of levels of measurement.

Doerschler, J. S., and Freeman, H. (1992) "A rule-based system for dense-map name placement." *Communications of the ACM* 35, no. 1:68–79.

An example of an approach for automated type placement based on serial computers. For an approach based on parallel computers, see Mower (1993).

Imhof, E. (1975) "Positioning names on maps." *The American Cartographer* 2, no. 2:128–144.

A classic reference on guidelines for positioning type on maps.

Kraak, Menno-J. (1988) *Computer-assisted Cartographical Three-Dimensional Imaging Techniques.* Delft, The Netherlands: Delft University Press.

An extensive treatment of methods for three-dimensional mapping.

MacEachren, A. M. (1994a) *Some Truth with Maps: A Primer on Symbolization and Design.* Washington, D.C.: Association of American Geographers.

Spatial arrangement of geographic phenomena, levels of measurement, and visual variables are covered in pages 13–34. A number of basic map design concepts are covered in Chapter 5.

Monmonier, M. S. (1978b) "Viewing azimuth and map clarity." *Annals, Association of American Geographers* 68, no. 2:180–195.

A basic problem in interpreting three-dimensional maps (e.g., a prism map) is the blockage of lower-valued areas. Monmonier describes an algorithm for finding a viewing angle that will minimize blockage.

Robinson, A. H., Morrison, J. L., Muehrcke, P. C., Kimerling, A. J., and Guptill, S. C. (1995) *Elements of Cartography.* 6th ed. New York: John Wiley & Sons.

See Chapter 22 for more information on the use of typography. Also see Chapter 14 of Dent (1996).

3

Statistical and Graphical Foundation

OVERVIEW

One purpose of this chapter is to provide a statistical and graphical foundation for the remainder of the text. For example, in this chapter we will consider basic statistical measures such as the mean and standard deviation. In the following chapter we will consider a method for classifying data that is based on these measures. Obviously, the logic of the classification method cannot be comprehended unless the concepts of mean and standard deviation are first understood.

A second purpose of this chapter is to introduce a range of nonmapping techniques (tables, graphs, and numerical summaries) that can be used along with maps to analyze spatial data. To illustrate the need for such techniques, consider Figure 1.6B, which used an unclassed choropleth map to show the distribution of foreign-owned agricultural land in the United States. Why is this map composed entirely of light tones of gray, with the exception of a solid black tone for the state of Maine? A graphical plot of the data along a number line would reveal that 47 of the 48 states were in the range 0–5.2, with a distinct ***outlier*** *at 14.1. Only by seeing such a graph would the reader develop a full appreciation of the spatial distribution shown on the map.*

Sections 3.1 and 3.2 briefly deal with the principles of population versus sample and descriptive versus inferential statistics. A ***population*** *is the total set of elements or things that could potentially be studied, while a* ***sample*** *is the portion of the population that is actually examined.* ***Descriptive statistics*** *are used to summarize the character of a sample or population, while* ***inferential statistics*** *are used to make a guess (or inference) about a population based on a sample. The focus of this chapter is on descriptive statistics because these are most useful in mapping.*

Section 3.3 covers a broad range of methods for analyzing data via tables, graphs, and numerical summaries. The section is split into three major parts: analyzing the distribution of individual variables (3.3.1), analyzing the relationship between two (or more) variables (3.3.2), and exploratory data analysis (3.3.3). Readers with coursework in statistics will find some of this material a review (such as ***histograms, measures of central tendency, correlation,*** *and* ***stem-and-leaf plots****), but other material will likely be new (for example,* ***hexagon bin plots,*** *the* ***scatterplot matrix,*** *the* ***reduced major axis,*** *and* ***scatterplot brushing****).*

There is one limitation of the numerical summaries covered in section 3.3: spatial location is not an integral part of the formulas used. In contrast, section 3.4 deals with numerical summaries in which spatial location is an essential element. Some methods covered in section 3.4 just analyze spatial location; for example, the formula for computing the ***centroid,*** *or balancing point for a geographic region, uses just the x and y coordinates bounding a region. Other methods consider both spatial location and the values of an attribute (or variable); for example,* ***spatial autocorrelation*** *measures the likelihood that similar attribute values occur near one another.*

To illustrate many concepts in this chapter, we will analyze the relationship between murder rate (number of murders per 100,000 people) and the following variables for 50 U.S. cities whose population was 100,000 or more in 1990: (1) percentage of families whose income was below the poverty level; (2) percentage of those 25 years and over who were at least high school graduates; (3) the drug arrest rate (number of arrests per 100,000 people); (4) population density (number of people per square *mile); and (5) total population (in thousands).* * *The raw data are shown in Table 3.1 (ordered on the basis of murder rate).*

One problem with these data is that an analysis at the city level may be inappropriate because it fails to account for the variation within a city; it might instead be desirable to look at finer geographic units, such as census tracts, or at individual murder cases. I chose the city level for analysis, however, because it is easier to relate to individual cities than to individual census tracts. Later in the chapter we will consider the effect of aggregation of enumeration units on measures of numerical correlation.

3.1 POPULATION AND SAMPLE

In statistics, a **population** is defined as the total set of elements or things that one could study, while a **sample** is the portion of the population that is actually examined (in this book, the number of elements in each is represented by N and n, respectively). Generally, scientists collect samples because they don't have the time or money to examine the entire population. For example, a geomorphologist studying the effect of wave behavior on beach development would collect data at a series of points along a shoreline, rather than examining the entire shoreline.

The data in Table 3.1 have characteristics of either a population or sample, depending on one's perspective of the data. Consider first a perspective from the standpoint of variables. Murder rate, drug arrest rate, population, and population density are all based on the entire population of each city (as defined by the census); for example, in the case of murder rate, all murders occurring within a city are considered relative to the entire population of that city. The other variables, percentage of families below the poverty level and percentage of high school graduates, are based on sampling approximately one of every six housing units (U.S. Bureau of the Census 1994a, A-2).

From the perspective of observations (cities, in this case), the 50 shown in Table 3.1 were sampled from 200 cities. Sampling was done in two stages. The first stage involved eliminating cities whose political boundaries extended beyond the limits of where most people live within those cities (using the "extended city" definition provided by the Census Bureau). This was done because one of the variables being analyzed, population density, was a function of city area. In the second stage, the remaining cities were ordered on the basis of total population and split into 10 classes using Jenks's (1977) optimal method (see Chapter 4). A proportional number of cities was sampled from each of the 10 classes in order to obtain a broad range of city sizes that would be representative of those found in the United States.

3.2 DESCRIPTIVE VERSUS INFERENTIAL STATISTICS

Statistical methods can be split into two types: descriptive and inferential. **Descriptive statistics** describe the character of a sample or population. For example, to assess the current president's job performance, you might ask a sample of 500 people, "Is the President doing an acceptable job?" The percentage responding yes, say 52 percent, would be an example of a descriptive statistic.

Inferential statistics are used to make an inference about a population from a sample. For example, based on the 52 percent figure given above, you might infer that 52 percent of the entire population thinks the president is doing a good job. You would be surprised, of course, if the 52 percent figure truly applied to the population because the figure is based on a sample. To correct for this problem, in inferential statistics it is necessary to compute a *margin of error* (for example, plus or minus 3 percent) around the sampled value; we often find such errors reported in daily newspapers.

3.3 GENERAL METHODS FOR ANALYZING SPATIAL DATA, IGNORING LOCATION

This section considers general methods for analyzing data, ignoring the spatial location of the data. The section is split into three parts: (1) analyzing the distribution of individual variables, (2) analyzing the relationship between two or more variables, and (3) exploratory data analysis.

3.3.1 Analyzing the Distribution of Individual Variables

Tables

Raw Table. The simplest form of tabular display is the **raw table** in which the data for a variable of interest are listed from lowest to highest value, as for the murder rate data in Table 3.1. Tabular displays are useful for providing specific information about particular places (for example, Buffalo, New York, had a murder rate of 11.3 per 100,000 people in 1990), but they can provide other information if they are examined carefully. First, note that the sorted values provide the

* The murder and drug arrest data were obtained from the *Sourcebook of Criminal Justice Statistics 1991* (Flanagan and Maguire 1992). The remaining data were taken from the *1994 City and County Data Book* (U.S. Bureau of the Census 1994a). All data were for either 1989 or 1990.

TABLE 3.1. Sample data for 50 U.S. cities (sorted on murder rate)

City	Murder Rate*	Families below Poverty Level (%)	High School Graduates (%)	Drug Arrest Rate†	Population Density‡	Total Population (in thousands)
Irvine,CA	0.0	2.6	95.1	780	2607	110
Cedar Rapids, IA	0.9	6.6	84.5	110	2034	109
Overland Park, KS	0.9	1.9	94.1	255	2007	112
Livonia, MI	1.0	1.7	84.7	665	2823	101
Lincoln, NE	1.6	6.5	88.3	294	3033	192
Madison, WI	1.6	6.6	90.6	57	3311	191
Glendale, CA	1.7	12.3	77.2	452	5882	180
Allentown, PA	1.9	9.3	69.4	1078	5934	105
Tempe, AZ	2.1	7.0	89.9	295	3590	142
Boise City, ID	2.4	6.3	88.6	512	2726	126
Lakewood, CO	2.4	5.2	88.2	216	3100	126
Mesa, AZ	3.1	6.9	84.8	223	2653	288
Pasadena, TX	3.4	11.1	69.8	370	2727	119
San Jose, CA	4.5	6.5	77.2	1289	4568	782
Waterbury, CT	4.6	9.9	66.8	1326	3815	109
Springfield, MO	5.0	11.6	77.0	446	2068	140
Chula Vista, CA	5.2	8.6	75.7	808	4661	135
St. Paul, MN	6.6	12.4	81.1	260	5157	272
Arlington, VA	7.0	4.3	87.5	758	6605	171
Alexandria, VA	7.2	4.7	86.9	834	7281	111
Portland, OR	7.6	9.7	82.9	1001	3508	437
Des Moines, IA	8.3	9.5	81.0	118	2567	193
Lansing, MI	8.7	16.5	78.3	780	3755	127
Pittsburg, PA	9.5	16.6	72.4	723	6649	370
Yonkers, NY	9.6	9.0	73.6	917	10403	188
Riverside, CA	9.7	8.4	77.8	1703	2916	227
Elizabeth, NJ	10.0	13.7	58.5	929	8929	110
Berkeley, CA	10.7	9.4	90.3	1569	9783	103
Buffalo, NY	11.3	21.7	67.3	580	8080	328
Raleigh, NC	11.5	7.7	86.6	634	2360	208
Sacramento, CA	11.7	13.8	76.9	1555	3836	369
Tacoma, WA	14.1	12.5	79.3	673	3677	177
Knoxville, TN	15.2	15.3	70.8	328	2137	165
Beaumont, TX	16.7	16.6	75.2	693	1427	114
Winston-Salem, NC	16.8	11.6	77.0	1343	2018	143
Montgomery, AL	18.2	14.4	75.7	131	1386	187
Waco, TX	21.2	19.7	68.4	400	1367	104
Jackson, MS	22.3	18.0	75.0	693	1804	197
Savannah, GA	23.9	18.5	70.1	707	2198	138
Norfolk, VA	24.1	15.1	72.7	624	4859	261
Los Angeles, CA	28.2	14.9	67.0	1391	7426	3485
Chicago, IL	30.6	18.3	66.0	1157	12251	2784
New York, NY	30.7	16.3	68.3	1255	23701	7323
Houston, TX	34.8	17.2	70.5	555	3020	1631
Newark, NJ	40.7	22.8	51.2	1751	11554	275
Baltimore, MD	41.4	17.8	60.7	2063	9108	736
Gary, IN	55.6	26.4	64.8	261	2322	117
Detroit, MI	56.6	29.0	62.1	1052	7410	1028
Atlanta, GA	58.6	24.6	69.9	2330	2990	394
Washington, DC	77.8	13.3	73.1	1738	9883	607

*Murders per 100,000 people.

†Arrests per 100,000 people.

‡Number of people per square mile.

minimum and maximum of the data (0 for Irvine, California, and 77.8 for Washington, D.C.). Second, with some mental arithmetic, the *range* can be calculated by simply subtracting the minimum from the maximum (77.8 − 0 = 77.8 in this case).

Third, raw tables can reveal any duplicate values in the data that may have special significance. For murder rate, duplicate values (for example, 0.9 for Cedar Rapids, Iowa, and Overland Park, Kansas) are unimportant, as they are a function of the number of significant digits reported.

For some variables, however, duplicates can be quite meaningful. For example, an examination of the distribution of major-league baseball player salaries would reveal numerous duplicates because several players on each team typically earn exactly the same amount, the minimum required salary for all of major-league baseball.

Fourth, raw tables may reveal **outliers**—values that are quite unusual or atypical (Barnett and Lewis 1994). For murder rate, no value is considerably different from the rest (although Washington, D.C., is clearly larger), but for the total population variable, note that several cities are quite large, with New York more than twice the value of any other city in the sample.

Although raw tables are useful for providing specific information, they are not very good at providing overviews of distributions. For example, in studying Table 3.1, note that roughly half the cities have murder rates below 10.0, but it is difficult to develop a feel for what the overall murder distribution really looks like. Grouped-frequency tables are more useful for this purpose.

Grouped-Frequency Table. **Group-frequency tables** are constructed by dividing the data range into equal intervals and then tallying the number of observations that fall in each interval. Although such tables generally are constructed using software (for example, SPSS and SAS), it is useful to consider the actual steps involved so that one has a clear understanding of what the resulting table reveals. (We will use a similar approach in the next chapter to create equal interval classes for choropleth maps.)

Step 1. Decide how many groups (or classes) you wish to display. When grouped-frequency tables were created manually, this was an important step because of the time and thought needed to construct the table. With the ready availability of software, it is reasonable to construct tables for various numbers of classes. For this discussion, presume that 15 classes will be chosen.

Step 2. Determine the width of each class (the class interval). The class interval is computed by dividing the range of the data by the number of classes. The following would be the computation for the murder rate data using 15 classes:

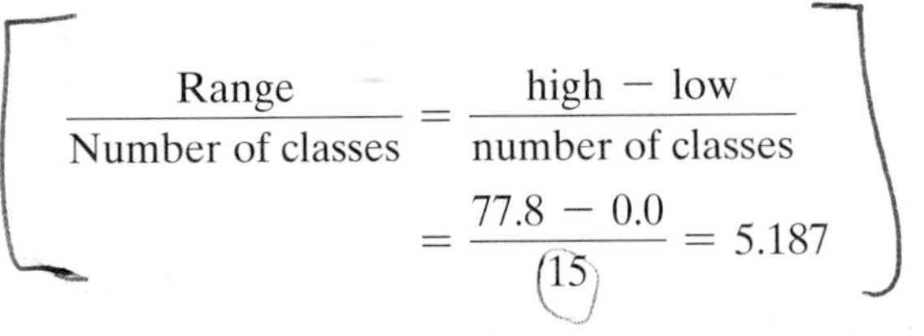

$$\frac{\text{Range}}{\text{Number of classes}} = \frac{\text{high} - \text{low}}{\text{number of classes}} = \frac{77.8 - 0.0}{15} = 5.187$$

Note that we use more decimal places for the class interval than appear in the actual data. We do so to avoid rounding errors that could cause the last class not to match the highest value in the data.

Step 3. Determine the upper limit of each class. The upper limit for each class is computed by repeatedly adding the class interval to the lowest value in the data. The result is the right-hand side of the Limits column in Table 3.2. Note that the highest calculated limit does not match the highest data value (77.8) exactly because the class interval is not a simple fraction; a more precise class interval would be 5.186666667, but even with this interval the match would not be exact (the highest value would be 77.80000001).

Step 4. Determine the lower limit of each class. Lower limits for each class are specified so that they are just above the highest value in a lower-valued class. This is done so that any observation can fall in only one class. For example, the lowest value in the second class is 5.188, which is .001 larger than 5.187, the highest value in class 1.

Step 5. Tally the number of observations falling in each class. These numbers are shown as Number in Class in Table 3.2. Also shown in the table are the percent and cumulative percent in each class.

In comparison to the raw table, the grouped-frequency table provides a somewhat better overview of the data. For example, note that 80 percent of the cities fall in the five lowest murder rate classes. Both raw and grouped-frequency tables, however, do not take full advantage of our visual processing powers; for this we turn to graphs.

Graphs

Point and Dispersion Graphs. In a **point graph** or **one-dimensional scatterplot** each data value is represented by a small point symbol plotted along the number line (Cleveland 1994, 133) (in Figure 3.1A an open circle is used). For the murder rate variable, the point graph shows that the data are concentrated at the lower

TABLE 3.2. Fifteen-class grouped-frequency table for the murder rate data

Class	Limits	Number in Class	Percent in Class	Cumulative Percent
1	0.000–5.187	16	32	32
2	5.188–10.374	11	22	54
3	10.375–15.561	6	12	66
4	15.562–20.748	3	6	72
5	20.749–25.935	4	8	80
6	25.936–31.122	3	6	86
7	31.123–36.309	1	2	88
8	36.310–41.496	2	4	92
9	41.497–46.683	0	0	92
10	46.684–51.870	0	0	92
11	51.871–57.057	2	4	96
12	57.058–62.244	1	2	98
13	62.245–67.431	0	0	98
14	67.432–72.618	0	0	98
15	72.619–77.805	1	2	100

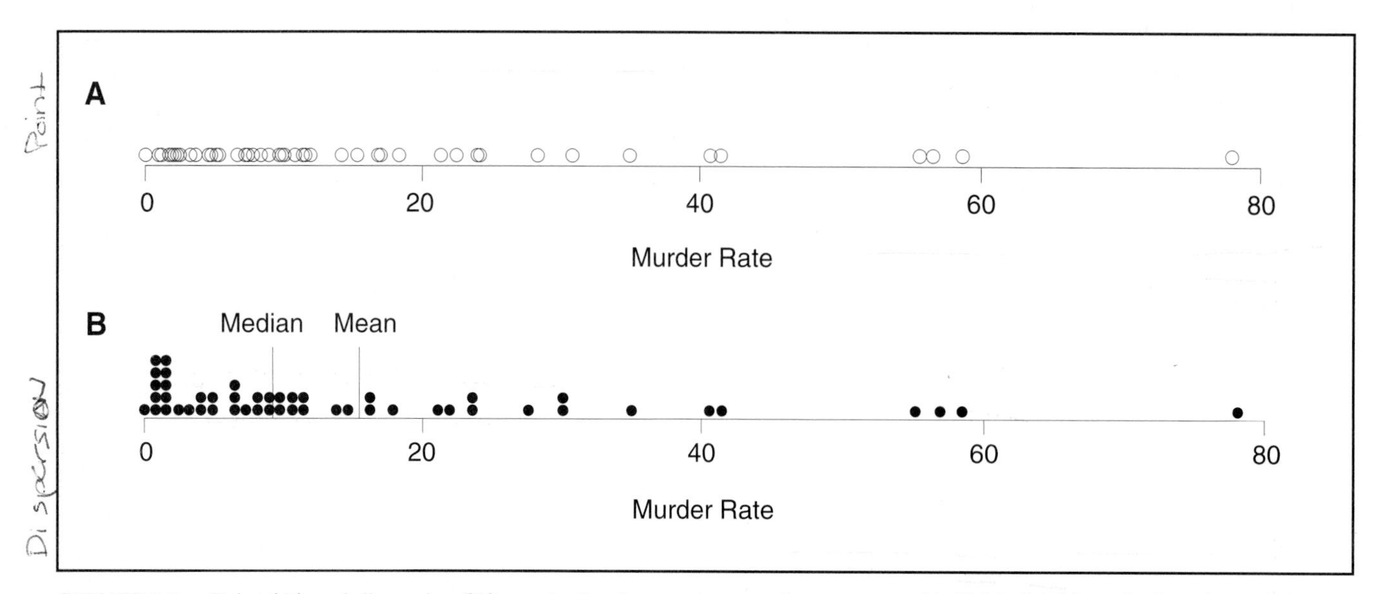

FIGURE 3.1. Point (A) and dispersion (B) graphs for the murder rate data presented in Table 3.1. Note the locations of the median and mean in (B).

end of the distribution (based on the graph, under a rate of about 12). One obvious problem with the point graph is that individual symbols may overlap, thus making the distribution difficult to interpret (note the "smearing" in the left-hand portion of Figure 3.1A); duplicate values, in particular, cannot be detected at all by this method.

An alternative to the point graph is the **dispersion graph** (Hammond and McCullagh 1978), in which data are grouped into classes, the number of values falling in each class are tallied, and dots are stacked at each class position (Figure 3.1B). Intervals for classes are defined in a fashion identical to the grouped-frequency table, except that a large number of classes is used. For example, 99 classes were used for Figure 3.1B. The dispersion graph for the murder rate data portrays a distribution similar to the point graph, except that potential confusion in the overlapping areas is eliminated.

Histogram. A **histogram** is constructed in a manner analogous to the dispersion graph, except that fewer classes are generally used and bars of varying height are used to represent the number of values in each class. Since the histogram is more commonly used than either the point or dispersion graph, all of the variables shown in Table 3.1 are graphed using the histogram in Figure 3.2. Looking first at the murder rate histogram, note that the up-and-down nature of the dispersion graph has been smoothed out; there is clearly a peak in the graph on the left, with a decreasing height in bars as one moves to the right.

Histograms are often compared with a hypothetical **normal** (or mound-shaped) **distribution** (Figure 3.3). For this reason normal curves have been overlaid on the histograms shown in Figure 3.2. Distributions lacking the symmetry of the normal distribution are said to be **skewed.** *Positively skewed* distributions have the tallest bars concentrated on the left-hand side (as for total population, murder rate, and population density), while negatively skewed distributions have the tallest bars on the right-hand side. (There is no distinctive example of a negatively skewed distribution in Figure 3.2, although percent with high school education has a slight negative skew).

Since many inferential tests require a normal distribution, raw data are often transformed to make them approximately normal. Transformation involves applying the same mathematical operation to each data value of a variable; for example, we might compute the $\log_{10}$ of each murder rate value, or alternatively we could compute the square root of each murder rate value.* Log_{10} and square root transformations are commonly used to convert a positively skewed distribution to a normal one; to illustrate, Figure 3.4 portrays such transformations for the murder and drug arrest data respectively.†

Numerical Summaries

Although tables and graphs are useful for interpreting data, they are prone to differing subjective interpretations, and the finite space of formal publications often

* The $\log_{10}$ for a number is the power that we would raise 10 to in order to get that number. For example, $\log_{10} 100 = 2$ because $10^2 = 100$.

† Technically, inferential statistics should be used to check data sets for normality (Stevens 1996, 244–247).

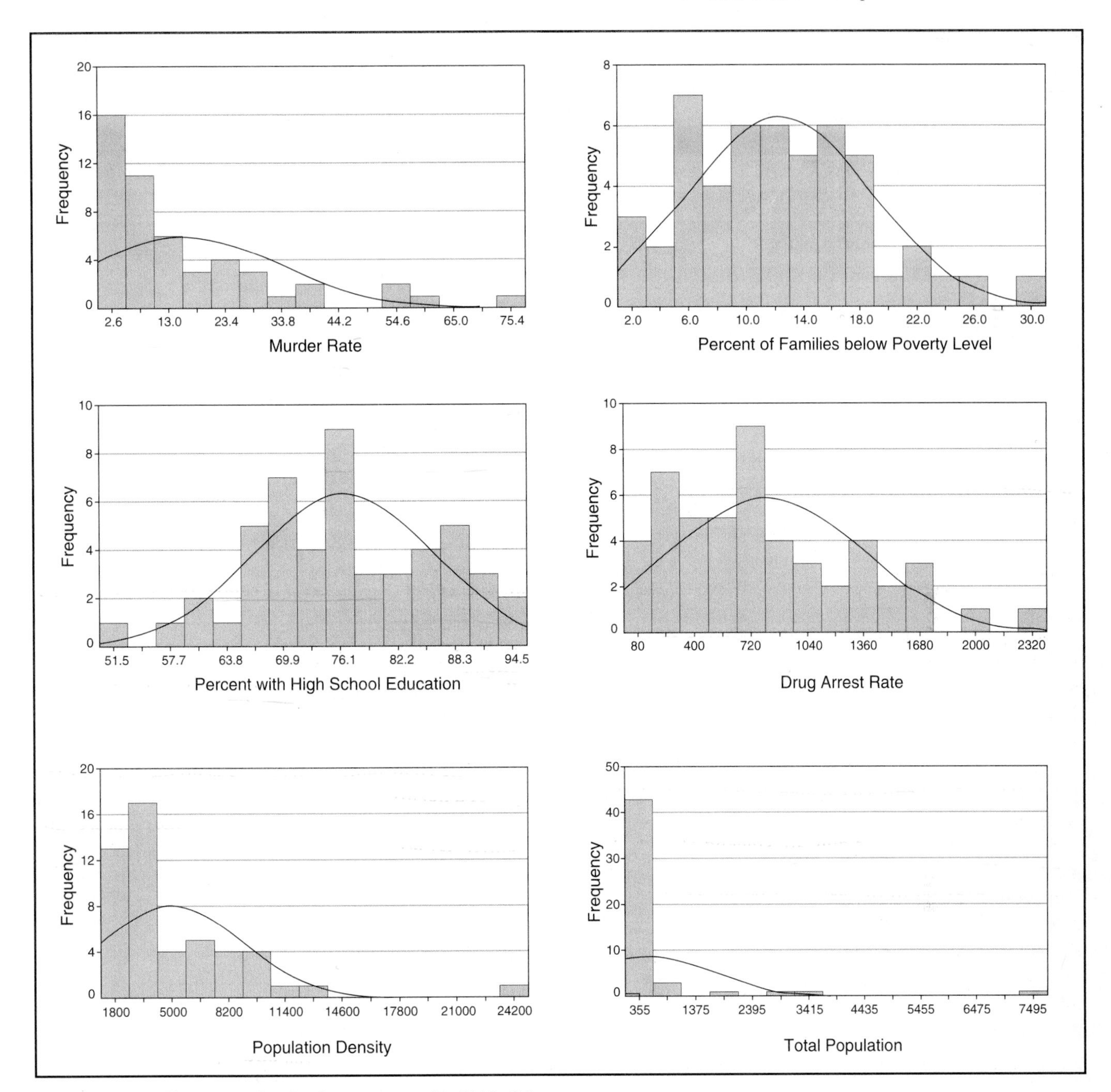

FIGURE 3.2. Histograms for the data presented in Table 3.1.

limits their use. As an alternative, statisticians frequently use numerical summaries, which typically are broken into two broad categories: measures of central tendency and measures of dispersion.

Measures of Central Tendency. **Measures of central tendency** are used to indicate a value around which the data are most likely concentrated. Three measures of central tendency are commonly recognized: mode, median, and mean. The **mode** is the most frequently occurring value, and is thus generally useful for only nominal data, such as on a land use/land cover map. The **median** is the middle value in an ordered set of data or, alternatively, the 50th percentile, since 50 percent of the data are below it. For the murder rate data, the median is 9.7. Note its location in Figure 3.1B.

The **mean** is often referred to as the "average" of the data and is calculated by summing all values and dividing by the number of values. Typically, separate formulas

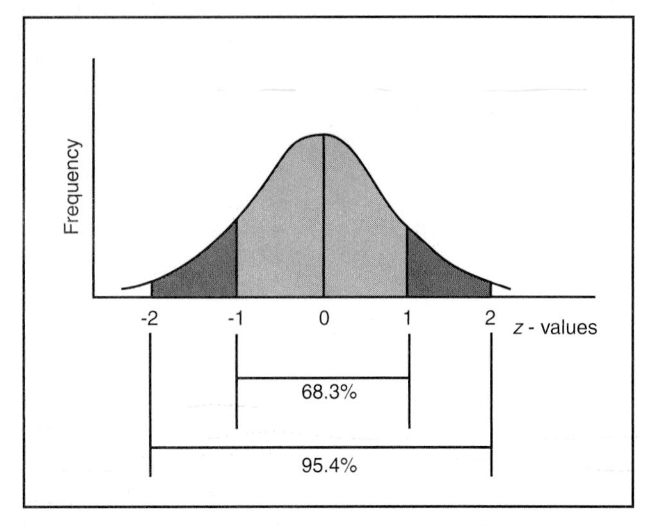

FIGURE 3.3. An example of a normal curve. Histograms will approximate this shape if the data are normal. For a perfectly normal data set, 68.3 percent and 95.4 percent of the observations will fall within 1 and 2 standard deviations, respectively, of the mean.

are given to distinguish mean values for the sample and population, as follows:*

Sample: $$\overline{X} = \frac{\sum_{i=1}^{n} X_i}{n}$$

Population: $$\mu = \frac{\sum_{i=1}^{N} X_i}{N}$$

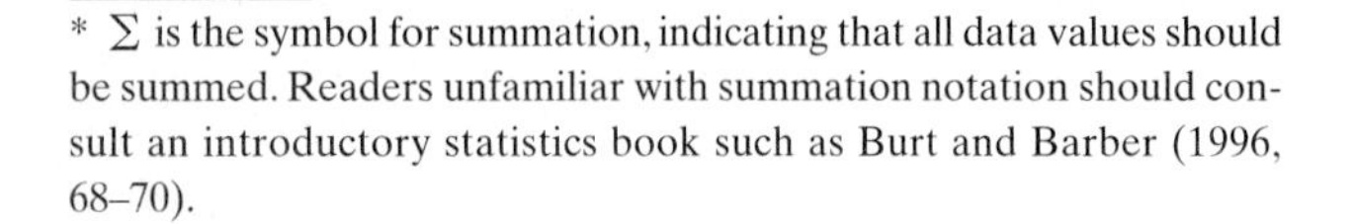

* Σ is the symbol for summation, indicating that all data values should be summed. Readers unfamiliar with summation notation should consult an introductory statistics book such as Burt and Barber (1996, 68–70).

where the X_i are the individual data values. Since the data in Table 3.1 are a sample from 200 cities, the left-hand formula is appropriate in this case. The mean for the murder rate data is 16.0. Note its location in Figure 3.1B.

One problem with the mean is that it is affected by either a skew or outliers in the data. In contrast, the median is resistant to these characteristics. We can see this in Figure 3.1B, where the median falls where most of the data are concentrated, while the mean is pulled to the right by the positive skew.

Measures of Dispersion. Measures of dispersion provide an indication of how data are spread along the number line. The simplest measure is the **range,** which was defined above as the maximum minus the minimum. Obviously, the range is of limited usefulness because it is based on only two values, the maximum and minimum of the data.

More useful measures of dispersion are the interquartile range and standard deviation, which should be used with the median and mean respectively. The **interquartile range** is the absolute difference between the 75th and 25th percentiles. For the murder rate data, the result is $|22.700 - 3.325| = 19.4$. An important characteristic of the interquartile range is that it, like the median, is unaffected by outliers in the data. For example, if the highest murder rate were replaced by a value of 150, the interquartile range would still be 19.4; the interquartile range is thus a measure of where the bulk of the data fall.

In a fashion similar to the mean, separate formulas normally are provided for the sample and population **standard deviation:**

Sample: $$s = \sqrt{\frac{\sum_{i=1}^{n} (X_i - \overline{X})^2}{n - 1}}$$

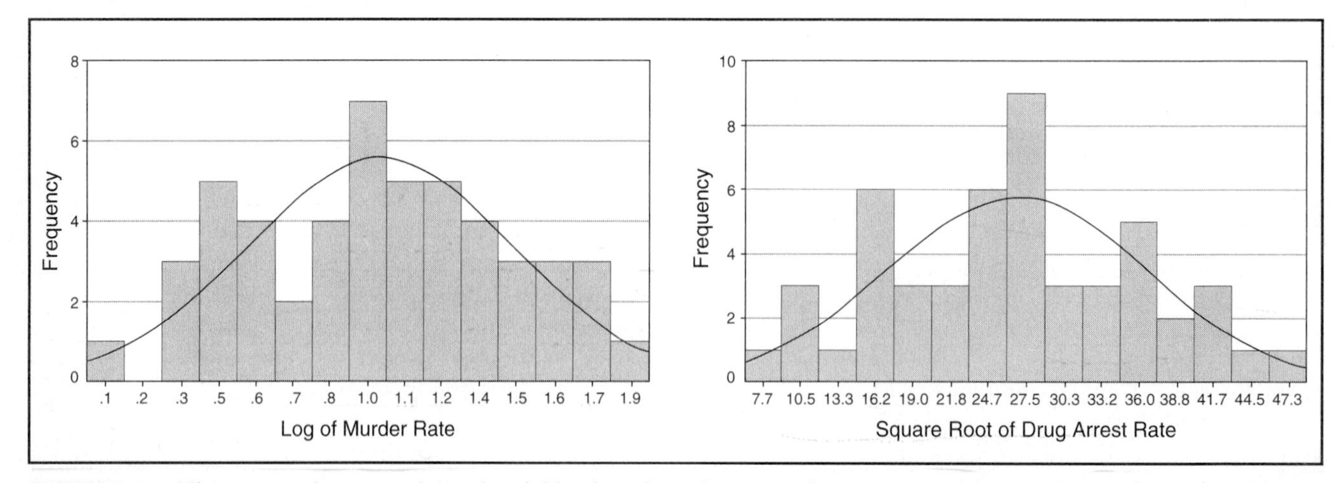

FIGURE 3.4. Histograms of two transformed variables: log of murder rate and square root of drug arrest rate (compare with corresponding histograms in Figure 3.2).

TABLE 3.3. Tabular relationship for selected cities between murder rate and percent of families below poverty level

	Percent Families below Poverty Level				
Murder rate	*1.70–7.16*	*7.17–12.62*	*12.63–18.08*	*18.09–23.54*	*23.55–29.00*
62.25–77.80			1		
46.69–62.24					3
31.13–46.68			2	1	
15.57–31.12		1	6	3	
0.00–15.56	13	14	5	1	

Population: $$\sigma = \sqrt{\frac{\sum_{i=1}^{N}(X_i - \mu)^2}{N}}$$

In comparing these formulas, note that they differ principally in the denominator: a 1 is subtracted in the sample case, but not in the population case. Subtracting a value of 1 is necessary because a sample estimate using just n tends to underestimate the population value (Burt and Barber 1996, 64). Using the sample formula, the standard deviation for the murder rate data is 17.5. In contrast to the interquartile range, the standard deviation is affected by outliers in the data. To illustrate, replacing the highest murder rate by a value of 150 results in a standard deviation of 24.4, an increase of nearly 40 percent.

3.3.2 Analyzing the Relationship between Two or More Variables

Tables

In the previous section, we saw that considerable information could be derived from raw tables when examining an individual variable. When trying to relate two variables, however, this task becomes difficult. To convince yourself of this, try using Table 3.1 to relate murder rate with percent families below the poverty level. In general, the variables appear related (low poverty values are associated with low murder rates and high poverty values are associated with high murder rates), but it is difficult to summarize the relationship. To simplify the process, we can class both variables using the grouped-frequency method and create a matrix of the result (Table 3.3). The same general relation between the variables is still apparent, but it is more easily seen; also, the matrix reveals that Washington, D.C. (in the highest murder rate class) does not fit the general trend of the data, which extends from the lower left to the upper right of the table.

Graphs

The **scatterplot** is used to examine the relationship of variables against one another in two-dimensional space. To illustrate, Figure 3.5 portrays scatterplots of murder rate and log of murder rate against percent of families below poverty level (also shown are best-fit regression lines, which we will consider shortly). On scatterplots, *dependent* and *independent variables* normally are plotted on the y and x axes, respectively.

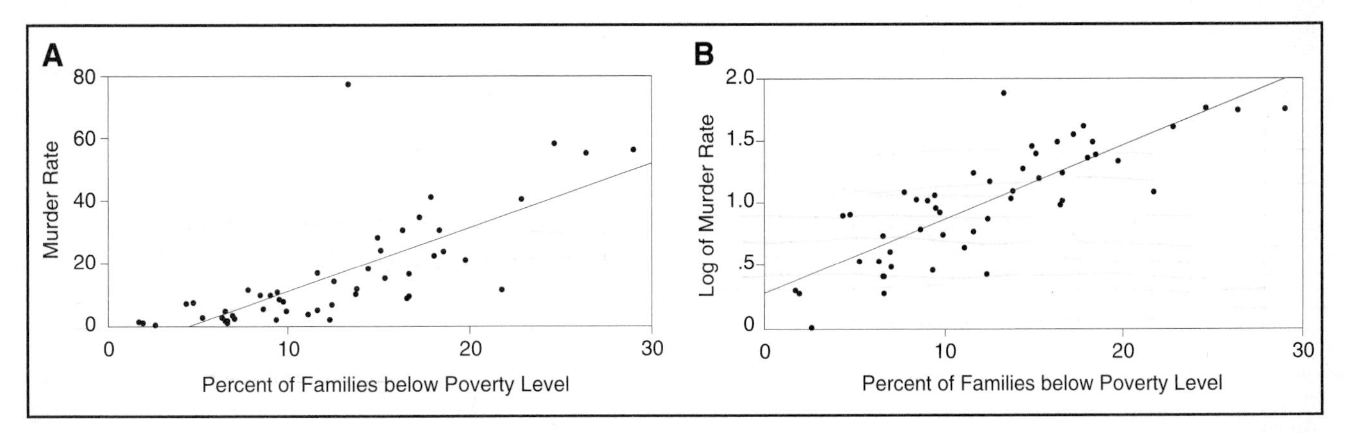

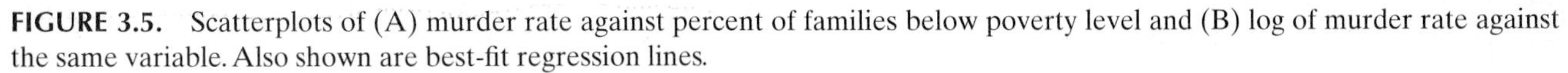

FIGURE 3.5. Scatterplots of (A) murder rate against percent of families below poverty level and (B) log of murder rate against the same variable. Also shown are best-fit regression lines.

Since it seems reasonable that the murder rate may depend, in part, on poverty, murder rate has been plotted on the y axis.

In examining Figure 3.5A, note the similarity of the distribution of points to the pattern of cells in Table 3.3. Also note that after transformation (Figure 3.5B), Washington, D.C., is not quite so different from the rest of the data; thus, data transformations affect not only the values for individual variables but also the relationship between variables.

When the number of observations is large, scatterplots may become difficult to interpret, just as smearing of dots made the point graph of murder rate hard to interpret. One solution to this problem is the **hexagon bin plot,** shown in Figure 3.6. Such a plot is created by making the size of hexagons shown on the right proportional to the number of dots falling within the hexagons on the left. Carr (1991) and Cleveland (1994) discuss several other approaches for handling a large number of observations.

To examine the relationship among three variables simultaneously, the scatterplot can be extended to a **three-dimensional scatterplot** by specifying x, y, and z axes. Many statistical packages have options for creating such plots, and even permit rotating them interactively. My experience is that these plots are difficult to interpret; moreover, they cannot be extended to handle more than three variables.

A method that can handle more than three variables is the **scatterplot matrix** (Figure 3.7). This graph may at first appear rather complex, but the principles underlying it are actually quite simple. Note first that the relation between any two variables is displayed twice, once above the diagonal of variable names and once below it. This approach enables each variable to be shown as either an independent or dependent variable in relation to other variables. By scanning a row, you can see what happens when a variable is considered dependent, and, by scanning a column, you can see what happens when it is considered independent.*

Numerical Summaries

Bivariate correlation-regression is the most widely used approach for summarizing the relationship between two numeric variables. **Bivariate correlation** is used to summarize the degree of fit, while **bivariate regression** determines the equation for a best-fit line passing through the data.

* Another method for visualizing multivariate data is the **parallel-coordinate plot,** in which variables are plotted on horizontal number lines parallel to one another and observations are connected by vertical lines (Bolorforoush and Wegman, 1988). The method appears satisfactory only when the number of observations is small (no more than, say, 10 to 20).

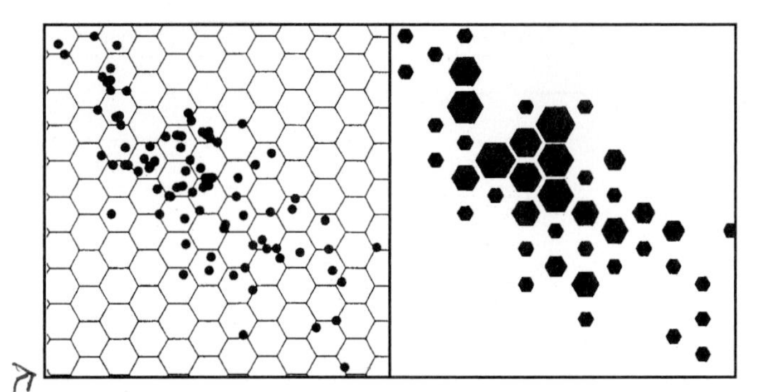

FIGURE 3.6. Hexagon bin plot: an alternative to the scatterplot. The size of hexagons shown on the right are proportional to the number of dots falling within the hexagons on the left. The x and y axes represent sulfate and nitrate deposition values for sites in the eastern United States (After Carr et al., 1992. First published in *Cartography and Geographic Information Systems* 19(4), p. 229. Reprinted with permission from the American Congress on Surveying and Mapping.)

Bivariate Correlation. One value commonly computed in bivariate correlation is the **correlation coefficient,** or r:

$$r = \frac{\sum_{i=1}^{n} (X_i - \overline{X})(Y_i - \overline{Y})}{\sqrt{\sum_{i-1}^{n} (X_i - \overline{X})^2}\sqrt{\sum_{i=1}^{n} (Y_i - \overline{Y})^2}}$$

Table 3.4 shows r values for the transformed sample data. (Note that the table is symmetric about a diagonal

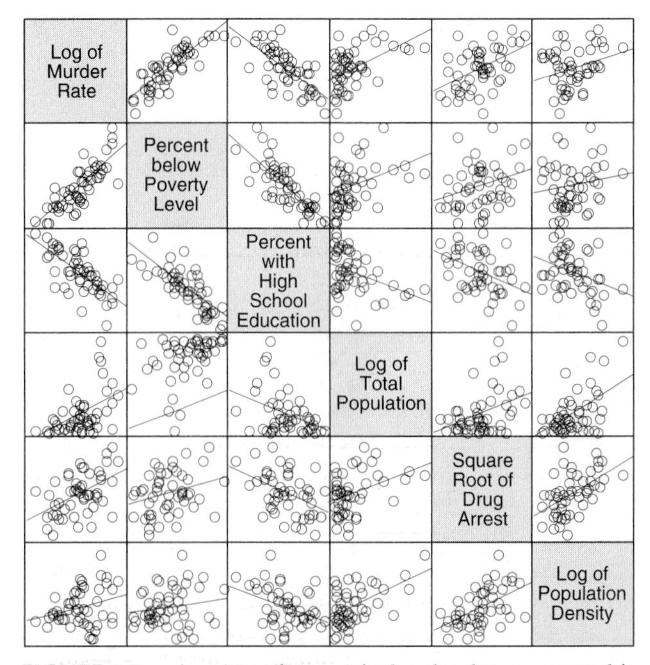

FIGURE 3.7. A scatterplot matrix for the data presented in Table 3.1. Also shown are best-fit regression lines.

TABLE 3.4. Matrix of correlation coefficients for the transformed city data

	Log of Murder Rate	*Percent below Poverty Level*	*Percent with High School Education*	*Log of Total Population*	*Square Root of Drug Arrest Rate*	*Log of Population Density*
Log of murder rate	1.00	.82	−.70	.52	.49	.27
Percent below poverty level	.82	1.00	−.79	.38	.29	.17
Percent with high school education	−.70	−.79	1.00	−.39	−.44	−.36
Log of total population	.52	.38	−.39	1.00	.39	.54
Square root of drug arrest rate	.49	.29	−.44	.39	1.00	.52
Log of population density	.27	.17	−.36	.54	.52	1.00

extending from upper left to lower right; this occurs because the correlation between variables A and B is identical to that between variables B and A.)

Extreme values for r range from -1 to $+1$. A positive r value indicates a *positive relationship* in which increasing values on one variable are associated with increasing values on another variable; for example, the r value for log of murder rate and percent below poverty level is .82. Note that for higher r values the points cluster more tightly about the best-fit line. (Compare Table 3.4 and Figure 3.7, and note that the graph of log of murder rate and percent below poverty level is more tightly clustered than the one for log of murder rate and square root of drug arrest.) Conversely, a negative r value indicates a *negative relationship* in which increasing values on one variable are associated with decreasing values on another variable; for example, log of murder rate and percent with high school education have a correlation of $-.70$.

Although r is a measure of the strength of relation between two variables, one must be careful in interpreting its magnitude, since an r value twice another does not equate to twice the strength of the relationship. To properly compute the relation between the strength of two r values, one must compute the coefficient of determination, or r^2, which measures the proportion of variation in one variable explained, or accounted for, by the other variable. For example, r^2 between log of murder rate and percent below poverty level is .67, indicating that 67 of the variation in murder rate can be accounted for by a linear function of the poverty variable.

Two admonitions are necessary with respect to correlation. The first is that high correlations do not necessarily imply a causal relationship. The high correlation between murder rate and poverty could be a result of chance (although it is unlikely, two random data sets could result in a correlation of this magnitude), or some other variable or set of variables may be influencing both of these variables.

The second admonition is that the magnitude of r may be affected by the level at which data have been aggregated (Clark and Hosking 1986; Barrett 1994). Generally, coarser levels of aggregation (for example, analyzing at the city level rather than at the tract level) will lead to higher r values because "aggregation reduces the between unit variation in a variable, making the variable seem more homogeneous" (Clark and Hosking, 1986, 405). For this reason, it could be argued that the data for cities (Table 3.1) should be broken down into census tracts or block groups, or even analyzed at the individual level (looking at individual murders and collecting data regarding the people involved), before any conclusions can be drawn. Analysis of aggregated data may, however, suggest variables that should be pursued through a more qualitative analysis (such as interviewing).*

Bivariate Regression. In general, the equation for any straight line is defined as $Y_i = a + bX_i$, where X_i, Y_i defines a point on the line and a and b are the y intercept and slope, respectively. In regression analysis, X_i and Y_i are raw values for the independent and dependent variables respectively, and the line-of-best-fit is defined as $\hat{Y}_i = a + bX_i$, where $\hat{Y}_i$ is a predicted value for the dependent variable. The best-fit line is found by minimizing the differences between the actual and predicted values

* An example of a study that minimizes aggregation effects is "The Project on Human Development in Chicago Neighborhoods," which is analyzing data from a variety of Chicago neighborhoods. The emphasis is on individual interviews and extensive field work (See "Research chases causes of crime: Chicago youths focus of study." (1994)).

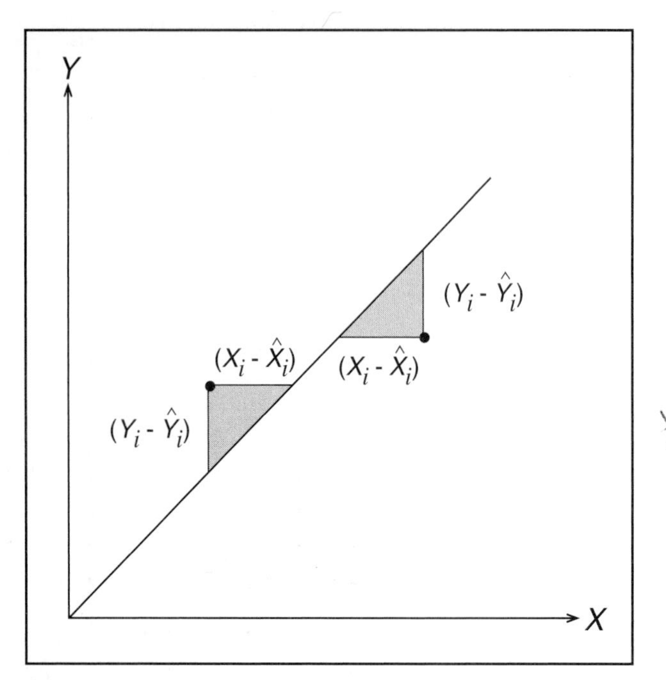

FIGURE 3.8. Possible approaches for determining best-fit lines for a set of data include (1) minimizing vertical distances $(Y_i - \hat{Y}_i)$; (2) minimizing horizontal distances $(X_i - \hat{X}_i)$; and (3) minimizing the area of the resulting triangles.

$(Y_i - \hat{Y}_i)$, the vertical distances shown in Figure 3.8. The values of a and b that minimize these values are:

$$a = \overline{Y} - b\overline{X}$$

$$b = \frac{n\sum_{i=1}^{n} X_iY_i - \sum_{i=1}^{n} X_i \sum_{i=1}^{n} Y_i}{n\sum_{i=1}^{n} X_i^2 - \left(\sum_{i=1}^{n} X_i\right)^2}$$

The best-fit lines shown in Figures 3.5 and 3.7 were derived using this approach.

Major Axis Approach. When one does not wish to specify a dependent variable (as when relating murder rate and race), it makes sense to minimize the vertical and horizontal distances shown in Figure 3.8 simultaneously, effectively minimizing the area of the triangles; this is known as the **reduced major-axis approach** (Davis 1986, 201). The equations for this approach turn out to be simpler than those for standard regression. The slope is just the ratio of the standard deviations of the variables, or s_Y/s_X, and the intercept is still $\overline{Y} - b\overline{X}$. In Chapter 12 we will consider how Eyton (1984a) has used the reduced major-axis approach to create a logical bivariate map.

Multiple Correlation-Regression and Other Multivariate Techniques. When the concept of bivariate correlation-regression is extended into the multivariate realm, it is termed **multiple regression.** In multiple regression, there is still a single dependent variable, but multiple independent variables are possible. In a fashion similar to the above, it is possible to perform correlation and regression analyses that summarize the relationship between the dependent and independent variables. A discussion of these and other multivariate techniques, such as **principal components analysis,** is beyond the scope of the present text. The reader should consult statistical texts, such as Clark and Hosking (1986), Stevens (1996), and Davis (1986), for related information.

3.3.3 Exploratory Data Analysis

Surely one of the most important advances in statistical analysis in the last 20 years was John Tukey's (1977) development of **exploratory data analysis.** In Chapter 1, it was suggested that rather than trying to make one "best" map, interactive graphics systems can provide multiple representations of a spatial data set. In much the same way, Tukey proposed that rather than trying to fit statistical data to standard forms (normal, poisson, binomial), data should be explored, much as a detective investigates a crime. In the process of exploring data, the purpose should not be to confirm what one already suspects, but rather to develop new questions or hypotheses.

One technique representative of Tukey's approach is the **stem-and-leaf plot,** which is depicted in Figure 3.9 using the murder rate data. To construct a stem-and-leaf plot, one first separates the digits of the data values into three classes: sorting digits, display digits, and digits not

Stem	Leaf
0 *	0111222222233
0 ·	5555777889
1 *	000011224
1 ·	5778
2 *	1244
2 ·	8
3 *	11
3 ·	5
4 *	11
4 ·	
5 *	
5 ·	679
6 *	
6 ·	
7 *	
7 ·	8

FIGURE 3.9. Stem-and-leaf plot of the murder rate data presented in Table 3.1.

displayed because of rounding.* For the murder rate data, I chose the 10s place as a sorting digit, the 1s place as the display digit, and did not display the 10ths place because I rounded to the nearest whole percent. Sorting digits are placed to the left of the vertical line shown in Figure 3.9 and are known as *stems,* while display digits are placed to the right and are known as *leaves.* For Figure 3.9, Tukey's conventional system of asterisks and dots was used to split the 10s place into two parts (leaf values of 0 to 4 are plotted on one row and 5 to 9 on the next row). For example, Norfolk's murder rate of 24.1 appears as the fourth leaf ("4") in the fifth row ("2 *"), while Los Angeles's rate of 28.2 appears as the only leaf ("8") in the sixth row ("2 ·").

If you mentally rotate the stem-and-leaf plot so that the stems are on the bottom, and then compare it to those graphical methods discussed previously for individual variables (Figures 3.1 and 3.2), you will note a great deal of similarity to the histogram. Both methods portray a peak on the left side of the graph with a distinct positive skew. Although the graphs are similar, the major advantage of the stem-and-leaf plot is that it is possible to determine approximate values for each observation. For the murder data, this may not be particularly useful, but for some data sets it can be. For example, Burt and Barber (1996, 542–544) describe a stem-and-leaf plot that portrays when houses were constructed in a neighborhood. The stem-and-leaf plot reveals that seven houses were built during the post–World War II period (1945–1947), but no houses were built from 1948 to 1950.

Another technique representative of Tukey's work is the **box plot** (Figure 3.10). Here, a rectangular box represents the interquartile range (the 75th minus the 25th percentile), and the middle line within the box represents the median, or 50th percentile. The position of the median, relative to the 75th and 25th percentiles is an indicator of whether the distribution is symmetric or skewed; for the murder rate data, the position indicates a positive skew (a tail toward higher values on the number line).

The horizontal lines outside the rectangular box represent the maximum and minimum values in the data. The relative positions of these lines also permit an examination of the symmetry of the distribution, and their position relative to the box indicates how extreme the maximum and minimum values are. In this case, their relative position again suggests a positive skew.

Box plots are most frequently used to compare two or more distributions having the same units of measurement. For example, Figure 3.11 compares murder rate data for cities falling in two distinct regions (the "Deep South" and "California," as defined by Birdsall and Florin 1992). Note that the California cities have a distinctly smaller range, a smaller maximum, and an interquartile range that is smaller and shifted toward lower murder rates.

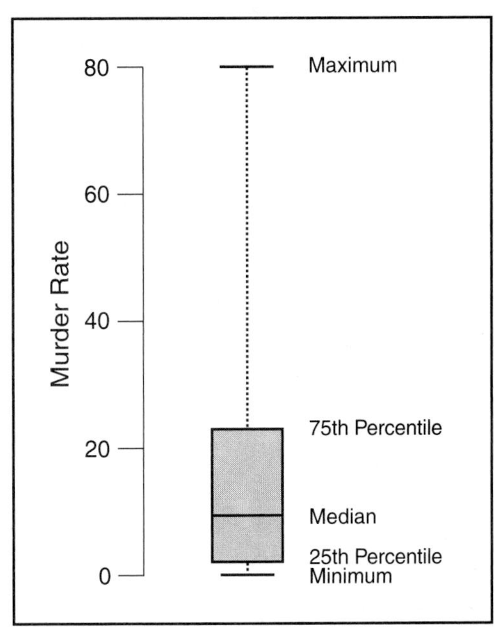

FIGURE 3.10. Box plot of the murder rate data presented in Table 3.1.

It is important to realize that many of the methods covered in previous pages can also be used in an exploratory manner. For example, in a fashion similar to the stem-and-leaf plot, a dispersion graph may uncover nuances in the data not revealed by a histogram. The scatterplot matrix can also be used in an exploratory fash-

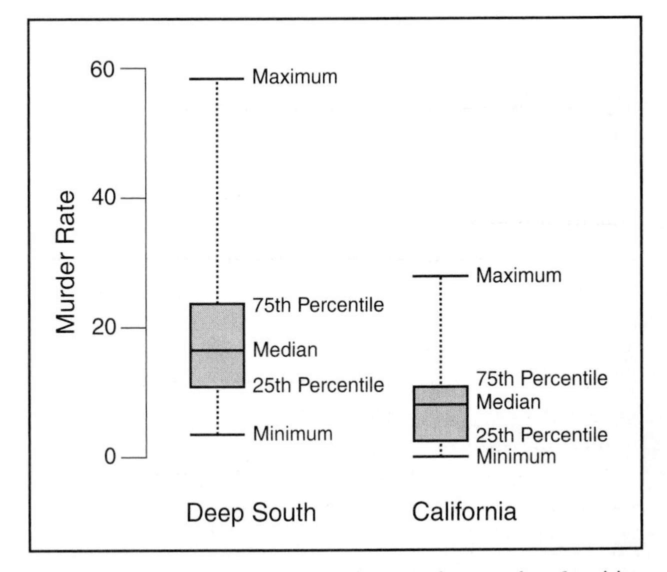

FIGURE 3.11. Box plots comparing murder rate data for cities in the Deep South and California. California cities appear to have a distinctly smaller range, a smaller maximum, and an interquartile range that is smaller and associated with lower murder rates.

* Tukey (1977) indicated that digits beyond the display digit could be used for rounding or simply ignored.

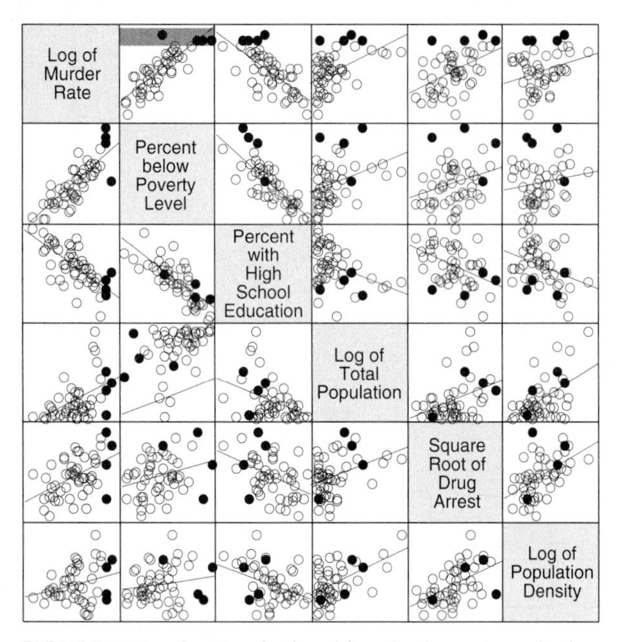

FIGURE 3.12. Scatterplot brushing. As the rectangular box (or brush) is moved, dots (cities) within it are highlighted in all scatterplots.

ion, especially if the number of variables is large. Remember that the key to exploratory analysis is to reveal hidden characteristics; the broadest possible range of approaches should be considered for achieving this.

A technique that fits especially well under the heading of exploratory data analysis is **scatterplot brushing,** which is illustrated in Figure 3.12 for the city data. Note that the log of murder rate–percent below poverty level scatterplot contains a gray rectangular box or "brush," and that all dots (cities) within this box are highlighted by a solid fill; these same cities are also highlighted within other scatterplots in the matrix. Brushing involves using an interactive graphics display to move the box; as the box moves, observations falling within it are highlighted within all scatterplots.

Monmonier (1989a) was responsible for integrating scatterplot brushing with maps. He indicated that as the brush is moved within the scatterplot, mapped areas corresponding to dots within the brush should also be highlighted (Figure 3.13). Monmonier also suggested the notion of a **geographic brush:** as areas of a map are brushed, corresponding dots in the scatterplot matrix should be highlighted. A number of Monmonier's ideas have been implemented in Tang's (1992) Visda software.

3.4. NUMERICAL SUMMARIES IN WHICH LOCATION IS AN INTEGRAL COMPONENT

This section considers numerical summaries in which spatial location is an integral component. Some of these methods analyze just spatial location by itself (formulas for the centroid and various shape indices), while others consider both spatial location and the values of an attribute or variable (such as spatial autocorrelation).

3.4.1 Analysis of Spatial Location

Centroid

The **centroid** is defined as

$$\overline{X}_r = \frac{\sum_{i=1}^{n}(X_iY_{i+1} - X_{i+1}Y_i)(X_i + X_{i+1})}{3\sum_{i=1}^{n}(X_iY_{i+1} - X_{i+1}Y_i)}$$

$$\overline{Y}_r = \frac{\sum_{i=1}^{n}(X_{i+1}Y_i - X_iY_{i+1})(Y_i + Y_{i+1})}{3\sum_{i=1}^{n}(X_{i+1}Y_i - X_iY_{i+1})}$$

where X_iY_i are a sequence of points defining the boundary of a region, assuming that the n + first point is identical to the first point (Bachi 1973). These formulas appear rather complicated, but the concept is really quite simple. Imagine cutting the outline of the 48 contiguous states out of a thin metal sheet and then attempting to balance the sheet on the eraser end of a pencil. The point at which the sheet balances (which happens to be located near Lebanon, Kansas) would be its centroid (or center of gravity). It should be recognized, however, that the centroid derived by the above formula may not necessarily fall within a region if the region is highly convoluted. As a result, it may be more appropriate to consider other measures for computing the center of region, such as the center of a rectangle surrounding a region—the so-called **bounding rectangle** (Carstensen 1987).

Indexes for Measuring Shape

Geographers have developed a wide variety of indexes for measuring the shape of geographic regions. One of the simplest is the **compaction index (CI)** (Hammond and McCullagh 1978, 69–70), which is defined as the ratio of the area of a shape to the area of a circumscribing circle (a circle that just touches the bounds of the shape). Values for CI range from 0 to 1, with 0 representing the least compact shape (for example, a narrow rectangular box) and 1 representing the most compact shape (a circle). As an example, consider the shapes of Tennessee and Arkansas relative to a circle (Figure 3.14); based on the compaction index, Arkansas is clearly more compact than Tennessee.

One limitation of the compaction index is that it does not differentiate some shapes well. For example, Figure 3.15 shows two regions whose shapes differ, but which have similar CI values. This sort of problem has led

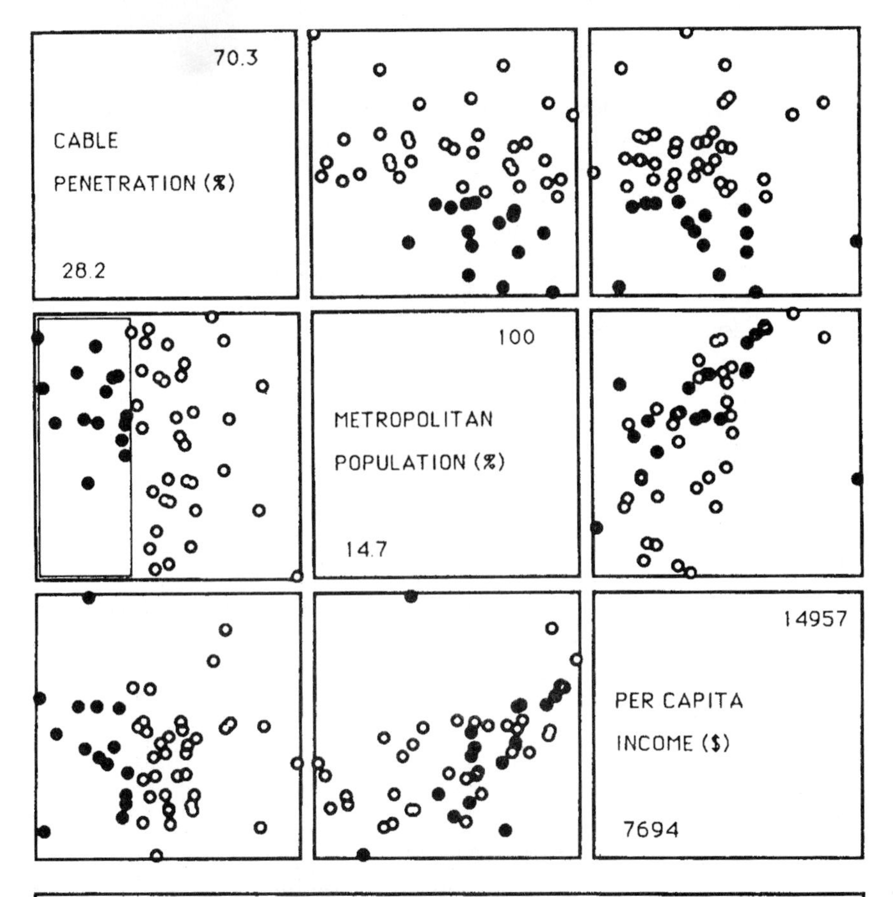

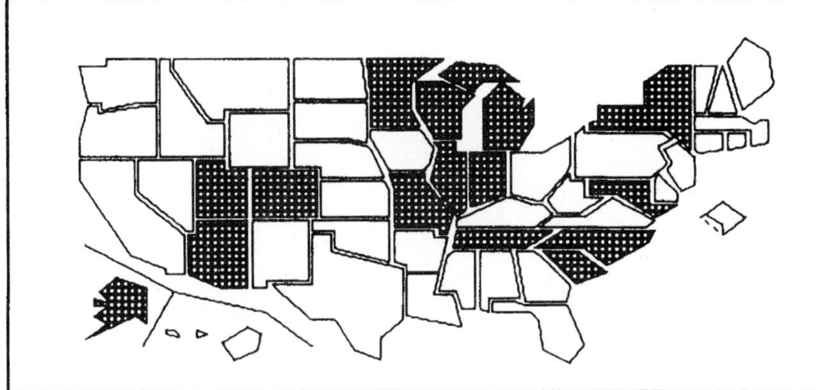

FIGURE 3.13. How scatterplot brushing can be integrated with a map. As the brush (the rectangular box) is moved in the scatterplot, dots within it are highlighted within the scatterplot and on the map (Monmonier, 1989a. Reprinted by permission from *Geographical Analysis,* Vol. 21, No. 1 (Jan 1989).

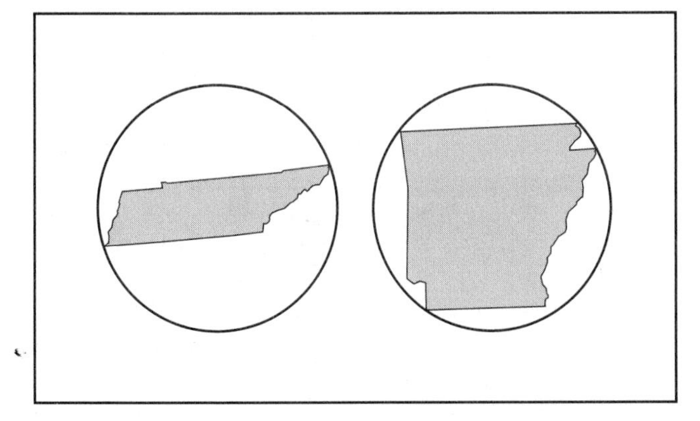

FIGURE 3.14. Computing the compaction index (CI) for two states. CI is computed as a ratio of the area of the region to a circle enclosing that region; the CI value for Arkansas clearly would be greater than for Tennessee.

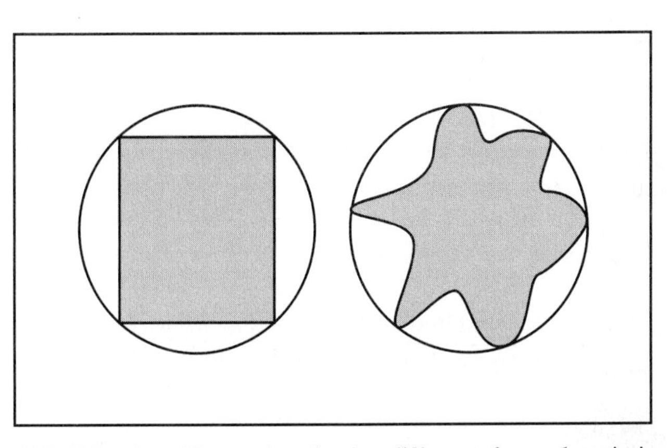

FIGURE 3.15. Two regions having different shapes, but similar CI values.

geographers to develop other methods for analyzing shape, many of which are more complex than the compaction index; examples include the Boyce-Clark index (Unwin 1981) and Moellering and Rayner's (1982) harmonic-analysis method.

One application of shape indices is in **redistricting,** which involves the combination of voting precincts to form legislative or congressional districts. This process must be done at each decennial census to account for migration and natural increases and decreases in population. The reader is probably familiar with the related term **gerrymandering,** in which districts are purposely structured for partisan benefit. A shape index can be used as an objective measure of the degree of gerrymandering (Morrill 1981).

3.4.2 Analyzing an Attribute in Association with Spatial Location

This section considers various methods for analyzing spatial data in which an attribute is linked with its spatial location. Topics covered include (1) weighting attribute data to account for sizes of enumeration units, (2) central tendency and dispersion for point data, and (3) spatial autocorrelation and map complexity measures.

Weighting Attribute Data to Account for Sizes of Enumeration Units

Robinson et al. (1995) argued that when data are associated with enumeration units, basic summary measures such as the mean and standard deviation should be weighted to account for the differing sizes of enumeration units. The formulas for a population are as follows:

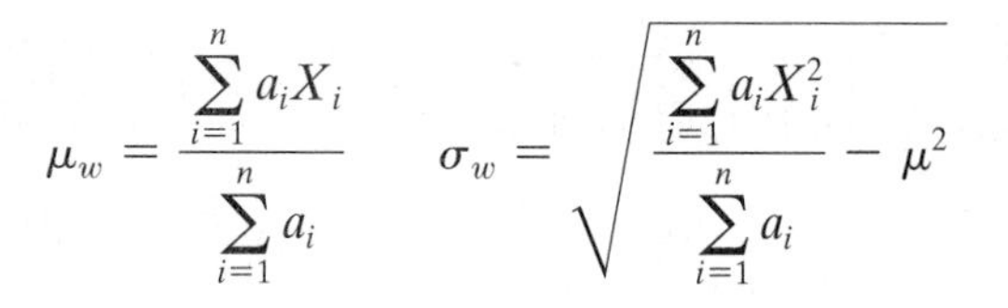

$$\mu_w = \frac{\sum_{i=1}^{n} a_i X_i}{\sum_{i=1}^{n} a_i} \qquad \sigma_w = \sqrt{\frac{\sum_{i=1}^{n} a_i X_i^2}{\sum_{i=1}^{n} a_i} - \mu^2}$$

where X_i is a value for an attribute for the ith enumeration unit, and a_i is the area of the ith enumeration unit.

These weighted measures are appropriate if the intention is to impart to the reader the visual impact that large enumeration units have on a distribution. One should realize, however, that different means and standard deviations will arise, depending on how enumeration units are defined. To illustrate, consider the single enumeration unit and its potential four subregions shown in Figure 3.16. Presume that within the single enumeration unit 39 out of 1300 people, or 3 percent, are college graduates, but that in three of the four subregions 13 out of 100 people, or 13 percent, are college graduates, while in the fourth subregion there are no college graduates. The average of the four subregions is 9.75 percent, a value that is quite different than the 3 percent for the aggregated data.

Another weighting procedure relevant to enumeration units would be to modify the formula for r to account for the sizes of the enumeration units—the logic being that r should be impacted more by larger enumeration units. The formula is as follows (Robinson 1956):

$$r_w = \frac{\sum_{i=1}^{n} a_i \sum_{i=1}^{n} a_i X_i Y_i - \sum_{i=1}^{n} a_i X_i \sum_{i=1}^{n} a_i Y_i}{\sqrt{\sum_{i=1}^{n} a_i \sum_{i=1}^{n} a_i X_i^2 - \left(\sum_{i=1}^{n} a_i X_i\right)^2} \times \sqrt{\sum_{i=1}^{n} a_i \sum_{i=1}^{n} a_i Y_i^2 - \left(\sum_{i=1}^{n} a_i Y_i\right)^2}}$$

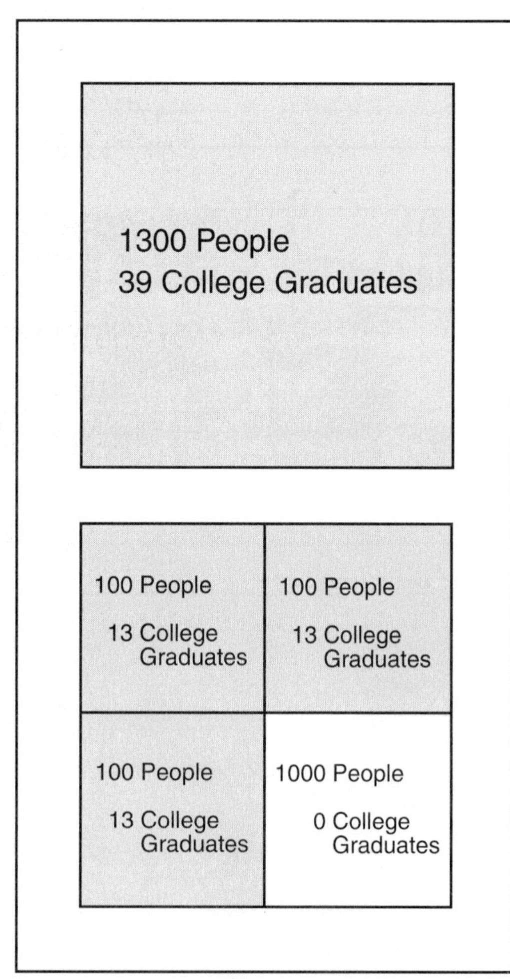

FIGURE 3.16. As explained in the text, the areally weighted mean is not necessarily a logical measure to use.

where a_i is the area of the ith enumeration unit. Thomas and Anderson (1965) showed that such a weighting procedure is inappropriate from a statistical perspective because unweighted correlation coefficients resulting from different arrangements of enumeration unit boundaries were not significantly different from one another. Although inappropriate from a statistical perspective, such a measure might be used as an indicator of the visual correlation between two maps (In Chapter 12 we will consider the problem of visual correlation in greater depth.)

Central Tendency and Dispersion for Point Data

Section 3.3.1 considered measures of central tendency and dispersion for attribute data, ignoring spatial location. For example, in computing the mean for the attribute murder rate, the location of individual cities was ignored. We now consider analogous measures of central tendency and dispersion for explicitly spatial data. Imagine a set of x and y coordinates defining the location of small towns in a portion of Minnesota. The central tendency for such a set of points is termed the **mean center,** and is defined as

$$\overline{X}_c = \frac{\sum_{i=1}^{n} X_i}{n} \qquad \overline{Y}_c = \frac{\sum_{i=1}^{n} Y_i}{n}$$

Note that these formulas simply apply the nonspatial mean formula twice, once for the x axis and once for the y axis. As with the nonspatial mean, there is some danger that the mean center will not fall where the data are concentrated. This problem can be handled using a spatial version of the median (Hammond and McCullagh 1978, 48–53).

A measure of dispersion associated with the mean center is **standard distance, SD,** which is computed as*

$$SD = \sqrt{\frac{\sum_{i=1}^{n} d_{ic}^2}{N}} = \sqrt{\sigma_X^2 + \sigma_Y^2}$$

where d_{ic} is the distance from the ith point to the mean center. The formula on the left indicates that SD is a measure of spread about the mean center, while the one on the right is valuable when using a calculator, which generally does not include functions for spatial statistics.

For some applications, it can be argued that measures of central tendency and dispersion for points should be weighted to reflect the magnitude of an attribute at the points. For example, imagine that you have population values for point locations (say, the centers of counties) and wish to find the mean center and standard distance for the population. The formulas for the *weighted mean center* and *weighted standard distance* are:

$$\overline{X}_{cw} = \frac{\sum_{i=1}^{n} w_i X_i}{\sum_{i=1}^{n} w_i} \qquad \overline{Y}_{cw} = \frac{\sum_{i=1}^{n} w_i Y_i}{\sum_{i=1}^{n} w_i}$$

$$SD_w = \sqrt{\frac{\sum_{i=1}^{n} w_i d_{ic}^2}{\sum_{i=1}^{n} w_i}}$$

Using formulas similar to these, the U.S. Bureau of the Census has found that the population mean center has moved from near Baltimore, Maryland, in 1790 to about 10 miles southeast of Steelville, Missouri, in 1990 (U.S. Bureau of the Census 1994b).

Spatial Autocorrelation and Measuring Spatial Pattern

Although maps allow us to visually assess spatial pattern, they have two important limitations: their interpretation varies from person to person; and there is the possibility that a perceived pattern is actually a result of chance factors, and thus not meaningful. For these reasons, it makes sense to compute a numerical measure of spatial pattern. In this section we will consider the notion of spatial autocorrelation and an associated numerical measure appropriate for choropleth maps. For a discussion of spatial pattern measures appropriate for other types of maps, see Unwin (1981) and Davis (1986).

Spatial autocorrelation is the tendency for like things to occur near one another in geographic space. For example, expensive homes likely will be located near other expensive homes, and soil cores with high clay content likely will be found near other soil cores with high clay content. Two measures used by statisticians to express the degree of spatial autocorrelation are the Moran coefficient (MC) and the Geary ratio (GR). We will consider only the Moran coefficient because it is statistically more powerful (Griffith 1993, 21), and it is more commonly used.

The formula for MC is

$$MC = \frac{\sum_{i=1}^{n}\sum_{j=1}^{n} w_{ij}(X_i - \overline{X})(X_j - \overline{X}) / \sum_{i=1}^{n}\sum_{j=1}^{n} w_{ij}}{\sum_{i=1}^{n} (X_i - \overline{X})^2 / n}$$

* The formulas given are for a population; Griffith and Amrhein (1991, 123) indicate that for standard distance there is little difference between sample and population formulas.

Enumeration Units

i	j	w_{ij}	$(x_i - \bar{x})(x_j - \bar{x})$	=	
1	2	1	(20–50)(40–50)	=	300
1	4	1	(20–50)(30–50)	=	600
2	1	1	(40–50)(20–50)	=	300
2	3	1	(40–50)(60–50)	=	–100
2	5	1	(40–50)(50–50)	=	0
3	2	1	(60–50)(40–50)	=	–100
3	6	1	(60–50)(70–50)	=	200
4	1	1	(30–50)(20–50)	=	600
4	5	1	(30–50)(50–50)	=	0
4	7	1	(30–50)(40–50)	=	200
5	2	1	(50–50)(40–50)	=	0
5	4	1	(50–50)(30–50)	=	0
5	6	1	(50–50)(70–50)	=	0
5	8	1	(50–50)(60–50)	=	0
6	3	1	(70–50)(60–50)	=	200
6	5	1	(70–50)(50–50)	=	0
6	9	1	(70–50)(80–50)	=	600
7	4	1	(40–50)(30–50)	=	200
7	8	1	(40–50)(60–50)	=	–100
8	5	1	(60–50)(50–50)	=	0
8	7	1	(60–50)(40–50)	=	–100
8	9	1	(60–50)(80–50)	=	300
9	6	1	(80–50)(70–50)	=	600
9	8	1	(80–50)(60–50)	=	300
					4000

$$\sum_{i=1}^{n}\sum_{j=1}^{n} w_{ij}(x_i - \bar{x})(x_j - \bar{x}) = 4000$$

$$\sum_{i=1}^{n}\sum_{j=1}^{n} w_{ij} = 24$$

$$\sum_{i=1}^{n}(x_i - \bar{x})^2/n = \sigma^2 = 333.333$$

$$MC = (4000/24)/333.333 = .50$$

FIGURE 3.17. Computation of the Moran coefficient (MC) for spatial autocorrelation.

where $w_{ij} = 1$ if enumeration units i and j are adjacent (or contiguous) and 0 otherwise.* Computations for MC for a hypothetical region consisting of nine enumeration units are shown in Figure 3.17. Within each enumeration unit, the upper left number is an identifier for that unit, while the lower right number is the value for a hypothetical attribute. Note that the formula involves multiplying a weight for two enumeration units (w_{ij}) times the product of the difference between the attribute values for the enumeration units and the mean of the data $(X_i - \overline{X})(X_j - \overline{X})$. Computations are shown only for adjacent enumeration units, because w_{ij} will be 0 for nonadjacent units and thus the product $w_{ij}(X_i - \overline{X})(X_j - \overline{X})$ also will be 0.

The formula for MC bears some resemblance to the formula for the correlation coefficient, r, discussed in section 3.3.2; for both equations, the denominator contains a measure of the variation in the attribute about the mean, while the numerator contains a measure of how adjacent enumeration units covary (the reader should compare the equation shown here with that for r in section 3.3.2). Thus, it is not surprising that MC also ranges between -1 and $+1$. A value close to $+1$ indicates that similar values are likely to occur near one another, while a value close to -1 indicates that unlike values are apt to occur near one another. Finally, a value near 0 is indicative of no autocorrelation, or a situation in which values of the attribute are randomly distributed. Griffith (1993, 2) indicates that "moderate" positive spatial autocorrelation is most frequently observed in the real world.

As an example of how the Moran coefficient might be used, consider the rates for respiratory cancer for white males in Louisiana counties (Figure 3.18). A visual assessment of this map suggests that there is a strong positive spatial autocorrelation, with high cancer rates occurring both in the southern part of the state and along an east-west strip in the northern part of the state. Odland (1988) computed an MC of .14 for this pattern, with a probability of less than 3 percent that this value occurred by chance. The value of .14 indicates that the pattern is not quite as strongly positively autocorrelated as a visual examination of the map suggests, but the associated probability indicates the pattern is significant and worthy of further exploration.

An important recent development in statistical geography is the incorporation of spatial autocorrelation into inferential statistics. For example, basic regression analysis can be modified to include an autocorrelation parameter; such an approach is essential because one of

* The weights can also be modified to account for the different sizes of enumeration units (Odland 1988, 29–31). I have used weights of 0 and 1 for computational simplicity.

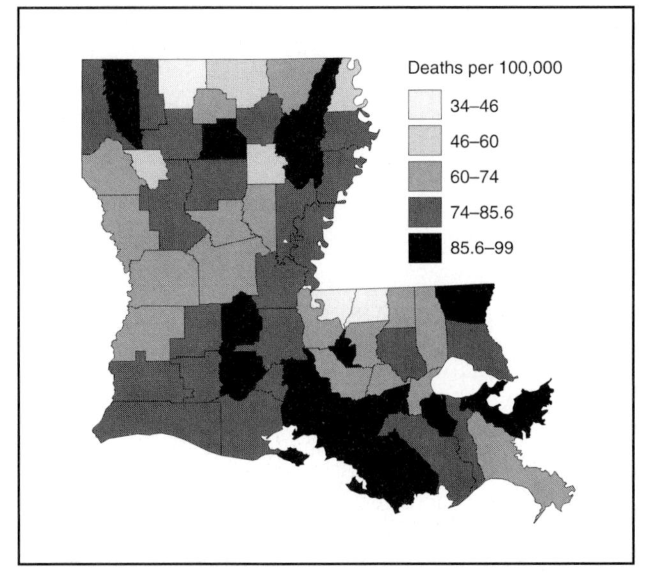

FIGURE 3.18. A map of rates for respiratory cancer for white males in Louisiana counties. The Moran coefficient (a measure of spatial autocorrelation) is .14, with a probability of less than 3 percent that this value occurred by chance. Thus, the pattern is significant and worthy of further exploration (Odland, *Spatial Autocorrelation,* p. 45, copyright © 1988 by Sage Publications. Reprinted by permission of Sage Publications, Inc.)

the basic assumptions in regression analysis is that the errors associated with the model (the vertical lines in Figure 3.8) are uncorrelated. Readers who wish to incorporate these concepts into their analysis should consult Griffith (1993).

MacEachren's Measures of Map Complexity

The notion of spatial autocorrelation and associated measures such as the Moran coefficient were developed by statistical geographers. Cartographers have also been interested in measures for describing spatial pattern. As an example, we'll consider the work of MacEachren (1982a), who developed what he called map complexity measures. MacEachren defined **map complexity** as "the degree to which the combination of map elements results in a pattern that appears to be intricate or involved" (p. 32). He argued that when a map is used for presentation, very complex maps might hinder the communication of information.

MacEachren's measures are based on Muller's (1974) application of graph theory to choropleth maps. To compute MacEachren's measures, the map is treated as a set of faces (enumeration units), edges (the boundaries of the units), and vertices (where the edges intersect) (Figure 3.19A). When a distribution is mapped, edges and vertices between like values are omitted (Figures 3.19B and 3.19C). Complexity measures are computed by simply taking the ratio of the number of faces, edges, and vertices on a base map and on the mapped distribution:*

$$C_F = \frac{\text{observed number of faces}}{\text{number of original faces}}$$

$$C_V = \frac{\text{observed number of vertices}}{\text{number of original vertices}}$$

$$C_E = \frac{\text{number of original edges remaining}}{\text{number of original edges}}$$

In the example illustrated in Figure 3.19, map C has the higher complexity measures, and so would presumably be more difficult for a user to interpret.

SUMMARY

In this chapter, we have examined a variety of nonmapping techniques (tables, graphs, and numerical summaries) that can be used along with maps for analyzing spatial data. We have seen that these techniques each have their advantages and disadvantages. In the univariate realm, **raw tables** are useful for providing specific information about places (for example, the murder rate for Atlanta, Georgia, in 1990 was 58.6 per 100,000 people), but they fail to provide an overview of the distribution (for example, that the murder rate for U.S. cities in 1990 had a distinct positive skew). **Grouped-frequency tables** do provide an overview of the distribution, but are not as effective as graphical methods (such as the **dispersion graph** and **histogram**). Potential weaknesses of graphical methods are the subjectivity of interpretation and the space they take up. Numerical summaries (such as the **mean** and **standard deviation**) are a solution to these problems. A weakness of numerical summaries, however, is that they hide the detailed character of the data; for example, an **outlier** in the data may be missed if only a numerical summary is examined. Because of these advantages and disadvantages, data exploration software (Chapter 13) should ideally include the capability to analyze spatial data using a broad range of nonmapping techniques.

When examining two or more variables at once, we saw that tabular displays are of limited use. Much more suitable are graphical displays, such as the **scatterplot** and **scatterplot matrix.** The numerical method of **bivariate correlation-regression** can also be useful if one wishes to summarize the relationship between two variables by a single number or equation. As with univariate data, care should be taken in using bivariate numerical summaries,

---

* Note that the edge measure is the "number of original edges left" rather than the "observed number of edges." MacEachren used this approach to attempt to account for the size of faces.

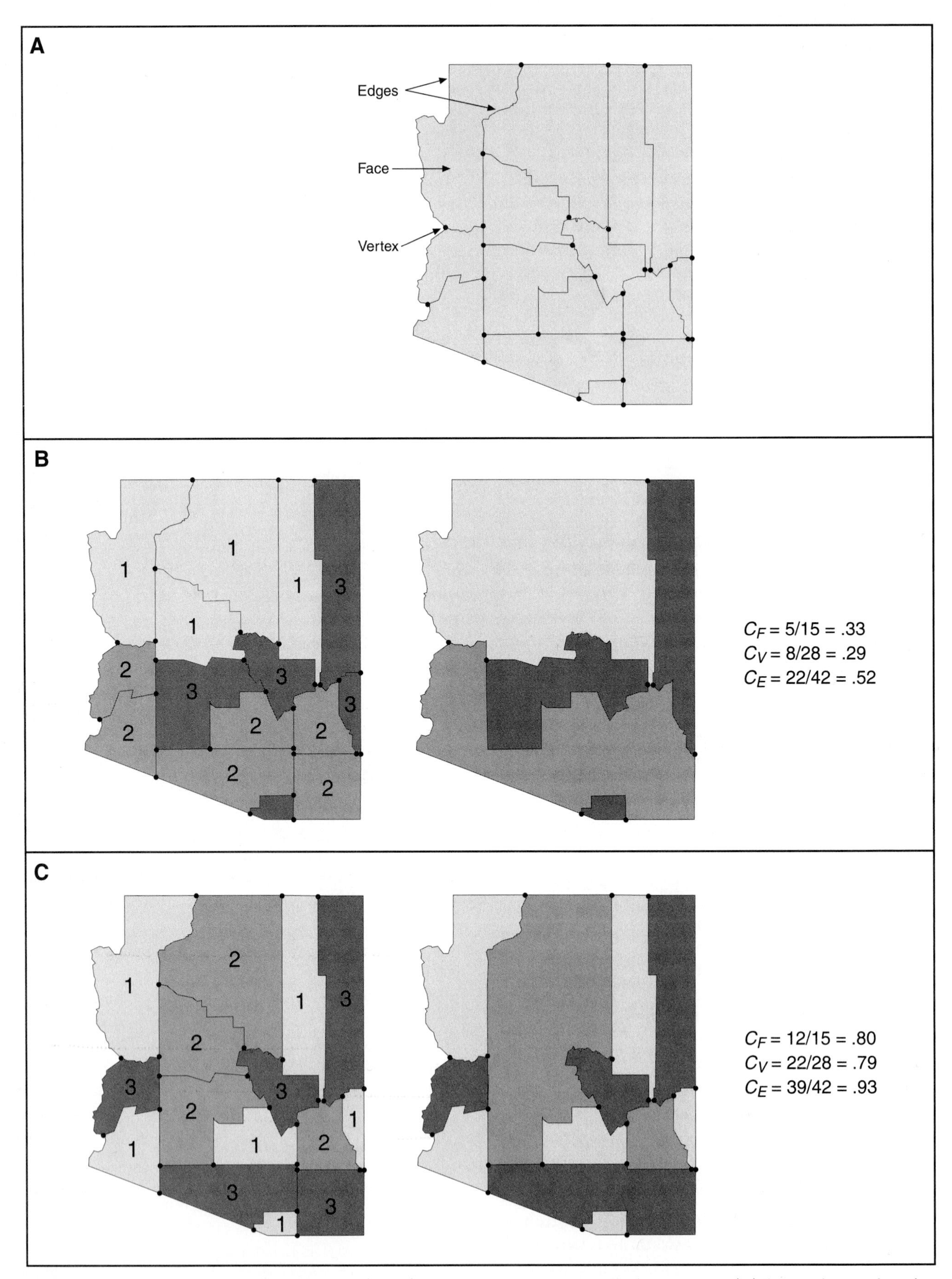

FIGURE 3.19. Computation of MacEachren's (1982a) face, vertex, and edge complexity measures. (A) Faces, edges, and vertices for a county-level map of Arizona; (B) three-class map for which low complexity values result; (C) three-class map for which high complexity values result.

as they hide the detailed character of the relationship between variables. An important problem with bivariate correlation is that it is a function of the level of aggregation (a different **correlation coefficient** might arise at the census tract level as opposed to the block-group level).

Given the emphasis on **data exploration** in this book, special emphasis in this chapter was placed on **exploratory data analysis,** which refers to the notion that data should be explored, much as a detective investigates a crime. We looked at three common methods for exploratory data analysis: the **stem-and-leaf plot,** the **box plot,** and **scatterplot brushing.** We saw that the notion of scatterplot brushing could be extended to a **geographic brush**—as areas on a map are highlighted, corresponding dots in a scatterplot matrix are highlighted. This is the sort of capability we would expect to find in data exploration software.

The latter section of the chapter considered several numerical summaries for which spatial location is an integral part of the formulas used; here we examined the **centroid, compaction index, mean center** (and associated **standard distance**), **spatial autocorrelation,** and **map complexity** measures. In the following chapter, we will make use of one of the map complexity measures when we consider a method for simplifying choropleth maps (section 4.3).

FURTHER READING

Barrett, R. E. (1994) *Using the 1990 U.S. Census for Research.* Thousand Oaks, Calif.: SAGE Publications.

Summarizes data available from the U.S. Bureau of the Census and discusses problems associated with using such data.

Burt, J. E., and Barber, G. M. (1996) *Elementary Statistics for Geographers.* 2d ed. New York: Guilford Press.

An introductory statistics book for geographers.

Carr, D. B. (1991) "Looking at large data sets using binned data plots." In *Computing and Graphics in Statistics,* ed. by A. Buja and P. A. Tukey, pp. 7–39. New York: Springer-Verlag.

Provides an in-depth discussion of the hexagon bin plot.

Clark, W. A. V., and Hosking, P. L. (1986) *Statistical Methods for Geographers.* New York: John Wiley & Sons.

Covers both introductory and advanced statistical methods for geographers.

Cleveland, W. S. (1994) *The Elements of Graphing Data.* rev. ed. Summit, N.J.: Hobart Press.

Covers a broad range of graphical methods for summarizing data. The text has more of a statistical emphasis than the Kosslyn text listed below. Also see Cleveland (1993).

Cleveland, W. S., and McGill, M. E. (eds.) (1988) *Dynamic Graphics for Statistics.* Belmont, Calif.: Wadsworth.

Contains 16 chapters describing various researchers' efforts to explore statistical data using interactive graphics.

Davis, J. C. (1986) *Statistics and Data Analysis in Geology.* 2d ed. New York: John Wiley & Sons.

Although intended primarily for geologists, this text covers a broad range of statistical methods, many of which are of interest to cartographers.

Dykes, J. A. (1994) "Visualizing spatial association in area-value data." In *Innovations in GIS,* ed. by M. F. Worboys, pp. 149–159. Bristol, Penn.: Taylor & Francis.

Considers problems of using a *mathematical measure* of spatial autocorrelation to represent the *perceived* degree of autocorrelation.

Griffith, D. A. (1993) *Spatial Regression Analysis on the PC: Spatial Statistics using SAS.* Washington, D.C.: Association of American Geographers.

Discusses spatial autocorrelation and how it can be used in various statistical methods (e.g., in regression analysis).

Kosslyn, S. M. (1994) *Elements of Graph Design.* New York: W. H. Freeman.

A how-to text on constructing graphs; the text is liberaly illustrated with *do* and *don't* examples.

MacEachren, A. M. (1982a) "Map complexity: Comparison and measurement." *The American Cartographer* 9, no. 1:31–46.

Discusses issues involved in measuring map complexity.

Monmonier, M. (1989a) "Geographic brushing: Enhancing exploratory analysis of the scatterplot matrix." *Geographical Analysis* 21, no. 1:81–84.

Briefly discusses geographic brushing, a form of exploratory data analysis.

Odland, J. (1988) *Spatial Autocorrelation.* Newbury Park, Calif.: SAGE Publications.

A primer on spatial autocorrelation.

Tufte, E. R. (1983) *The Visual Display of Quantitative Information.* Cheshire, Conn.: Graphics Press.

Covers a broad range of techniques (both graphs and maps) for representing numerical data.

Tukey, J. W. (1977) *Exploratory Data Analysis.* Reading, Mass.: Addison-Wesley.

The classic reference on exploratory data analysis.

Unwin, D. (1981) *Introductory Spatial Analysis.* London: Methuen.

Covers a broad range of methods for analyzing spatial data (that is, the data have a distinct spatial component and thus can be mapped).

4

Data Classification

OVERVIEW

A common problem faced by mapmakers is whether raw data should be classed into groups, and if so, which method of classification should be used. Although this problem is relevant to a wide variety of maps, research and discussion within the cartographic community has been confined largely to choropleth maps. Traditionally, cartographers argued for classed maps on two grounds: readers' inability to discriminate among many differing areal symbols, and the difficulty of creating unclassed maps using traditional photomechanical procedures. Today the latter constraint has been eliminated as numerous forms of computer hardware are capable of producing unclassed maps; as a result, many cartographers now question whether data classification is a necessity.

In this chapter, we will consider three issues relevant to data classification: common methods of classification, the issue of classed versus unclassed maps, and how spatial context can be used to simplifying the appearance of choropleth maps. Section 4.1 considers six common methods of data classification: ***equal intervals, quantiles, mean-standard deviation, maximum breaks, natural breaks,*** *and* ***optimal.*** *A common problem confronting the mapmaker is determining which of these methods should be used. Criteria that can assist in selecting a classification method include: (1) whether the method considers how data are distributed along the number line, (2) ease of understanding the method, (3) ease of computation, (4) ease of understanding the legend, (5) whether the method is acceptable for ordinal data, and (6) whether the method can assist in selecting an appropriate number of classes.*

Many cartographers have promoted the optimal method because it does the best job of considering how data are distributed along the number line (by minimizing differences within classes and maximizing differences between classes). The optimal method, however, does not score well on some of the other criteria listed above; for example, the legend is difficult to understand and the method is unacceptable for ordinal data. Furthermore, we'll see in Chapter 12 that it is inappropriate for map comparison tasks. As a result, it is important that mapmakers learn about the advantages and disadvantages of other methods of classification.

Section 4.2 addresses the issue of classed versus unclassed maps. Two criteria that may assist in choosing between classed and unclassed maps are: (1) whether or not the mapmaker wishes to portray correct numerical data relations, and (2) whether the intention is to present or explore data. If the intent is to maintain correct data relations, then unclassed maps are appropriate because they provide a spatial expression of numerical relations in the data (for example, if there is a skew in the data, the map will provide a spatial expression of that skew). In contrast, classed maps are constructed using ***maximum contrast shades*** *(shades that maximally differ from one another), and thus numerical relations are not maintained.*

When a map is constructed for presentation (that is, using the five steps of the communication model discussed in Chapter 1), the mapmaker normally is limited to one map and must make a choice between a classed and unclassed map. If data exploration software is used, however, it should be possible to show the data in both classed and unclassed forms. Section 4.2 also summarizes the results of experimental studies dealing with classed and unclassed maps. The results of these studies may assist the mapmaker in choosing between a classed and unclassed map when only one map can be presented.

One limitation of the classification methods described in section 4.1 is that they do not consider the spatial con-

text of the data. If the purpose of classification is to simplify the appearance of the map, it can be argued that spatial context also should be considered. In light of this, section 4.3 covers two approaches for simplifying choropleth maps that do consider spatial context. The first uses the optimal classification method, but incorporates a spatial constraint by requiring that values falling in the same class be contiguous, while the second employs no classification approach, but rather smooths the raw data by changing values for enumeration units as a function of values of neighboring units.

4.1 COMMON METHODS OF DATA CLASSIFICATION

This section considers six common methods of data classification: equal intervals, quantiles, mean-standard deviation, maximum breaks, natural breaks, and optimal. These methods range from those that do not consider how data are distributed along the number line (such as equal intervals) to those that do (such as optimal). Advantages and disadvantages of each method are pointed out, and a general set of criteria by which the methods can be evaluated is developed. For simplicity, we will assume that the intent is to analyze a single distribution (as opposed to comparing distributions, which will be discussed in Chapter 12).

For illustrative purposes, we will work with two variables: (1) the wheat data used to create Figure 2.8A (the percentage of land area from which wheat was harvested in Kansas in 1993), and (2) the percentage foreign-born for counties in Florida in 1990. Raw data and dispersion graphs for both variables are given in Tables 4.1 and 4.2, and Figures 4.1 and 4.2, respectively. In examining the dispersion graphs, note that both variables have a positive skew, with the foreign-born variable having a distinctive outlier (Dade County, which includes the city of Miami).

Before attempting any form of data classification, it is essential to consider the kind of numerical data being mapped (whether it is bipolar, balanced, or unipolar). Remember from section 2.2 that bipolar data have a natural or meaningful dividing point that can be used to partition the data. For example, a data set of "percent population change" has a natural dividing point of zero, which can be used to create two classes: values at or above zero and values below zero. Once bipolar data have been split in such a fashion, it may be appropriate to apply one of the methods discussed in this section to each subset of data.

For balanced and unipolar data, there is generally no natural dividing point, and thus the data can be classified directly using one of the methods discussed here. With such data, however, it may be desirable to create a meaningful dividing point prior to classifying. For example, we might compute a mean percentage value for wheat harvested in the 105 counties of Kansas and then split the data into values above and below the mean.

An important consideration in any method of classification is selecting an appropriate *number of classes.* Owing to map readers' inability to discriminate between differing areal symbols, cartographers traditionally have recommended no more than approximately five to seven classes, regardless of the method of classification. The maximum recommended number of classes can be increased slightly (say, up to 9) if hue and lightness are combined (as in a yellow-orange-dark red color scheme). In order to compare methods of classification in this section, we will assume five classes, which are easy to discriminate on the gray-tone maps shown here.

4.1.1 Equal Intervals

In the **equal intervals** (or **equal steps**) method of data classification each class occupies an equal interval along the number line. As a result, this method is identical to creating a grouped-frequency table (see section 3.3.1), except that cartographers commonly distinguish between the calculated class limits and the limits actually used for mapping. The steps for computation are as follows:

Step 1. Determine the ***class interval,*** *or width that each class occupies along the number line.* This is computed by dividing the range of the data by the number of classes. The results are as follows:

FOR WHEAT

$$\frac{\text{Range}}{\text{Number of classes}} = \frac{\text{high} - \text{low}}{\text{number of classes}} = \frac{58.5 - 0.7}{5} = 11.56$$

FOR FOREIGN-BORN

$$\frac{\text{Range}}{\text{Number of classes}} = \frac{\text{high} - \text{low}}{\text{number of classes}} = \frac{45.1 - 0.6}{5} = 8.90$$

Note that for both data sets no rounding is necessary; this differs from the murder rate data used to compute grouped-frequency tables in section 3.3.1.

Step 2. Determine the upper limit of each class. The upper limit for each class is computed by repeatedly adding the class interval to the lowest value in the data.

TABLE 4.1. Percentage of land area from which wheat was harvested in Kansas Counties in 1993

Observation	*County*	*% Wheat*	*Observation*	*County*	*% Wheat*
1	Wyandotte	0.7	54	Ellsworth	21.1
2	Greenwood	1.5	55	Wallace	21.2
3	Jefferson	2.5	56	Jewell	21.6
4	Elk	2.8	57	Stevens	21.8
5	Miami	2.9	58	Smith	22.0
6	Lyon	2.9	59	Ottawa	22.4
7	Wabaunsee	2.9	60	Cheyenne	22.7
8	Chase	3.0	61	Lincoln	23.0
9	Pottawatomie	3.1	62	Ellis	23.1
10	Doniphan	3.5	63	Decatur	23.3
11	Bourbon	3.6	64	Marion	23.3
12	Johnson	3.7	65	Rawlins	23.3
13	Leavenworth	3.8	66	Cloud	23.4
14	Chautauqua	3.8	67	Clay	23.6
15	Franklin	3.9	68	Gove	23.7
16	Shawnee	4.1	69	Seward	24.0
17	Linn	4.1	70	Norton	24.7
18	Jackson	4.1	71	Hodgeman	24.8
19	Osage	4.2	72	Ness	24.9
20	Atchison	4.4	73	Kiowa	25.1
21	Riley	4.5	74	Hamilton	25.1
22	Douglas	5.2	75	Kearny	25.9
23	Woodson	5.8	76	Logan	26.1
24	Nemaha	6.7	77	Saline	26.4
25	Geary	6.8	78	Finney	27.0
26	Brown	7.3	79	Edwards	27.3
27	Coffey	7.5	80	Sherman	27.4
28	Butler	8.7	81	Dickinson	28.1
29	Anderson	9.8	82	Stanton	30.7
30	Morris	10.2	83	Stafford	31.5
31	Crawford	10.8	84	Pawnee	31.8
32	Allen	10.8	85	Ford	32.2
33	Neosho	12.2	86	Mitchell	32.7
34	Rooks	12.5	87	Wichita	33.1
35	Montgomery	13.1	88	Grant	33.5
36	Wilson	14.2	89	Rush	33.5
37	Clark	14.9	90	Lane	33.8
38	Washington	15.3	91	Thomas	34.3
39	Labette	15.7	92	Gray	34.3
40	Marshall	16.0	93	Haskell	34.4
41	Morton	16.7	94	Rice	34.4
42	Graham	16.8	95	Barton	34.6
43	Russell	17.0	96	Pratt	34.7
44	Cowley	17.3	97	Greeley	35.3
45	Phillips	18.1	98	Reno	35.9
46	Comanche	18.2	99	Harvey	36.2
47	Osborne	18.8	100	Scott	37.1
48	Meade	19.3	101	Kingman	39.1
49	Barber	20.6	102	McPherson	39.3
50	Cherokee	20.6	103	Sedgwick	40.6
51	Trego	20.8	104	Harper	54.9
52	Sheridan	21.0	105	Sumner	58.5
53	Republic	21.0			

TABLE 4.2. Percentage of foreign-born population in Florida Counties in 1990

Observation	*County*	*% Foreign-Born*	*Observation*	*County*	*% Foreign-Born*
1	Madison	0.6	35	Leon	3.7
2	Calhoun	0.8	36	Lafayette	4.1
3	Dixie	0.8	37	Okaloosa	4.3
4	Baker	0.8	38	Glades	4.5
5	Taylor	0.9	39	Highlands	4.6
6	Jefferson	1.0	40	Citrus	4.9
7	Liberty	1.0	41	Lee	5.2
8	Bradford	1.0	42	Brevard	5.3
9	Wakulla	1.2	43	Manatee	5.4
10	Gadsden	1.2	44	DeSoto	5.5
11	Gilchrist	1.3	45	Hernando	5.5
12	Gulf	1.3	46	Volusia	5.8
13	Holmes	1.4	47	Pasco	5.9
14	Suwannee	1.5	48	Alachua	5.9
15	Hamilton	1.6	49	Sarasota	6.0
16	Walton	1.6	50	Indian River	6.1
17	Nassau	1.6	51	Okeechobee	6.3
18	Columbia	1.7	52	Seminole	6.3
19	Franklin	1.8	53	St. Lucie	6.3
20	Sumter	1.9	54	Charlotte	6.3
21	Jackson	2.2	55	Hardee	6.3
22	Putnam	2.2	56	Martin	6.8
23	Sana Rosa	2.2	57	Osceola	7.1
24	Levy	2.3	58	Pinellas	7.1
25	Union	2.4	59	Orange	7.5
26	Washington	2.4	60	Hillsborough	7.6
27	Escambia	2.7	61	Flagler	8.3
28	Clay	3.1	62	Monroe	10.1
29	Bay	3.4	63	Collier	10.5
30	Duval	3.5	64	Palm Beach	12.2
31	Lake	3.5	65	Hendry	14.6
32	St. Johns	3.6	66	Broward	15.8
33	Polk	3.6	67	Dade	45.1
34	Marion	3.6			

(For the wheat data, adding the class interval 11.56 to 0.7 yields a value of 12.26.) The result is the right-hand set of numbers in the Calculated Limits column in Table 4.3.

Step 3. Determine the lower limit of each class. Lower limits for each class are specified so that they are just above the highest value in a lower-valued class (for the wheat data, the lower limit of class 2 is 12.27, which is .01 more than the upper limit of class 1). Up to this point the equal-intervals method is the same as the grouped-frequency method.

Step 4. Specify the class limits actually shown in the legend. The class limits actually shown in the legend should (1) reflect the precision of the data used to create the classification, and (2) be easy for the map reader to use. Given that both raw data sets were reported to the nearest 10th (see Tables 4.1 and 4.2), the first of these criteria suggests that the data also should be reported to the nearest 10th. Values to the nearest 10th of a percent should also be easy for readers of this text to work with, although it could be argued that rounding class limits to the nearest whole percent would make the map easier to read. Legend limits expressed to the nearest 10th of a percent are shown in Table 4.3.

TABLE 4.3. Class limit computations for equal-intervals classification

Class	*Calculated Limits*	*Legend Limits*
	Kansas Wheat Data	
1	0.70 to 12.26	0.7 to 12.3
2	12.27 to 23.82	12.4 to 23.8
3	23.83 to 35.38	23.9 to 35.4
4	35.39 to 46.94	35.5 to 46.9
5	46.95 to 58.50	47.0 to 58.5
	Florida Foreign-Born Data	
1	0.6 to 9.5	0.6 to 9.5
2	9.6 to 18.4	9.6 to 18.4
3	18.5 to 27.3	18.5 to 27.3
4	27.4 to 36.2	27.4 to 36.2
5	36.3 to 45.1	36.3 to 45.1

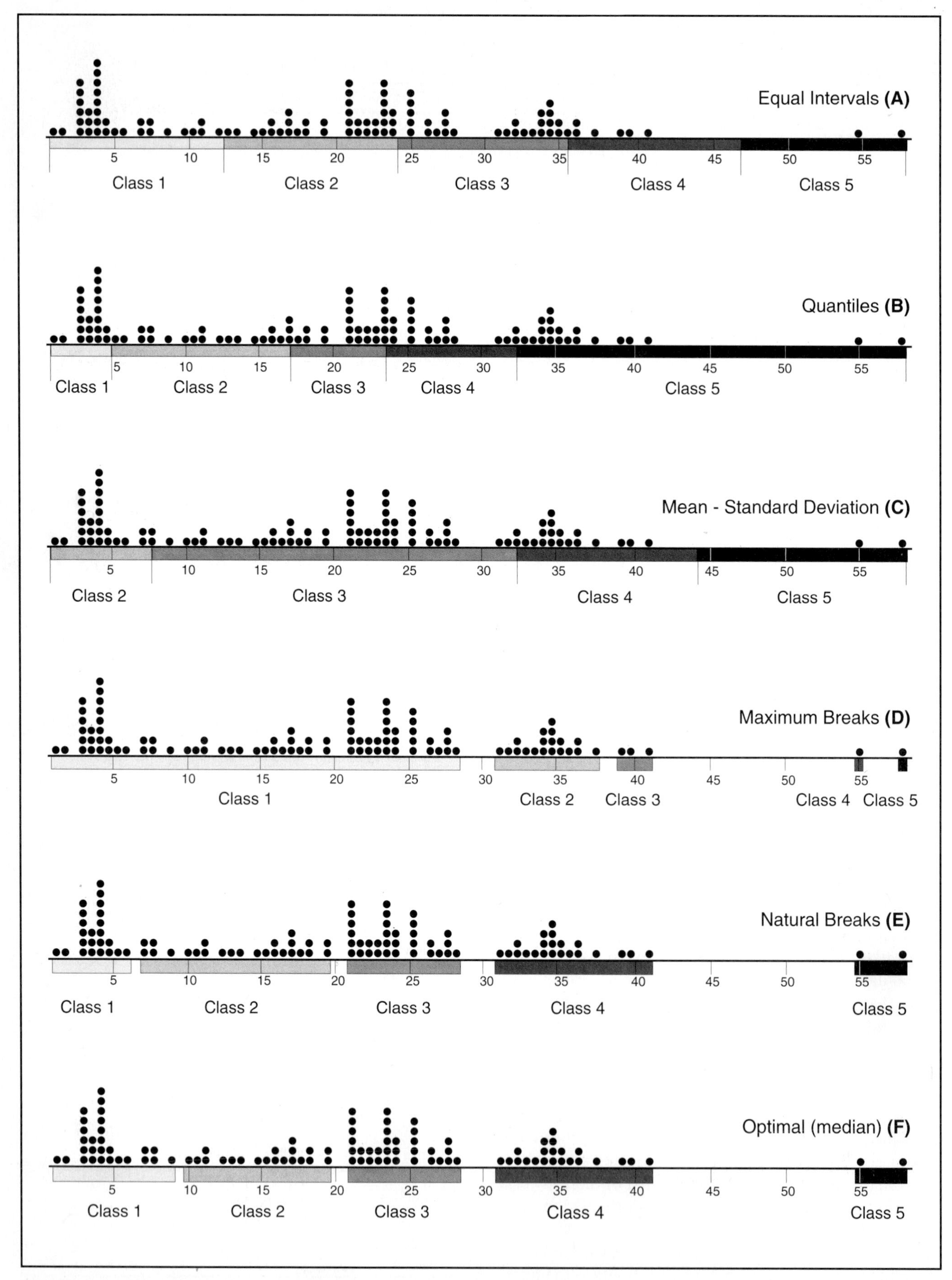

FIGURE 4.1. Dispersion graph of the wheat data shown in Table 4.1 combined with the various methods of data classification. The gray-shaded boxes represent the class limits for each method of classification.

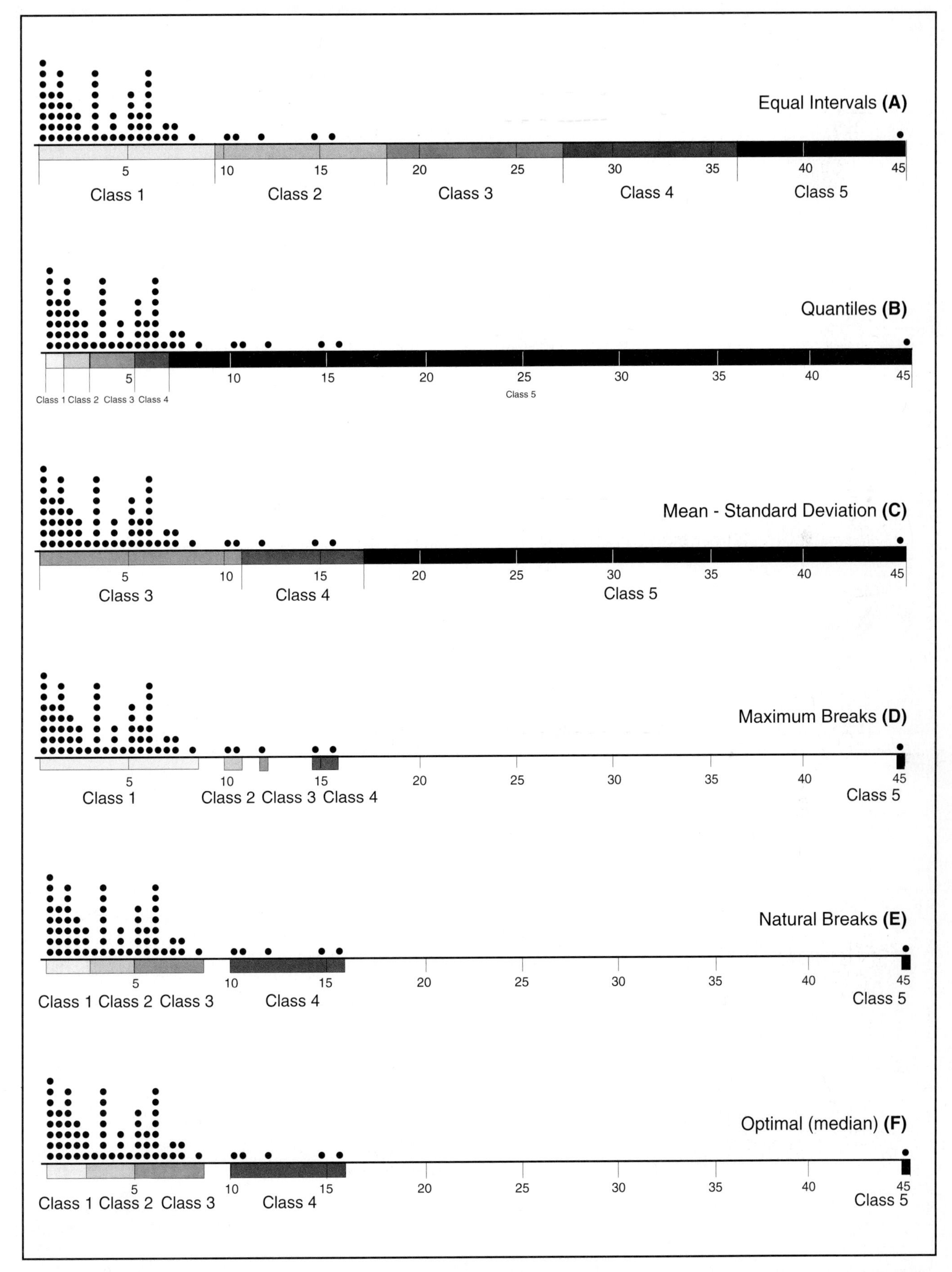

FIGURE 4.2. Dispersion graph of the foreign-born data shown in Table 4.2 combined with the various methods of data classification. The gray-shaded boxes represent the class limits for each method of classification.

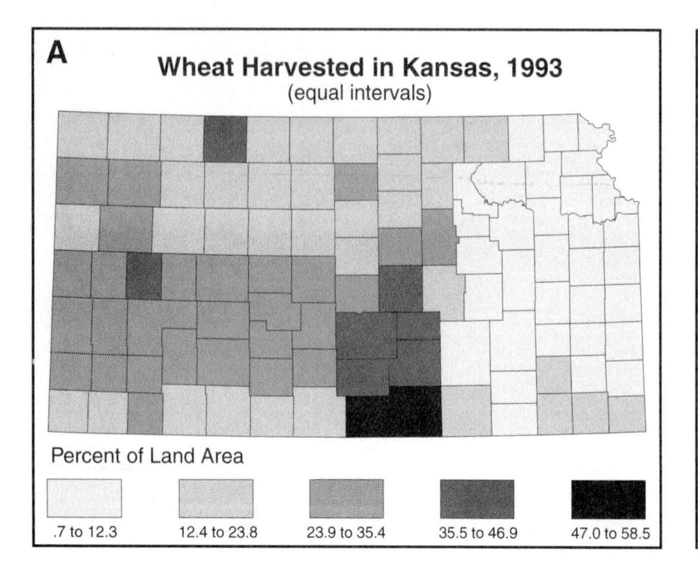

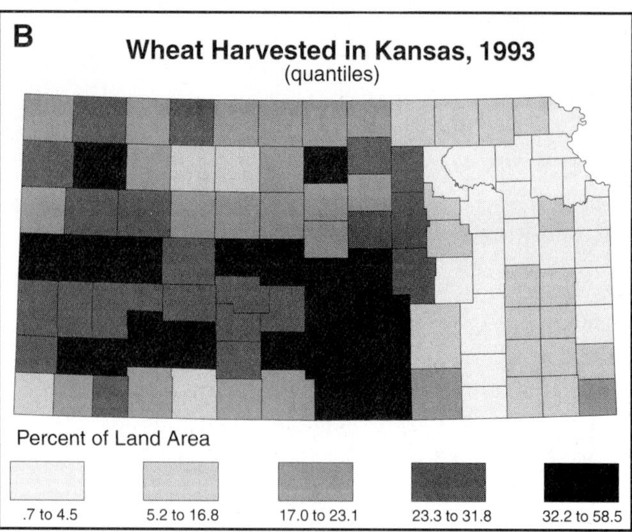

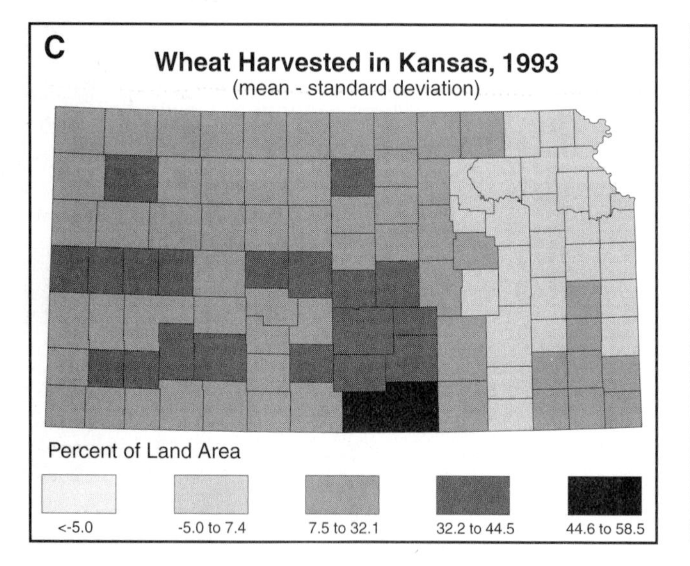

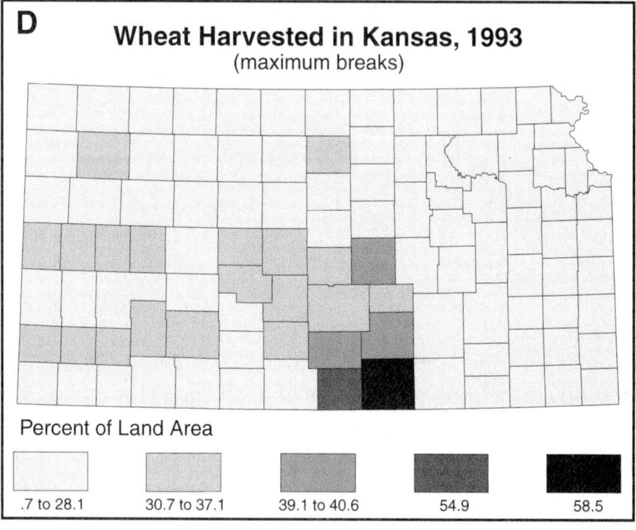

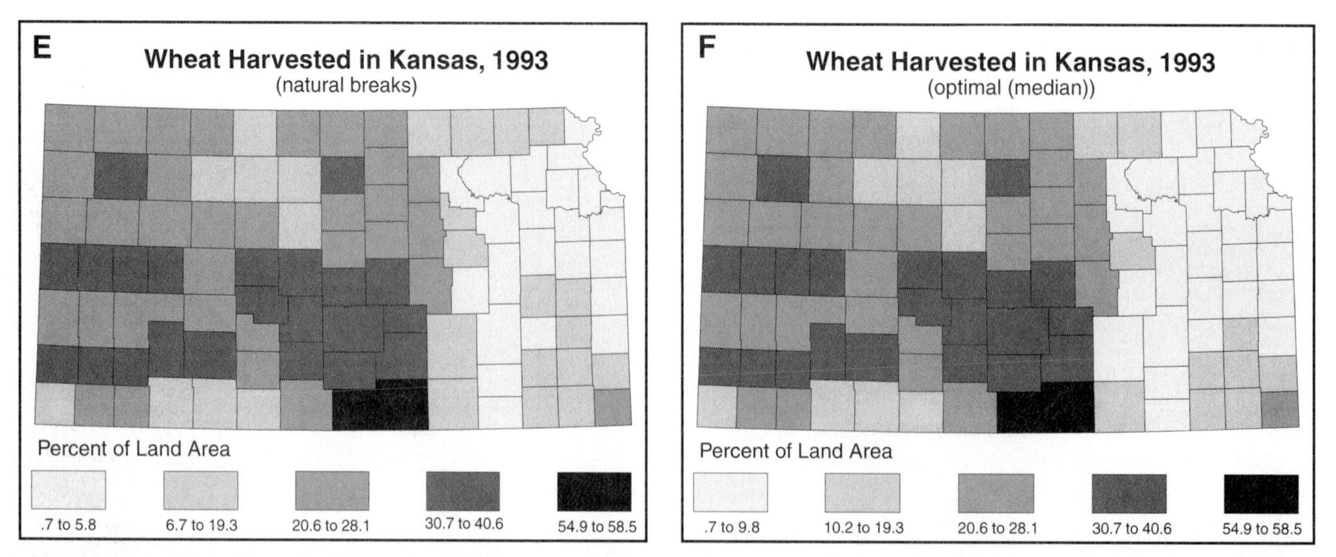

FIGURE 4.3. Maps illustrating various methods of data classification for the wheat data shown in Table 4.1. (A) Equal intervals, (B) quantiles, (C) mean–standard deviation, (D) maximum breaks, (E) natural breaks, and (F) optimal, based on medians.

Step 5. Determine which observations fall in each class. This involves simply comparing the raw values with the legend limits from step 4. Figures 4.1 and 4.2 present these observations in graphic form and maps of the classified data appear in Figures 4.3A and 4.4A.

An obvious advantage of equal intervals is that the above steps can be completed using a calculator, or even pencil and paper. As a result, this method was often favored before mapping software became available. A second advantage is that the resulting equal intervals will, in some cases, be easy for map users to interpret. For example, if you were making a five-class map of "percent urban population," and the data ranged from 0 to 100, the resulting classes would be convenient rounded multiples (0–20, 21–40, 41–60, 61–80, and 81–100). For the wheat and foreign-born data, however, note that the mapped limits do not readily reveal what the class interval might be.

A third advantage of equal intervals is that the legend limits contain no missing values (or gaps): for both data sets, the difference between the upper value in a class and the lower value in the next class is 0.1, which is the precision of the data. Gaps, which we will see occur in other classification methods (such as quantiles), may cause a reader to wonder why some of the data are missing. A related advantage for equal intervals is that the legend limits can be simplified so that only the lowest and highest values in the data and the upper limit of each class are shown (Figure 4.5). This approach should permit faster map interpretation, but it may also create confusion concerning the bounds of each class (for example, the reader may wonder whether 12.3 falls in the first or second class).

The major disadvantage of equal intervals is that the class limits fail to consider how data are distributed along the number line (Figures 4.1 and 4.2). For example, if you inspect the dispersion graph for the wheat data, you will note that the boundary between classes 4 and 5 falls in a blank area; as a result, the legend limits for classes 4 and 5 are not particularly meaningful. This problem is even more serious for the more sharply skewed foreign-born data, where the third and fourth classes have no members.

4.1.2 Quantiles

In the **quantiles** method of classification, data are rank-ordered and an equal number of observations is placed in each class. Different names for this method are used, depending on the number of classes; for example, four- and five-class quantiles maps are referred to as *quartiles* and *quintiles,* respectively.

To compute the number of observations in a class, the total number of observations is divided by the number of classes.

WHEAT

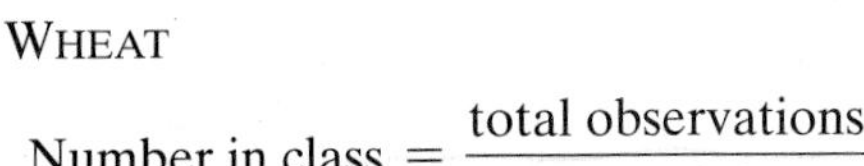

$$\text{Number in class} = \frac{\text{total observations}}{\text{number of classes}} = 105/5 = 21$$

FOREIGN-BORN

$$\text{Number in class} = \frac{\text{total observations}}{\text{number of classes}} = 67/5 = 13.4$$

To determine particular observations to be placed in each class, one simply progresses through the rank-ordered data until the desired number of members in a class is obtained. For example, for the wheat data, the first 21 observations in Table 4.1 would be considered members of the first class.

Identifying class membership for the foreign-born data is more complicated because the number of observations desired in each class, 13.4, is not an integer value. In such a situation, one should attempt to place approximately the same number of observations in each class. Thus, I placed 13 observations in the first class and 14 observations in the second class so that the total for the first two classes was 27 (approximately 13.4×2).

Since identical data values should not be placed in different classes, ties can complicate the quantiles method. This can be seen for the wheat data if a class break is attempted between the 63rd (21×3) and 64th numeric positions. Defining a break here doesn't make sense because the data value for both positions is 23.3. Since the value 23.3 also occurs at the 65th numeric position, I chose a class break between the 62nd and 63rd positions in order to minimize the number of tied values that would have to be moved to another class. (As an exercise, the reader should consider whether similar manipulations need to be made with the foreign-born data.)

A review of basic cartographic texts indicates that two approaches are used for defining legend limits for the quantiles method. One is to specify the lowest and highest values of members in a class, and this is the approach I have taken here; for example, for the wheat data the 21 members of the first class range from 0.7 to 4.5, so these limits are shown in the legend. The other approach is to compute a class boundary as an average of the highest value in a class and the lowest value of the next class; using this approach for the wheat data, the upper limit for the first class would be $(4.5 + 5.2)/2$, or 4.9. I prefer the former approach because it more accurately reflects the data values falling in a class. An advantage of the latter approach, however, is that only the lowest and highest values in the data and the upper limit of each class would need to be shown, as was done for the equal-intervals method in Figure 4.5.

As with equal intervals, an advantage of quantiles is that class limits can be computed manually. A second advantage is that if enumeration units are approximately

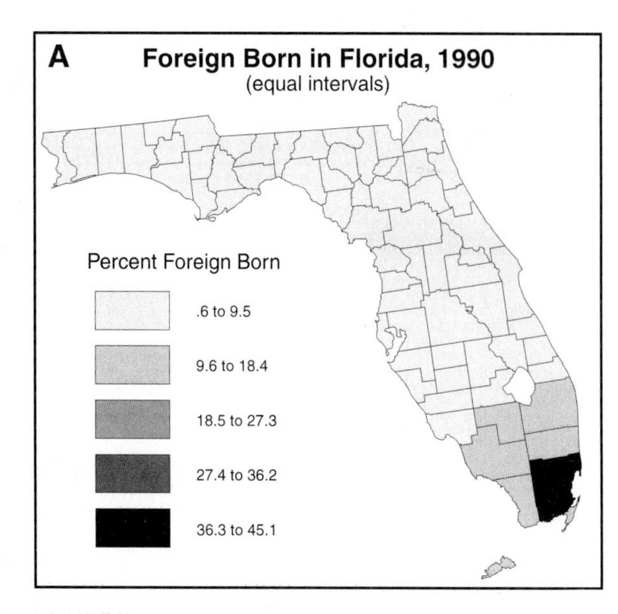

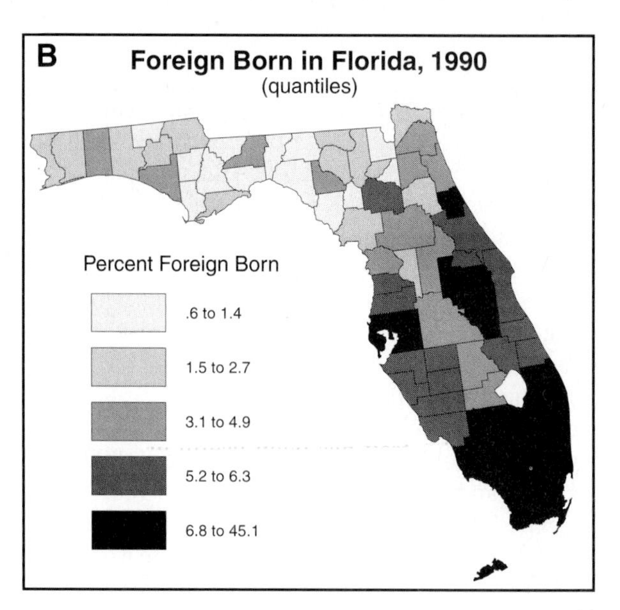

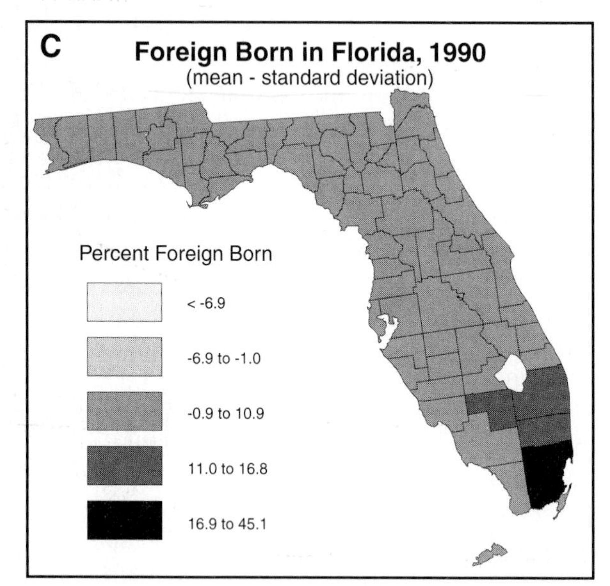

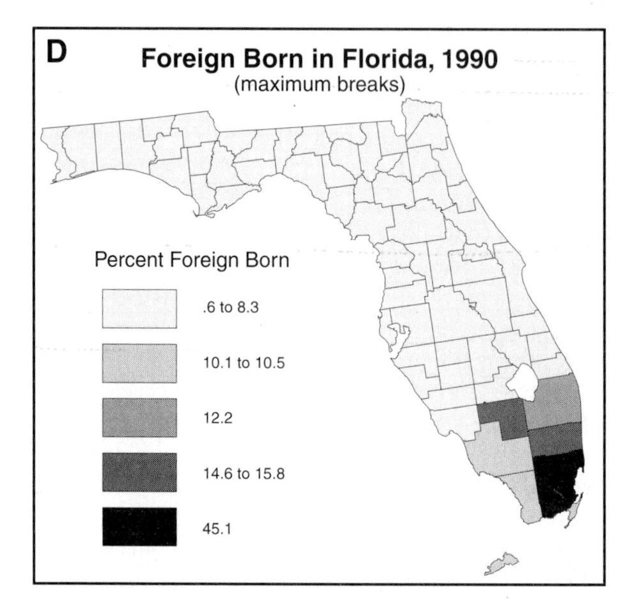

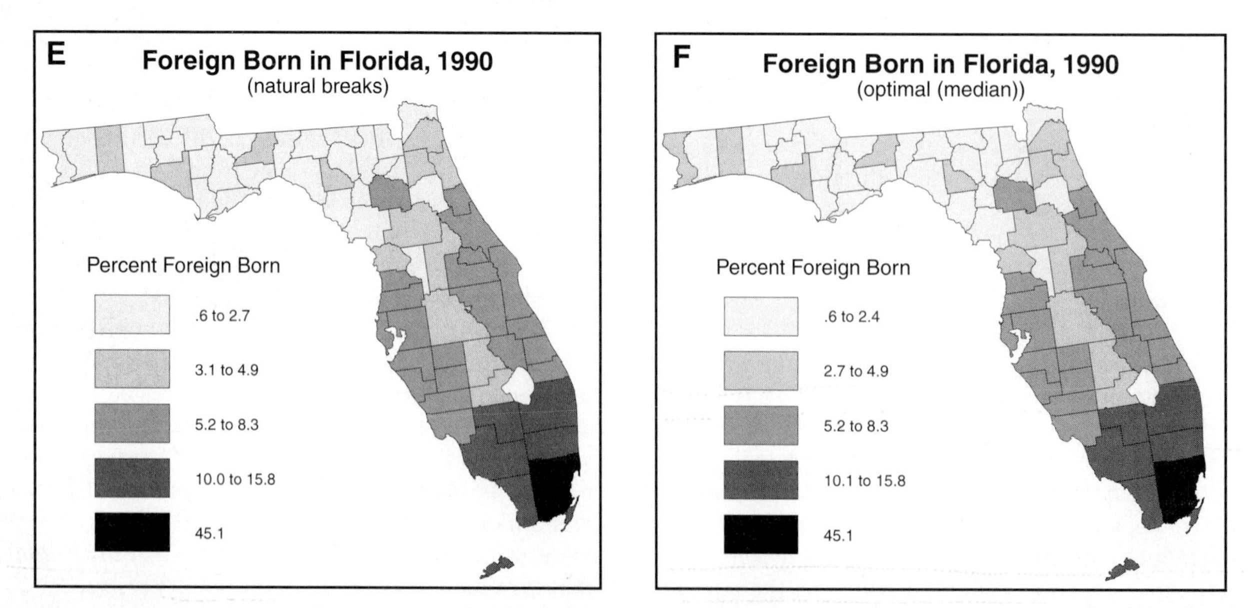

FIGURE 4.4. Maps illustrating various methods of data classification for the foreign-born data shown in Table 4.2. (A) Equal intervals, (B) quantiles, (C) mean–standard deviation, (D) maximum breaks, (E) natural breaks, and (F) optimal, based on medians.

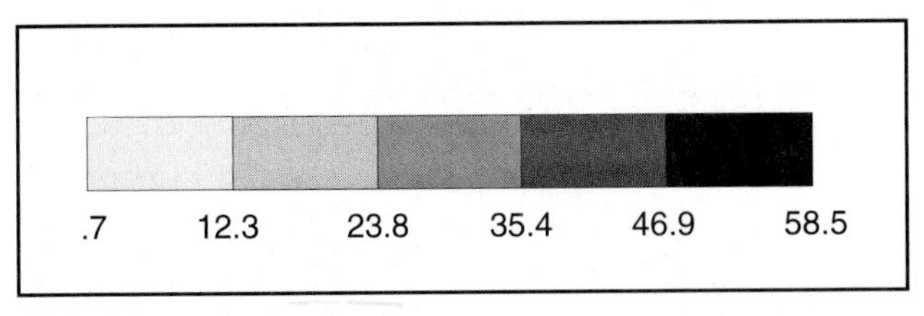

FIGURE 4.5. A legend for the equal intervals method that takes advantage of the continuous nature of the class limits.

the same size, each class will have approximately the same map area; this trait has been useful in comparing maps (see Chapter 12). A third advantage is that since class assignment is based on rank order, quantiles are useful for ordinal-level data. For example, if the 50 states of the United States were ranked on "quality of life," the resulting ranks could be split into five equal groups: no numeric information would be necessary to create the classification.

A disadvantage of equal intervals is that gaps result (for the wheat data, the highest value in class 1 is 4.5 and the lowest value in class 2 is 5.2, resulting in a gap of 0.7) (Figure 4.3B). Gaps are problematic because the reader may wonder why they occur. However, they do permit the legend to reflect the range of data actually occurring on the map.

The quantiles method also shares the major disadvantage of equal intervals: it fails to consider how the data are distributed along the number line. For example, note that for the foreign-born data, the outlier for Dade County is included in the same class with values of considerably lower magnitude (Figure 4.2).

4.1.3 Mean-Standard Deviation

The **mean–standard deviation** method is one of several classification techniques that do consider how data are distributed along the number line. In this method, classes are formed by repeatedly adding or subtracting the standard deviation from the mean of the data, as shown in Table 4.4.* As with the equal intervals method, both calculated limits and legend limits can be computed. Calculated limits are computed by using the mean and standard deviation values listed in column 2, Normal Distribution Limits. To create legend limits, the calculated limits are adjusted so that identical values cannot fall in two different classes and the upper limit of the highest class reflects the highest value in the data (or, alternatively, the lower limit of the lowest class reflects the lowest value in the data).

* This section uses mean and standard deviation formulas appropriate for a sample; because of the relatively large number of observations in each case, similar results would be obtained with population formulas.

TABLE 4.4. Class limit computations for mean–standard deviation classification

Class	*Normal Distribution Limits*	*Calculated Limits*	*Legend Limits*
Kansas Wheat Data: $\bar{x} = 19.7$, $s = 12.4$			
1	$< \bar{x} - 2s$	<-5.0	<-5.0
2	$\bar{x} - 2s$ to $\bar{x} - 1s$	−5.0 to 7.4	−5.0 to 7.4
3	$\bar{x} - 1s$ to $\bar{x} + 1s$	7.4 to 32.1	7.5 to 32.1
4	$\bar{x} + 1s$ to $\bar{x} + 2s$	32.1 to 44.5	32.2 to 44.5
5	$> \bar{x} + 2s$	>44.5	44.6 to 58.5
Florida Foreign-Born Data: $\bar{x} = 4.96$, $s = 5.94$			
1	$< \bar{x} - 2s$	<-6.9	<-6.9
2	$\bar{x} - 2s$ to $\bar{x} - 1s$	−6.9 to −1.0	−6.9 to −1.0
3	$\bar{x} - 1s$ to $\bar{x} + 1s$	−1.0 to 10.9	−1.1 to 10.9
4	$\bar{x} + 1s$ to $\bar{x} + 2s$	10.9 to 16.8	11.0 to 16.8
5	$> \bar{x} + 2s$	>16.8	16.9 to 45.1

A major disadvantage of the mean–standard deviation method is that it works well only with data that are normally distributed. This is particularly evident with the foreign-born data, in which the two lowest classes contain solely negative values and therefore have no members (Figure 4.4). Even with the less skewed wheat data, there is still an empty class at the low end of the distribution (Figure 4.3). One solution to this problem is to transform the data (as described in section 3.3.1), but this is obviously inappropriate if the intention is to examine the *raw* data. Another disadvantage is that the mean–standard deviation method requires an understanding of some basic statistical concepts; a message on the map or in the text indicating that "classes were developed based on the mean and standard deviation" would not be meaningful if one had no statistical training.

A distinct advantage, however, is that if the data are normally distributed (or near normal), the mean serves as a useful dividing point, enabling a contrast of values above and below it. This is most effectively accomplished if an even number of classes is used; for example, a six-class map could consist of the positive classes ($\bar{x}$ to $\bar{x} + s$, $\bar{x} + s$ to $\bar{x} + 2s$, and $> \bar{x} + 2s$) and the negative classes ($\bar{x}$ to $\bar{x} - s$, $\bar{x} - s$ to $\bar{x} - 2s$, and $< \bar{x} - 2s$). (For the five-class maps shown in Figures 4.3 and 4.4, the two middle classes were combined.) Another advantage is that there are no gaps in the legend that might confuse the reader.

4.1.4 Maximum Breaks

Although the mean–standard deviation method considers how data are distributed along the number line, it does so in a holistic sense by trying to fit a normal distribution to the data. An alternative approach is to look at individual values along the number line and group those that are similar to one another (or, alternatively, avoid grouping values that are dissimilar). The **maximum-breaks** method is a simplistic means for accomplishing this.* In this method, raw data are ordered from low to high, the differences between adjacent values are computed, and the largest of these differences serve as class breaks. The dispersion graph provides a visual expression of computations performed in the maximum-breaks method, as larger numeric differences create larger blank spaces (Figures 4.1 and 4.2).

A distinct advantage of using maximum breaks is that it clearly does consider how data are distributed along the number line. This is illustrated by the outlier for the foreign-born data, which ends up in a class by itself (Figure 4.2). Another advantage of maximum breaks is that it is relatively easy to compute, simply involving subtracting adjacent values.

A disadvantage of maximum breaks is that by paying attention only to the largest breaks, the method seems to miss natural clusterings of data along the number line. For example, for the wheat data, maximum breaks places the two highest values (54.9 and 58.5) in separate classes, but the distance of these values from the rest of the data suggests that they should be in the same class.

4.1.5 Natural Breaks

The **natural-breaks** method is one solution to the failure of maximum breaks to consider natural groupings of data. In natural breaks, graphs (such as the dispersion graph or histogram) are examined visually to determine logical breaks (or, alternatively, clusterings) in the data. Stated another way, the purpose of natural breaks is to minimize differences between data values in the same class and maximize differences between classes. We will see that this is also the objective of the optimal method, but with the optimal method, the classification is done using a mathematical measure of classification error, while with natural breaks, the classification is subjective.

To illustrate the computation of natural breaks, consider how we might divide the foreign-born data into five classes (the reader should visually examine the dispersion graph associated with the natural-breaks method in Figure 4.2). The highest value in the data (45.1) appears to be quite different from the rest of the data, so we will place it in a class by itself. Our next decision is how to handle the five values ranging from 10.1 to 15.8. We could divide these into two separate classes (10.1 to 12.2 and 14.6 to 15.8), but then we would have to divide the remaining data into two classes. It seems easier to divide the remaining data into three classes, roughly corresponding to the three peaks in the dispersion graph, and thus we group the values ranging from 10.1 to 15.8 in one class. We can see in this example that an obvious problem with natural breaks is that decisions on class limits are subjective, and therefore may vary among mapmakers.*

* I have borrowed the term "maximum breaks" from the teaching of George Jenks.

4.1.6 Optimal

The **optimal** classification method is a solution to the limitations noted for maximum and natural breaks. The optimal method places like data values in the same class by minimizing an objective measure of classification error. To illustrate, consider how a small hypothetical data set of nine values would be classified by the quantiles and optimal methods (Table 4.5). The quantiles method assigns the same number of observations to each class (three, in this case), and thus places similar values in different classes (14 appears with 31 and 32 in class 2, even though 14 is clearly more similar to 11, 12, and 13, the members of class 1). In contrast, the optimal method places only similar values in the same class (the first class consists of 11, 12, 13, and 14, while the second class consists of 31, 32, and 33).

One measure of classification error commonly used in the optimal method is the sum of absolute deviations about class medians (ADCM). Computing this measure involves calculating the median of each class and the sum of absolute deviations of class members about the class median, and then adding the resulting sums of absolute deviations. For example, for quantiles, the median in the first class is 12 (remember, the median is simply the middle value in an ordered set). The sum of absolute deviations for the class is $|11 - 12| + |12 - 12| + |13 - 12| = 2$. Adding the sums of absolute deviations for all classes for quantiles, we compute $2 + 18 + 67 = 87$. Note that ADCM for the optimal method is 7, which is obviously a smaller value and thus indicative of a better classification.

* It is also interesting to note that natural breaks suggests that six classes might be appropriate for the foreign-born data.

TABLE 4.5. Computing the sum of absolute deviations about class medians (ADCM)

Raw Data: 11, 12, 13, 14, 31, 32, 33, 99, 100					
Quantiles Classification			Optimal Classification		
Class	*Values*	*Error*	*Class*	*Values*	*Error*
1	11, 12, 13	2	1	11, 12, 13, 14	4
2	14, 31, 32	18	2	31, 32, 33	2
3	33, 99, 100	67	3	99, 100	1
		ADCM = 87			ADCM = 7

The data for this hypothetical example were selected so that the results would be clear-cut. In the real world, the desired minimum-error classification is normally not obvious, so researchers have developed automated methods for analyzing possible solutions. Two major automated methods have been developed, which we will refer to as the Jenks-Caspall and Fisher-Jenks algorithms.

The Jenks-Caspall Algorithm

The **Jenks-Caspall algorithm,** developed by Jenks and Caspall (1971), is an empirical solution to the problem of determining optimal classes. It is based on minimizing the sum of absolute deviations about class means (as opposed to medians). The algorithm begins with an arbitrary set of classes (say, the quantiles classes shown in Table 4.5), calculates a total map error analogous to ADCM (but involving the mean), and attempts to reduce this error by moving observations between adjacent classes.

Movements are accomplished using what Jenks and Caspall termed reiterative and forced cycling. In *reiterative cycling,* movements are accomplished by computing the difference between an observation in one class and the mean of another class; for example, for the quantiles data in Table 4.5, the value 14 is closer to the mean of class 1 (12) than is 13 to the mean of class 2 (25.7), so 14 would be moved to the first class. Movements based on the relation of observations to class means are repeated until no further reductions in total map error can be made.

In *forced cycling,* individual observations are moved into adjacent classes, regardless of the relation between the mean value of the class and the moved observation. After a movement, a test is made to determine whether any reduction in total map error has occurred. If error has been reduced, the new classification is considered an improvement and the movement process continues in the same direction. Forcing is done in both directions (from low to high classes and from high to low classes). At the conclusion of forcing, the reiterative procedure described above is repeated to see whether any further reductions in error are possible. Although this approach does not guarantee an optimal solution, Jenks and Caspall (1971, 236) indicated that they were "unable to generate, either purposefully or by accident, a better . . . representation in any set of data."

In addition to developing an automated algorithm for determining optimal classes, Jenks and Caspall also introduced three general criteria that might be used to select a "best" classification. They introduced these criteria by posing three questions (Jenks and Caspall 1971, 225):

1. Which map provides the reader with the most accurate intensity values for specific places?
2. Which map creates the most accurate overview?
3. Which map contains boundaries that occur along major breaks in the statistical surface?

Corresponding to these questions, Jenks and Caspall discussed three kinds of error: tabular, overview, and boundary.

To understand these forms of error, it is helpful to consider a three-dimensional prism map, such as the one shown in Figure 2.1B, which shows no error due to classification because each county is raised to a height proportional to the data. If the data were classed, error would arise as a result of counties in the same class being raised to the same height. The difference in *height* between corresponding prisms would constitute **tabular error,** while **overview error** would be the difference in *volume* between corresponding prisms. Tabular error is equivalent to the error measure described above (sum of absolute deviations about the mean of each class), while overview error is weighted to account for the size of enumeration units.

Jenks and Caspall used the term **boundary error** to describe the error occurring along the boundary between two enumeration units on a classed map. They argued that the highest cliffs appearing on the unclassed prism map (such as Figure 2.1B) should ideally appear on the classed prism map. They computed a measure of boundary accuracy by dividing the n actual cliffs used on the classed map by the n largest cliffs occurring in the raw data.

Of these three kinds of error, cartographers (including Jenks himself) have focused on tabular error, primarily because of its simplicity. I have presented the three different kinds of error, however, in order to provide the reader with a broader perspective on the classification problem.

The Fisher-Jenks Algorithm

In contrast to the empirical approach used by Jenks and Caspall, the **Fisher-Jenks algorithm** has a mathematical foundation that guarantees an optimal solution. Fisher (1958) was responsible for developing the mathematical foundation, while Jenks (1977) introduced the idea to cartographers. Cartographers generally have chosen to recognize only Jenks for this contribution, so the reader may find the algorithm referred to as "Jenks's optimal method."

To understand the Fisher-Jenks algorithm, it is worthwhile to consider how an optimal solution might be computed using brute force. Imagine that you wanted to develop an optimal two-class map of the data 1, 3, 7, 11, and 22. With such a small data set, it is easy to list all possible two-class solutions and compute associated error measures (Table 4.6). If the process is so simple with a small data set, it would seem that for large data sets a computer could be used to determine an optimal solution by simply considering all possibilities. Unfortunately, for large data sets, the number of possible solutions becomes prohibitively large; for example, Jenks and Caspall (1971, 232) calculated that for the 102 counties of Illinois there would be over 1 billion possible seven-class maps.

Rather than consider all solutions, the Fisher-Jenks algorithm takes advantage of the mathematical foundation provided by Fisher, which states that any optimal partition is simply the sum of optimal partitions of subsets of the data. We will illustrate this concept by considering some initial steps for handling the small data set just mentioned (Table 4.7). For computational simplicity, we will use the median (and associated sum of absolute deviations); another version of the algorithm uses the mean (and associated sum of squared deviations about the mean).

Step 1 involves computing the sum of absolute deviations about the class median for any ordered subset of the raw data, ignoring how these subsets might fit into a particular classification. For example, the sum of absolute deviations for the first through third observations (the subset 1, 3, and 7) is $|1 - 3| + |3 - 3| + |7 - 3| = 6$. This result appears in row 1, column 3, of the matrix shown in step 1 of Table 4.7. The resulting sums of absolute deviations are commonly termed the *diameter (D)*, and are represented by $D(i, j)$, where i and j identify the observations in step 1 (Hartigan 1975, Chapter 6); thus, $D(i, j)$ for this example would be $D(1, 3) = 6$.

In step 2, the optimal solution for a two-class map of the complete data set is computed, along with optimal two-class solutions for subsets of the data. Together these are termed the optimal two partitions. Calculations for the optimal two-class map for the complete data set are shown in part (a) of step 2 (Table 4.7), while some of the results for subsets of the data are shown in part (b) of step 2. Although the subset calculations are not used in determin-

TABLE 4.6. Computing ADCM for all potential two-class maps

Raw Data: 1, 3, 7, 11, 22

Solution 1			Solution 2		
Class	*Values*	*Error*	*Class*	*Values*	*Error*
1	1	0	1	1,3	2
2	3, 7, 11, 22	23	2	7, 11, 22	15
		ADCM = 23			ADCM = 17

Solution 3			Solution 4		
Class	*Values*	*Error*	*Class*	*Values*	*Error*
1	1, 3, 7	6	1	1, 3, 7, 11	14
2	11, 22	11	2	22	0
		ADCM = 17			ADCM = 14

Solution 4 is optimal because it has the smallest total error (or ADCM).

TABLE 4.7. Initial steps in the Fisher-Jenks Algorithm for optimal data classification

Raw Data: 1, 3, 7, 11, 22

Step 1. Compute the sum of absolute deviations about the class median for all ordered subsets of the data.

The following matrix shows the sum of absolute deviations about the median for the *i*th through the *j*th observation; for example, if $i = 1$ and $j = 3$, then the sum is $|1 - 3| + |3 - 3| + |7 - 3| = 6$. (Note that this result appears in the *first* row and the *third* column of the matrix.) The resulting sums of absolute deviations are commonly termed the *diameter (D)* and are represented by $D(i, j)$; for this example $D(1, 3) = 6$.

		*j*th observation				
		1	2	3	4	5
*i*th observation	1	0	2	6	14	29
	2		0	4	8	23
	3			0	4	15
	4				0	11
	5					0

Step 2. Compute all optimal two partitions.

a. The results for the optimal two-class map of the complete data set are as follows:

1 | 3 7 11 22
D(1, 1) + D(2, 5) = 0 + 23 = 23
1 3 | 7 11 22
D(1, 2) + D(3, 5) = 2 + 15 = 17
1 3 7 | 11 22
D(1, 3) + D(4, 5) = 6 + 11 = 17
1 3 7 11 | 22
D(1, 4) + D(5, 5) = 14 + 0 = 14

b. The following are some results for optimal two-class partitions of subsets of the data:

1 | 3 7 11
D(1, 1) + D(2, 4) = 0 + 8 = 8
1 3 | 7 11
D(1, 2) + D(3, 4) = 2 + 4 = 6
1 3 7 | 11
D(1, 3) + D(4, 4) = 6 + 0 = 6

1 | 3 7
D(1, 1) + D(2, 3) = 0 + 4 = 4
1 3 | 7
D(1, 2) + D(3, 3) = 2 + 0 = 2

Ideally, all optimal two partitions would be computed!

Step 3. Compute all optimal three partitions (let Opt-2 represent an optimal two partition.)

a. The results for the optimal three-class map of the complete data set would be calculated. Some of the calculations follow. The question marks represent optimal two-class partitions not computed above.

1 | 3 7 11 22 = 0 + ?
D(1, 1) Opt-2
1 3 | 7 11 22
D(1, 2) + Opt-2 = 2 + ?
1 3 7 | 11 22
Opt-2 + D(4, 5) = 2 + 11 = 13
1 3 7 11 | 22
Opt-2 + $D(5, 5) = 6 + 0 = 6$

b. Calculate all other optimal three-class partitions.

ing the optimal two-class map for the complete data set, they are used to determine optimal classifications for maps with a greater number of classes, such as a three-class map, for which some calculations are shown in step 3.

Advantages and Disadvantages of Optimal Classification

The obvious advantage of the optimal method is that it does consider how data are distributed along the number line. It is the "best" choice for classification when the intention is to place like values in the same class (and unlike values in different classes) based on the position of values along the number line.

Another advantage is that the optimal method can assist in determining the appropriate number of classes. When the median is used as the measure of central tendency, this is accomplished by computing the **goodness of absolute deviation fit (GADF)**, which is defined as:

$$\text{GADF} = 1 - \frac{\text{ADCM}}{\text{ADAM}}$$

where ADCM is the sum of absolute deviations about class medians for a particular number of classes, and ADAM is the sum of absolute deviations about the median for the entire data set. An analogous measure can be computed when the mean is used as the measure of central tendency, and is known as the **goodness of variance fit (GVF)** (Robinson et al. 1984, 363).

GADF ranges from 0 to 1, with 0 representing the lowest accuracy (a one-class map) and 1 the highest accuracy. If there are no ties in the data, then a GADF value of 1 will result only when each observation is a separate class (an n-class map will be required, where n is the number of classes). Ties can, however, considerably reduce the number of classes needed to achieve a GADF of 1. For example, in the case of the foreign-born data, only 46 classes were needed (there were 67 data values).

It is important to note that the n-class map mentioned here is equivalent to the "unclassed map" that will be discussed in the next section. This may be confusing because the term *unclassed* suggests no classes, while there are actually n classes on an n-class map. "Unclassed" is commonly used to indicate that no classing or grouping has been applied to the data. (It is not necessary to run the optimal program to create an unclassed map.)

GADF calculations can assist in selecting the appropriate number of classes in two ways. One approach is to construct a graph of the number of classes against the GADF value and look for a point at which the curve begins to flatten out. In the case of the foreign-born data (Figure 4.6), the curve appears to flatten out at about five classes. A flattening at this point indicates that a larger number of classes would not contribute significantly to a reduction in the classification error. (One should also bear in mind that a map with more classes would be more difficult to use.)

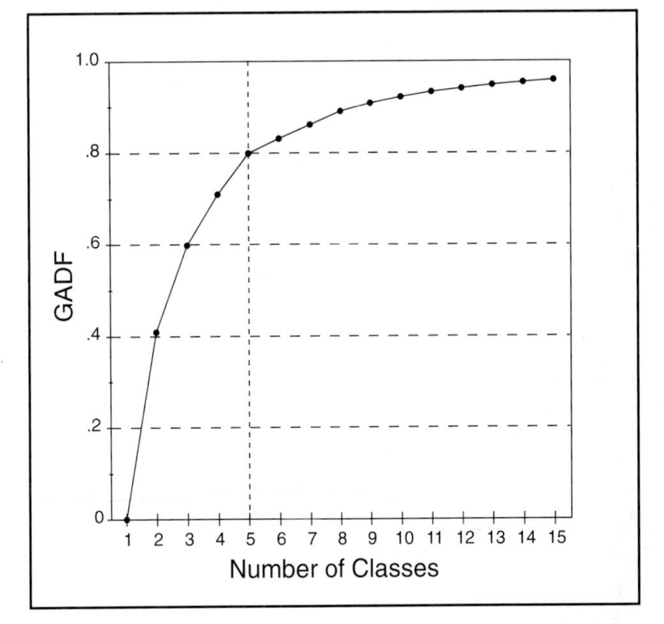

FIGURE 4.6. Graph of the number of classes plotted against GADF values. The curve appears to flatten out at about five classes, indicating that a greater number of classes would not significantly reduce the error on the map.

Another approach is to determine the number of classes for which the GADF first exceeds a certain value, say, 0.80 (the accuracy is 80 percent). For the foreign-born data, this approach yields a six-class map (Figure 4.6). Note, however, that if a more stringent value is used, say, 0.90, a nine-class map would be required. Admittedly, both of these approaches are subjective, but they are an improvement over choosing an arbitrary number of classes.

In addition to assisting in determining an appropriate number of classes for the optimal method, the GADF can also be used to determine whether another classification method is appropriate. For example, if you were to compute similar GADF values for both optimal and quantiles, you might choose the quantiles method because it would be easier for the user to understand how the class limits were created. Alternatively, GADF values might be computed for various numbers of classes for quantiles to assist in determining an appropriate number of classes for that method.

Disadvantages of the optimal method include the difficulty of understanding the concept and the appearance of gaps in the legend. Another traditional disadvantage was that software packages generally did not include the optimal method as an option, but this appears to be changing (for example, the popular package ArcView now includes such an option).

4.1.7 Criteria for Selecting a Classification Method

In discussing the common methods of classification, I have pointed out numerous criteria that might be used to judge their usefulness. Figure 4.7 summarizes these criteria and rates each classification method as "very good," "good," or "poor" on each measure ("acceptable" or "unacceptable" in the case of "acceptable for ordinal data"). One problem with any rating system is that it will be a function of the computer environment the mapmaker has available and the knowledge of the map user. For Figure 4.7, I assume that the computer software is capable of creating all of the classification methods we have considered and that the map user is a college-level student with a basic foundation in introductory statistics.

Note that "ease of understanding legend" is a function of whether or not there are gaps in the legend: remember that gaps between class limits can make the legend difficult to understand. The equal-intervals method receives a very good rating on this criterion because not only are there no gaps, but the rounded intervals may be very easy to understand (such as 0–25, 26–50, etc.). Some mapmakers may wish to avoid the problem of gaps by creating continuous legends for all classification methods (as in Figure 4.5). Remember, though, that this approach will not indicate the actual range of values falling in a class (the latter is dealt with in the criterion "legend values match range of data in a class").

An analysis of Figure 4.7 reveals that there is no single "best" method of classification. Although the optimal approach is often touted as the best method, it is best only in terms of grouping like values together (as a function of their position along a number line) and in selecting an appropriate number of classes. Clearly, there are several other criteria for which it is not the best. Ultimately, a mapmaker creating a display for presentation purposes must consider the purpose of the map and the knowledge of the intended audience before selecting a classification method.

	Equal Interval	Quantiles	Mean SD	Maximum Breaks	Natural Breaks	Optimal
Considers distribution of data along a number line	P	P	G[a]	G	VG[b]	VG
Ease of understanding concept	VG	VG	VG	VG	VG	G[c]
Ease of computation	VG	VG	VG	VG	VG	G[d]
Ease of understanding legend	VG[e]	P	G	P	P	P
Legend values match range of data in a class	P	VG	P	VG	VG	VG
Acceptable for ordinal data	U	A	U	U	U	U
Assists in selecting number of classes	P	P	P	P	G	VG

P = Poor **G** = Good **VG** = Very Good **A** = Acceptable **U** = Unacceptable

[a] Rating would be poor if data are not normal.
[b] Although breaks are subjectively determined, the results are often similar to those obtained by the optimal method.
[c] Only a good rating is assigned because of the fairly complex nature of the algorithm.
[d] When the Fisher-Jenks algorithm is used, only about 1000 observations can be handled; this problem does not occur with the Jenks-Caspall algorithm.
[e] Only a good rating would be appropriate if round numbers are not used.

FIGURE 4.7. Criteria for selecting a method of classification.

4.2 CLASSED VERSUS UNCLASSED MAPPING

In the past, virtually all choropleth maps were produced using classification methods such as those discussed in the preceding section. Cartographers argued that classed maps were essential because of the limited ability of the human eye to discriminate shades for areal symbols; also, practically speaking, classed maps were the only option because of the large time and effort required to produce unclassed maps using traditional photomechanical procedures. In 1973, the latter constraint was eliminated when Tobler introduced a method for creating unclassed maps using a line plotter (Figure 2.9C could have been produced using such a device). Today, unclassed maps can be created using a variety of hardware devices.

The development of unclassed mapping led to a hotly contested debate on its merits and demerits (Dobson, 1973; Dobson 1980b; Peterson 1980; Dobson 1980a; and Muller 1980a), and numerous experimental studies (Peterson 1979; Muller 1979, 1980b; Gilmartin and Shelton 1989; Mak and Coulson 1991; MacEachren 1982b; Mersey 1990). Although the results of the experimental studies can be helpful in selecting between classed and unclassed maps, two criteria should be considered first: (1) whether the mapmaker wishes to maintain numerical data relations, and (2) whether the map is intended for presentation or exploration. In this section, we will consider these two criteria, and then appraise the results of some of the experimental studies.

4.2.1 Maintaining Numerical Data Relations

To illustrate the concept of maintaining numerical data relations, consider the classed and unclassed maps shown in Figure 4.8 for foreign-owned agricultural land and high school graduation. (The optimal classification method was used to create the classed maps because it does the best job of minimizing classification error.) Shades for the classed maps were selected using a conventional maximum-contrast approach in which tones are perceptually equally spaced from one another. In contrast, shades on the unclassed maps were made directly proportional to the values falling in each enumeration unit, thus maintaining the numerical relations among the data (see section 6.5.1 for appropriate equations).

Clearly, there is an apparent difference between each pair of maps, with the difference most distinct for the agricultural maps. The difference for the agricultural maps is a function of the severe skew in the data (examine Figure 4.9 and note the concentration of data on the left, with Maine a notable outlier at 14.1). The unclassed agricultural map is a spatial expression of this skew; the numerous low-valued gray tones correspond to the closely spaced low values on the dispersion graph, while the black tone for Maine indicates that it is quite different. In contrast, on the classed map Maine does not appear quite so different from the rest of the data.

Although unclassed maps do a better job than classed maps of portraying correct data relations, a disadvantage is that for skewed distributions the ordinal relations in much of the data may be hidden. For example, in the case of foreign-owned agricultural land it is difficult to determine whether states bordering the Great Lakes have lower or higher values than Deep South states. When the data are classed, these differences become more obvious.

4.2.2 Presentation versus Data Exploration

When a map is intended for presentation, it is generally only possible to show one map of a distribution. In such a situation, one must make a choice between the classed and unclassed map. If, however, data exploration software is available, then numerous options are possible. One option is to compare a variety of classification approaches visually; for example, Color Plate 4.1, taken from the program ExploreMap (Chapter 13) compares four methods of classification. With the appropriate software, such a comparison might also include an unclassed map. Another option is to apply unclassed shading to a subset of the data. For example, for the agricultural data, Maine can be assigned a unique symbol, with the remainder of the data displayed using the unclassed method (Figure 4.10). Note how this approach makes it easier to contrast the pattern of states bordering the Great Lakes with those from the Deep South.

The latter approach is similar to Brassel and Utano's (1979) **quasi-continuous tone method,** in which the lightest and darkest shades were reserved for outlying values and some subset of the remaining shades was used to represent the data. Rather than use the darkest shade for the outlier (Maine), I used a unique symbol so that it would be clearly distinguished from other values. If color were used, it might be useful to portray outlying values with a subdued gray tone and apply a continuous color scheme (such as shades of red) to the remaining data. In an interactive environment it is, of course, possible to make such changes dynamically, focusing on any subset of the data desired; we will consider this notion in more detail in Chapter 13.

4.2.3 Summarizing the Results of Experimental Studies

This section summarizes the results of some experimental studies that may assist in determining whether a classed or unclassed map is appropriate. The summary

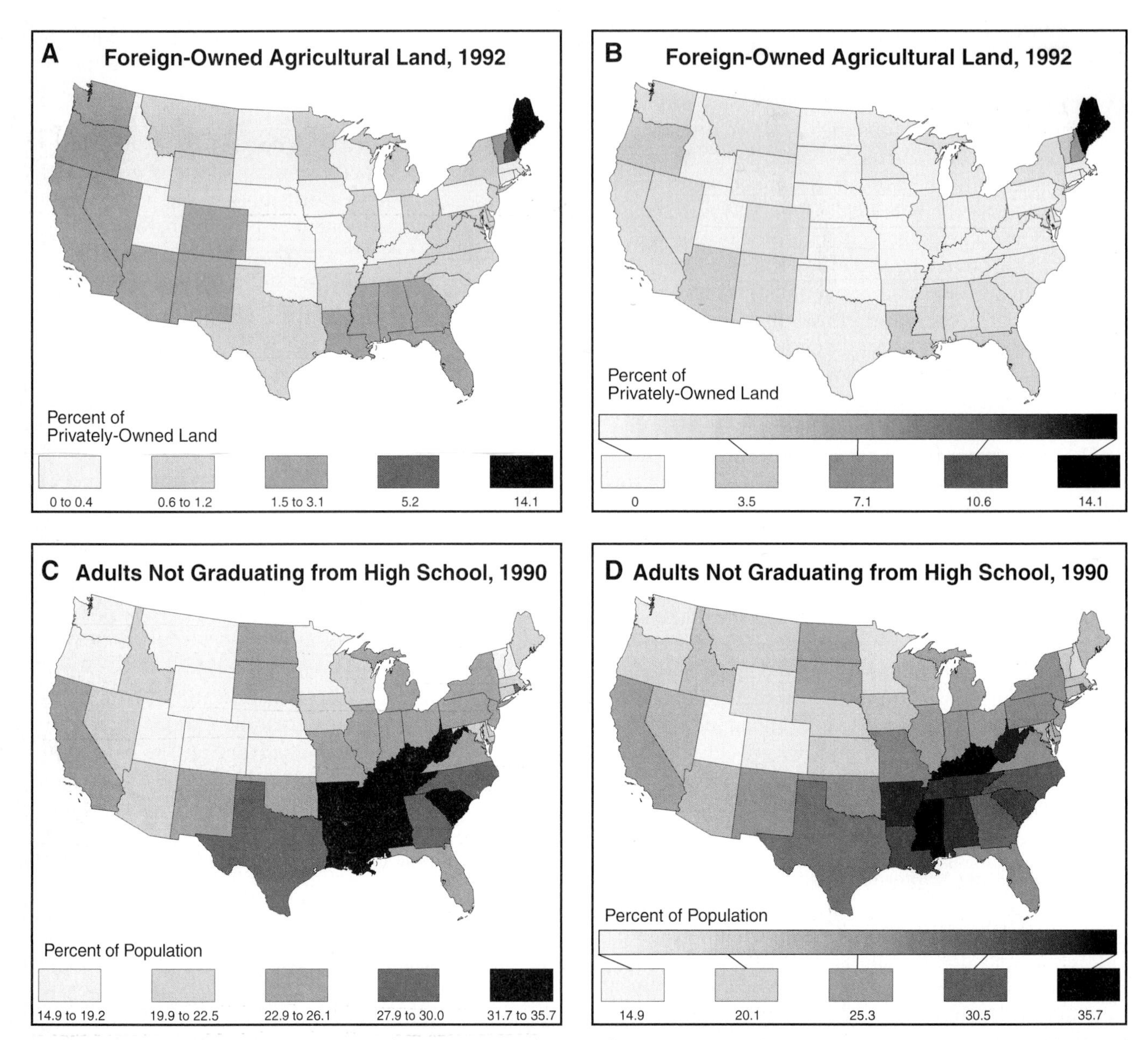

FIGURE 4.8. Optimally classed (A and C) and unclassed (B and D) maps of two variables: foreign-owned agricultural land and high school education.

is based on studies by Muller (1979), MacEachren (1982b), Gilmartin and Shelton (1989), Mersey (1990), and Mak and Coulson (1991). The studies of Peterson (1979) and Muller (1980b) are not considered here because they deal with map comparison, a topic to be covered in Chapter 12. One problem in summarizing the studies is that two of them (MacEachren and Mersey) did not use unclassed maps, but rather varied the number of classes (3–11 and 3–9, respectively); thus, it is useful conceptually to substitute "maps with few classes" for classed maps and "maps with many classes" for unclassed maps in the following discussion. The results of the studies can be conveniently summarized under the types of information that readers acquire and recall from maps: specific and general.

Specific Information

For the *acquisition* of specific information, studies have generally found classed maps more effective. This result may seem surprising because unclassed maps are usually touted as being more accurate (because data are not grouped into classes). The high accuracy of unclassed maps is, however, mathematical, not perceptual. Visually matching a shade on an unclassed map with a shade in the legend is difficult because there are so many shades

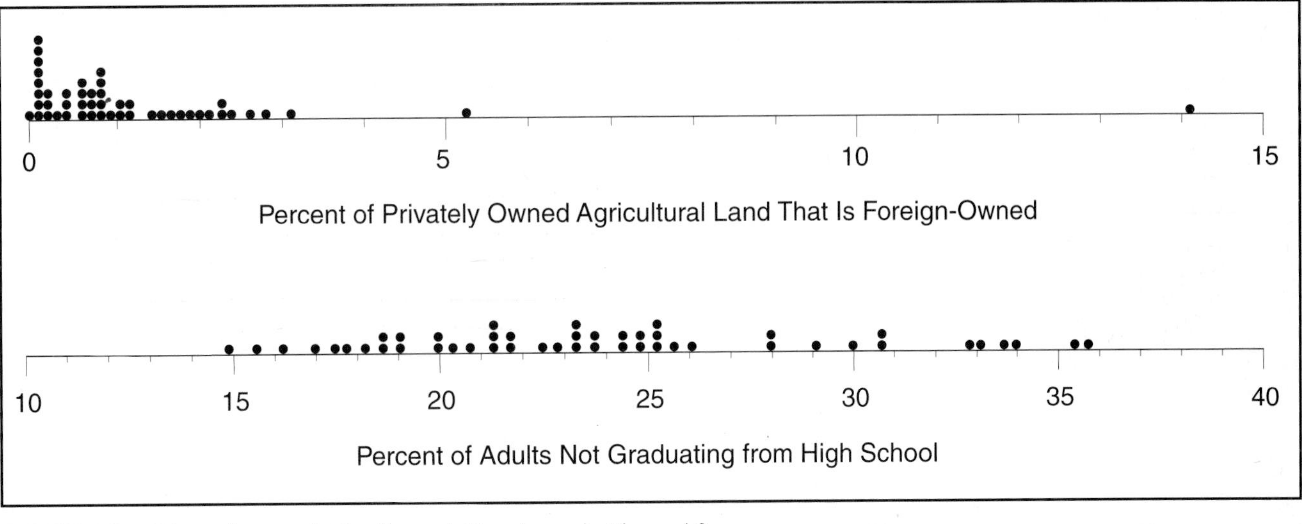

FIGURE 4.9. Dispersion graphs for the variables shown in Figure 4.8.

and their appearance is affected by simultaneous contrast (see section 5.1.4).

Although, in general, studies have shown that classed maps are more effective for acquisition of specific information, there are some tasks for which unclassed maps are more effective. For example, consider ascertaining the ordinal relation of enumeration units, as on the education map shown in Figure 4.8. On the unclassed map it is possible to determine ordinal relations; for example, Montana appears to have a higher value than Wyoming, which in turn is higher than Colorado. On a classed map, such ordering is impossible to determine when the data fall in the same class (as for these three states). Although this task is better performed with unclassed maps, note that it will be difficult if the enumeration units are separated by a considerable distance (and thus likely appear in different contexts), or if a distribution is highly skewed (making shades difficult to discriminate).

It also must be realized that the results for specific information will depend on the number of classes. For example, on a two-class map the error resulting from classification clearly will be greater than the error resulting from incorrect estimation on an unclassed map. Although experimental studies do not suggest a clear choice for the ideal number of classes, I suspect that the error resulting from classification will be unacceptably large on maps with four or fewer classes.

For the *recall* of specific information, results are inconclusive. MacEachren found that maps with fewer classes were more effective, while Mersey found that although a three-class map was most effective, five- and seven-class maps did not elicit as accurate a response as those for a nine-class map. Unfortunately, Mersey did not speculate why such a result occurred.

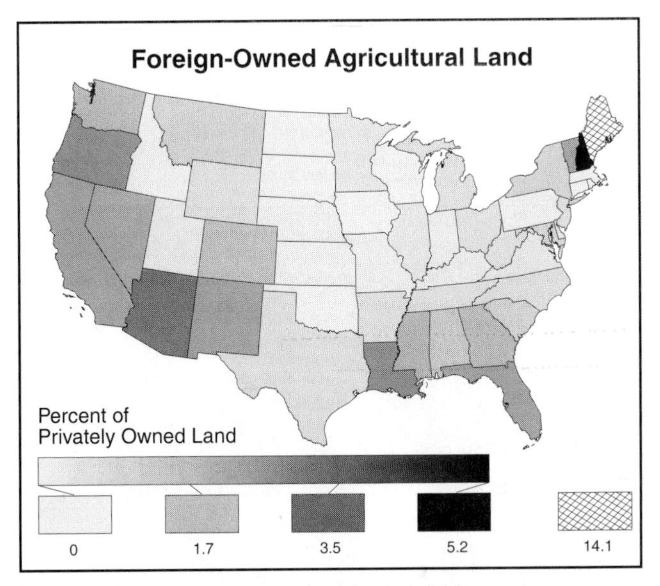

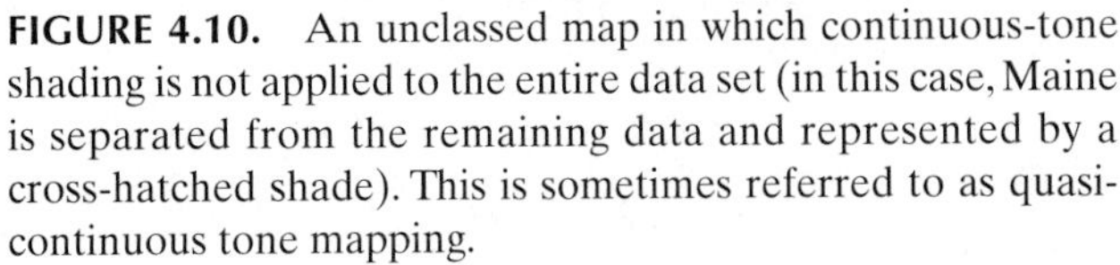

FIGURE 4.10. An unclassed map in which continuous-tone shading is not applied to the entire data set (in this case, Maine is separated from the remaining data and represented by a cross-hatched shade). This is sometimes referred to as quasi-continuous tone mapping.

General Information

For *acquisition* of general information, studies generally have revealed no significant difference for classed and unclassed maps. Only the Mak and Coulson study found any significant differences, and this occurred only when individual regions on each map were analyzed; they concluded that "for more complex classed maps (six to eight classes) classification may have a distinct advantage over unclassed maps" (p. 121).

For *recall* of general information, MacEachren found no significant relationship between the number of classes and subjects' ability to recall data, while Mersey found that a greater number of classes decreased recall effectiveness. Based on his findings, MacEachren suggested that classification may be unnecessary for general tasks. In contrast, Mersey noted:

> A large number of map symbols, even on a map with regular surface trends, may create a notion in the mind of the user that the distribution is more complex than in fact it is. Using many categories to provide the user with more detailed specific information may serve as "noise" when the same user attempts to reconstruct the thematic distribution from memory. (p. 125)

One problem with interpreting the results of these studies is that usually there was no direct comparison between classed and unclassed maps. For example, Mak and Coulson asked subjects to divide each map into five ordinal-level regions (ranging from "low" to "high"). Rather than comparing the resulting regions directly on classed and unclassed maps, they analyzed the consistency for each type of map (classed and unclassed). As a result, it is possible that the consistency was similar for both but the locations of regions differed.

Only one of the studies (Gilmartin and Shelton) measured processing time. Measuring processing time is important because classed and unclassed maps could provide the same perceived information, but one map could take longer to process. (In such a situation, the more rapidly processed map is clearly more desirable.) If classed maps are processed more rapidly, as traditional classification proponents such as Dobson (1973) argue, and the information acquired (or recalled) is identical, an argument can be made for classing the data. If, however, the information acquired or recalled varies for classed and unclassed maps, faster processing is a moot point: in this case, the unclassed map might be argued for on the grounds that it provides a more correct portrayal of the overall distribution.

4.3 USING SPATIAL CONTEXT TO SIMPLIFY CHOROPLETH MAPS

One limitation of the classification methods described in section 4.1 is that they do not consider the spatial context of the data. If the purpose of classification is to simplify the appearance of the map, it can be argued that spatial context also should be considered. This section considers two approaches that do consider spatial context: one that begins with the optimal classification approach but incorporates a spatial constraint, and one that works solely with spatial context.

4.3.1 Spatial Constraint with Optimal Classification

To illustrate how a spatial constraint can be combined with the optimal classification approach, consider Figure 4.11. The top portion of the figure lists raw data for 16 hypothetical enumeration units, while the bottom portion portrays two maps of these data. The map on the left is an optimal classification; note that breaks for this map occur between 6 and 9 and between 13 and 16. The GADF for this map is .76, and the complexity using the C_F face measure (see section 3.4.2) is 10/16 or .63.

A spatial constraint can be applied by starting with the optimal solution and then allowing data values to shift between classes so that the map pattern is simplified. One potential solution is shown in the right-hand map, where the values 9 and 16 have been shifted from the second and third classes to the first and second classes, respectively. The result is a slightly lower GADF (.70 as opposed to .76), but a considerably simpler complexity (.38 as opposed to .63).

Declercq (1995) has developed an algorithm that will find the least complex map for a particular number of classes, assuming that one is willing to shift a certain percentage of observations between classes. Obviously, one problem with this approach is that there is a tradeoff between complexity and accuracy: as complexity decreases, accuracy will also decrease. Consequently, the mapmaker must use some subjectivity in determining what magnitude of GADF is acceptable for a particular application.

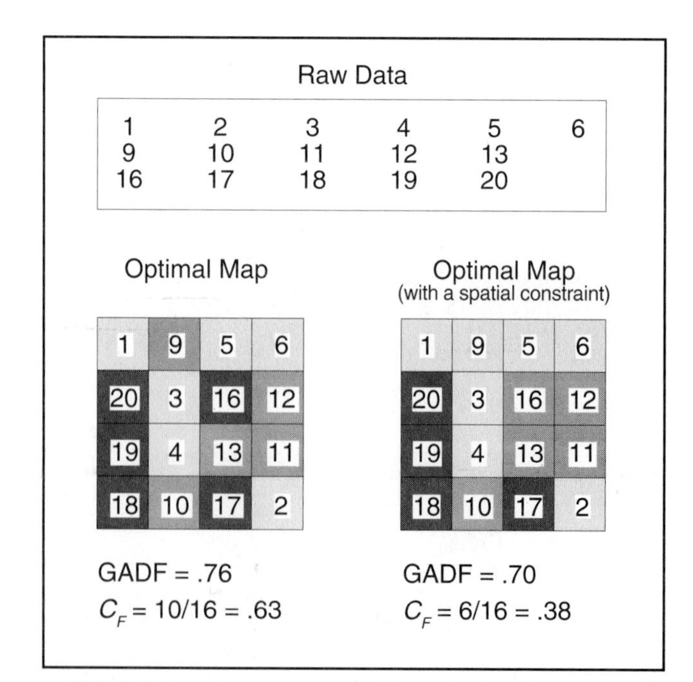

FIGURE 4.11. Applying a spatial constraint to the optimal classification approach.

4.3.2 Spatial Context Only

Another approach for simplifying map pattern is to adjust the values of enumeration units as a function of values of surrounding units, an idea developed by Tobler (1973) and promoted by Herzog (1989). This notion is based on the concept of **random error:** if we repeatedly measure a value for an enumeration unit, we will likely get a different value each time; therefore, some change in the value is permissible (Clark and Hosking 1986, 14). Moreover, the spatially autocorrelated nature of geographic data (see section 3.4.2) provides a mechanism for making adjustments: nearby units can assist in determining appropriate values for a particular enumeration unit because similar values are likely to be located near one another.

Although this approach could be used with a wide variety of data, it is easiest to implement for proportion data (such as the proportion of adults who smoke cigarettes) because the statistical theory for such data is well known. To illustrate, consider the hypothetical portion of a map shown in Figure 4.12, where it is assumed we wish to change the value for the central enumeration unit as a function of the surrounding units. The simplest formula for determining the value of the central unit would be to average all four values, but such a formula would not consider two factors: (1) that we have presumably taken some care in collecting the data for the central unit (and thus would like to place greater weight on it), and (2) that those units having a longer common boundary with the central unit should have greater impact. Thus, a more appropriate formula is

$$V_e = W_c V_c + W_s\left(\sum_{i=1}^{n} \frac{L_i}{L_T} V_i\right)$$

where V_e = the estimated value for the central unit
V_c = the original value for the central unit
V_i = the original value for the ith surrounding unit
W_c = the weight for the central unit
W_s = is the weight for the surrounding units
L_i = the length of the boundary between the ith unit and the central unit
L_T = the total length of the central unit boundary
n = the number of surrounding units

If we assume W_c and W_s values of .67 and .33 respectively, then for Figure 4.12 the formula would be calculated as follows:

$$V_e = .67(.15) + .33\left(\frac{1}{6}(.26) + \frac{2}{6}(.22) + \frac{3}{6}(.19)\right) = .17$$

Herzog indicated that a more complete algorithm should involve several other considerations. First, some

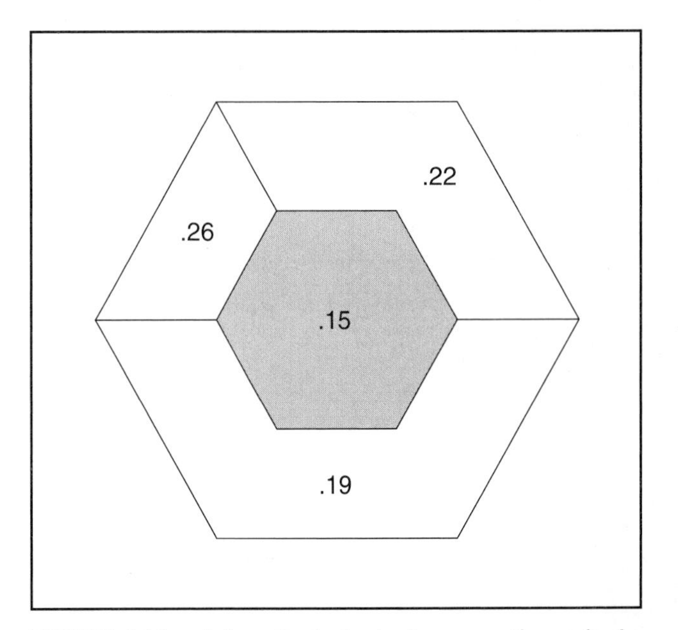

FIGURE 4.12. A hypothetical set of enumeration units for which a simplified value is desired for the central enumeration unit.

limit should be placed on how much change is permitted to the central unit. For proportion data, Herzog suggested that the central value should not be moved outside of a 95 percent confidence interval around its original value (see Burt and Barber 1996, 272–274, for a discussion of confidence intervals for proportions). Second, surrounding values differing dramatically from the central value should not be used in the calculations. For proportion data, Herzog suggested a difference of proportions test (see Burt and Barber 1996, 322–324), with proportions differing significantly from the central unit not being used. Third, the process should be implemented in an iterative fashion, meaning that the entire map should be simplified several times; for example, this process could continue until all enumeration units reached the bounds of their confidence intervals or the largest change made in a unit was less than a small tolerance. Finally, the map resulting from the iterative process should be further smoothed by combining areas with nearly equal values.

Figure 4.13 illustrates the net effect of using Herzog's approach for the percentage of students in each commune of Switzerland who study at the University of Zurich. Here the top map is an original (unsimplified) unclassed map, while the bottom map is the simplified result. Although the bottom map is still unclassed, note that it is much simpler to interpret than the top map.

It should be noted that the two methods we have examined for simplifying map appearance are both relatively new and have not yet been incorporated in commercial software. Assuming that such methods do become readily available, it will be interesting to see to what extent they are used, and what impact this has on

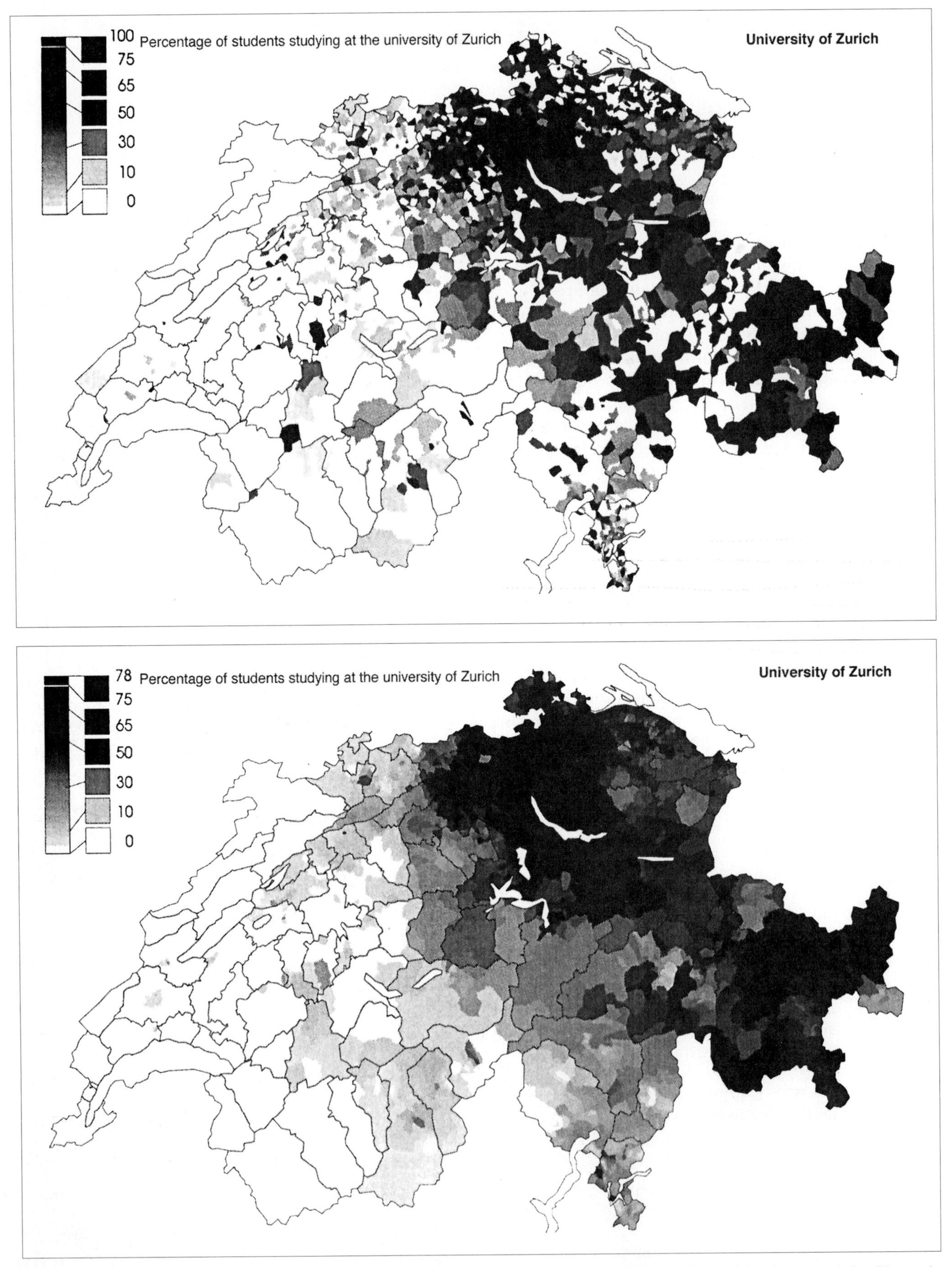

FIGURE 4.13. A comparison of an original unclassed map (*top*) and a simplified map (*bottom*) resulting from applying Herzog's method for simplifying choropleth maps. (Courtesy of Taylor & Francis.)

the use of more traditional classification methods that do not consider spatial context.

SUMMARY

In this chapter, we examined several data classification methods and evaluated criteria for selecting among them. We concluded that methods that do consider how data are distributed along the number line (such as **natural breaks** and **optimal**) are desirable because they place similar data values in the same class (and dissimilar data values in different classes). Methods that do not consider the distribution of data along the number line (such as **equal intervals** and **quantiles**) may, however, also be desirable because they satisfy other criteria. For example, equal intervals is desirable because it is easy to understanding, is easy to compute, and has an easily understood legend (at least when rounded percentage data are used, such as 0–100 percent urban). In Chapter 12, we will see that certain methods (such as quantiles) may also be useful in map comparison.

Now that unclassed maps can be readily constructed using modern computer hardware, the cartographer is faced with the question of whether a classed or unclassed map should be created. In this chapter, we saw that two criteria can be used to assist in selecting between classed and unclassed maps. The first is whether or not the mapmaker wishes to portray correct numerical relationships among the data. If the intent is to maintain correct data relations, then unclassed maps are appropriate because they provide a spatial expression of numerical relations in the data. (For example, if there is a skew in the data, the map will provide a spatial expression of that skew.)

A second criterion that can be used to assist in selecting between classed and unclassed maps is whether the intention is to present or explore data. For presentation purposes, it generally is only possible to show one map of a distribution, and thus one must make a choice between the classed and unclassed map. If, however, data exploration software is available, then it is possible to display the data in a variety of ways: as a classed map, as an unclassed map, or possibly focus on a subset of the data. The notion of data exploration will be considered more fully in Chapter 13.

Experimental studies for evaluating the effectiveness of classed and unclassed maps have revealed mixed results. For example, classed maps have been effective for the acquirement of specific information, while neither mapping technique has performed consistently better in terms of the acquirement of general information. The experimental studies appear to have been limited by a failure to directly compare classed and unclassed maps, and by not considering the time for readers to process the map; thus, it appears that more research is needed in comparing the effectiveness of classed and unclassed maps.

One limitation of traditional classification methods (such as equal intervals and optimal) is that they fail to consider the spatial context of the data. If the purpose of classification is to simplify the appearance of the map, it can be argued that spatial context should also be considered. In this chapter, we looked at two approaches that consider spatial context when simplifying a choropleth map. The first used the optimal classification method but incorporated a spatial constraint by requiring that values falling in the same class be contiguous, while the second employed no classification approach, but rather smoothed the raw data by changing values for enumeration units as a function of values of neighboring units. It will be interesting to see to what extent such novel models are adopted by commercial software vendors.

FURTHER READING

Brassel, K. E., and Utano, J. J. (1979) "Design strategies for continuous-tone area mapping," *The American Cartographer* 6, no. 1:39–50.

Describes various methods for symbolizing unclassed maps.

Coulson, M. R. C. (1987) "In the matter of class intervals for choropleth maps: With particular reference to the work of George F. Jenks," *Cartographica* 24, no. 2:16–39.

Summarizes data classification methods, with an emphasis on Jenks's work.

Cromley, R. G. (1995) "Classed versus unclassed choropleth maps: A question of how many classes," *Cartographica* 32, no. 4:15–27.

Describes a method for optimally classifying data by treating classification as an integer programming problem; argues that this approach can assist in choosing an appropriate number of classes.

Cromley, R. G. (1996) "A comparison of optimal classification strategies for choroplethic displays of spatially aggregated data," *International Journal of Geographical Information Systems* 10, no. 4:405–424.

Describes a variety of methods for optimally classifying data by treating classification as an integer programming problem.

Dobson, M. W. (1980a) "Perception of continuously shaded maps," *Annals, Association of American Geographers* 70, no. 1:106–107.

An example of some arguments against unclassed maps. Read in association with Muller (1979; 1980a).

Evans, I. S. (1977) "The selection of class intervals," *Transactions, Institute of British Geographers* (new series) 2, no. 1:98–124.

A classic article on methods of data classification.

Herzog, A. (1989) "Modeling reliability on statistical surfaces by polygon filtering." In *Accuracy of Spatial Databases,* eds. M. Goodchild and S. Gopal, pp. 209–218. London: Taylor & Francis.

I introduced Herzog's method for simplifying patterns on choropleth maps in section 4.3.2. This article presents a more detailed discussion of the approach.

Jenks, G. F., and Caspall, F. C. (1971) "Error on choroplethic maps: Definition, measurement, reduction," *Annals, Association of American Geographers* 61, no. 2:217–244.

The classic article on optimal data classification.

Kennedy, S. (1994) "Unclassed choropleth maps revisited: Some guidelines for the construction of unclassed and classed choropleth maps," *Cartographica* 31, no. 1:16–25.

Summarizes the debate concerning classed and unclassed choropleth maps and provides some guidelines for constructing both kinds of maps.

Lindberg, M. B. (1990) "Fisher: A Turbo Pascal unit for optimal partitions," *Computers & Geosciences* 16, no. 5:717–732.

Describes a computer implementation of the Fisher-Jenks optimal method of data classification.

Mak, K., and Coulson, M. R. C. (1991) "Map-user response to computer-generated choropleth maps: Comparative experiments in classification and symbolization," *Cartography and Geographic Information Systems* 18, no. 2:109–124.

An example of a study that compared readers' ability to interpret classed and unclassed maps.

Monmonier, M. S. (1982) "Flat laxity, optimization, and rounding in the selection of class intervals," *Cartographica* 19, no. 1:16–27.

Argues that round-number breaks should be used with optimal data classification in order to create a more readable map.

Muller, Jean-C. (1979) "Perception of continuously shaded maps," *Annals, Association of American Geographers* 69, no. 2:240–249.

An early article describing a method for creating unclassed maps, along with a study of readers' ability to interpret classed and unclassed maps.

Muller, Jean-C. (1980a) "Perception of continuously shaded maps: Comment in reply," *Annals, Association of American Geographers* 70, no. 1:107–108.

An example of some arguments for unclassed maps.

Peterson, M. P. (1992) "Creating unclassed choropleth maps with PostScript," *Cartographic Perspectives* no. 12:4–6.

Describes an approach based on PostScript that permits 9,999 different gray tones to be created for unclassed choropleth maps. This is in contrast to programs such as Freehand that only permit 99 gray tones. Maps in the present book were created using Freehand.

Rowles, R. A. (1991) Regions and Regional Patterns on Choropleth Maps. Unpublished Ph.D. dissertation, University of Kentucky, Lexington, Kentucky.

Compares regions on an optimally classified choropleth map with those formed by a cluster analysis method. Understanding this work requires some background in cluster analysis.

Tobler, W. R. (1973) "Choropleth maps without class intervals?" *Geographical Analysis* 5, no. 3:262–265.

The paper that arguably initiated the debate concerning classed and unclassed maps.

5

Principles of Color

OVERVIEW

The increasing use of color in cartography (for example, maps shown on television and via the Internet) means that mapmakers have a greater need to understand how color should be used. To assist you in developing this understanding, chapters 5 and 6 deal with color issues in mapping. Chapter 5 covers a number of basic color principles, including how color is processed by the human visual system, hardware considerations in creating color maps, models for specifying colors used on maps, and methods for disseminating maps to users (and the implications this has for color maps). Chapter 6 covers color usage on two of the more common forms of thematic map: choropleth and isarithmic.

How color is processed by the human visual system is the topic of section 5.1. Here we learn about the nature of ***visible light,*** *the structure of the eye, theories of color perception (focusing on* ***opponent-process theory****),* ***simultaneous contrast*** *(how color perception is influenced by its surround),* ***color vision impairment,*** *and the visual processing that takes place beyond the eye. Some of this material (such as details of the structure of the eye) may seem to diverge from cartography, but developing rules for good map design requires knowledge of how our visual system processes map information.*

Section 5.2 describes fundamental hardware aspects related to the production of color maps. In the realm of soft-copy maps, we will focus on how ***cathode ray tubes (CRTs)*** *are able to produce millions of colors by using the* ***additive colors*** *red, green, and blue* ***(RGB).*** *With respect to hard-copy maps, we will consider some of the basic kinds of printing techniques (****ink-jet, laser,*** *and* ***dye sublimation****) that have been developed in recent years. In contrast to the additive colors used to create soft-copy maps, we will see that the* ***subtractive colors*** *cyan, magenta, yellow, and black* ***(CMYK)*** *are used to create printed maps. Hardware technology, especially for printers, is changing rapidly, and readers should consult popular computer magazines such as* Byte *(http://www.byte.com) to learn about the most recent developments.*

*An important problem facing cartographers is color specification. For example, as a mapmaker, you might like to create a smooth progression of colors extending from a light desaturated green to a dark saturated green. Section 5.3 covers numerous models that have been developed for color specification. Some of these are hardware-oriented (RGB and CMYK), and therefore of limited use to the mapmaker. Others are user-oriented (****HSV, Munsell,*** *and Tektronix's* ***HVC****) and thus permit color specification in terms mapmakers are apt to be familiar with, such as hue, lightness, and saturation. This section also considers the* ***CIE*** *model, which, in theory, allows the mapmaker to reproduce colors specified by other mapmakers (for example, a friend tells you that a particular color progression is very effective, and you want to be sure that you use the same progression).*

The last section of the chapter touches on methods for disseminating maps to users, and some of the implications this has for color maps. The section is split between paper and nonpaper dissemination approaches. In the paper realm, we consider the use of digital printers, xerography, and ***offset printing.*** *In the nonpaper realm, we consider television, videocassette, and computer-based approaches (CD-ROM and the Internet). Traditionally, when many copies were to be disseminated, paper maps were most frequently used, but today the Internet has become an important alternative. From the standpoint of color, this is problematic because the actual colors the user sees will be a function of the particular printer or graphic display on which the map is produced or viewed.*

5.1 HOW COLOR IS PROCESSED BY THE HUMAN VISUAL SYSTEM

5.1.1 Visible Light and the Electromagnetic Spectrum

We see maps as **visible light,** whether it is reflected from a paper map or emitted from a computer screen. Visible light is a type of **electromagnetic energy,** which is a wave form having both electrical and magnetic components (Figure 5.1).* The distance between two wave crests is known as the **wavelength of light.** Because visible wavelengths are small, they are typically expressed in nanometers (nm). Visible wavelengths range from 380 to 760 nm; 1 nm equals 1 billionth of a meter. Figure 5.2 relates visible light to other forms of electromagnetic energy that humans deal with; the complete continuum of wavelengths is called the **electromagnetic spectrum.**

We have all seen or read about how a prism splits sunlight into the color spectrum (red, orange, yellow, green, blue, indigo, and violet). This phenomenon occurs because the visible portion of sunlight consists of a broad range of wavelengths, rather than being concentrated at a particular wavelength. Different colors arise in a prism as a function of how much each wavelength is bent, with shorter wavelengths (such as blue) bent more than longer wavelengths (such as red). Note that the colors in the visible portion of the spectrum in Figure 5.2 are arranged from short to long wavelength (from violet to red), and thus match the nature of colors we might see using a prism.

* Light also consists of photons (packets of energy), which behave as particles when light strikes a surface (Birren 1983, 20).

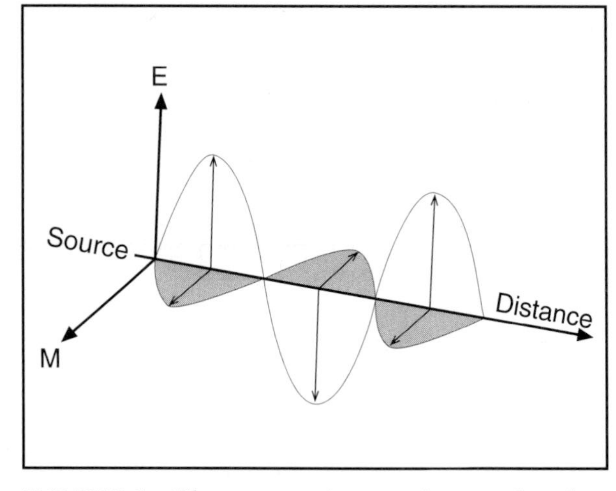

FIGURE 5.1. Electromagnetic energy is a waveform having both electrical (E) and magnetic (M) fields. Wavelength is the distance between two crests.

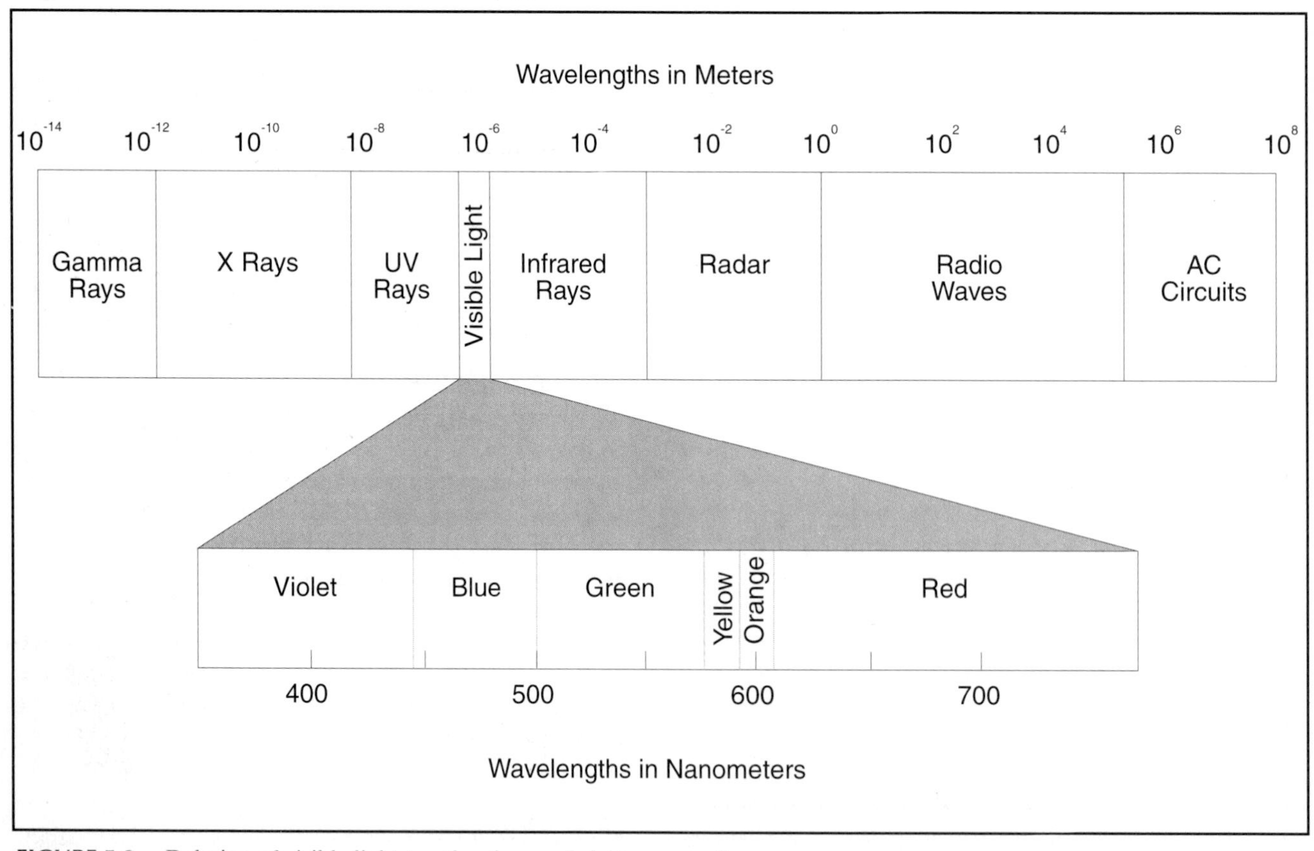

FIGURE 5.2. Relation of visible light to other forms of electromagnetic energy.

5.1.2 Structure of the Eye

The basic features of the eye of concern to cartographers are shown in Figure 5.3. After passing through the **cornea** (a protective outer covering) and the **pupil** (the dark area in the center of our eye), light reaches the **lens,** which focuses it on the **retina.** Images are focused by changing the shape of the lens, an automatic process known as **accommodation.** As we age, our lens becomes more rigid, and our ability to accommodate thus weakens. Generally, around the age of 45, our ability to accommodate becomes so weak that a corrective lens (glasses or contacts) is necessary. The **fovea** is the portion of the retina where our visual acuity is the greatest. The **optic nerve** carries information from the retina to the brain.

A term used to describe the size of an image projected onto the retina is **visual angle,** the angle formed by lines projected from the top and bottom of an image through the center of the lens of the eye (ϕ_1 in Figure 5.4). (Note that ϕ_1 is identical to ϕ_2; thus the degrees of coverage in the visual field corresponds to degrees of coverage on the retina.)

An enlargement of the retina is shown in Figure 5.5. Note that it consists of three major layers of nerve cells (rods and cones, bipolar cells, and ganglion cells), along with two kinds of connecting cells (*horizontal* and *amacrine cells*), which enable cells within the major layers to communicate with one another. **Rods and cones** are specialized nerve cells that contain light-sensitive chemicals called **visual pigments,** which generate an electrical response to light. The concentration of cones is greatest at the fovea, while the highest concentration of rods is about 20 degrees on either side of the fovea. Overall there are about 120 million rods and 6 million cones.

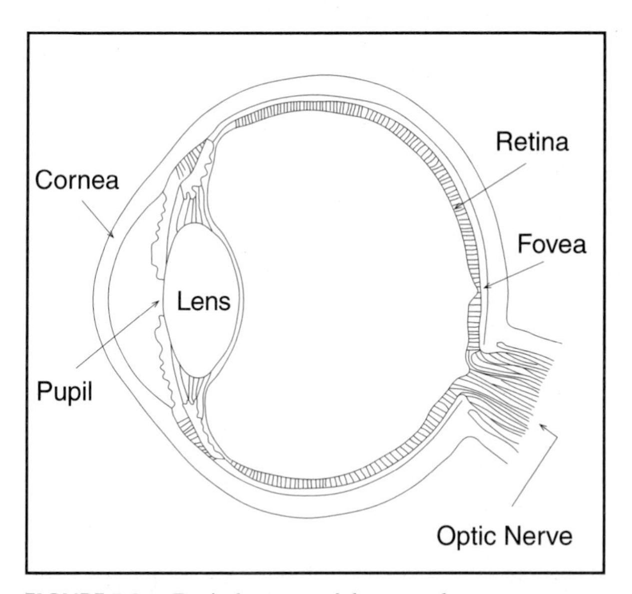

FIGURE 5.3. Basic features of the eye relevant to cartography.

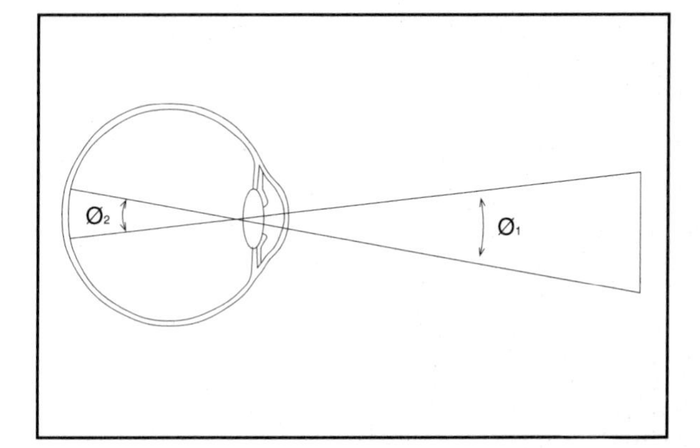

FIGURE 5.4. The projection of an image onto the back of the eye. A visual angle is computed by measuring the angle ϕ_1 formed by lines projected from the top and bottom of the image passing through the center of the lens of the eye (see Figure 5.3). Note that ϕ_1 is identical to ϕ_2, and thus the degrees of coverage in the visual field will correspond to degrees of coverage on the retina.

Cones function in relatively bright light and are responsible for color vision, while rods function in dim light and play no role in color vision. The cones are of primary interest to cartographers because most maps are viewed in relatively bright light (an exception would be maps viewed in the dim light of an aircraft cockpit). Physiological examination of cones taken from the eye of a person with normal color vision reveals three distinct kinds based on the wavelength to which they are most sensitive: short (blue), medium (green), and long (red) (Bowmaker and Dartnall 1980).* MacEachren (1995, 56) notes that the distribution and sensitivity of these three kinds of cones varies in the retina: although blue cones cover the largest area, they are least sensitive, thus making blue inappropriate for small map features.

The major function of the **bipolar** and **ganglion cells** (Figure 5.5) is to merge the input arriving from the rods and cones. Although there are about 126 million rods and cones, there are only about 1 million ganglion cells. Considerable convergence must take place between the rods and cones and the ganglion cells; each single ganglion cell corresponds to a group of rods or cones, or what is termed a **receptive field.** These receptive fields are circular in form and overlap one another.

5.1.3 Theories of Color Perception

Psychology textbooks (for example, Goldstein 1989) generally consider two major theories of color perception: the trichromatic and opponent process theories. The **trichromatic theory,** developed by Young (1801) and

* Hubel (1988, 163–164) indicates that technically the terms *violet, green,* and *yellowish-red* would probably be more appropriate.

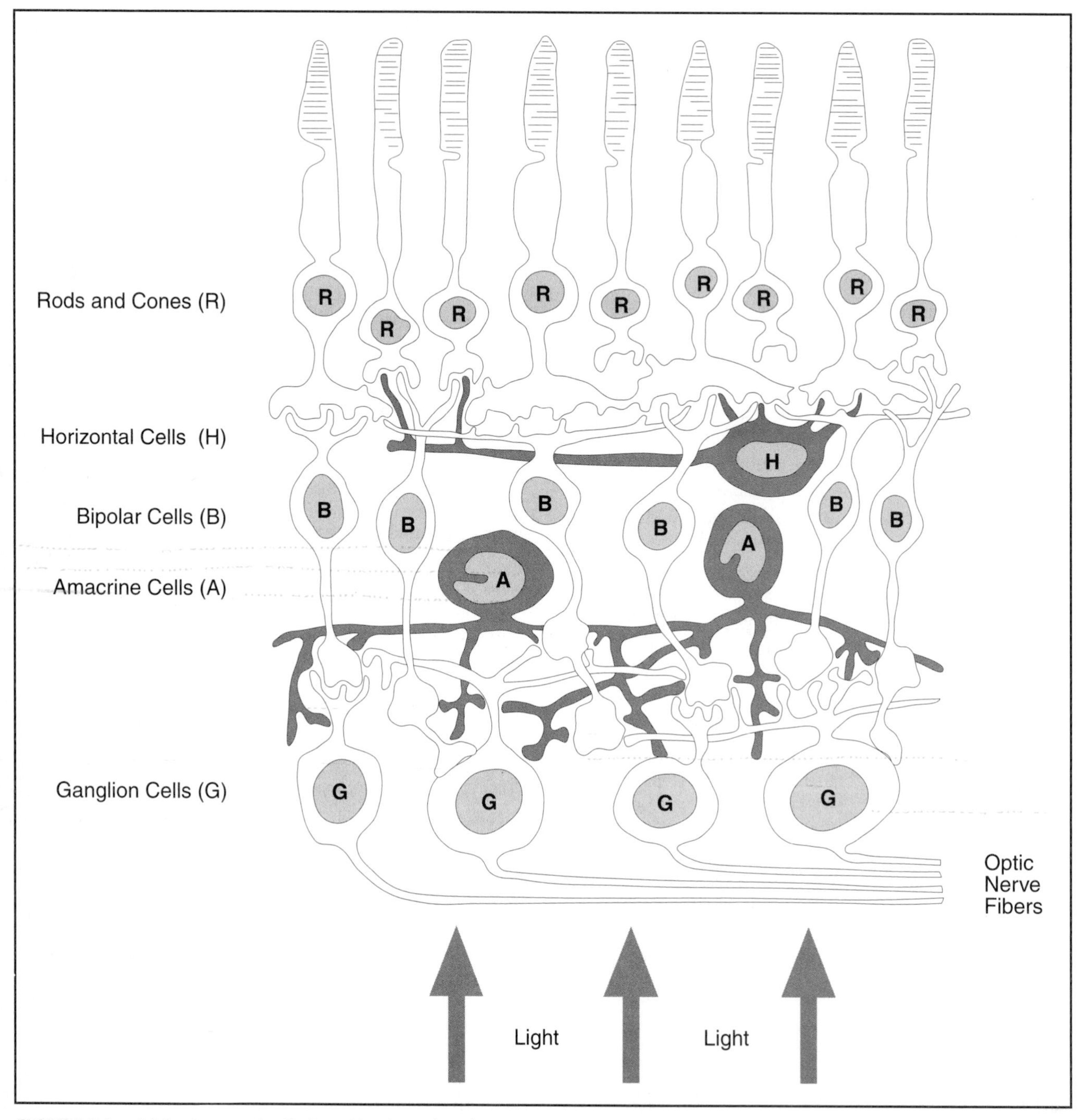

FIGURE 5.5. Major layers of cells found in the retina. (Adapted from J. E. Dowling and B. B. Boycott, 1966, "Organization of the primate retina: electron microscopy," *Proceedings of the Royal Society of London,* 166, Series B, pages 80–111.)

championed by Helmholtz (1852), presumes that color perception is a function of the relative stimulation of the three types of cones (blue, green, and red). If only one cone type is stimulated, then that color is perceived (for example, a red light would stimulate primarily red cones, and thus red would be perceived). Other perceived colors would be a function of the relative ratios of stimulation (a yellow light would stimulate the green and red cones and yellow would be perceived).

The **opponent-process theory,** originally developed by Hering (1878), states that color perception is based on a lightness-darkness channel and two opponent color channels: red-green and blue-yellow. Colors within each opponent color channel are presumed to work in opposition to one another, meaning that we do not perceive mixtures of red and green or blue and yellow; rather, we see mixtures of pairs from each channel (red-blue, red-yellow, green-blue, and green-yellow).

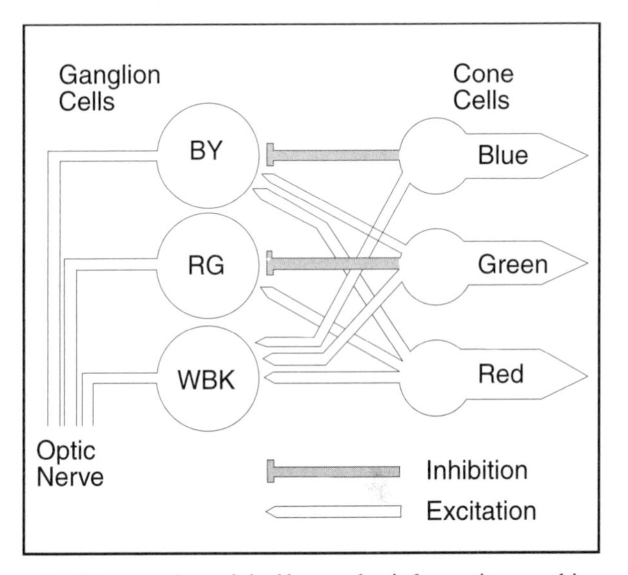

FIGURE 5.6. A model of how color information reaching the cones may be converted to opponent processes. (After Eastman, 1986. First published in *The American Cartographer,* 13(4), p. 326. Reprinted with permission from the American Congress on Surveying and Mapping.)

For many years proponents of the two theories of color perception hotly debated their merits, presuming that only one theory could be correct. It is now apparent, however, that both can help explain the way we see color. The trichromatic theory is correct in the sense that our color vision is based on three types of cones and that information from these cones combines to produce the perception of color. The manner, however, in which information from the cones combines is based on opponent-process theory.

There is both psychophysical and physiological evidence in support of opponent-process theory. The psychophysical evidence comes from the seminal work of Hurvich and Jameson (1957), who showed that a color of an opposing pair could be eliminated by adding light for the other color in the pair; for example, when adding yellow light to blue light, the blue eventually disappears. The physiological evidence is based on an analysis of how electrical signals pass through cells in the nervous system. In this regard, an important concept is that nerve cells fire at a constant rate even when they are not stimulated. Firing above this constant rate is termed *excitation,* while firing below it is termed *inhibition.* By studying electrical activity in cells, physiologists have noted linkages between the blue, green, and red cones and the bipolar and ganglion cells; for example, a red light might excite red cones, which in turn excite bipolar and ganglion cells (Derrington et al. 1983; De Valois and Jacobs 1984).

Although experts in human vision are reasonably sure that certain colors are in opposition to one another and that excited and inhibited nerve cells play a role, the precise linkage between the cones and bipolar and ganglion cells is unknown. One model that has been suggested is shown in Figure 5.6. In this model the blue-yellow channel is excited by green and red cones and inhibited by blue cones; the red-green channel is excited by red cones and inhibited by green cones; and the lightness-darkness channel is stimulated by red, green, and blue cones. For our purposes, the precise nature of the model is not as important as its applications, which we will consider in Chapter 6.

5.1.4 Simultaneous Contrast

One problem sometimes encountered when reading maps is that the perceived color of an area may be affected by the color of the surrounding area, a problem known as **simultaneous contrast,** or **induction** (Brewer 1992). This concept is illustrated for lightness in Figure 5.7. Here the gray tones in the central boxes are physically identical, but the one on the left appears lighter. This occurs because a gray tone surrounded by black shifts toward a lighter tone, while the tone surrounded by white shifts toward a darker tone. Note that in this case the shifts are toward the opposite side of the lightness-darkness channel in the opponent-process model.

FIGURE 5.7. An illustration of simultaneous contrast for a black-and-white image. The central gray strips are physically identical, but the one surrounded by black appears lighter.

When different hues are used, the apparent color of an area will tend to shift toward the opponent color of the surround color. For example, Color Plate V of Hurvich (1981) illustrates an example in which a gray tone is surrounded by either green or blue. When the surround is green, the gray tone appears reddish; in contrast, when the surround is blue, the gray tone appears yellowish.

Simultaneous contrast is believed to be due to the receptive fields mentioned above. Receptive fields have not been found to be uniform; rather, they are characterized by distinctive centers and surrounds, with visual information in the surround having an impact on the information found in the center. For a detailed discussion of how simultaneous contrast operates, see Hurvich (1981).

5.1.5 Color Vision Impairment

Up to this point, we have assumed map readers with normal color vision. Actually, a significant percentage of the world's population has some form of color vision impairment. The highest percentages are found in the United States and Europe (approximately 4 percent, primarily males); the lowest incidence (about 2 percent overall) appears in the Arctic and the equatorial rain forests of Brazil, Africa, and New Guinea (Birren 1983).

The color vision–impaired can be split into two broad groups: **anomalous trichromats and dichromats.** These groups are distinguished on the basis of the number of colors that must be combined to match any given color; anomalous trichromats use three colors, while dichromats use two. For both groups, the most common problem is distinguishing between red and green; for anomalous trichromats, there is some difficulty, while for dichromats the colors cannot be distinguished.

Two hypotheses have been proposed for color vision impairment: (1) a change in the colors to which cone cells are sensitive and (2) changes in one of the opponent-process channels (normally the red-green one). Assuming that changes in the cones cells are the cause, the two major groups have been divided into subgroups: protanomalous and deuteranomalous for anomalous trichromats and protanopes and deuteranopes for dichromats. The two subgroups differ on the basis of the types of cones affected; for example, protanopes and deuteranopes are presumed to be missing red and blue cones, respectively. In Chapter 6 we will consider how color schemes for maps can be adjusted to account for color vision impairment.

5.1.6 Beyond the Eye

It is important to realize that the eye is part of the larger visual processing system shown in Figure 5.8. Note first that information leaving the eyes via the **optic nerves** crosses over at the *optic chiasm;* up to this point information from each eye is separate, but pathways beyond this point contain information from both eyes. After passing through the optic chiasm, each pathway enters the *lateral geniculate nucleus (LGN).* Physiological experiments with animals reveal that opponent cells similar to those found within the retina are also found here (De Valois and Jacobs 1984).

Interpretation of the visual information begins in the **primary visual cortex,** the first place where all of the information from both eyes is handled. As with the LGN, our knowledge of processing in this area is largely a function of physiological experiments with animals. Probably the most significant of these is the work of David Hubel and Torsten Wiesel, who received the 1981 Nobel Prize for their efforts. They found three kinds of specialized cells in the primary visual cortex: *simple cells,* which respond best to lines of particular orientation; *complex cells,* which respond to bars of particular orientation that move in a particular direction; and *end-stopped cells,* which respond to moving lines of a specific length or to moving corners or angles. Not only did they discover these different kinds of cells, but they also mapped out where they occur within the primary visual cortex (Goldstein 1989).

Although such findings are certainly significant, specialists still have not provided an explanation of how the brain handles a complex real-world situation, such as a map. As a result, I concur with MacEachren (1995, 64), who argues that "from a cartographic point of view . . . we are interested in . . . how the brain processes visual signals not because this knowledge is likely to tell us how maps work, but because these processes put limits on what symbolization and design variations might work."

5.2 HARDWARE CONSIDERATIONS IN PRODUCING COLOR MAPS

This section considers some hardware aspects of producing color maps in both soft-copy and hard-copy form. The term **graphic display** is commonly used to describe the computer screen (and associated color board) on which a map is displayed in soft-copy form; examples include cathode ray tubes (CRTs), liquid crystal displays (LCDs), plasma displays, and electroluminescent displays. We will focus on CRTs because they are still the most commonly used; LCDs, however, are becoming increasingly important because of their use in portable microcomputers and overhead computer projection systems.

Both **printers** and **plotters** are used to create hard-copy maps. A wide range of printing technologies have been developed, including dot matrix, electrostatic, ink-jet, thermal-wax transfer, laser, and dye sublimation methods. We will consider only the latter four because they appear to be the most common technologies in

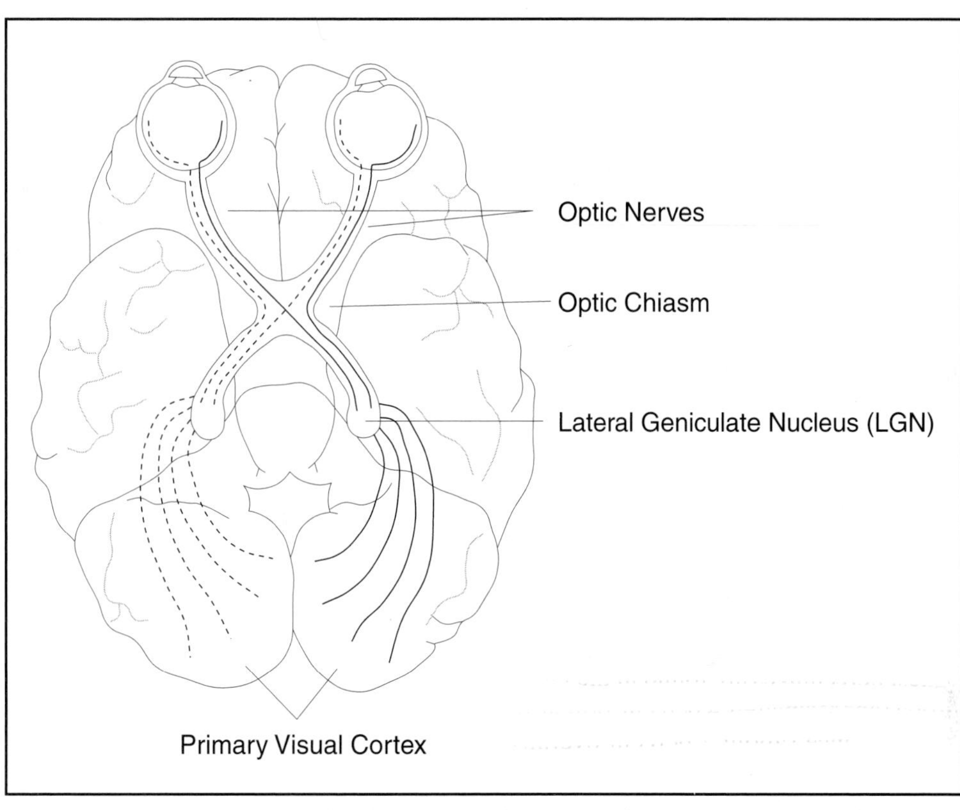

FIGURE 5.8. An overview of the visual processing system viewed from underneath the brain. (From *Sensation and Perception,* by E. B. Goldstein. Copyright © 1999, 1996, 1989, 1984, 1980 Brooks/Cole Publishing Company, Pacific Grove, CA 93950, a division of International Thomson Publishing Inc. By permission of the publisher.)

today's marketplace. In terms of plotters, readers may be familiar with the Hewlett Packard series of line plotters, which simulate traditional pen-and-ink cartography. We will not consider plotters in detail because it appears that this technology will soon disappear.

5.2.1 Vector versus Raster Graphics

Images on graphic displays and printers (or plotters) can be generated using two basic hardware approaches: vector and raster. In the **vector** approach images are created much like we would draw a map by traditional pen-and-ink methods: the hardware moves to one location and draws to the next location. In contrast, in the **raster** approach the image is composed of **pixels** (or **picture elements**), which are created by scanning from left to right and from top to bottom (Figure 5.9). Prior to about 1980, the vector approach was more common, but today raster is dominant; virtually all graphic displays use a raster approach, as do all printers.

5.2.2 CRTs

Images on CRT screens are created by firing electrons from an **electron gun** at *phosphors,* which emit light when they are struck. Monochrome CRTs contain a single electron gun, while color CRTs contain three guns, normally designated as R (red), G (green), and B (blue). The names for the guns have nothing to do with the type of electrons they fire, but are a function of which type of phosphor the electrons strike on the screen. Two common arrangements of electron guns and phosphors (*delta* and *in-line*) are shown in Figure 5.10.

Different colors on a CRT screen result from the principle of **additive color:** the colored phosphors are visually added (or combined) to produce other colors. This principle is normally demonstrated with overlap-

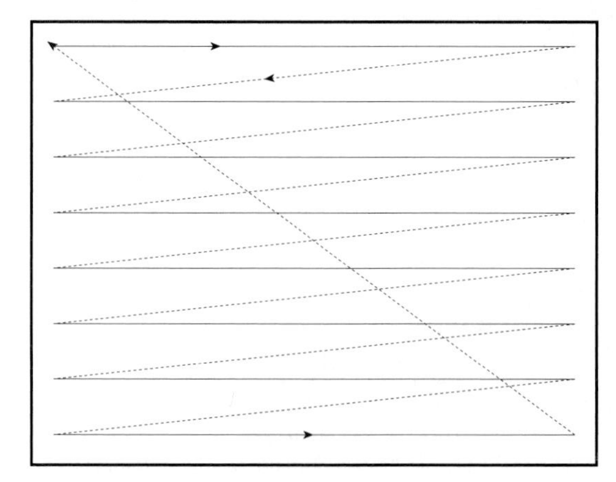

FIGURE 5.9. The order of refresh on a CRT.

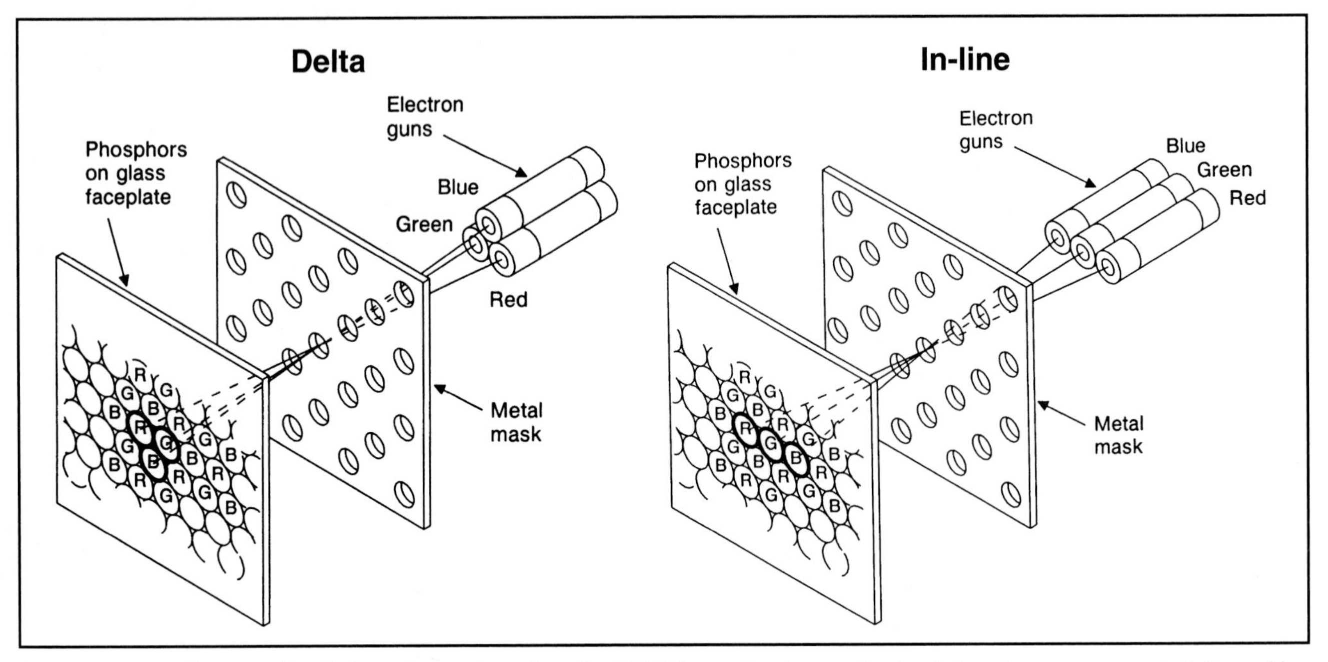

FIGURE 5.10. Cross-sectional view of a portion of a color CRT illustrating two methods of phosphor arrangement: delta and in-line. (J Foley/S Feiner/J Hughes/A Van Dam, COMPUTER GRAPHICS: PRINCIPLES AND PRACTICE, (figures 4.14 and 4.15 from pages 159 and 160). © 1990 Addison Wesley Longman Inc. Reprinted by permission of Addison Wesley Longman.)

ping colored circles (as in Color Plate 5.1), but phosphors on a CRT screen do not actually overlap. Rather, we see a mixture of color because our eye cannot resolve the very fine detail of individual phosphors; the concept is analogous to pointillism techniques used in 19th-century painting.

Together three phosphors compose a pixel. One measure of the **resolution** of a monitor is the number of *addressable pixels,* normally specified as the number of pixels displayable horizontally and vertically.* A common resolution for microcomputers has been 640 × 480, or 640 pixels horizontally by 480 pixels vertically. Some problems with using this relatively coarse resolution include **jaggies** (or a staircase appearance to straight lines), the inability to smoothly vary the size of small symbols, and the difficulty of creating crisp text.† Higher resolutions, such as 1024 × 768 and 1280 × 1024, have been available for workstations for some time, and are now becoming common for microcomputers.

In a later section, we will see that in order to compare the resolution of monitors and printers, it is useful to calculate the number of pixels per inch. For example, for a 17-inch monitor with 1024 × 768 displayable pixels, there are approximately 75 pixels per inch.

Since phosphors on a CRT screen have a low *persistence* (stay lit only briefly), the screen must be **refreshed** constantly. Refresh takes place by scanning across the screen from left to right and top to bottom (Figure 5.9). Two types of refresh are possible: interlaced and noninterlaced. In the *interlaced* method, every odd-number scan line is refreshed in the first 1/60 of a second, and every even-number scan line in the next 1/60 of a second, for a total refresh rate of 1/30 of a second. If information on adjacent scan lines is similar, then this approach is acceptable, but if the information is different (as on a map with horizontal political borders 1 pixel thick) the screen will appear to flicker. Although flicker is undesirable, interlacing is sometimes used on high-resolution systems (such as 1280 × 1024) because so many pixels must be addressed. Fortunately, the trend is toward *noninterlaced* systems in which the entire screen is refreshed from top to bottom in 1/60 of a second or less.

The **frame buffer** is an area of memory that stores a digital representation of colors appearing on the screen. We will consider the frame buffer in detail because it determines the number of colors available for cartographic applications. Consider first a monochrome system in which only black-and-white pixels are possible. In such a system, we can think of the frame buffer as an area of memory in which each bit corresponds to a pixel on the screen; bits can have values of either 0 or 1, corresponding to either lit or unlit pixels (Figure 5.11). The resulting layer of bits is termed a *bit-plane.*

Now consider a monochrome system in which shades of gray are possible. In this case there is a layer of bit-planes, with the lightness of a pixel a function of whether

* See Peddie (1994, pp. 7–10) for a variety of definitions for resolution.
† Jaggies can be handled using antialiasing routines (Foley et al., 1996, 132–142), but such routines are not common in mapping and design software.

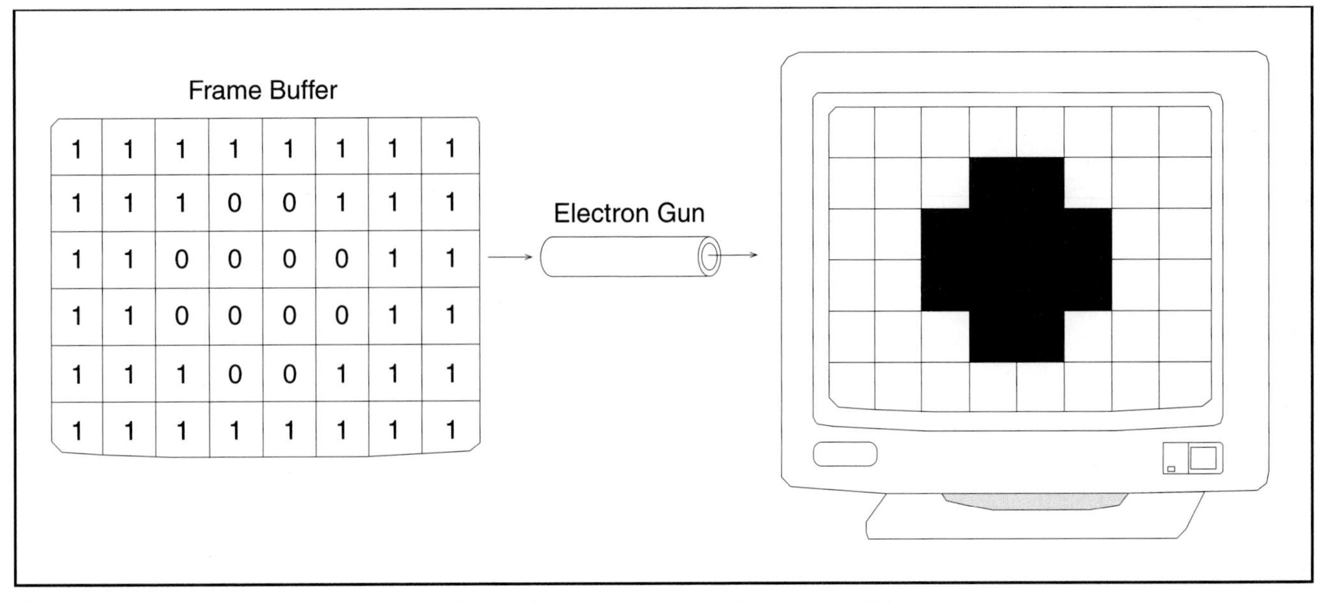

FIGURE 5.11. A monochrome system in which only black and white pixels are possible.

the bits associated with that pixel are set to 0 or 1 (Figure 5.12). Since each bit-plane corresponds to a power of 2, the lightness value for any pixel is the sum of the products of the frame buffer values times the corresponding power of 2 (for the upper right pixel in Figure 5.12, the result is $0 \times 2^0 + 1 \times 2^1 + 0 \times 2^2 = 2$. In general, if n is the number of bit planes, then 2^n different values will be possible (in Figure 5.12, n is 3, so $2^n = 8$ possible values), and the values will range from 0 to $2^n - 1$ (0 to 7 in Figure 5.12).

Also note in Figure 5.12 that a **digital-to-analog converter (DAC)** is placed between the frame buffer and the electron gun. The purpose of the DAC is to convert the digital information stored in the frame buffer into the analog signal produced by the electron gun.

Now let's consider color display systems. In sophisticated color display systems, a set of bit-planes is assigned to each color gun (Figure 5.13). Values for individual color guns are computed in a manner identical to the monochrome system just described. Thus, in Figure 5.13 each gun can fire with eight intensities. The total number of colors possible for a pixel is the product of the number of intensities for each color gun, or $2^n \times 2^n \times 2^n$; in Figure 5.13 the result is $8 \times 8 \times 8 = 512$. Typically, sophisticated color display systems have 8 bit-planes per color gun, meaning

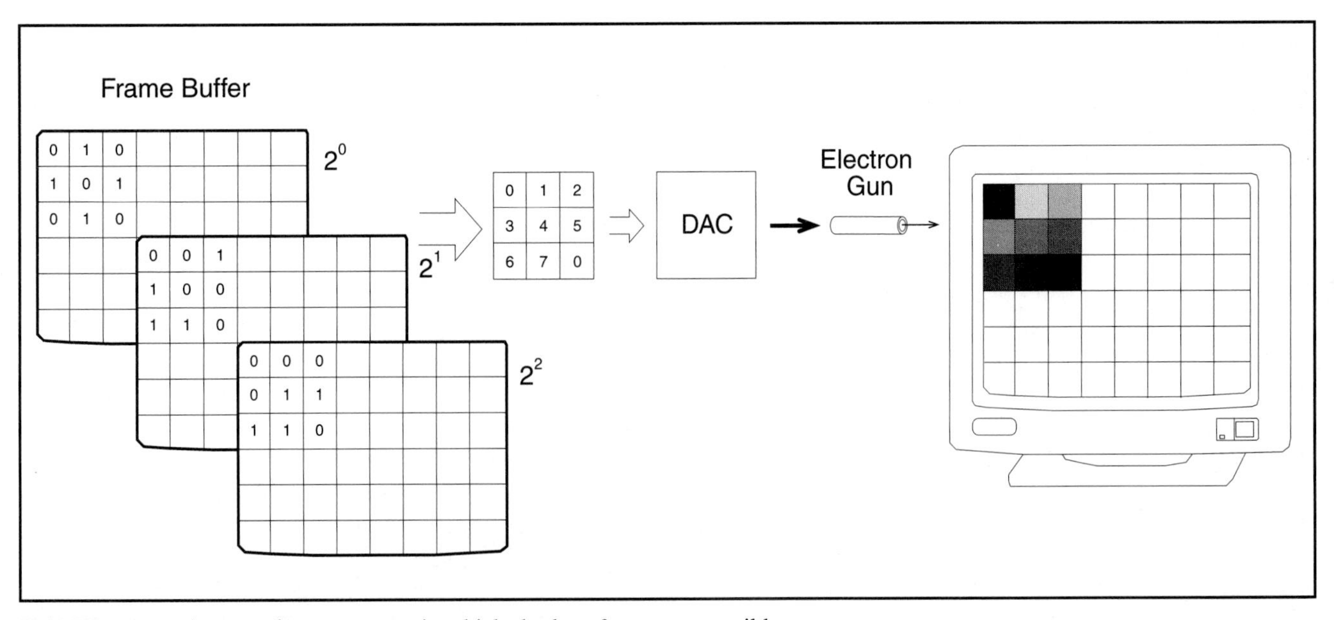

FIGURE 5.12. A monochrome system in which shades of gray are possible.

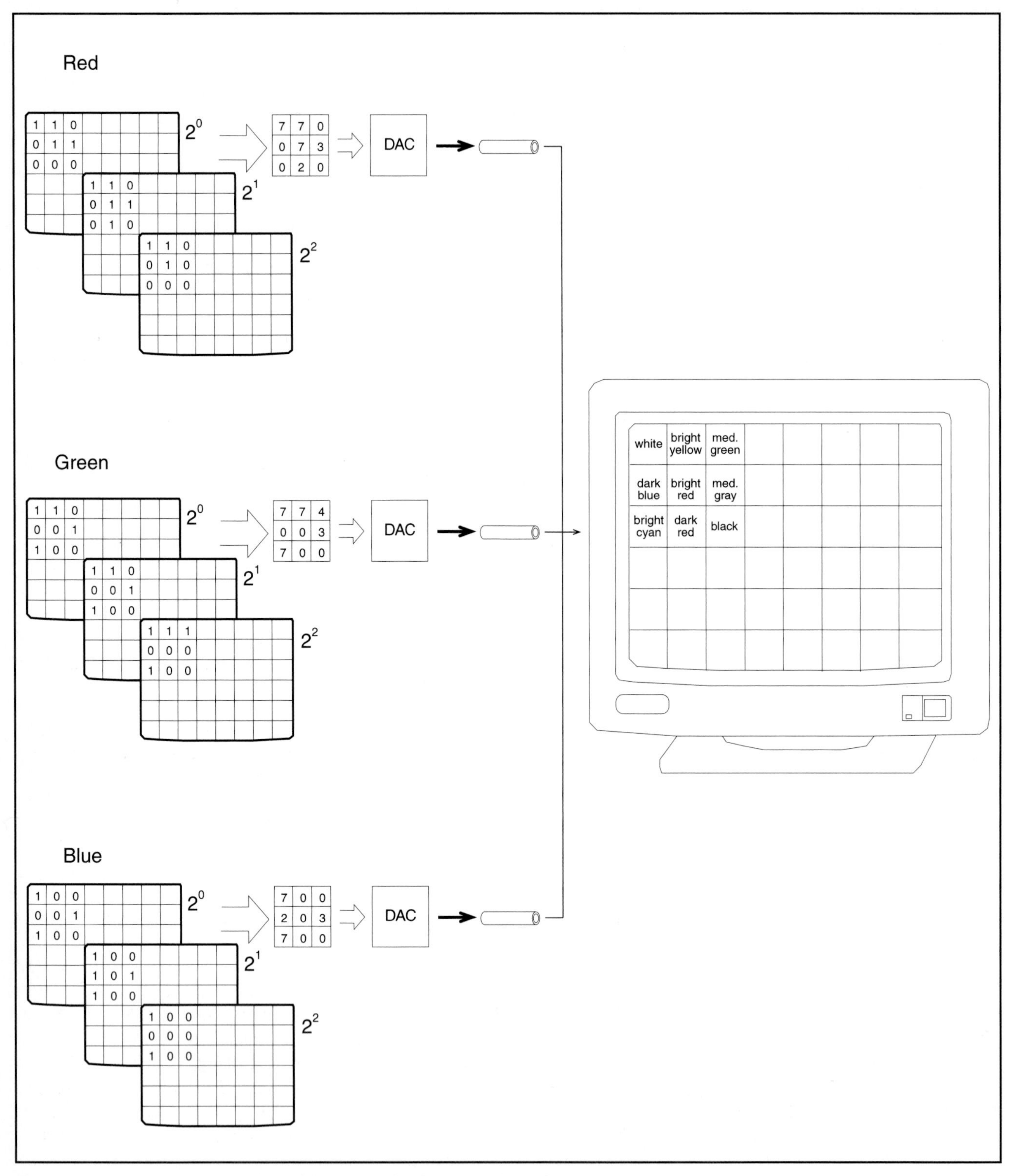

FIGURE 5.13. A color system in which several bit-planes are allotted to each color gun.

they can, in theory, display $2^8 \times 2^8 \times 2^8 = 16.8$ million colors.

Although 16.8 million colors may seem excessive, three points must be kept in mind. First, we can distinguish a much larger number of colors than is commonly recognized; Goldstein (1989, 111) suggests an upper limit of 7 million. Second, the differences between light intensities for a color gun are not constant throughout the range of possible digital values. For example, a change in intensity between 0 and 1 may not be the

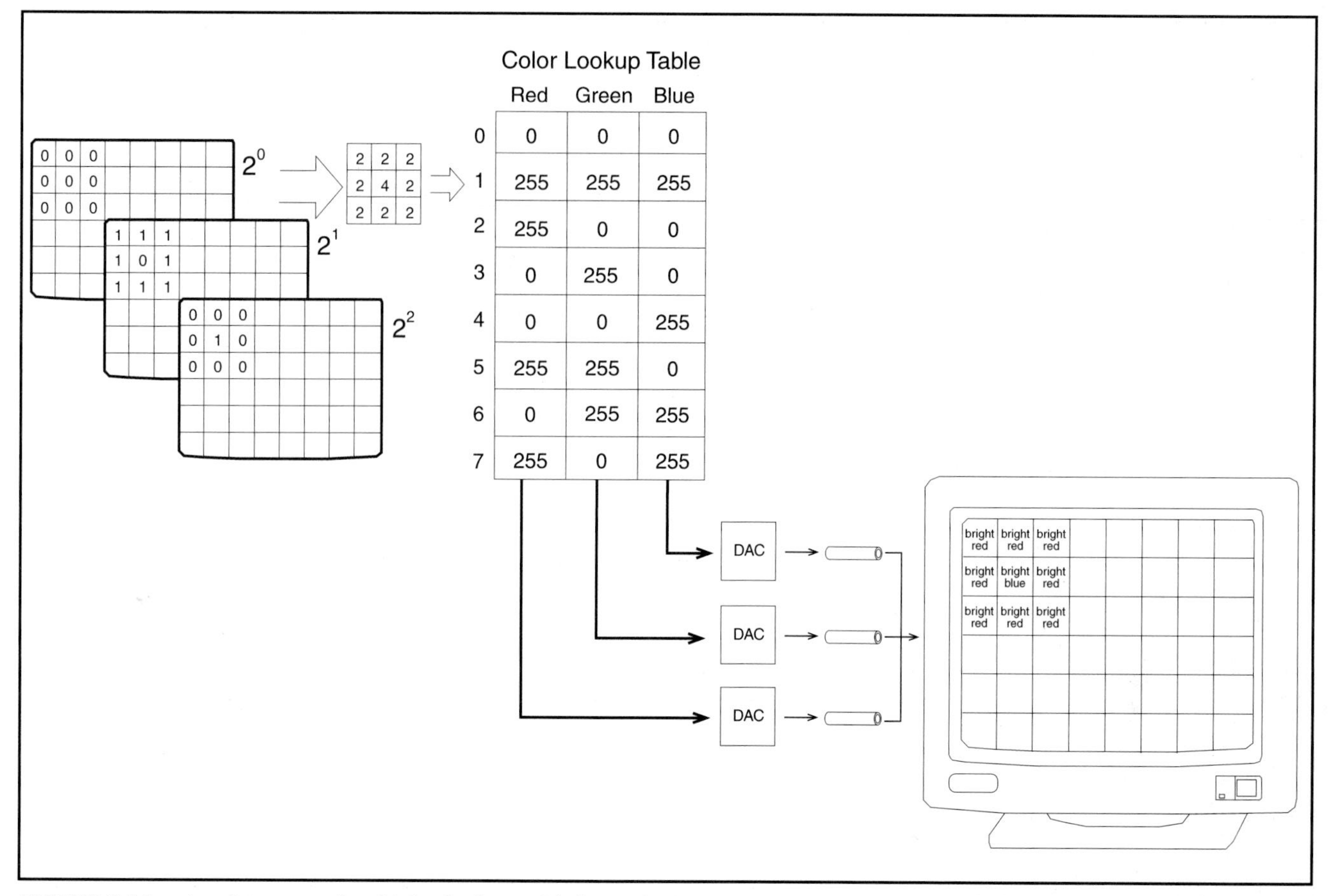

FIGURE 5.14. A color system in which a lookup table is used.

same as between 200 and 201; some changes may be so small as to be indistinguishable. Third, the actual intensities used do not account for the fact that our eye-brain system discriminates among lighter shades better than darker ones.

Unfortunately, display systems in which a set of bitplanes is assigned to each color gun are beset by two problems: (1) large memory requirements and (2) the slow speed of changing a geographic region of uniform color (as might be desired in cartographic animation). The solution to such problems is **color lookup tables,** in which values in the frame buffer serve as indexes to a table that provides the actual values sent to the color guns. Figure 5.14 shows a simplified example for a table with 8 rows (typically, tables consist of 256 rows). Note that the bright red pixels on the screen could be changed to a moderate red simply by replacing the 255 for "Red" at lookup position 2 in the table with a value of 127.

Dithering is another approach that conserves frame buffer memory and permits the display of additional colors. In dithering, new colors are created by presuming that readers will perceptually merge colors displayed in adjacent pixels. For example, Color Plate 5.2 shows a dithered orange *(A)* and the individual yellow and red pixels *(B)* used to create that orange. One problem with dithering is that the resulting colors often exhibit pattern (a qualitative characteristic), which is inappropriate for use in a quantitative series.

5.2.3 CRTs versus LCDs

Readers of this book may be disturbed to find little mention of **liquid crystal displays (LCDs),** which are commonly used for color display on personal laptop computers. The justification for doing so is twofold: First, CRTs are still the dominant technology for personal computers. In an extensive review of graphic display systems, Peddie (1994, 190) stated that "CRTs are still the most prevalent display technology and will continue to be so for the next decade. . . . They offer brightness, contrast, colors, resolution, and reliability, as well as a wide viewing angle—all at a low cost." Peddie, however, noted a number of disadvantages of CRTs: they take up space, are heavy, consume considerable power, and produce x-rays and low-frequency magnetic fields, which are suspected of causing health hazards. Since LCDs do not have these disadvantages, it is apparent that we will see continued improvements in them.

The second justification for not focusing on LCDs is that the associated technology is dynamic. In his review, Peddie evaluates a broad range of **flat-panel displays (FPDs)** that could conceivably replace CRTs; he lists 16 types of LCDs along with 5 other major types of FPDs. At this point, it is unclear which of these displays will ultimately dominate the marketplace. I encourage readers interested in LCD and related technologies to examine Peddie's review, and to peruse recent issues of popular computing magazines to obtain the most recent developments.

One limitation of both LCDs and CRTs is that they cannot handle large map displays (such as an entire USGS topographic sheet). The largest CRT screens available generally do not exceed 21 inches (along a diagonal), and the largest marketed LCDs are considerably smaller. One alternative to displaying an entire large-format map is, of course, to use *pan, scroll,* and *zoom* functions, but there are times when the reader would rather examine the entire map at once or compare a variety of maps simultaneously; for example, imagine comparing 50 maps showing changes in wheat production for Kansas on a yearly basis over a 50-year period. Although large, low-cost map displays are not available today, it will be interesting to see what becomes available as technology continues to evolve.

5.2.4 Printers

The basic picture element of CRTs is a pixel; the corresponding element for printers is a **dot.** Resolution for printers is generally expressed in *dots per inch (dpi),* and is generally higher than the resolution for CRTs. For example, printers usually have resolutions ranging from 300 to 1200 dots per inch (dpi), while a typical 17-inch monitor (assuming 1024 × 768 resolution) displays only about 75 pixels per inch.* The high resolution of printers means that linework will exhibit fewer of the jaggies characteristic of graphic displays.

Since printed maps are based on reflected (as opposed to emitted) light, they create color using a subtractive (as opposed to an additive) process. The three basic **subtractive colors** are cyan, magenta, and yellow; black is often overprinted when a true black is desired (Color Plate 5.1). In the past, printers were able to display only a limited number of colors for a dot (the three subtractive colors and their various combinations at 100 percent ink), so dithering was used to create a greater variety of colors. As noted with CRTs, dithering results in colors that often exhibit a pattern. Fortunately, some printers now avoid dithering by changing the color of a dot in a continuous fashion (actually changing the amount of cyan, magenta, or yellow printed for an individual dot).

An important characteristic of printers is whether they are PostScript-compatible. PostScript is a **page description language (PDL)** that describes an image in terms of its vector and raster components. The major advantage of PDLs is that they take advantage of the maximum resolution of a particular output device. For example, a PostScript-based map generated on a 100-dpi printer would appear coarse (have jaggies), while the same map generated on a 600-dpi printer would be much smoother. Although there are several page description languages, "PostScript has become the *de facto* standard imaging model for printed text and graphics in the publishing industry" (DiBiase 1990a, 17).

Four types of color printers are commonly available in the marketplace today: ink-jet, laser, thermal-wax transfer, and dye-sublimation.* **Ink-jet printing** involves squirting microscopic droplets of subtractive inks onto paper. A traditional problem with this method was that ink fluid enough to form droplets tended to seep into fibers of the paper; as a result, specialized paper was required. Today, refined approaches have been developed for drying the ink or controlling its spread once it reaches the paper, and specialized paper is no longer essential. As a result, the ink-jet technology is now challenging dot matrix technology as an affordable low-quality form of color printing. Although dithering is normally used to create multiple colors, more expensive ink-jet devices can produce continuous color.

Laser printing is probably most familiar to readers because of its common use for printing black-and-white text documents. The basic process involves passing a laser beam over a charged belt. When the laser beam is turned on, a corresponding location on the belt is discharged. After the laser beam has passed over the belt for a particular subtractive color, the belt passes under the appropriate toner-developer unit, and the toner sticks to the discharged areas. This process is repeated for all four subtractive inks, and the image on the belt is then transferred to paper. Laser printing provides higher quality than ink-jet printing, but the initial purchase price is also higher. Traditionally, laser printing was accomplished using only dithering, but a form of continuous-tone printing is now available (by adjusting the size of individual dots), a process referred to as *contone.* Distinct advantages of laser printers are speed and ease of networking.

* In general, black-and-white printers have the highest resolutions because their memory can be used to store the greater number of picture elements, as opposed to color information.

* References used for this section included Zeis (1993), Weiss (1995), and the October 21, 1997, issue of *PC Magazine.* Since printer technology is rapidly changing, I advise readers to keep up-to-date by examining recent issues of popular computing magazines.

In **thermal-wax transfer printers,** blank paper is placed in contact with a banded, multicolor sheet (bands would be used for cyan, magenta, yellow, and black, respectively) along a pinch roller, where a thermal print head is also located. To get a dot at a location, the print head is turned on or off to either melt the ink or resolidify it onto the paper. As with laser printers, color traditionally was created using dithering, but it is expected that continuous-tone color will soon be available. Printing quality is generally higher than for ink-jet printers, but the initial purchase cost is somewhat greater, and best results are obtained only with special print media. On the plus side, thermal-wax transfer printers produce excellent transparencies.

Dye-sublimation printers (or *thermal-dye printers*) are similar to thermal-wax transfer printers in that they use heat to move colorant from a ribbon to paper. They differ in several respects, however. Dye-sublimation printers use a special dye, as opposed to a color-impregnated wax or plastic. Another difference is that dye-sublimation printers use a sublimation process (converting the solid ink to gas), as opposed to simply melting the wax or plastic. Advantages of this process are that 256 different intensities of each color can be created at a dot (continuous color is possible), and the resulting image cannot be scratched easily. Dye-sublimation printers produce the highest quality images, but the initial purchase price of the printer is the highest of those discussed here, and expensive specialized paper is required.

There are a host of criteria involved in selecting a printer. Some of those mentioned above include the initial purchase price, the resolution, the number of colors generated, and the need for special paper. Other criteria include costs for toner, ease of replacing toner cartridges, precision of dot placement (a high resolution won't necessarily be desirable if dots are not consistently placed in the proper row-and-column position), the consistency with which color is applied at each dot (a device that claims to produce continuous color may not actually produce the identical color each time that color is specified), and maintenance costs (because the technology is changing very rapidly these costs are frequently unknown).

5.2.5 Color Management Systems: Matching Graphic Display and Printer Colors

A common problem confronting mapmakers is getting graphic display and printer colors to match. Ideally, when creating a map on a graphic display, we would like to push a button and print a map having the same colors on the hard copy that appear on the graphic display. Unfortunately, this generally has not been possible because the color production methods of graphic displays and printers are dissimilar; consequently the **color gamuts** (ranges of colors produced) differ. Fortunately, some companies have begun to provide specialized software known as **color management systems** to assist in this process.

Color management systems work with either specific software applications or are included as part of the computer operating system. Examples of the former include the Kodak Precision CMS used in Adobe Systems' PageMaker 6.0 and EfiColor XTension bundled with QuarkXPress 3.2. The latter includes Apple's ColorSync, Microsoft's ICM (*I*ndependent *C*olor *M*atching) for Windows 95, and the PANTONE Personal Color Calibrator, which works with either MacIntosh or Windows 95 operating systems. Including color management as part of the operating system is the most flexible approach because it guarantees that different applications will be compatible in terms of color use (Sugihara 1995).

Unfortunately, since color management systems are a recent phenomenon, their quality, ease of use, and cost vary. Reviews by Sugihara (1995) and Hilliard (1995) suggest that the systems bundled with specific applications are the most sophisticated. An indication of the difficulty of use and high cost is that specialized hardware may be necessary to calibrate the management software properly. Although it is clear that cartographers ultimately will find color management systems useful, it is also apparent that reviews of software (and any associated hardware) by cartographers will be necessary to determine which is most appropriate for mapping purposes.

5.3 MODELS FOR SPECIFYING COLOR

This section considers six models that have been used for specifying colors appearing on maps: **RGB, CMYK, HSV, Munsell, HVC,** and **CIE.** The RGB and CMYK models are termed hardware-oriented because they are based on hardware specifications of red, green, and blue color guns, and cyan, magenta, yellow, and black ink, respectively. In contrast, the HSV, Munsell, and HVC models are user-oriented because they are based on how we perceive colors (using variables such as hue, lightness, and saturation). The CIE system is neither hardware- nor user-oriented; however, it is "optimal" in the sense that if you provide me with the CIE coordinates of a color, then I should be able to create exactly the same color.

5.3.1 The RGB Model

In the RGB model, users define color by specifying the intensity of the red, green, and blue color guns described in section 5.2. The range of intensities for these guns can be represented as a cube, which appears in schematic form in Figure 5.15. In Figure 5.15A, gray tones (or completely desaturated colors) are found along the diagonal

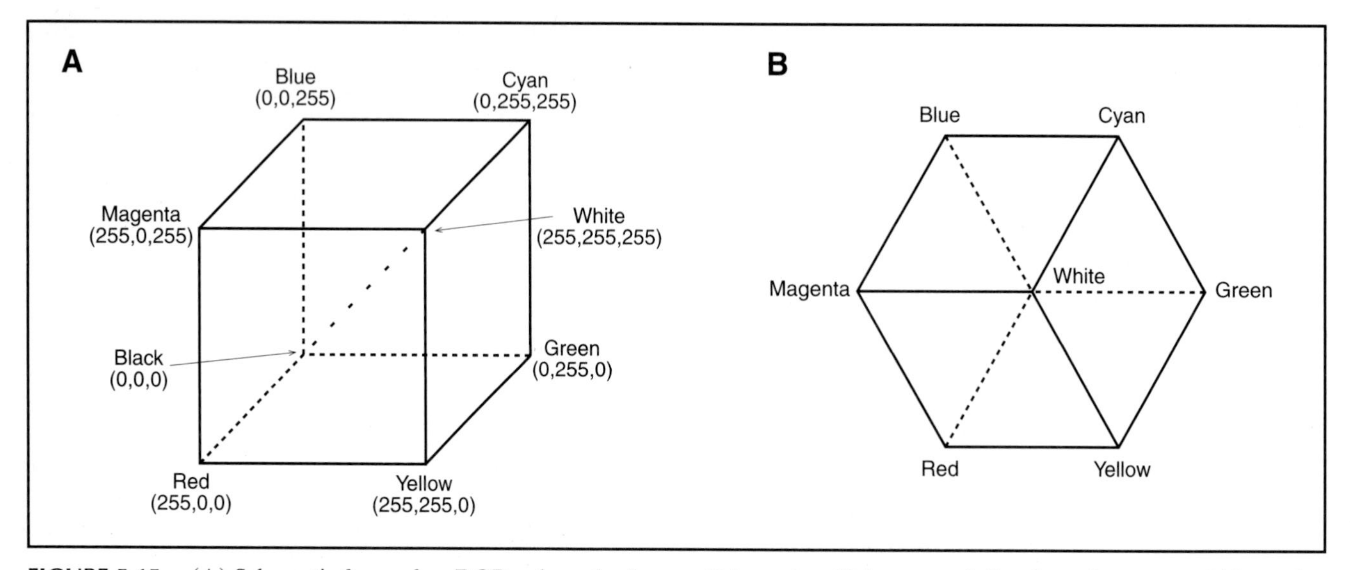

FIGURE 5.15. (A) Schematic form of an RGB color cube for specifying color. (It is assumed that the red, green, and blue color guns each have a maximum value of 255. (B) The cube viewed looking directly down the diagonal from white to black.

extending from "white" to "black"; in general, lighter colors are found around the white point of the cube, while darker colors are found around the black point of the cube. As you move away from the white-black line, you move toward saturated colors; for example, at the "red" point you would be at the maximum saturation of red. Finally, note that hues are arranged in a hexagonal fashion around the white-black diagonal line. The latter can be seen most easily if you look directly down the diagonal from white to black, as seen in Figure 5.15B.

The RGB color cube has the advantage of relating nicely to the method of color production on graphic displays, but it has two major disadvantages. One is that common notions of hue, saturation, and lightness are not inherent to the system. Another disadvantage is that equal steps in the color space do not correspond to equal visual steps; for example, a color gun value of 125, 0, 0 will not appear to fall midway between values of 0, 0, 0 and 250, 0, 0. Typically, you will find that an incremental change in low RGB values represents a smaller visual difference than the same incremental change in high RGB values. In spite of these disadvantages, RGB values are commonly used as an option for color specification in many software packages, probably because of their long history and the consequent familiarity that users have with them.

5.3.2 The CMYK Model

Section 5.2.4 indicated that printed maps are created by a combination of cyan, magenta, yellow, and black. If we think of cyan, magenta, and yellow as analogous to the red, green, and blue color guns, then it is also possible to conceive of the CMY portion of CMYK as a cube; a certain percentage of cyan, magenta, and yellow would correspond to a particular point in the cube. Black would, of course, need to be added to create true shades of gray within the cube. Given the analogy to the RGB color cube, it makes sense that CMYK will share the same disadvantages: lack of relation to common color terminology and colors that are unequally spaced visually.

5.3.3 The HSV Model

In contrast to RGB and CMYK, HSV is more intuitive from a map design standpoint because it allows users to work directly with hue, saturation, and value (lightness). Color space in HSV is represented as a hexcone, as shown in Figure 5.16. The logic of the hexcone can be seen if you compare it with the color cube shown in Figure 5.15B; note that the hexagonal structure of the hues in the cube are retained in the hexagonal structure at the base of the hexcone. Value changes occur as you move from the apex of the cone to its base, and saturation changes occur as you move from the center to the edge of the cone. The intuitive notions of hue, saturation, and value in HSV have led to its common use in software.

Although HSV is commonly used, it too has disadvantages. One is that different hues having the same value in HSV will not all have the same perceived value. As an example, consider the base of the cone, where the highest-value green and red are found. If you create such colors on your monitor, green will appear lighter than red. In a similar fashion, different hues having identical saturations in HSV will not have the same perceived saturations (Brewer 1994b). HSV also shares the disadvan-

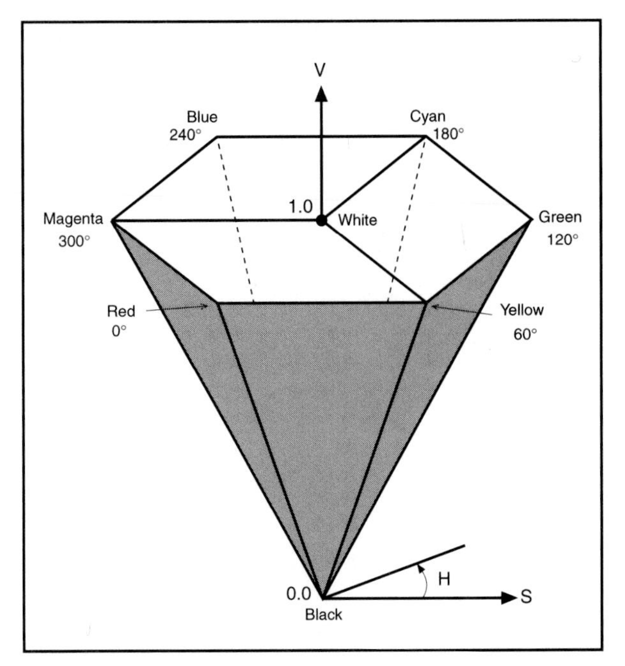

FIGURE 5.16. The HSV (hue, saturation, and value) color system represented as a hexcone.

tage noted for RGB: selecting a color midway between two colors will not result in a color that is perceived to be midway between the colors.

5.3.4 The Munsell Model

The Munsell color model is a user-oriented system that was developed prior to the advent of computers. Munsell colors are specified using the terms *hue, value* (for lightness), and *chroma* (for saturation). The general structure of the model (Figures 5.17 and 5.18 and Color Plate 5.3) is similar to HSV (hues are arranged in a circular fashion around the center, chroma increases as one moves outward from the center, and value increases from bottom to top). Note, however, that in contrast to HSV, the Munsell model is asymmetrical; for example, if you were to hold the model in your hands, you would note that the lightest green would be higher on the model than the lightest red. The asymmetry occurs because the model is perceptually based (the lightest possible green does appear brighter than the lightest possible red).

Ten major Munsell hues are recognized, and these are split into five principal (represented by a single letter, such as Y for yellow) and five intermediate (represented by two letters, such as YR) hues (Figures 5.17 and 5.18). Each major hue is also split into 10 subhues (consider the 10 subhues shown for 5R in Figure 5.18). Values range from 0 to 10 (darkest to lightest) and chromas from 0 to 16 (least to most saturated), respectively. Due to the

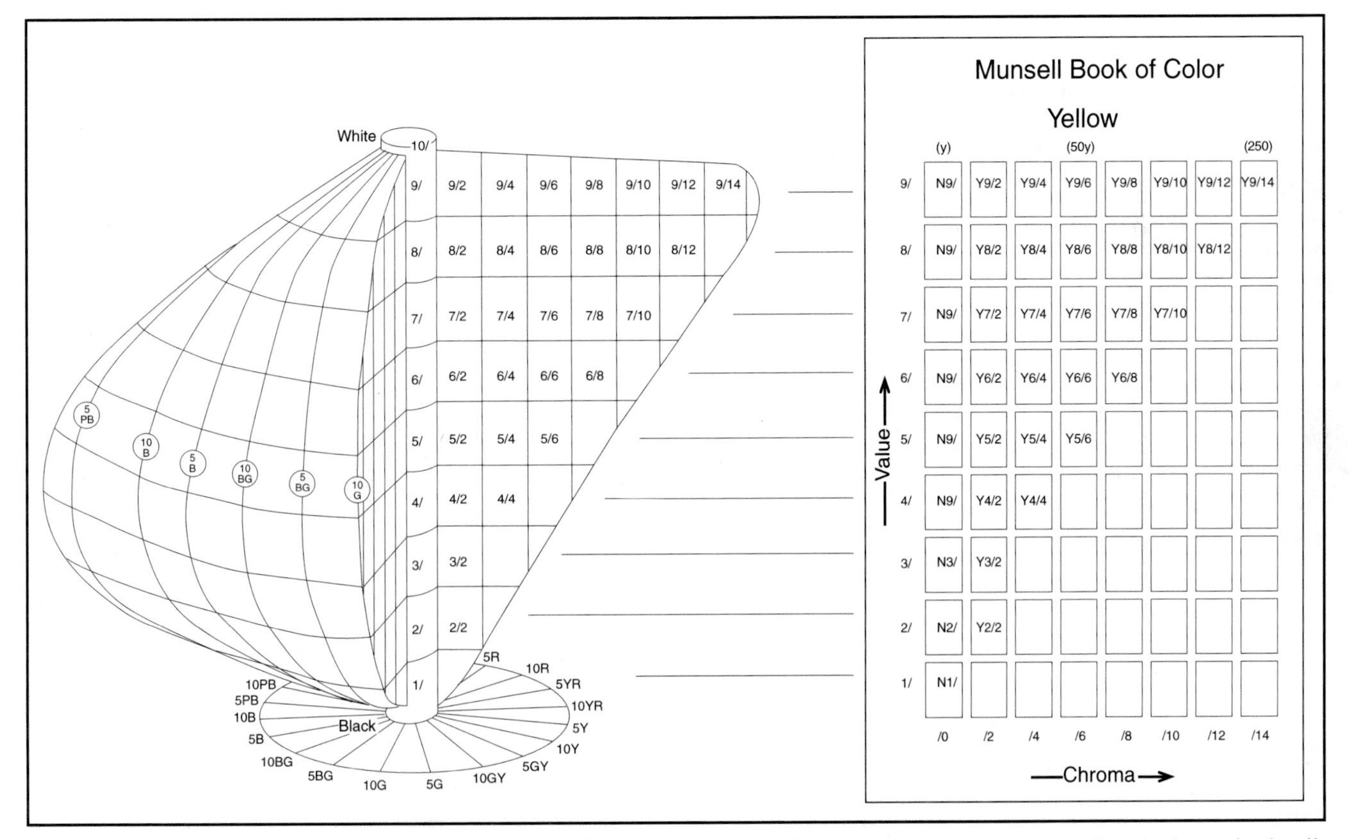

FIGURE 5.17. Three-dimensional representation of the Munsell color solid. A vertical slice through yellow is shown in detail. (From Hurvich, *Color Vision,* p. 274. Copyright © 1981 Sinauer Associates, Inc. Reprinted by permission of Sinauer Associates.)

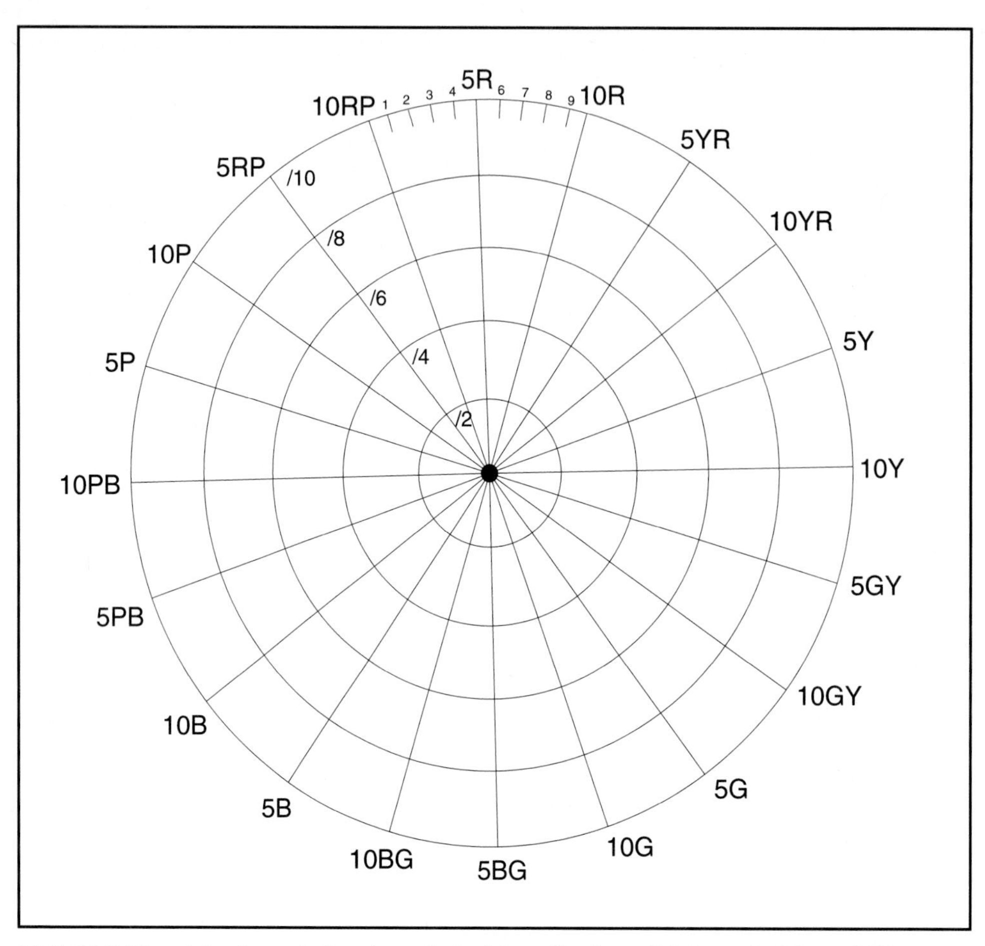

FIGURE 5.18. A horizontal slice through the Munsell color solid shown in Figure 5.17.

asymmetry of the model, not all values and chromas will occur for each hue. In general, colors are represented symbolically as H V/C. Thus, 5R 5/14 would represent a red of moderate value and high saturation (crimson).

An important characteristic of Munsell is that equal steps in the model represent equal perceptual steps. Thus, a color that is numerically midway between two other colors will appear to be perceptually midway between those colors. For example, the color 5R 5/5 should appear midway between 5R 2/2 and 5R 8/8. Remember that this was not a characteristic of the other color models described thus far.

5.3.5 The HVC Model

The HVC (hue, value, and chroma) color model developed by Tektronix was an attempt to duplicate the Munsell system on computer graphic displays. The similarity of HVC and Munsell is seen in the use of the same three terms (hue, value, and chroma), and the irregular shape of the color space (compare Figures 5.17 and 5.19). The two differ, however, in color notation. Munsell uses the

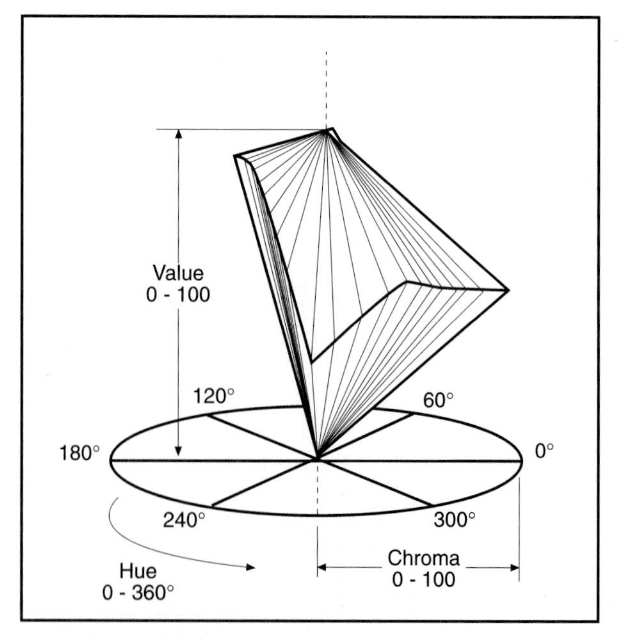

FIGURE 5.19. The HVC color system developed by Tektronix. (Taylor et al., 1991. First published in *Information Display,* 7 (4/5), p. 21. Courtesy of Tektronix, Inc.)

H V/C notation, while in HVC, hue is specified in degrees counterclockwise from 0 degrees (red), value varies vertically from 0 to 100, and chroma varies from 0 to 100 from the central axis to the edge of the model. Although the literature (for example, Taylor et al., 1991) suggests that HVC is effective, unfortunately, associated software is no longer distributed. Apparently, many designers (outside the field of cartography) either do not require or do not see the advantage of specifying colors that are equally spaced in the visual sense, so Tektronix has chosen not to support the software. I have included the HVC model here because it is illustrative of the type of model that developers should consider including in cartographic software.

5.3.6 The CIE Model

CIE is an abbreviation for the French Commission International de l'Eclairage (International Commission on Illumination). In theory, careful color specification in CIE means that anyone in the world should be able to recognize and reproduce a desired color. CIE colors can be specified in several ways (*Yxy*, L*u*v*, L*a*b*), but in all cases a combination of three numbers is used. We will consider the *Yxy* model (commonly referred to as the 1931 CIE model) first because it forms the basis of all CIE methods.

In the *Yxy* model, the *x* and *y* coordinates define a two-dimensional space within which hue and saturation vary (Figure 5.20). Note that hues are arranged around a central *white point* (or *equal-energy point*) and saturation increases as one moves outward from the white point. The *Y* portion of the model provides the third dimension—the lightness or darkness component (Color Plate 5.4).

The structure of the *Yxy* model is similar to both HVC and Munsell (all have hues arranged in a circular fashion, desaturated colors in the middle, and a vertical-lightness axis), but note that in CIE the hues and saturations are not related in a simplistic fashion to the *x* and *y* axes. The

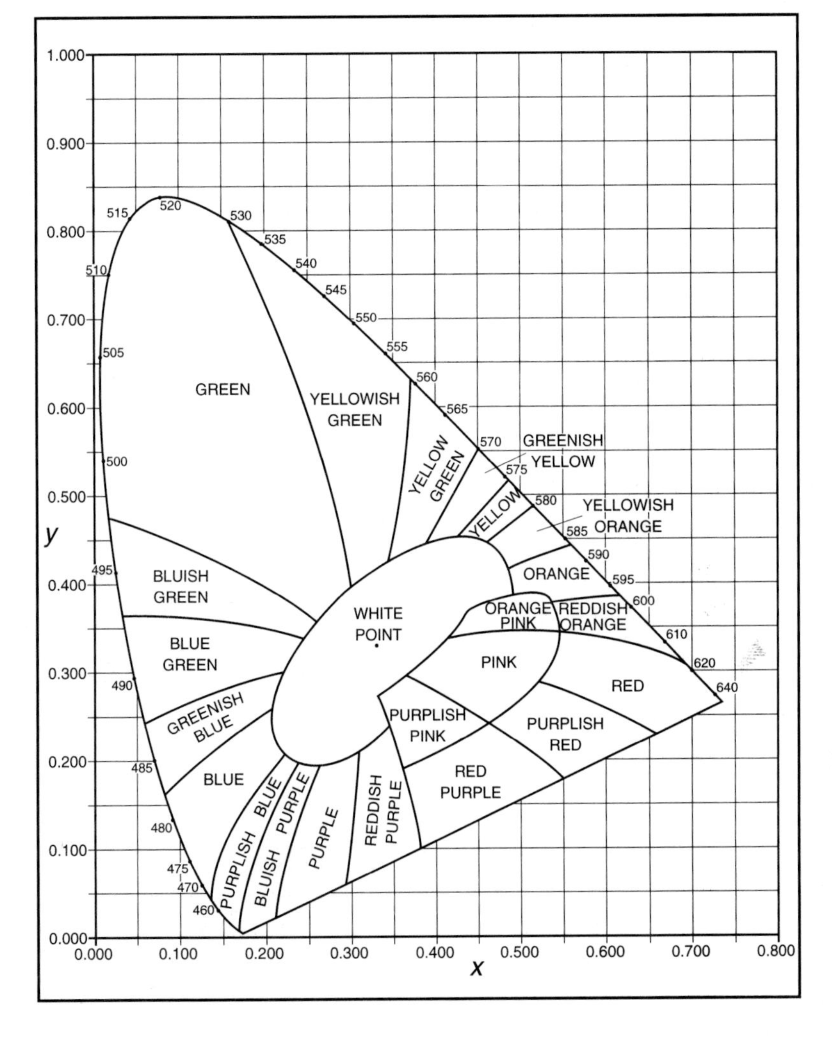

FIGURE 5.20. The *Yxy* 1931 CIE system. Hues are arranged in a continuum around a central white point. Saturation is a maximum on the edge of the horseshoe and a minimum at the white point. Numerical values on the edge of horseshoe represent wavelengths in nanometers. Differing lightnesses cannot be shown by this diagram; a three-dimensional diagram is required, as in Color Plate 5.4. (After Kelly, K. L. (1943) "Color designations for lights," *Journal of the Optical Society of America*, 33, no. 11:627–632.)

reason for this can be found in the manner in which CIE was established.

CIE was developed using the notion that most colors can be defined by a mixture of three colors (roughly speaking, we can call these red, green, and blue). The appropriate combination of three colors needed to match selected colors was determined using human subjects (the average response of the subjects was termed the *standard observer*). To understand the matching process, imagine that you were asked to view a screen on which a single circle was projected using a standard light source. In the top portion of the circle a test color appeared, while in the bottom portion you manipulated the three colors in order to produce a color identical to the test one. If you repeated this task for many test colors, you would discover that various amounts of the three colors would be required to make the appropriate matches.

Results of actual CIE matching experiments are shown in Figure 5.21. The three curves correspond to the three colors combined in the experiments. The x axis represents the wavelength of the test color and the y axis represents the relative magnitudes of the three colors needed to match the test color. For example, a test color at 530 nm would require .005 of blue, .203 of green, and −.071 of red (Wyszecki and Stiles 1982, 750); these are known as *tristimulus values.* The negative value for red is necessary because in some cases you would find that you could not match the test color with any combination of the three colors; to achieve a match, you would have to mix one of the three with the test color, and this is recorded as a minus value in the graph. To avoid having to work with negative values, the developers of the CIE model transformed the results shown in Figure 5.21 to purely positive values, which are commonly referred to as $X, Y,$ and Z.

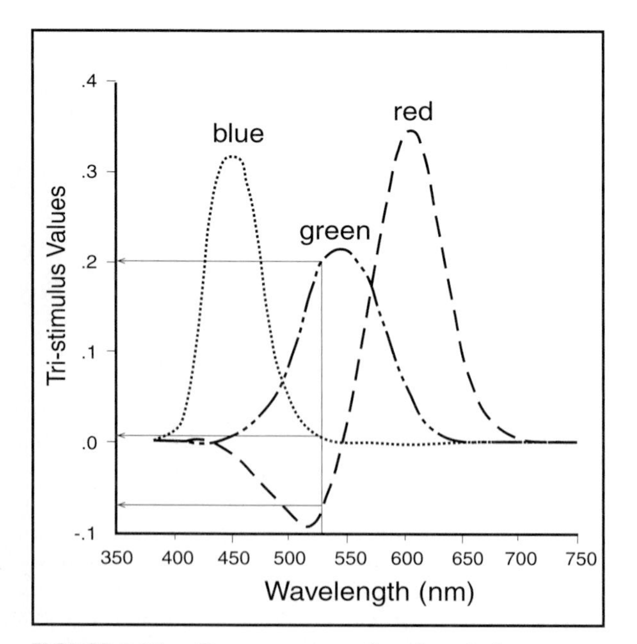

FIGURE 5.21. Curves representing the relative proportion of CIE primaries needed to match test colors at various wavelengths.

To get coordinates for the *Yxy* system, the $X, Y,$ and Z values were converted to proportional values:

$$x = X/(X + Y + Z)$$
$$y = Y/(X + Y + Z)$$
$$z = Z/(X + Y + Z)$$

Since these proportions add to 1, it is not necessary to plot z in the *Yxy* system (z would be $1 - (x + y)$).

One problem with plotting proportional values is the elimination of information about lightness or darkness. (We get the same proportions when $X, Y,$ and Z are all 10 units as we do when they are all 20 units.) This problem was handled in CIE by arbitrarily assigning the lightness information to Y and plotting this as the third dimension, as shown in Color Plate 5.4 (Hurvich 1981, 284).

It should also be noted that the CIE *Yxy* coordinate values can be adjusted to account for the lighting conditions under which the colors are viewed. For example, you might want to consider the potential effects of natural sunlight versus fluorescent room light. This is a capability that is not generally included in other color models.

One problem with the 1931 CIE diagram is that colors are not equally spaced in a visual sense, as was true for RGB, CMYK, and HSV. Fortunately, two **perceptually uniform color models** (or simply **uniform color models**) have been developed based on CIE: L*u*v* (CIELUV) and L*a*b* (CIELAB). CIELUV is appropriate for graphic displays, while CIELAB is appropriate for printed material. We will consider the potential use of the CIELUV system on choropleth maps in the subsequent chapter.

5.3.7 Discussion

Given the variety of color models presented here, it is natural to ask which models mapmakers need to be familiar with. At present, the RGB, CMYK, and HSV models are most frequently used in mapping software. This is problematic because the RGB and CMYK models do not relate easily to our common notions of hue, saturation, and lightness, and all three models do not produce equally spaced colors (in the visual sense). The Munsell model is not generally found in mapping software, but it is useful to understand because it relates well to our notion of hue, saturation, and lightness, and stresses the notion of creating equally spaced colors. The HVC model is useful for illustrating how the Munsell model can be included in software. The CIE model, although thus far not commonly used in mapping software, is important to understand because you may see reference to CIE colors when reading the results of cartographic research; for example, recent articles by Olson and Brewer (1997) and Brewer et al. (1997) have reported colors in CIE coordinates.

5.4 METHODS FOR DISSEMINATING MAPS TO USERS

A basic problem facing mapmakers is deciding how maps should be disseminated to potential users. We use the term **dissemination** rather than reproduction because dissemination encompasses the distribution of maps in both paper and nonpaper form. Consequently, this section is broken into paper and nonpaper dissemination subsections.

5.4.1 Paper Dissemination

One obvious possibility for disseminating multiple copies of paper maps is to use one of the printers discussed in section 5.2.4. For relatively few copies (say, under 20), this approach is feasible, but it is clearly inappropriate when numerous copies are desired because each copy can take several minutes to print, the cost per copy can be high (a sophisticated dye-sublimation print can cost several dollars per copy), repeated copying will limit the lifespan of the printer, and the printer will be unavailable to others while the copying is being done. As technology changes, however, it is conceivable that this approach may become more practical, as evidenced by "some printer makers . . . including add-ons, such as collators and staplers, to their high-end printers" (Mello 1993, 95).

A second possibility for disseminating paper maps is *xerography* (or *photocopying*). In the past, xerography was generally unacceptable because fine map tones could not be maintained and color copiers were uncommon. Today the quality of black-and-white xerography has improved and color copiers are more common, but there is still some loss of quality in the duplication process. An additional problem is the expense of color copying (generally $1 or more per page). As with printers, it will be interesting to see how technological change affects the quality and price of xerography in coming years.

Although printers and xerography may eventually be used for more than just limited copies of a map, for the immediate future offset printing is still the method of choice for a large number of copies. **Offset printing** is a mechanical (as oppoed to digital) method of reproduction in which lithographic plates containing the map image are placed on drums. Areas on the plates representing the image are inked, and when paper is passed over the drums, the image is transferred to paper. This approach is commonly used in the printing and publishing industry, for example, to print this book.

Black-and-white maps can be produced using a single lithographic plate and drum. The number of plates and drums for color maps is a function of the type of color reproduction method used. The most common method used by cartographers is **four-color process printing,** in which four color separations (or plates) are used. Three of these separations are based on the subtractive primaries (cyan, magenta, and yellow), while the fourth is black. (Remember that cyan, magenta, and yellow will not produce true grays.) Different intensities of color are achieved on the separations by using screen tints composed of very small dots of varying sizes. Fortunately, most mapmakers do not need to know the details of this process since illustration programs such as Freehand automatically determine the appropriate separations once the mapmaker has selected desired colors for map features.

A less common method for offset printing is the **spot color** technique (Robinson et al. 1995, 581) in which preblended PANTONE colors are used. The advantage of this approach is that the mapmaker (and ultimately the printer) has greater control over the colors, since they are preblended. A disadvantage is that separate lithographic plates are required for each color used, thus making printing costs excessive when a large number of colors is desired. (For example, imagine printing an unclassed map in which both hue and value change within each class.)

Traditionally, cartographers created film negatives of color separations (or desired PANTONE colors) in their darkrooms and provided these to printers, who made the necessary lithographic plates. Today mapmakers who publish in color print media typically send digital files resulting from illustration programs to a **service bureau,** which then produces the necessary film negatives using an **imagesetter.** This approach has the advantage of eliminating the need for a darkroom, but it also means that the mapmaker loses some control over the map. As Waldorf (1995, 171) notes:

> When film and proofs are released to clients and/or printers, cartographers need not worry about significant changes to scale, aspect ratio, color or text. However, experience has shown us that in untrained hands, a digital map can be seriously compromised (intentionally or not).

In a similar fashion, Mattson (1990) described a number of characteristics of imagesetters that may lead to unanticipated results on the film separations. For example, he found that a specified percent area inked in the PostScript file did not lead to that same value on the resulting negative. Such problems suggest that those using service bureaus need to work very closely with them to insure a high-quality result.

5.4.2 Nonpaper Dissemination

Dissemination via Television and Videocassette

Television was the first medium where maps were disseminated widely in nonpaper form. Most common has been the use of maps (particularly animated) for the weather segment of newscasts, even small stations spend thousands of dollars for such technology. In the late 1980s, Monmonier (1989b, 216) indicated that weather stations were spending $20,000 to $50,000 on software related to weather newscasts. A conversation with a local television weather reporter suggests that these figures may be even higher today. Although commonly used in weather reporting, maps are less frequently used in other areas, such as national and international news. One reason is that the home television screen imposes a number of constraints on image viewing. Since television is a form of CRT display, some of these constraints are identical to those found in CRTs associated with personal computers. For example, the relatively low resolution of television (525 scan lines in the United States) means that small type and symbols are not possible and straight lines may appear jagged. This problem will, however, be less of an issue if **high-definition television (HDTV)** becomes widely accepted.

Other constraints are a function of the television medium itself. Particularly important is the limited time viewers have to examine maps, with the average news map appearing for less than 15 seconds (Caldwell 1979, 180). Another constraint is that there is often little time to prepare a graphic for a late-breaking news story; in today's fast-paced world people expect news about an event as soon as it happens (White and Rosenquist 1995).

An important consideration in creating color maps for television is that the signal used for broadcast transmission (NTSC in the United States) was developed to handle both color and black-and-white television. One result is that some colors (such as highly saturated reds and pale pastels) do not transmit correctly (Caldwell 1981). Another is that color maps have often had to be designed with the presumption that some people would view them in black-and-white. The latter problem is, of course, becoming less of an issue as black-and-white televisions disappear.

In recent years, an important mode of map dissemination (particularly map animations) associated with television has been the *videocassette*, which is played on videocassette recorders (VCRs). We will consider several examples of these in Chapter 14. (See Appendix E for a more complete listing.)

Computer-Based Dissemination

As indicated in Chapter 1, improvements in computer technology have drastically changed the discipline of cartography. Computers are now being used to produce the majority of maps, and the resulting maps are being disseminated via diskette and CD-ROM, and more recently via the Internet. From the standpoint of color, an important consideration in dissemination is that the characteristics of graphic displays and printers vary, and thus the mapmaker cannot anticipate exactly which colors the map reader will see. This problem may eventually be solved using color management systems (such as the PANTONE Personal Color Calibrator), but at present the nature of color produced is often a function of the particular graphic display or printer used. For example, if I design a map on my graphics monitor and send you the resulting map via the Internet, there is no guarantee that you will see the same colors I saw. Your monitor brand or settings may vary from mine, or the color board used to create colors on your monitor may differ from mine.* Unfortunately, this problem has yet to be fully addressed by cartographers.

SUMMARY

In this chapter, we have examined a number of basic principles of color that will assist in understanding the use of color to be discussed elsewhere in the text. For example, in the next chapter we will examine the use of color on choropleth and isarithmic maps, while Chapter 12 will consider the role of color in bivariate and multivariate mapping.

In considering how color is processed by the human visual system, we found that there are three types of **cones** in the **retina** of the eye—red, green, and blue—each of which are sensitive to certain wavelengths of light; for example, red cones are sensitive to long-wavelength red light. Our perception of color is not simply a function of which of these cones are stimulated, however, but to the extent that the cones excite or inhibit other nerve cells (for example, the **bipolar** and **ganglion** cells). The net result is that we perceive color according to the **opponent-process theory,** which states that three channels are involved in color perception: a lightness-darkness channel and two opponent-color channels (red-green and

* A **color board** is a piece of hardware that determines the number and range of colors displayed on a graphics display device.

blue-yellow). Perceived colors are a function of a mix of colors in the opponent channels; for example, we see orange as a mixture of red and yellow.

It is important to keep in mind that not everyone sees color in the same way. About 4 percent of the population in the United States and Europe (mostly males) have some form of **color vision impairment;** color-impaired readers cannot discriminate between certain colors (such as red and green). Fortunately, as we will see in the next chapter special color schemes have been developed for color-impaired readers. Another characteristic of color perception that we must keep in mind is **simultaneous contrast:** our perception of a color may be a function of the colors that surround it. In Chapter 16 we will look at some recent color research that has used this knowledge to create useful color schemes for choropleth and isarithmic maps.

In terms of the hardware used to produce color maps, we focused on **CRTs** and **printers.** We noted that CRTs create color using an **additive** process, while printers utilize a **subtractive** process. Important in producing color on CRTs is the **frame buffer,** an area of memory that stores a digital representation of colors appearing on the screen; in sophisticated color display systems, the frame buffer is divided into a series of *bit-planes* for each of the three **electron guns** (red, green, and blue). Although CRTs are dominant today and are likely to remain so for the next few years, **LCDs** or a similar technology may eventually become dominant. In a similar fashion, it must be recognized that common printers today (such as the **ink-jet** and **laser**) may be replaced by other technologies. Recently, **color management systems** have been developed to attempt to "match" the colors produced on graphic displays and printers.

We considered a variety of models that can be used by mapmakers to specify the colors used on maps. Some of these are hardware-oriented (**RGB** and **CMYK**), and of limited use to the mapmaker. Others are user-oriented (**HSV, Munsell,** and Tektronix's **HVC**), and thus permit color specification in terms mapmakers are apt to be familiar with, such as hue, lightness, and saturation. We also examined the **CIE** model, which, in theory, allows the mapmaker to reproduce colors specified by other mapmakers. At present, the RGB, CMYK, and HSV models are most frequently used in mapping software. This is problematic because the RGB and CMYK models do not relate easily to our common notions of hue, saturation, and lightness, and all three models do not produce equally spaced colors (in the visual sense).

Maps can be **disseminated** to potential users in both paper and nonpaper form. In the paper realm, **offset printing** is still the method of choice when a large number of copies must be made. The technologies for *xerography* and digital **printers** have improved, however, and are now being used for modest numbers of copies, and when high quality is not of paramount importance. Mapmakers using offset printing must beware of losing control over the quality of digital map files if those files are handled by a **service bureau.** In the past, nonpaper map dissemination was done principally via television (and via VCR, especially for map animations). More recently, computer-based dissemination has become increasingly common, particularly via the Internet. From the standpoint of color, an important consideration in dissemination is that the characteristics of graphic displays and printers vary; thus the mapmaker cannot anticipate exactly which colors the map reader will see. Sophisticated **color management systems** may, however, alleviate this problem.

FURTHER READING

Birren, F. (1983) *Colour.* London: Marshall Editions.

A general text on color and its varied uses.

Byte 20, no. 1, 1995 (entire issue).

Contains several articles dealing with color management.

Caldwell, P. S. (1979) Television News Maps: An Examination of Their Utilization, Content, and Design. Unpublished Ph.D. dissertation, University of California at Los Angeles, Los Angeles, California.

A dated but extensive treatment of maps in television. Also see Caldwell (1981).

Computer Graphics 31, no. 2, 1997 (entire issue).

The bulk of this issue deals with novel graphic displays such as HDTV and virtual reality (the latter is considered in chapter 16 of the present text).

Foley, J. D., van Dam, A., Feiner, S. K., and Hughes, J. F. (1996) *Computer Graphics: Principles and Practice.* 2d ed. Reading, Mass.: Addison-Wesley.

A standard reference on computer graphics. Chapter 4 covers hardware issues, while Chapter 13 covers issues related to color models.

Goldstein, E. B. (1989) *Sensation and Perception.* 3d ed. Belmont, Calif.: Wadsworth.

A standard text on the perception of color. Chapters 3 and 4, especially, are a useful supplement to section 4.1 of the present text.

Hubel, D. H. (1988) *Eye, Brain, and Vision.* New York: Scientific American Library.

A very readable treatise on how the eye-brain system functions. Based on the work of Hubel, a Nobel Prize winner.

Hunt, R. W. G. (1987a) *Measuring Colour.* Chichester, England: Ellis Horwood.

A thorough treatment of color models.

Hunt, R. W. G. (1987b) *The Reproduction of Colour in Photography, Printing & Television.* Tolworth, England: Fountain Press.

A good reference text on color vision.

Hurvich, L. M. (1981) *Color Vision.* Sunderland, Mass.: Sinauer Associates, Inc.

Covers the details of how we perceive color, with an emphasis on opponent-process theory.

Monmonier, M. S. (1989b) *Maps with the News: The Development of American Journalistic Cartography.* Chicago, Ill.: University of Chicago Press.

Chapter 5 covers the use of maps in electronic media.

Peddie, J. (1994) *High-resolution Graphics Display Systems.* New York: Windcrest/McGraw-Hill.

A comprehensive treatment of graphic display systems (for example, CRTs and LCDs).

Robinson, A. H., Morrison, J. L., Muehrcke, P. C., Kimerling, A. J., and Guptill, S. C. (1995) *Elements of Cartography.* 6th ed. New York: John Wiley & Sons.

Chapters 19 and 20 cover color-related issues. Also see pp. 292–302 of Dent (1996).

Travis, D. (1991) *Effective Color Displays: Theory and Practice.* London: Academic Press.

Chapters 1 to 3 deal with computer hardware, the human visual system, and color models, respectively.

Weiss, M. (1995) "Final output," *Byte* 20, no. 1:109–110, 112, 114–115.

Covers some of the technical aspects of printers. Also see Mello (1993), Zeis (1993), and the entire issue of *PC Magazine* 16, no 18.

Wyszecki, G., and Stiles, W. S. (1982) *Color Science: Concepts and Methods, Quantitative Data and Formulae.* 2d ed. New York: John Wiley & Sons.

A detailed reference on technical aspects of color.

6

Color Schemes for Univariate Choropleth and Isarithmic Maps

OVERVIEW

A major problem faced by mapmakers is the selection of appropriate **color schemes** *for univariate choropleth and isarithmic maps. For example, a mapmaker depicting the distribution of median family income by county for the United States may wonder whether lightnesses of a single hue (say, green) or a combined hue-lightness scheme (say, bright yellow to dark red) would be more appropriate. The purpose of this chapter is to provide an answer to this sort of question by examining a variety of factors that mapmakers should consider in selecting color schemes. The focus is on univariate choropleth and isarithmic maps because of their common use; many of the concepts covered will, however, have wider applicability. Later chapters cover color issues relevant to other topics; for example, Chapter 12 considers the role of color in bivariate choropleth mapping.*

Sections 6.1 to 6.3 consider three of the more important factors for selecting appropriate color schemes: map use tasks, kind of data, and type of map. Map use tasks *refers to whether the map is used to obtain specific or general information, and whether this information is acquired while looking at the map or recalled from memory. Mersey (1990) revealed that unordered hue-based schemes worked best for acquiring specific information, while ordered lightness-based schemes worked best for acquiring general information. For recalling general information, Mersey found that a combined hue-lightness scheme worked best; in fact, she found that the hue-lightness scheme worked best overall, contradicting the traditional cartographic recommendation to use lightnesses of a single hue.*

Kind of data *refers to the three kinds of numerical data: bipolar, balanced, and unipolar. Bipolar data have either natural or meaningful dividing points: for example, percentage of population change has a natural dividing point of zero. Balanced data are characterized by two phenomena that coexist in a complementary fashion: for example, the percentage of English and French spoken in Canadian provinces. Unipolar data have no dividing points and do not involve two complementary phenomena: for example, per capita income in African countries.*

The discussion of kind of data will be based on recommendations developed by Brewer (1994a). For unipolar data, Brewer recommends ***sequential schemes,*** *in which lightness predominates, while for bipolar data, she recommends* ***diverging schemes,*** *in which two hues diverge away from a common light hue or a neutral gray. For balanced data, she recommends either sequential schemes (to emphasize one end of the data), or alternatively, a diverging scheme (to emphasize the midpoint of the balanced data).*

Type of map *refers to the general method of symbolization used: choropleth or isarithmic. A broader set of color schemes is possible on isarithmic maps because of the smooth continuous nature of the phenomenon underlying these maps. We will consider contributions by Eyton (1990, 1994), which indicate that a* **spectral color scheme** *(colors that span the visual portion of the electromagnetic spectrum) can be used on isarithmic maps if the colors are shown in association with contours and/or hill shading. In this approach long-wavelength colors (such as red) appear closer to the viewer than short-wavelength colors (such as blue); the effect is particularly dramatic if the map is viewed with specialized viewing lenses.*

Section 6.4 considers additional factors in selecting color schemes: (1) color associations, (2) aesthetics, (3) color vision impairment, (4) age of the intended audience, (5) whether the map is intended for presentation or data exploration, (6) economic limitations, and (7) client requirements. In selected instances, these factors will be more important than those already discussed; for example, an ideal scheme from a map use standpoint may be inappropriate for someone with impaired color vision.

Section 6.5 describes various approaches that can be used to specify particular colors. For example, a mapmaker may decide that shades of blue varying in lightness are appropriate to represent unipolar data, but how should specific shades of blue be selected? Should RGB or CMYK notation be used, or is another approach appropriate? Separate systems of specification have been developed for black-and-white printed maps, colored printed maps, and maps for soft-copy display.

For ease of comparison, color schemes used in this chapter will be mapped using an education data set for the 48 contiguous U.S. states: the percentage of adults not graduating from high school in 1990. Those wishing to create similar color schemes will find CMYK values for the mapped colors in Appendix A. Since color schemes recommended by cartographers have changed in recent years, and continue to change as a result of new research findings, we will consider some additional recommendations for color use in Chapter 16.

6.1 MAP USE TASKS

Prior to about 1990, most cartographers recommended using shades of a single hue when mapping numerical data. For example, Robinson et al. (1984) stated in the widely used *Elements of Cartography:* "Several tests have shown that a progression of values of one hue is the most efficient in conveying the magnitude message of a simple graded series" (p. 186).*

A study by Mersey (1990) changed this view. When Mersey analyzed the schemes shown in Color Plate 6.1, she found that the most effective color scheme was a function of the type of map use. She intended these schemes to represent a range of ordering, from the highly unordered (A) to the highly ordered (F). The data for Mersey's maps were unipolar (the percent of black population in counties of Kentucky), as are the education data used here to illustrate her schemes.

Mersey divided map use tasks into specific and general ones, and considered both information acquisition and memory. For specific acquisition tasks, she found that the unordered hues (A) worked best; this result is not surprising given that shades for ordered schemes are difficult to discriminate and this discrimination is complicated by simultaneous contrast. For general acquisition tasks, Mersey found that ordered schemes, such as (E) and (F), performed better, thus supporting the traditional thinking on using shades of a single hue.

Since the memory tasks were largely general in nature, Mersey focused just on general tasks in her discussion of results for memory. Although lightness-based schemes outperformed hue-based schemes, the hue-lightness scheme was the best overall. According to Mersey, "Perhaps hue variations, when combined in a logical manner with regular [lightness] changes, serve to imprint in the mind of the map viewer a clearer mental construct of the distribution" (p. 125).

Overall, Mersey's results lead to the conclusion that the ideal color scheme depends on the nature of the map use task required. When the manner in which the map will be used is uncertain, however, the hue-lightness scheme is a reasonable compromise since it scored highest on four of the ten tasks she tested and was a close second or third on the remaining tasks.

One limitation of Mersey's work is that she did not provide a theoretical foundation for why the hue-lightness scheme was successful. Eastman's (1986) work with opponent process theory may provide such a theoretical foundation. Opponent process theory posits that our perception of color is based on a lightness-darkness channel and two opponent color channels: red-green and blue-yellow. The theory further supposes that we see mixtures of color from each of the opposing channels; for example, that we visualize red and yellow merging to form orange. A diagram of opponent colors and their corresponding mixtures is shown in Figure 6.1.

Eastman's major objective was to test the role that opponent process theory might play in selecting appropriate color schemes for balanced and bipolar data. His most significant result, however, was that two color schemes spanning the sides of the diagram in Figure 6.1 were successful for unipolar data: a yellow-red sequence and a green-yellow sequence, both of which involved hue and lightness changes. The similarity of the yellow-red scheme to Mersey's most effective scheme suggests that the sides of the opponent process diagram shown in Figure 6.1 might be used for color specification for unipolar data sets, and that opponent process theory may explain, in part, the success of Mersey's hue-lightness scheme.

6.2 KIND OF DATA

Brewer's recommendations for using *kind of data* as a factor in selecting color schemes are not based on experimental studies, but on "personal experience, cartographic convention and the writings and graphics of

* The "tests" referred to were experimental studies performed by Cuff (1972a, 1972b, 1973, 1974a, 1974b).

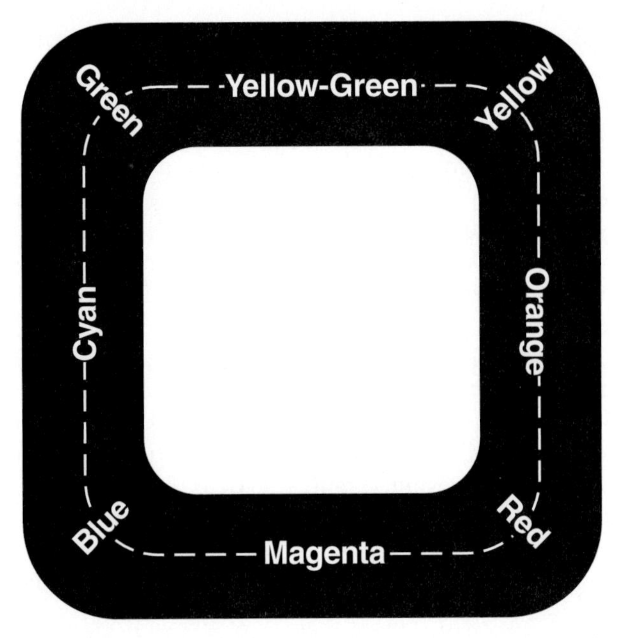

FIGURE 6.1. A simple model of how colors from the red-green and blue-yellow opponent-color channels might be combined. Colors in the middle of each side of the model should be seen as mixtures of the colors at the end points of that side; for example, a magenta should be seen as a mixture of blue and red. (After Eastman 1986, 326.)

others" (p. 124). She is, however, beginning to test some of her ideas, as we will see in Chapter 16. She has formulated recommendations not only for numerical univariate data (which is the focus of this chapter), but also for qualitative univariate data and for numerical bivariate and multivariate data. We will consider her recommendations for bivariate and multivariate data in Chapter 12.

For unipolar data, Brewer recommends that sequential steps in the data should be represented by sequential steps in lightness; she terms this a **sequential scheme.** An example of this would be the black-and-white series used in Mersey's study (Color Plate 6.1F). Concurring with earlier cartographers' recommendations (such as Robinson et al. 1984), Brewer advocates using light colors for low data values and dark colors for high data values. She argues that if the opposite is done (as is sometimes the case on graphic displays), a clear legend is essential. These recommendations are supported by studies by McGranaghan (1989, 1996).

Although Brewer argues that lightness differences should predominate for sequential schemes, she stresses that visual contrast can be enhanced if saturation differences are also used. This can be seen in Color Plate 6.2 in which a pure lightness scheme is contrasted with a combined lightness-saturation scheme. In 6.2A only lightness changes—from a light to a dark green; in 6.2B lightness again changes from light to dark green, but saturation also changes, increasing from the first to the third class and then decreasing for the latter two classes. Although the increase and decrease in saturation in this case does not correspond logically to the increase in the data, Brewer (1994a, 137) indicates this is acceptable "if high saturation colors do not overemphasize unimportant categories."

Brewer indicates that hue difference can also be used for sequential schemes, but that such differences should be subordinate to lightness differences. Paralleling Mersey's findings, she advocates a yellow-orange-red scheme, but a scheme she illustrates spans a greater variety of hues, extending from yellow to green to purple. Color Plate 6.3A illustrates the middle five of the seven classes Brewer uses. She does not recommend a greater range of hues than this, although she states that "sequential schemes may be constructed that use the entire color circle" (as in Color Plate 6.3B). In my opinion, the latter implies qualitative differences, and therefore should not be used for numerical data.

For bipolar data, Brewer (1994a, 139) recommends using a **diverging scheme,** in which two hues diverge away from a common light hue or a neutral gray. An example converging on a light hue would be a dark green-greenish yellow-yellow-orange-dark orange scheme, while an example converging on a neutral would be a dark red-light red-gray-light blue-dark blue scheme. (Color Plate 6.4 illustrates the latter.) Brewer argues that the greatest flexibility in construction can be achieved if diverging schemes are thought of as two sequential schemes running end to end, with the lowest value toward the middle.

Brewer discusses balanced data in association with bivariate color schemes because balanced data technically involve two variables. We consider it here because the legend appears only slightly more complicated than a legend for a univariate map (on a five-class map, percent English might be labeled above the legend boxes and percent French below the legend boxes). Brewer indicates that although balanced schemes can be created by overlaying two sequential schemes—she mixes magenta and cyan to create a purple—she recommends using standard sequential schemes to emphasize the high end of the data, or alternatively a diverging scheme to emphasize the midpoint of the balanced data.

6.3 TYPE OF MAP

Although not normally emphasized by cartographers, *type of map* (choropleth or isarithmic) can be an important factor in selecting an appropriate color scheme because of the different kinds of data that are mapped. In the case of isarithmic maps, the underlying data are presumed to be part of a smooth continuous surface, meaning that symbols of similar value must occur adjacent to one another. In contrast, on choropleth maps quite different values may occur next to one another. Therefore, it seems reasonable to conclude that color schemes should be easier to interpret on isarithmic maps: on the choropleth map the reader will have to consult the legend repeatedly in order to understand the scheme, while on the isarithmic map the order of colors within the map pattern match the order in the legend, and thus less frequent consultation of the legend is required.

One color scheme that particularly illustrates the role that map type may play is a **spectral scheme,** in which colors span the visual portion of the electromagnetic spectrum (this is sometimes referred to as a ROYGBIV scheme, for the first letter of each color). The logic of the spectral scheme is that colors from the red portion of the spectrum should appear slightly nearer than colors from the blue portion, primarily because the lens of the eye refracts light as a function of wavelength (see Travis 1991, 135–139, for a more detailed explanation). The resulting appearance is sometimes referred to as the **color stereoscopic effect.** Cartographers have contended that in a map context this effect is minimal, and have argued against using it for anything other than elevation data; for example, Robinson et al. (1984) described the approach as "graphically illogical with little to recommend it both because the connotative associations of some of the hues detract from the map's effectiveness and because the [lightness] changes result in marked variations in the perceptibility of other data being shown" (p. 186). Brewer also recommends against using a spectral scheme, primarily because yellow (an inherently light color) is in the middle of the spectrum.

Eyton has written two interesting papers describing ways in which spectral colors *can* be appropriately used on isarithmic maps. In the first paper, Eyton (1990) focuses on how spectral colors should be generated and how they might be combined with contours and hill shading to enhance the color stereoscopic effect. He indicates that, ideally, the subtractive CMYK process should not be used to create spectral colors because the process is not purely subtractive. He states that "subtractive ink dots . . . adjacent to each other produce desaturated colors that are formed additively" (p. 21). As one solution to this problem, Eyton created maps on a computer graphics display and then produced hard copies by recording the images onto film. As another solution, he printed maps using **fluorescent inks,** which produced "brilliant, intense color" (pp. 23–24). Maps resulting from his use of fluorescent ink can be found in Plates 1 to 4 of his paper.

In spite of taking these careful approaches, Eyton found little color stereoscopic effect when spectral colors were used alone. With respect to the fluorescent inks, he noted that the effect could be enhanced with a reading magnifier, but the desired order of the colors was distorted (green appeared above yellow and orange, instead of below them). He did find, however that when contour lines were added to the display, a distinct color stereoscopic effect was perceived by users. He noted that an explanation for the enhancement caused by the contours cannot be found in the literature, but hypothesized that the rapid change in slope display associated with closer contour spacing might provide depth cues.

A particularly dramatic color stereoscopic effect was achieved when Eyton combined spectral colors with a hill-shaded display. **Hill shading** is a process in which terrain is shaded as a function of its orientation and slope relative to a presumed light source, typically from the northwest. Eyton indicated that many viewers found a combination of spectral colors, contours, and hill shading to be the most effective for enhancing the color stereoscopic effect.

In his second paper, Eyton (1994) described how the color stereoscopic effect could be enhanced by viewing spectral colors with special viewing glasses. Originally developed by Steenblik (1987), these glasses are now available as holographic films fitted to cardboard (known as ChromaDepth lenses), and are marketed as Valiant Vision (from Valiant comics) and Jumping Colours (from Crayola). These glasses can produce dramatic three-dimensional images when spectral colors are combined with contours or hill shading. I encourage the reader to acquire such a set of glasses and examine Color Plate 6.5, which was created using a continuous spectral scheme developed by Eyton (1990, 24–26).

Out of curiosity, I also used the specialized viewing glasses to examine choropleth maps based on a spectral color sequence recommended by Eyton for isarithmic maps. However, I found no distinct ordering of the colors in three-dimensional space. It appears the color stereoscopic effect can be achieved only on *isarithmic maps* in which contours or hill shading are used.

6.4 OTHER FACTORS IN SELECTING COLOR SCHEMES

The preceding sections focused on three of the more important factors that should be considered in selecting appropriate color schemes. This section considers seven additional factors: color associations, aesthetics, color vision impairment, age of the intended audience, whether the map is intended for presentation or data exploration, economic limitations, and client requirements. In some situations, these factors will be more important than those already discussed; for example, an ideal scheme from a map use standpoint may be completely inappropriate for someone with impaired color vision.

6.4.1 Color Associations

Mapmakers may wish to take advantage of people's associations of color and mapped phenomena (for example, green for money in the United States), but they should be aware that these associations are cognitive in nature, and thus may change over time and be inconsistent among map users. A good illustration of the potential for change over time is the use of blue and red for cool and warm temperatures. In an early study, Cuff (1973) concluded these colors were not effective, but a more recent study by Bemis and Bates (1989) found the opposite. Bemis and Bates suggested that the different results may be a function of the more common use of this scheme since the time of Cuff's study and thus users' greater familiarity with it.

6.4.2 Aesthetics

The aesthetics of a color scheme are an important consideration, regardless of how effective that scheme might be for certain map use tasks. As an example, Stephen Egbert and I (Slocum and Egbert 1993) compared traditional static maps (the map is viewed all at once) with sequenced maps (the map is built while the reader views it). Our study was split into two major parts: a formal one in which subjects viewed one of the two types of maps (static or sequenced) and performed various map use tasks, and an informal one in which subjects viewed both types of maps and were asked to comment on them. We used a hue-lightness scheme similar to Mersey's (yellow-orange-red) in the formal portion, and a lightness-based blue scheme in the informal portion. Although the major purpose of the informal portion was to have subjects contrast the method of presentation (static or sequenced), they often commented on the color schemes, indicating their preference for the blue one over the supposedly more appropriate yellow-orange-red one.

Interestingly, studies of color preference outside the discipline of cartography have also found the color blue appealing. For example, using subjects' rating of Munsell color chips, Guilford and Smith (1959) found that colors were preferred in the order blue, green, purple, red, and yellow. More recently, McManus et al. (1981) found that people's preferences were more variable than Guilford and Smith implied, but that blue and yellow were still the most and least preferred, respectively. One limitation of such studies is that they tend to focus on a particular culture group. For example, McManus et al. studied only "undergraduate members of the University of Cambridge" (p. 653). One wonders what the results might have been if the study were done in other areas of the world. Since color preference varies among individuals, and there are likely to be differences among culture groups, the key point to remember is that color schemes *you* find attractive may not be attractive to others.

6.4.3 Color Vision Impairment

The various forms of color vision impairment can be represented on CIE diagrams by a series of confusion lines. For example, Figure 6.2 illustrates confusion lines for protanopes and deuteranopes (two subgroups of dichromats, people who cannot distinguish red from green). Colors running along the confusion lines will be confused by these groups, while colors running roughly perpendicular to the lines should be distinguishable.

Using CIE diagrams like the one shown in Figure 6.2, Olson and Brewer (1997) developed sets of confusable and adjusted colors for various color schemes. As an example, Figure 6.3 portrays their results for a diverging color scheme representing gains or losses in the number of manufacturing jobs. For the confusable rendition, gains are represented by a yellow-green to dark green scheme, while losses are represented by a yellow-orange to red scheme. Note that these colors all fall along the same set of confusion lines in the CIE diagram. In contrast, the adjusted scheme ranges from light blue to dark blue and

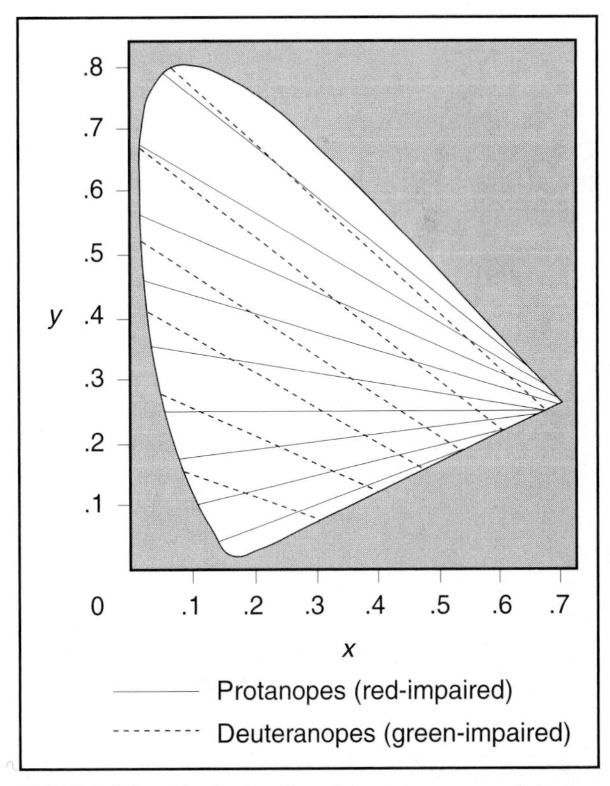

FIGURE 6.2. Confusion lines for protanopes and deuteranopes drawn on a *Yxy* 1931 CIE diagram. Colors of similar lightness will be difficult to discriminate if they are placed along the same confusion line. (After Olson and Brewer, 1997, p. 108.)

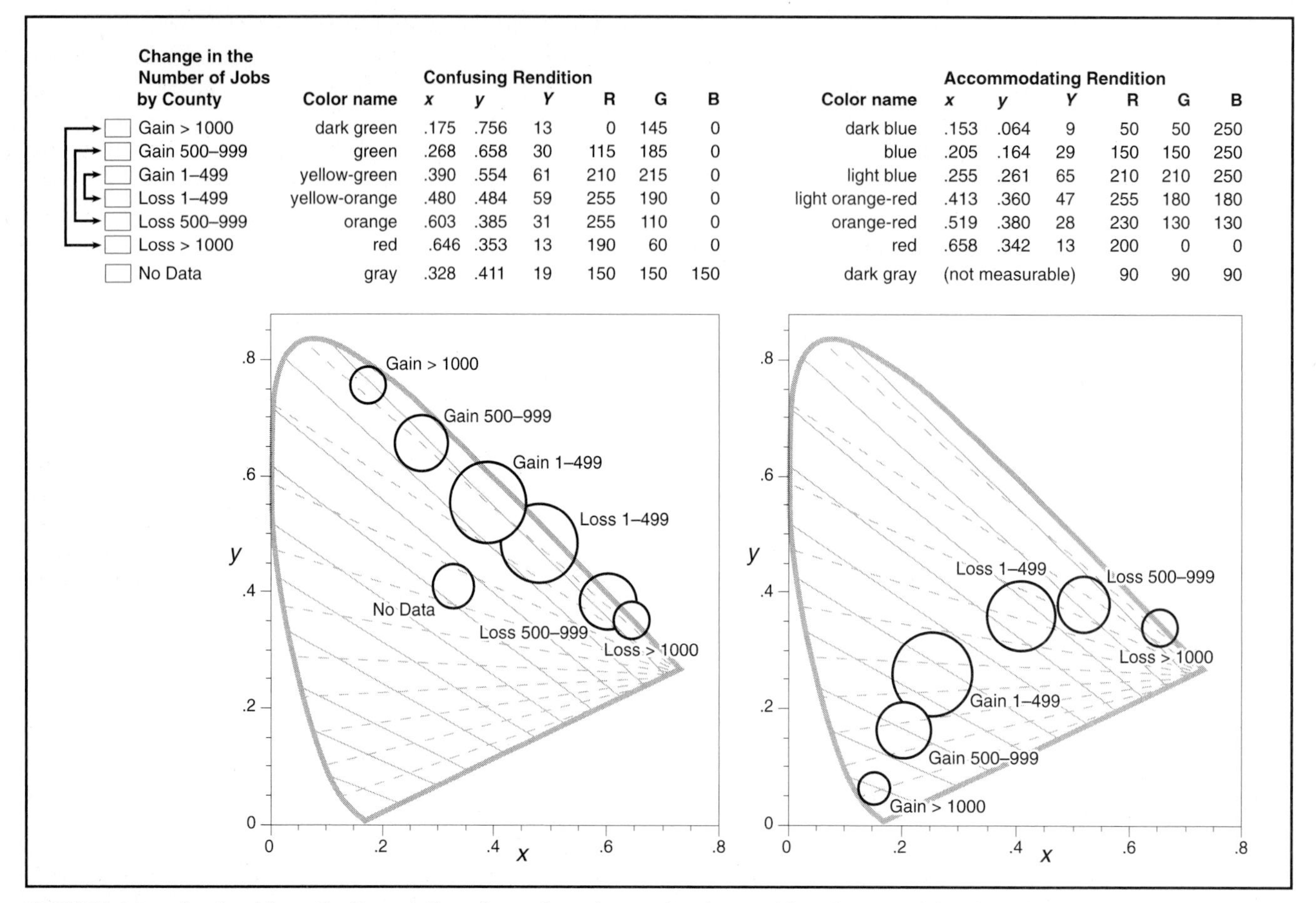

Change in the Number of Jobs by County	Confusing Rendition: Color name	x	y	Y	R	G	B	Accommodating Rendition: Color name	x	y	Y	R	G	B
Gain > 1000	dark green	.175	.756	13	0	145	0	dark blue	.153	.064	9	50	50	250
Gain 500–999	green	.268	.658	30	115	185	0	blue	.205	.164	29	150	150	250
Gain 1–499	yellow-green	.390	.554	61	210	215	0	light blue	.255	.261	65	210	210	250
Loss 1–499	yellow-orange	.480	.484	59	255	190	0	light orange-red	.413	.360	47	255	180	180
Loss 500–999	orange	.603	.385	31	255	110	0	orange-red	.519	.380	28	230	130	130
Loss > 1000	red	.646	.353	13	190	60	0	red	.658	.342	13	200	0	0
No Data	gray	.328	.411	19	150	150	150	dark gray	(not measurable)			90	90	90

FIGURE 6.3. Confusable and adjusted diverging color schemes for those with red-green vision impairment. Note that colors for the confusable scheme are arranged along similar confusion lines, while colors for the adjusted scheme cross the confusion lines. (From Olson and Brewer, 1997. First published in *Annals, Association of American Geographers,* 87(1), p. 131. Reprinted with permission of Association of American Geographers.)

light orange-red to red, and is oriented perpendicular to the confusion lines. Not surprisingly, Olson and Brewer found that adjusted schemes were much more easily interpreted by color-impaired readers.

Although it is difficult to justify separate maps for those with normal and impaired vision within a published book or atlas, modern interactive graphics systems make it an almost trivial task to produce separate maps on-screen or on color printers. To assist those wishing to create schemes for the color vision–impaired, Olson and Brewer provide *Yxy* and RGB values for the various color schemes they tested (an example is shown in Figure 6.3). Care should be taken in using their RGB values directly because identical RGB values will not produce the same colors on all monitors.

6.4.4 Age of the Intended Audience

The age (and presumably experience) of the intended audience is an important consideration in selecting colors. For example, young children would probably not be familiar with the blue-to-red scheme commonly used for representing temperature data. Research by Trifonoff (1994, 1995) suggests that young children have a strong preference for color over black-and-white maps: in comparing four methods of symbolization (gray tones, proportional circles, proportional circles coded redundantly with gray tones, and shades of red), children overwhelmingly chose the colored symbolization (shades of red) as most desirable. Although this finding might be true for only young children, a recent study by Brewer et al. (1997) suggests that adults may also have a strong preference for color maps, as a black-and-white choropleth scheme was decidedly less preferred than various colored schemes.

6.4.5 Presentation versus Data Exploration

To this point in the chapter, we assumed that color schemes were selected for a map to be *presented* to others (for *communication* purposes). If, instead, the map is used for data exploration, then a wider range of color schemes is appropriate. Those exploring data might use "recommended" color schemes to obtain an initial overview of the distribution, but employ a "nonrecommended" color scheme to examine detailed aspects of that distribution. For example, Berton (1990, 112) argues:

> The continuous blending of color is not conducive to highlighting artifacts in the interior of the data set. While it is usually critical that a palette not "double-back" on itself... certain kinds of banding effects which violate the traditional order of the color wheel can provide excellent markers for transitional areas in the data.

6.4.6 Economic Limitations and Client Requirements

In an ideal world, there would be no economic limitations and the mapmaker would have sole responsibility for selecting colors. In the real world, of course, this is frequently not the case. One obvious economic limitation is the expense of color reproduction in book or journal form: although color may communicate information more effectively than black-and-white, it may not be feasible from an economic standpoint. As an example, academic journals generally require authors to pay for color reproduction, which can cost $500 or more per page.

Another economic limitation is the cost of sophisticated hardware and software, and the time needed to learn how to use it. With respect to hardware, I encountered a very practical problem when producing colored maps for this book. When trying to create maps using Brewer's (1989) method described in section 6.5.2, the colors produced by an inexpensive ink-jet printer were quite different from those appearing in her published work. But when the maps were printed on a more expensive dye-sublimation printer, color differences were negligible.

The mapmaker often does not have sole responsibility for the design, but rather must respond to client requirements or desires. Although the mapmaker may suggest an ideal, or "optimal," set of colors, clients may reject these because they find other colors more pleasing or have traditionally used other colors. For example, the director of a cartographic production laboratory told me that "the school district made us make a boundary map for attendance areas in hot pink and blue to 'match their old map.' " Although mapmakers may try to dissuade clients from choosing such schemes, they must bear in mind that clients are keeping the laboratory in business.

6.5 DETAILS OF COLOR SPECIFICATION

In the previous sections, we considered how hue, lightness, and saturation should be used in general to select color schemes, but we did not consider the details of color specification. For example, if a mapmaker selects a light yellow to dark red scheme, how should colors be specified in RGB or CMYK notation? This section describes a variety of methods for detailed color specification; for ease of discussion, it is convenient to split the methods into those for black-and-white printed maps, colored printed maps, and maps for soft-copy display.

6.5.1 Black-and-White Printed Maps

Because of the high cost traditionally associated with producing color maps, the bulk of research dealing with the selection of appropriate areal symbols for choropleth and isarithmic maps was done with black-and-white maps (see Kimerling 1985 for a comparison of various studies). Such research has resulted in **gray scales** or **gray curves,** which express the relation between printed area inked and perceived blackness. Separate gray curves have been developed for smooth and coarse (or "textured") shades, respectively.

Smooth Shades

For smooth shades, or those in which the pattern of marks making up the shade is not apparent, the **Munsell curve** is the most widely accepted gray scale.* In examining this curve (Figure 6.4), note that at the light end, a small difference in percent area inked leads to a large difference in perceived blackness, while at the dark end, a large difference in percent area inked is required to achieve a small difference in perceived blackness. Using the Munsell scale is a four-step process.

Step 1. Pick your smallest and largest desired perceived blackness values. Let's assume that you select values of 12 and 100 as initial choices. (A value of 12 equates to 6 percent area inked and so will produce a gray tone that can be seen as a figure against the white background typically used on printed maps. A value of 100 will produced a solid black.)

Step 2. Determine the contrast between each pair of perceived values, assuming the perceived values are equally spaced from one another. This is accomplished by dividing the range of perceived values by the number of classes minus 1. Assuming five classes, and the above perceived values, the result is:

$$\frac{PB_{max} - PB_{min}}{NC - 1} = \frac{100 - 12}{4} = \frac{88}{4} = 22$$

where PB_{max} and PB_{min} are the perceived maximum and minimum blackness, and NC is the number of classes.

* Research by Castner and Robinson (1969) indicated that shades having 75 or more marks per inch would be perceied as gray values without any pattern.

Step 3. Interpolate the intermediate perceived values. This involves simply adding the contrast value derived in step 2 to each perceived value, beginning with the lowest. For perceived values of 12 and 100, and a contrast value of 22, we have: 12 + 22 = 34; 34 + 22 = 56; 56 + 22 = 78; and 78 + 22 = 100.

Step 4. Determine the percent area inked values corresponding to the perceived values. Using Figure 6.4, this can be accomplished by drawing a horizontal line from the perceived blackness value to the Munsell curve, and from this point drawing a vertical line to the percent area inked axis. For perceived values of 12, 34, 56, 78, and 100, you should find percent area inked values of approximately 6, 20, 41, 67, and 100.

Although the Munsell curve is the most widely accepted scale for smooth shades, the **Stevens curve** (Figure 6.4) has been argued more appropriate for unclassed maps (Kimerling 1985, 141; MacEachren 1994a, 105). Note that the Stevens curve is displaced above the Munsell curve; thus, at the light end a small difference in percent area inked leads to a very large difference in perceived blackness, while at the dark end, a very large difference in percent area inked is required to achieve a small difference in perceived blackness.

To understand why the Munsell and Stevens curves might be appropriate for classed and unclassed maps, respectively, it is necessary to consider the methods by which these curves were constructed. The Munsell curve was constructed using **partitioning,** in which a user places a set of areal shades between white and black such that the resulting shades will be visually equally spaced. For example, in a study that essentially duplicated Munsell's curve, Kimerling (1975) had subjects place seven gray tones (chosen from a set of 37) between white and black. It can be argued that the tones resulting from such an approach will have maximum contrast with one another and thus be appropriate for a classed choropleth map.

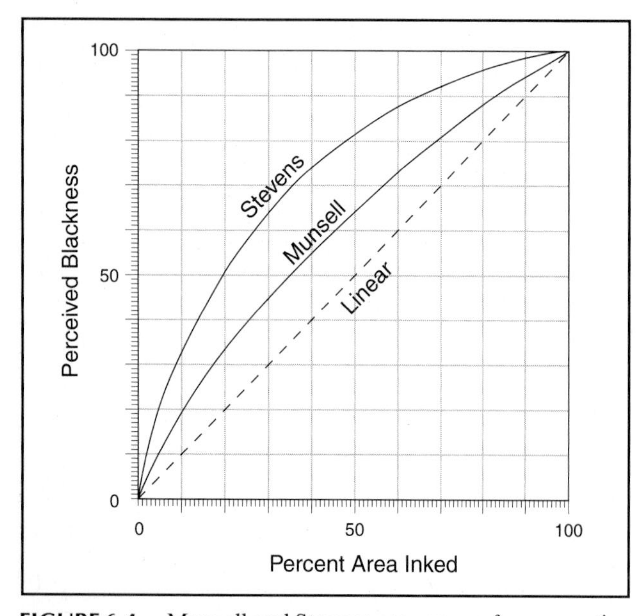

FIGURE 6.4. Munsell and Stevens gray curves for converting desired perceived blackness to percent area inked. The curves are intended for fine-toned black-and-white areal symbols. (After Kimerling 1985, 137.)

The Stevens curve was created using **magnitude estimation,** in which a user estimates the lightness or darkness of one shade relative to another. Stevens and Galanter (1957) described the approach as follows:

> Typically a particular gray was shown to [the observer] and he was told to call it by some number. It was then removed and the stimuli were presented twice each in irregular order. [The observer] was told to assign numbers proportional to the apparent lightness. (p. 398)

Some cartographers argue that this process is similar to how readers compare individual areas on an unclassed map. Readers do not, of course, use a number system, but they might attempt to conceive of ratios of gray tones (that a given gray tone is say five times as dark as another). It seems more likely, however, that readers consider only ordinal relations (one tone is considered darker or lighter than another or "much" darker or lighter than another). If this is the case, the Stevens curve would seem to be inappropriate for maps.

One of the problems with applying either the Munsell or Stevens curve to unclassed maps is that the four-step process described above is tedious with a large number of classes. To solve this problem, I digitized the curves and used the technique of curvilinear regression to fit the digitized data (Davis 1986, 186–195).* The results for Munsell and Stevens respectively were:

$$PA = .287029 + .329157PB + .008753PB^2 - 6.24654 \times 10^{-7}PB^4 + 4.16010 \times 10^{-11}PB^6$$

$$PA = .465370 + .013841PB^2 - 1.26098 \times 10^{-4}PB^3 + 8.38227 \times 10^{-11}PB^6$$

where PB represents a desired perceived blackness and PA represents the percent area ink necessary to achieve that level of blackness. The equations produced an excellent fit in the case of Munsell (the maximum error between predicted and digitized values for percent area inked was 0.3 percent) and a reasonably good fit in the case of Stevens (the maximum error was 3.4 percent).

* In a personal communication, Kimerling indicated this approach was appropriate since he fit the curves shown in Figure 6.4 to sample data points by eye (the process is described in Kimerling 1985). I also experimented with a power function equation, but the results were less successful.

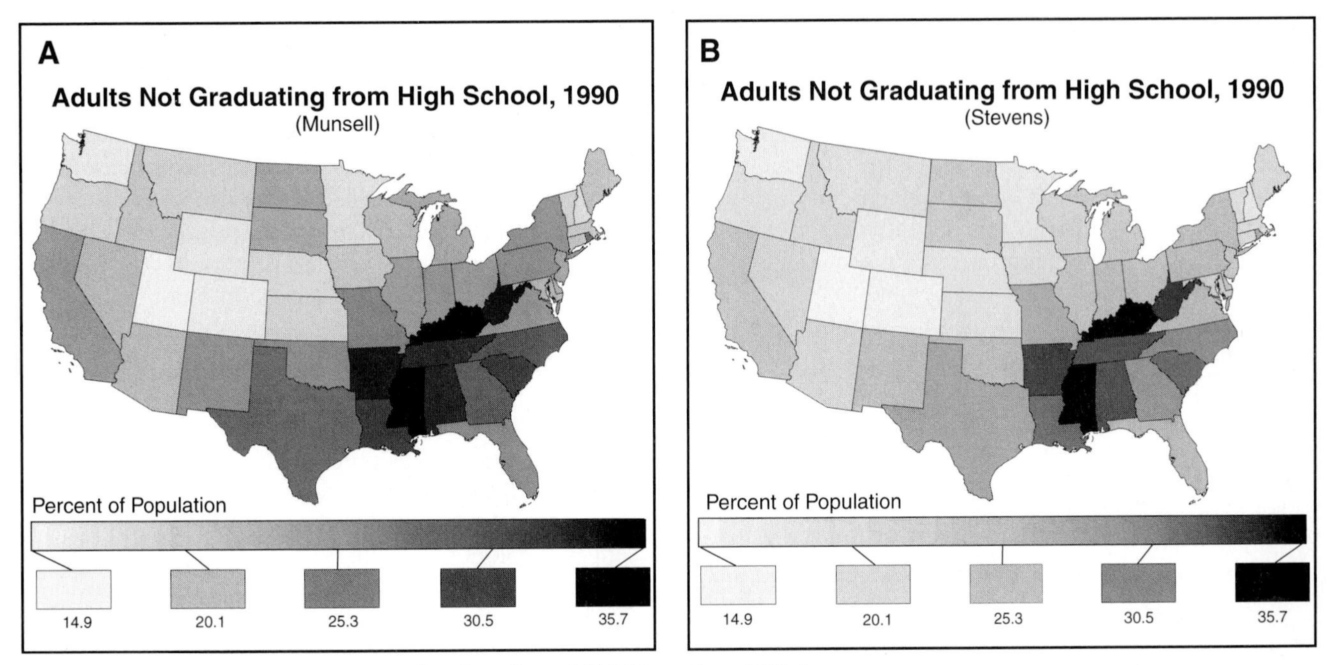

FIGURE 6.5. Unclassed maps illustrating the effect of (A) Munsell and (B) Stevens gray curves.

Although the above equations may appear complex, they can be incorporated into a spreadsheet in the following three-step procedure:

Step 1. Determine the proportion of the data range represented by each data value. This is accomplished by subtracting the minimum of the data from each value, and dividing by the range of the data.

$$Z_i = \frac{X_i - X_{\min}}{X_{\max} - X_{\min}}$$

where X_i is a raw data value, $X_{\min}$ is the minimum of the raw data, $X_{\max}$ is the maximum of the raw data, and Z_i is the proportion of the data range. The resulting Z values will range from 0 to 1.

Step 2. Convert the values calculated in step 1 to perceived blackness. This involves computing

$$PB_i = Z_i(PB_{\max} - PB_{\min}) + PB_{\min}$$

where Z_i, $PB_{\min}$, and $PB_{\max}$ are defined as before, and PB_i is the perceived blackness for an individual data value. (To achieve the 6 to 100 percent area inked used in the classed example, you would use values of 13 and 100 [for Munsell], and 22 and 100.8 [for Stevens].)

Step 3. Insert the perceived blackness values into whichever equation (Munsell or Stevens) you wish to use.

Figure 6.5 provides a comparison of the effect of the Munsell and Stevens curves for the unclassed education data set. Note that when the entire distribution is considered, the Munsell curve produces a slightly darker map. On a more detailed level, note that the Munsell curve enables greater discrimination for low values, while the Stevens curve permits greater discrimination for high values.

Coarse Shades

For coarse shades, or those in which the pattern of marks making up the shade is apparent, the **Williams curve** (Figure 6.6) is usually recommended (see, for example,

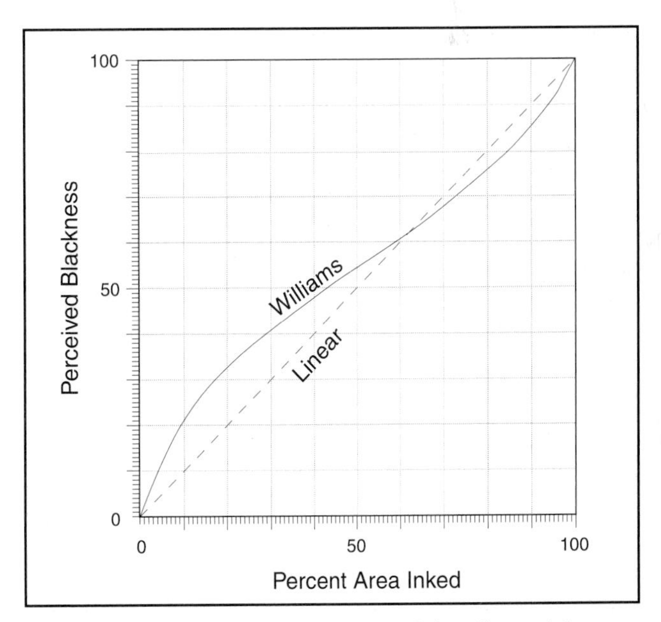

FIGURE 6.6. Williams gray curve traditionally used for converting desired perceived blackness to percent area inked. The curve is intended for coarse black-and-white areal symbols. (After Kimerling 1985, 139.)

Robinson et al. 1995, 394–396). Note that this curve is similar to the Munsell and Stevens curves near the origin, but that it crosses over a hypothetical 45-degree line connecting the lower left and upper right corners of the diagram (compare Figures 6.4 and 6.6). To calculate percent-area-inked values for the Williams curve, a four-step procedure similar to that already described for classed maps would be appropriate. In a manual environment, the Williams curve was applied to coarse **Zip-a-Tone** shades. Although similar shades can be created using the computer, the trend seems to be toward smoother tones, probably because they are considered more aesthetically pleasing.

It is important to realize that the Williams curve was intended for symbols having a relatively constant number of marks per inch. As a result, it should not be used for a symbol in which the number of marks per inch varies considerably, such as Tobler's (1973) crossed-line shading. Subjects tend to view crossed-line and gray-tone shades as having a similar percent area inked as equivalent, thus making the Munsell or Stevens scales appropriate for crossed-line shading (Slocum and McMaster 1986).

More recently, Leonard and Buttenfield (1989) have developed a gray scale for 300 dpi laser printers. I have chosen not to illustrate their curve because it was developed using percent reflectance and was not converted to a percent ink equivalent.* Their paper does, however, offer useful suggestions for generating maximum contrast maps using laser printers.

Practical Application of Gray Scales

Regardless of which gray scale is used, the mapmaker must be aware that the results in a real-world production environment are likely to differ from those suggested by the curves; that is, if the mapmaker specifies a particular percent area ink using software, there is no guarantee that the hardware will actually produce the same percent area ink. (This should be obvious to anyone who has used laser printers for word processing: once a toner cartridge has been used heavily, words will not appear as a solid black.) The only real solution to this problem is to analyze available hardware carefully to insure that percent-area-inked specifications match software specifications, and that black marks are actually produced.

Even if one has taken a great deal of care in producing maps within one's own shop, consideration must also be given to what will happen if the map is reproduced (for example, using a fax machine, a photocopy machine, or offset lithography). Most of us are familiar with how degraded an image becomes as a result of using fax or copy machines. In the case of offset lithography, the basic problem is that individual marks making up an areal symbol may be larger on the printed map (than suggested by numerical specifications in a digital file) because of "overinking or too much pressure between the paper and blanket cylinder" (Monmonier 1980, 25). Although this problem might be handled by applying a correction factor similar to that described above for gray scales, it is more realistic to find a high-quality printer, have a good working relationship with that printer, and thus try to avoid (or at least limit) this sort of problem.

6.5.2 Colored Printed Maps

Colors on printed maps are normally specified in the CMYK system. This is problematic because the CMYK system is not perceptually based; for example, a 40 percent cyan will not appear midway between 30 and 50 percent cyan colors. One solution to this problem is to use an equivalency that Brewer (1989) developed between CMYK and the perceptually based Munsell system.

To illustrate Brewer's approach, consider Color Plate 6.6, which portrays the equivalency between CMYK and Munsell hue 5R. Munsell value (lightness) and chroma (saturation) specifications are given in the upper right of the plate; equivalent CMYK specifications are given in the upper left. A particular CMYK specification will produce a color close to a corresponding Munsell specification. I say "close" because the CMYK colors will vary as a function of the method and quality of printing. Presumably, the colors will be very close if one uses a high-quality offset printer.

At the bottom of Color Plate 6.6, you will note a series of color progressions that might be appropriate for numerical data: A represents a lightness progression; B, a saturation progression; C, a combined lightness-saturation progression in which lightness decreases and saturation increases; and D a combined lightness-saturation progression in which lightness decreases, while saturation increases and then decreases. Progressions similar to A and D were used for the maps A and B respectively in Color Plate 6.2, except hue 5G was used (as opposed to 5R).

In total, Brewer provides CMYK-Munsell equivalencies for 10 major Munsell hues (those with a 5 preceding the letter designating the color). Since these hues are equally spaced from one another visually, progressions of them also can be created. For example, the colors 5Y 7/10, 5YR 7/10, and 5R 7/10, should appear as equally spaced yellow, orange, and red (Remember that the numbers after the letter represent Munsell value and chroma.) Hue changes can also be combined with value and saturation changes; for example, it should be possible to create a light desaturated yellow to dark saturated red progression by specifying 5Y 8/8, 5YR 6/10, and 5R 4/12.

One problem with Brewer's method is that if one wishes to use a narrow range of hues and many classes, some interpolation of CMYK values will be necessary.

* See MacEachren (1994a, 103–105) for a discussion and graph of the Leonard-Buttenfield curve.

This is problematic because there is no guarantee that the interpolated values will also be equally spaced in the visual sense. My experience, however, is that a simple linear interpolation between values often yields reasonable results. For example, to create the five-class "hue" map shown in Color Plate 2.4B, I first specified Munsell colors for the lowest, middle, and highest classes as 5Y 7/10, 5YR 7/10, and 5R 7/10 and determined the CMYK equivalents for these colors from Brewer's work. I then interpolated the remaining two classes using CMYK notation.

Another problem with Brewer's method is that within a hue the number of lightnesses for a particular saturation (or alternatively the number of saturations for a particular lightness) is limited. For example, a five-class map of saturation 12 is not possible for the 5R hue (see Color Plate 6.6). Brewer indicated that such restrictions were due to her desire to simulate existing Munsell charts. Although she indicated that her charts might be extended through a broader range of CMYK colors, she argued that "perceptual order would be compromised" (Brewer, 276).

6.5.3 Soft-Copy Maps

The preceding two sections considered approaches for specifying color on hard-copy (or printed) maps. This section deals with potential approaches for specifying color on soft-copy (or graphic-display) devices. The approaches include (1) using Brewer's method introduced in the preceding section to create CMYK colors, and then converting these colors to their RGB equivalents; (2) color ramping (which involves simple interpolation between RGB values for two colors); and (3) using the CIE color model. Although the latter approach will, in theory, produce the most effective color sequences, the second approach appears to be most commonly used in commercially available map design software.

Converting between CMYK and RGB

Since most software packages contain automatic methods for converting from CMYK to RGB values, one potential approach for specifying color on soft-copy maps is to enter Brewer's CMYK specifications and have the software convert these to their presumed RGB equivalents. The problem with this approach is that the different procedures for creating hard-copy and soft-copy maps (subtractive versus additive color processes) do not produce identical colors (as discussed in section 5.2.5). In preparing illustrations for this book, this was readily apparent when comparing Brewer's on-screen colors with their supposed printed equivalent: the on-screen colors did produce reasonably logical progressions, but they were distinctly different from the colors appearing in Brewer's printed work.

Color Ramping

A simple and common approach for soft-copy color specification is **color ramping.** In this approach, users select endpoints for a desired scheme from a color palette, and the computer automatically interpolates values between the endpoints on the basis of RGB values. As an example, imagine that a user selects end points of white (RBG values of 255, 255, 255) and black (RBG values of 0, 0, 0). Simple interpolation between these end points yields RGB values of: 204, 204, 204; 153, 153, 153; 102, 102, 102; and 51, 51, 51. The problem with this approach is that a simple arithmetic increase in a color gun does not correspond to an arithmetic increase in perceived lightness, and thus the resulting shades likely will not appear to be equally visually spaced. Although interpolation algorithms might be modified to account for this discrepancy, they generally are not. As a result, color ramping must be used with caution.

Using CIE

A more sophisticated approach for specifying colors is to use the CIE color model. This can be accomplished in two ways. One is to use equivalencies that have been established between the perceptually uniform Munsell color space and the 1931 *Yxy* CIE colors. For example, a Munsell 2.5R 9/6 has been determined to be equivalent to the following *Yxy* CIE coordinates: .7866, .3665, and .3183. (Travis 1991, 214–270, provides a complete set of equivalent colors.) The resulting CIE colors can be converted to RGB values using equations provided by Travis (pp. 93–97).

A second approach is to specify end points for a color scheme in RGB form, convert these to the 1976 CIE L*u*v* uniform color space (in which colors are equally spaced from one another in the visual sense), interpolate colors in the uniform color space, and then convert the interpolated colors back to RGB. Equations for this process are provided in Appendix B.

One limitation of both of these approaches is that they require you to establish the gamma function for each color gun on your monitor. **Gamma function** refers to the relation between the voltage of a color gun and the associated luminance of the CRT display. Equations for converting between CIE and RGB assume this relation is linear, but generally it is not, having the form $L = D^{\gamma}$, where D is the digital value sent to a color gun, L is the luminance measured on the display, and γ is an exponent describing the nonlinearity.* (The term *gamma function* comes from the use of γ as the exponent.) Travis (1991, 154–160) describes how the gamma function can be lin-

* L and D are standardized so that they both range from 0 to 1; thus no constant is necessary to convert between the different units of measurement for luminance and digital color gun values.

earized; the basic result is a lookup table equating standard RGB values (those that would normally be used in the absence of any correction) with gamma-corrected RGB values.

Fortunately, these sorts of approaches are now being programmed into color management systems. In section 5.2.5 the emphasis was on how a color created on one device (say, a graphics display) could be duplicated on another (say, an ink-jet printer). From the discussion presented here, it is clear that color management systems should ideally also be capable of specifying colors that are, perceptually, uniformly spaced between two arbitrary end points.

SUMMARY

In this chapter we have learned about several factors that can be used to select color schemes for choropleth and isarithmic maps. One important factor is the *map use task:* whether the map is used to obtain specific or general information, and whether this information is acquired while looking at the map or recalled from memory. We found that an unordered hue-based scheme (for example, Color Plate 6.1A) worked best for acquiring specific information, while a lightness-based scheme (such as light red to dark red) worked best for acquiring general information. For recalling general information, a hue-lightness scheme (such as bright yellow to dark red) is recommended. The hue-lightness scheme should also be used when the map is to be used for a variety of purposes (for both specific and general information).

A second factor for selecting color schemes is the *kind of data:* unipolar (there is no dividing point in the data), bipolar (there is a dividing point), or balanced (two phenomena coexist in a complementary fashion). For unipolar data, **sequential schemes** are recommended, in which lightness predominates (the light red to dark red and the yellow to dark red schemes mentioned above). For bipolar data, **diverging schemes** are recommended, in which two hues diverge away from a common light hue or a neutral gray (for example, dark blue to gray to dark red). For balanced data, either sequential or diverging schemes can be used; the former emphasize one end of the data, while the latter emphasize the midpoint of the data.

A third factor useful in selecting color schemes is the *type of map:* choropleth or isarithmic. A broader set of color schemes is possible on isarithmic maps because of the smooth continuous nature of the phenomenon underlying these maps. A **spectral color scheme** (colors that span the visual portion of the electromagnetic spectrum) is useful on isarithmic maps, especially when the scheme is shown in association with contours or hill shading. In the spectral scheme, long-wavelength colors (such as red) appear closer to the viewer than short-wavelength colors (such as blue); the effect is most dramatic if the map is viewed with specialized viewing lenses.

Other factors that should be considered in selecting color schemes include: (1) color associations, (2) aesthetics, (3) color vision impairment, (4) age of the intended audience, (5) whether the map is intended for presentation or data exploration, (6) economic limitations, and (7) client requirements. These seven factors, together with the three discussed above, provide a useful checklist for a mapmaker designing a color scheme. In selecting color schemes, it is also important to recognize that cartographers' recommendations have changed over recent years, and continue to change as a result of new research findings; we will consider some of these recent research findings in Chapter 16.

In this chapter, we have also considered several methods for specifying *particular* colors in a scheme. We divided the approaches into those for black-and-white printed maps, colored printed maps, and maps for soft-copy display. For black-and-white printed maps, we saw how **gray scales** could be used to select a set of gray tones that would appear to be equally spaced from one another. For colored printed maps, shades are normally specified in CMYK notation, which unfortunately does not provide colors that appear equally spaced from one another. We saw that this problem could be handled by using an equivalency that Brewer has established between the CMYK and the Munsell color models.

The most common approach for specifying colors on soft-copy displays is simple interpolation in RGB notation. This approach is deficient, however, in that the resulting colors may not appear to be equally spaced. A better technique is to use the CIE color model, either through an equivalency that has been developed between CIE and Munsell or by employing the 1976 CIE uniform color space. Although equations for working with CIE are rather complex, **color management systems** are now beginning to include this capability.

FURTHER READING

Bergman, L. D., Rogowitz, B. E., and Treinish, L. A. (1995) "A rule-based tool for assisting colormap selection." Proceedings, *Visualization '95,* October 29–November 3, Atlanta Georgia, pp. 118–125. (See also http://www.almaden.ibm.com/dx/vis96/proceedings/PRAVDA/index.htm.)

Describes the development of logical color schemes within the IBM Visualization Data Explorer (DX).

Brewer, C. A. (1989) "The development of process-printed Munsell charts for selecting map colors." *American Cartographer* 16, no. 4:269–278.

Describes how an equivalency was established between the Munsell and CMYK color models. Also provides Munsell and CMYK equivalencies for a large range of colors.

Brewer, C. A. (1994a) "Color use guidelines for mapping and visualization." In *Visualization in Modern Cartography,* ed. A. M. MacEachren and D. R. F. Taylor, pp. 123–147. Oxford: Pergamon Press.

Discusses color schemes appropriate for various kinds of data (unipolar, bipolar, and balanced).

Cuff, D. J. (1973) "Colour on temperature maps." *Cartographic Journal* 10, no. 1:17–21.

An example of an early study that supported using shades of a single hue when mapping numerical data.

Eastman, J. R. (1986) "Opponent process theory and syntax for qualitative relationships in quantitative series." *American Cartographer* 13, no. 4:324–333.

Introduces the notion that opponent process theory may be useful in selecting color schemes and presents an experiment designed to evaluate the theory.

Eyton, J. R. (1990) "Color stereoscopic effect cartography." *Cartographica* 27, no. 1:20–29.

Describes methods for creating spectral colors and illustrates how these colors can be combined with contours and hill shading to enhance the color stereoscopic effect.

Heyn, B. N. (1984) "An evaluation of map color schemes for use on CRT's." Unpublished M.S. thesis, University of South Carolina, Columbia, South Carolina.

A study similar to Mersey's (1990) in that it supported the role that map use tasks should play in selecting color schemes.

Kimerling, A. J. (1975) "A cartographic study of equal value gray scales for use with screened gray areas." *American Cartographer* 2, no. 2:119–127.

An example of an experimental study for developing a gray scale.

Kimerling, A. J. (1985) "The comparison of equal-value gray scales." *American Cartographer* 12, no. 2:132–142.

Compares various gray scales that have been developed.

Leonard, J. J., and Buttenfield, B. P. (1989) "An equal-value gray scale for laser printer mapping." *American Cartographer* 16, no. 2:97–107.

A study on gray scales appropriate for laser printers.

McManus, I. C., Jones, A. L., and Cottrell, J. (1981) "The aesthetics of color." *Perception* 10, no. 6:651–666.

A study that evaluated peoples' preference for various colors.

Mersey, J. E. (1990) "Colour and thematic map design: The role of colour scheme and map complexity in choropleth map communication." *Cartographica* 27, no. 3:1–157.

An extensive study of the role of map use tasks in selecting color schemes.

Monmonier, M. S. (1980) "The hopeless pursuit of purification in cartographic communication: A comparison of graphic-arts and perceptual distortions of graytone." *Cartographica* 17, no. 1:24–39.

Discusses the effect that map reproduction can have on area symbols (e.g., gray tones) and presents several models for simulating its effect. Also see Monmonier (1979).

Olson, J. M., and Brewer, C. A. (1997) "An evaluation of color selections to accommodate map users with color-vision impairments." *Annals, Association of American Geographers* 87, no. 1:103–134.

An experimental study of color schemes for color-vision impaired readers. Includes RGB and *Yxy* values for a number of suitable schemes.

Robinson, A. H., Morrison, J. L., Muehrcke, P. C., Kimerling, A. J., and Guptill, S. C. (1995) *Elements of Cartography.* 6th ed. New York: John Wiley & Sons.

Chapter 21 deals with the appropriate use of color. Also see pp. 303–311 of Dent (1996).

Travis, D. (1991) *Effective Color Displays: Theory and Practice.* London: Academic Press.

Contains a variety of useful information related to color, including formulas for converting between CIE and RGB (pp. 89–97), discussion of the gamma function (pp. 154–160), and specifications for matching Munsell and CIE colors (Appendix 3).

7

Proportional Symbol Mapping

OVERVIEW

This chapter examines the ***proportional symbol map*** *(or* ***graduated symbol map****), which is used to represent numerical data associated with point locations. Section 7.1 discusses two basic types of point data that can be displayed with proportional symbols: true and conceptual.* ***True point data*** *can actually be measured at a point location; an example would be the production of an oil well.* ***Conceptual point data*** *are collected over an area (or volume), but the data are conceived as being located at a point for the purpose of symbolization; an example would be the number of oil wells in each state. Section 7.1 also considers how data standardization can be applied to proportional symbol maps (for example, the murder rate for cities generally is a more useful measure than the raw number of murders).*

Section 7.2 introduces two general kinds of proportional symbols: ***geometric*** *and* ***pictographic*** *(respective examples would be circles and caricatures of people). Although geometric symbols have been more frequently used, pictographic symbols are becoming common because of the ease with which they can be created with software. Circles have been the most frequently used geometric symbol because they are visually stable, users prefer them, and they conserve map space. Three-dimensional symbols (such as spheres) have traditionally been frowned upon because of the difficulty readers have in estimating their size, but they can produce attractive eye-catching graphics and can be useful for representing a large data range.*

Section 7.3 deals with three methods of scaling (or sizing) proportional symbols: mathematical, perceptual, and range grading. In ***mathematical scaling,*** *symbols are sized in direct proportion to the data; thus, a data value 20 times another is represented by an area (or volume) 20 times as large. In* ***perceptual scaling,*** *a correction is introduced to account for visual underestimation of larger symbols; thus, larger symbols are made bigger than would normally be specified by mathematical scaling. In* ***range grading,*** *data are grouped into classes, and a single symbol size is used to represent all data falling in a class.*

Section 7.4 considers the problem of designing legends for proportional symbol maps. Basic decisions include deciding how symbols should be arranged in the legend and how many and what size of symbols should be used. In the ***nested-legend arrangement,*** *smaller symbols are drawn within larger symbols, while in the* ***linear-legend arrangement,*** *symbols are placed adjacent to each other in either a horizontal or vertical orientation. A nested arrangement conserves map space, while the linear arrangement permits a solid fill, which enhances* ***figure-ground contrast*** *(symbols tend to appear as a figure against the background of the rest of the map).*

Section 7.5 addresses the issue of symbol overlap. Small symbols cause little or no overlap, potentially resulting in a map devoid of spatial pattern; in contrast, large symbols can create a cluttered map, making it difficult to interpret individual symbols. In this context, this section considers two issues: deciding how much overlap there should be, and how the overlap should be symbolized. Two basic solutions for symbolizing overlap are transparent and opaque symbols. ***Transparent symbols*** *enable readers to see through overlapping symbols, while* ***opaque symbols*** *display smaller symbols stacked on top of larger symbols. Transparent symbols ease the problem of estimating symbol size, while opaque symbols enhance figure-ground contrast. Other solutions for handling overlap include* ***inset maps*** *portraying an enlarged scale of a congested area, the use of a* ***zoom function*** *in an interactive graphics environment, and the possibility of moving symbols slightly away from the center of a congested area.*

Finally, section 7.6 describes how redundant symbols can be used on proportional symbol maps. ***Redundant symbols*** *represent a single attribute (say, magnitude of oil production) by two visual variables (for example, size and value), thus enabling users to perform specific tasks more easily and accurately. Since redundant symbols are arguably less effective for portraying general information, they are best used in a data exploration framework.*

7.1 SELECTING APPROPRIATE DATA

Proportional symbols can be used for two forms of point data: true and conceptual. **True point data** can actually be measured at a point location; examples include the number of calls made from a telephone booth and temperature at a weather station. Although we generally map a continuous phenomenon such as temperature using contour maps (see Chapter 8), we may use proportional symbols to focus on the raw data, which are collected at point locations. **Conceptual point data** are collected over an area (or volume), but the data are conceived of as being located at a point (such as the centroid of an area) for purposes of symbolization. An example is the number of microbreweries and brewpubs in each state (Figure 7.1A).

Some data are not easily classified as either true or conceptual. For example, data associated with cities are collected over an area, but they normally are treated as occurring at point locations because at typical mapping

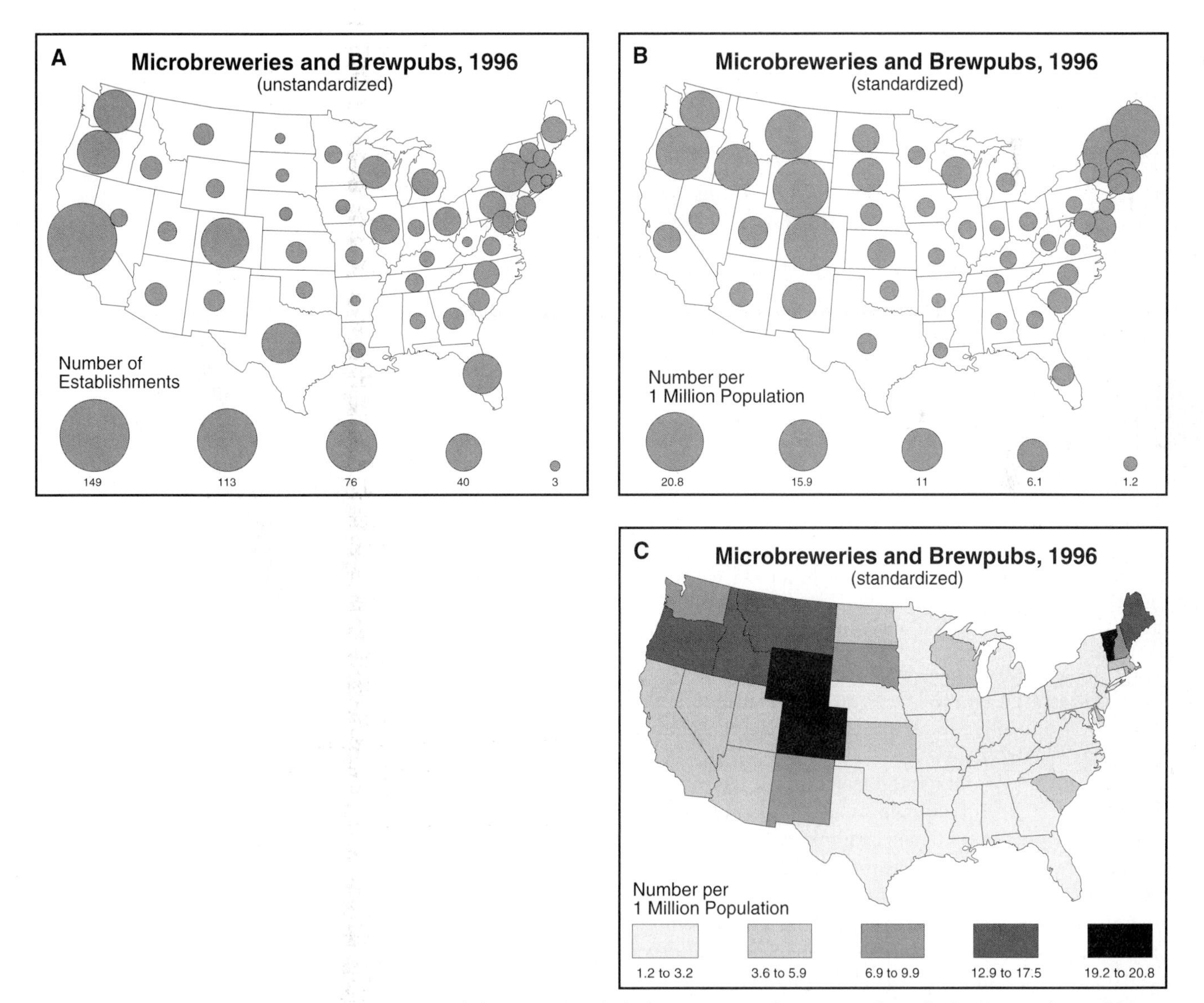

FIGURE 7.1. Mapping conceptual point data: (A) proportional circles represent the raw number of microbreweries and brewpubs in each state; (B) proportional circles represent standardized data—the number of microbreweries and brewpubs per 1 million people; (C) a choropleth map representing the standardized data shown in part B. (Data Source: http://www.beertown.org/.)

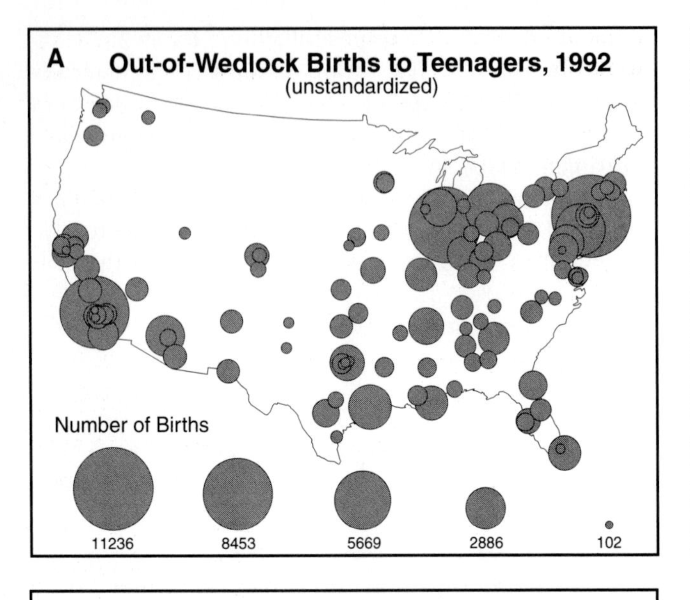

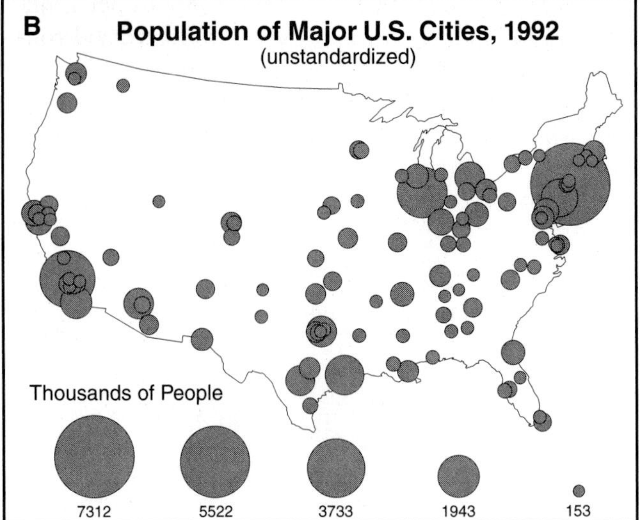

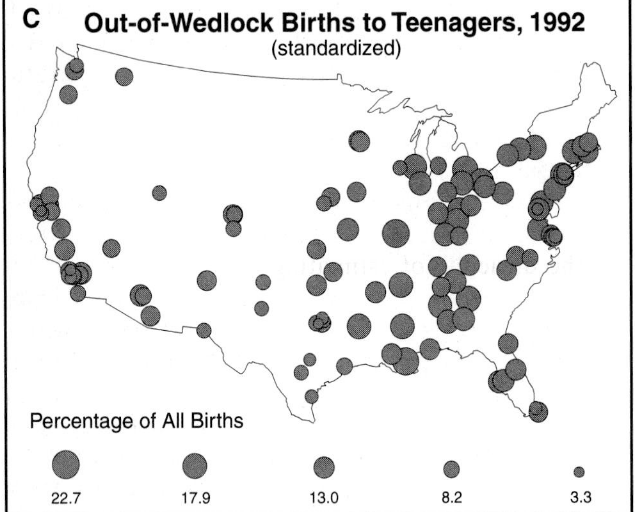

FIGURE 7.2. The data, which were collected over areas (cities), can easily be conceived as points, given the scale of the map: (A) raw number of out-of-wedlock births for U.S. cities with a population of 150,000 or more; (B) the population of those cities (note that this map correlates highly with A); (C) standardized map obtained by dividing the number of out-of-wedlock births by the total number of births. (Data Source: National Center for Health Statistics, 1995.)

scales cities are depicted as points. An example is the number of out-of-wedlock births to teenagers in major U.S. cities (Figure 7.2A).

The concept of data standardization introduced in Chapter 2 for choropleth maps is also applicable to proportional symbol maps. As an example, consider the data on out-of-wedlock births. The unstandardized map (Figure 7.2A) is useful for showing the sheer magnitude of out-of-wedlock births, but care must be taken in interpreting any spatial patterns on this map because cities with a large population are likely to have a large number of out-of-wedlock births. (Note the strong visual correlation between Figure 7.2A and 7.2B, the population of U.S. cities.)

One method of standardization is to compute the ratio of two raw-count variables. In the case of the birth data, we can compute the ratio of out-of-wedlock births to the total number of births, obtaining the proportion of out-of-wedlock births in each city (this is shown in percentage form in Figure 7.2C). This map has a markedly different appearance than the unstandardized map, in large part because of a much narrower range of data: 3.3 to 22.7 percent, as opposed to 102 to 11,236 births.

One can also argue that the microbrewery and brewpub data should be standardized, as the number of microbreweries and brewpubs is likely to be greater in more populous states. In this case, a useful standardization is the number of microbreweries and brewpubs per 1 million people (Figure 7.1B). Although standardized conceptual point data can be represented with proportional symbols, a choropleth map is more commonly used (Figure 7.1C) because the data are associated with areas.

7.2 KINDS OF PROPORTIONAL SYMBOLS

Proportional symbols can be divided into two basic groups: geometric and pictographic. **Geometric symbols** (circles, squares, spheres, and cubes) generally do not mirror the phenomenon being mapped, while **pictographic symbols** (heads of wheat, caricatures of people, diagrams of barns) do. Prior to the development of digital cartography, geometric symbols predominated because templates were readily available for manually constructing common geometric shapes. Today, digital mapping has made the development of pictographic symbols much easier: one can create the basic design for a symbol by hand, scan the symbol into a computer file, and then use design software to duplicate that symbol at various sizes. The map of beer mugs shown in Figure 7.3 was created using this approach. Alternatively, one can find numerous pictographic symbols already in digital form in **clip art** files.

The ease with which readers can associate pictographic symbols with the phenomenon being mapped, their eye-catching appeal, and their increasingly greater ease of construction suggest that these symbols will appear more commonly on maps. Pictographic symbols, however, are not without problems. One is that when symbols overlap, they may be more difficult to interpret than geometric symbols (compare the northeastern portions of Figures 7.1A and 7.3). Another problem is that it may be more difficult for readers to judge the relative sizes of pictographic symbols (for example, judging size relations among beer mugs is arguably more difficult than judging size relations among circles).

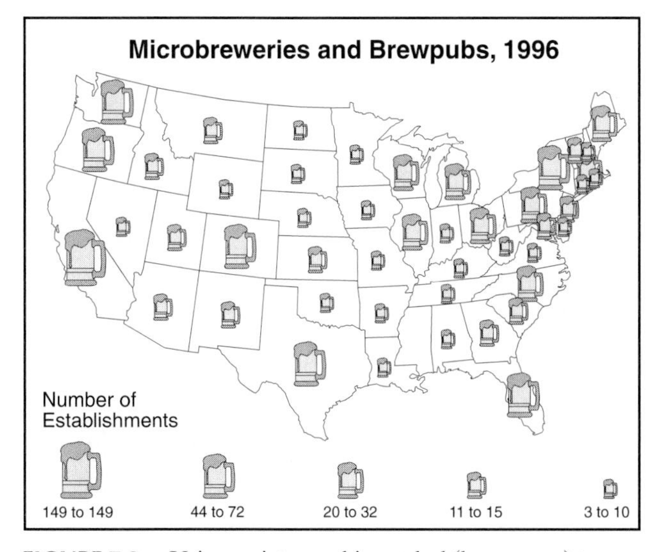

FIGURE 7.3. Using a pictographic symbol (beer mugs) to represent the number of microbreweries and brewpubs in each state. The data were range-graded and the sizes of symbols chosen by "eye."

Within the realm of geometric symbols, circles have been most frequently used. These are the arguments traditionally offered for using circles.

1. Circles are considered visually stable.
2. Users prefer circles over other geometric symbols.
3. Circles (as opposed to bars) conserve map space.
4. When constructing circles, it is easy to determine appropriate relations between data and circle size.
5. Circles are easy to construct.

The latter two advantages have largely disappeared with the rise of digital cartography: formulas for calculating the sizes of various symbols can now be implemented in spreadsheets, and software can be developed to automatically position symbols on a map. (For example, squares can be automatically aligned relative to one another and to the map border.)

Traditionally, three-dimensional geometric symbols (spheres, cubes, and prisms) have not been used because of the difficulty of estimating their size and the difficulty of constructing them. Like pictographic symbols, however, these symbols can produce attractive eye-catching graphics (Figure 7.4). Additionally, three-dimensional geometric symbols can be useful for representing a large range in data; for example, in Figure 7.5 the small three-dimensional symbol is easily detected, while the corresponding two-dimensional symbol nearly disappears.

7.3 SCALING PROPORTIONAL SYMBOLS

7.3.1 Mathematical Scaling

Mathematical scaling sizes areas (or volumes) of point symbols in direct proportion to the data; thus, if a data value is 20 times another, the area (or volume) of a point symbol will be 20 times as large (Figure 7.6). We now consider some formulas for calculating symbol sizes; we will deal with circles first because of their common use.

Remembering from basic math that the area of a circle is equal to πr^2, we can establish the relation

$$\frac{\pi r_i^2}{\pi r_L^2} = \frac{v_i}{v_L}$$

where r_i = radius of the circle to be drawn
r_L = radius of the largest circle on the map

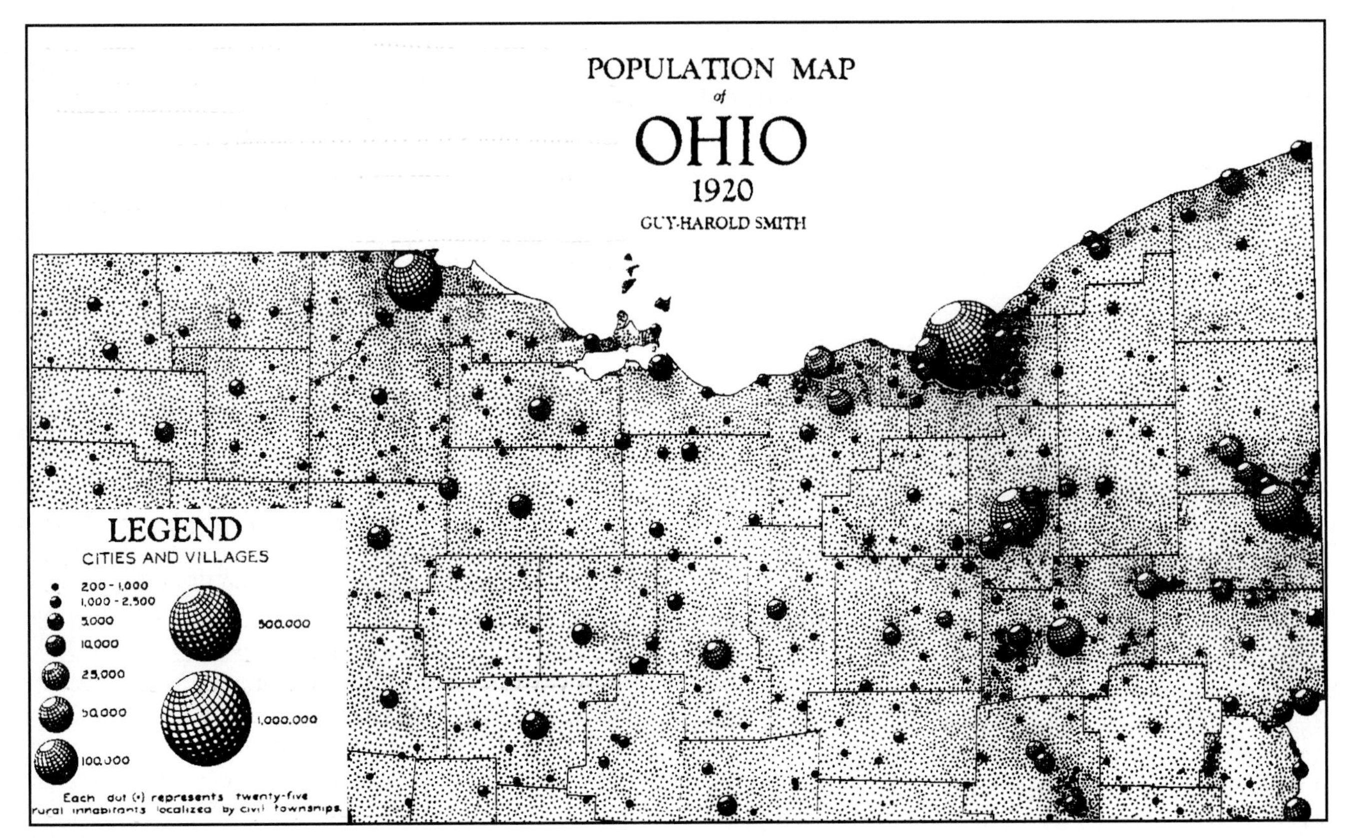

FIGURE 7.4. An eye-catching map created using three-dimensional geometric symbols. (After Smith, 1928. First published in *The Geographical Review,* 18(3), plate 4. Reprinted with permission of the American Geographical Society.)

v_i = data value for the circle to be drawn
v_L = data value associated with the largest circle

Note that this formula specifies circle areas in direct proportion to corresponding data values. The relation uses the largest radius (and largest data value) for one of the circles because proportional symbol maps are often constructed by beginning with a largest symbol size to minimize the effect of symbol overlap.

Since the values of π cancel, this equation reduces to

$$\frac{r_i^2}{r_L^2} = \frac{v_i}{v_L}$$

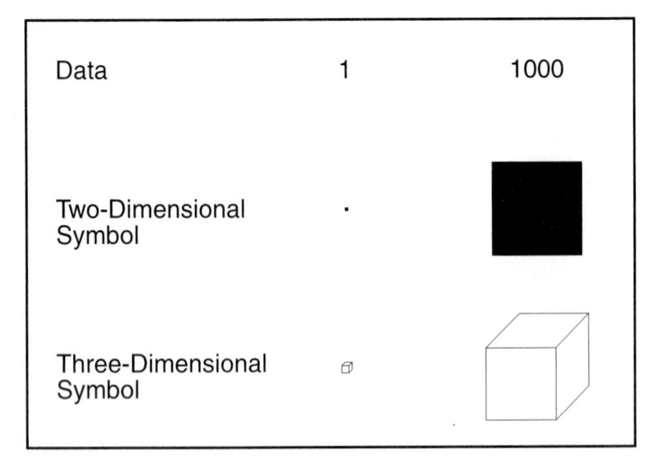

FIGURE 7.5. Portraying a large range of data. The smallest cube (a three-dimensional symbol) is more easily seen than the smallest square (a two-dimensional symbol).

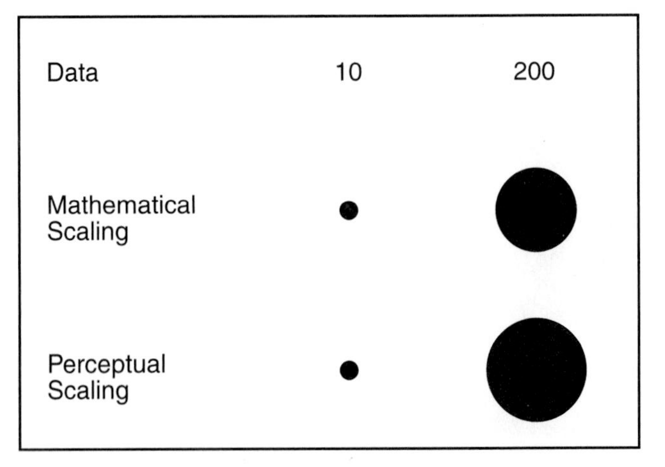

FIGURE 7.6. Mathematical versus perceptual scaling. In mathematical scaling, the areas of the circles are constructed directly proportional to the data; in perceptual scaling, the size of the larger circle is increased (using Flannery's exponent) to account for underestimation.

Taking the square root of both sides, we have

$$\frac{r_i}{r_L} = \left(\frac{v_i}{v_L}\right)^{0.5}$$

Finally, solving for r_i, we compute

$$r_i = \left(\frac{v_i}{v_L}\right)^{0.5} \times r_L$$

To illustrate the application of this formula, consider how the radius for a circle representing Los Angeles was computed for Figure 7.2A. After some experimentation, it was decided that the largest circle (New York) should have a radius of .2125 inches (.5398 cm). To determine the radius for Los Angeles, the number of out-of-wedlock births for Los Angeles and New York (8,507 and 11,236, respectively) were inserted into the formula, along with the largest radius as follows

$$r_{\text{Los Angeles}} = \left(\frac{v_{\text{Los Angeles}}}{v_{\text{NewYork}}}\right)^{0.5} \times r_{\text{New York}}$$

$$r_{\text{Los Angeles}} = \left(\frac{8507}{11236}\right)^{0.5} \times .2125 = .1849$$

Formulas for other geometric symbols could be derived in a similar fashion. The results would be as follows:

SQUARES: $$s_i = \left(\frac{v_i}{v_L}\right)^{0.5} \times s_L$$

where s_i is the length of a side of a square to be drawn and s_L is the length of a side of the largest square;

BARS: $$h_i = \frac{v_i}{v_L} \times h_L$$

where h_i is the height of a bar to be drawn and h_L is the height of the tallest bar;

SPHERES: $$r_i = \left(\frac{v_i}{v_L}\right)^{1/3} \times r_L$$

where r_i is the radius of a sphere to be drawn and r_L is the radius of the largest sphere;

CUBES: $$s_i = \left(\frac{v_i}{v_L}\right)^{1/3} \times s_L$$

where s_i is the length of a side of a cube to be drawn and s_L is the length of a side of the largest cube.

Since these formulas typically involve taking the square or cube root of the ratio of data (raising the ratio to either the 1/2 or 1/3 power), mathematical scaling is also sometimes referred to as *square root* or *cube root scaling*. We will term the power to which the ratio of data is raised the *exponent for symbol scaling*. Although some design and mapping programs have these formulas imbedded in software, users will sometimes have to enter the formulas in spreadsheets along with the raw data.

It is important to recognize that the above formulas produce unclassed maps, as each differing data value is represented by a differing symbol size. Classed, or **range-graded maps,** can be created by classing the data and letting a single symbol size represent a range of data values, but *unclassed* proportional symbol maps are more common. This may seem surprising given the frequency with which *classed* choropleth maps are used. The difference probably stems from the ease with which unclassed proportional symbol maps could be created in manual cartography (for example, an ink compass could be used to draw circles of numerous sizes; alternatively, a template with numerous circle sizes could be used).

Some software (such as MapViewer) permits arbitrarily specifying the smallest and largest symbol sizes, with other symbols scaled proportionally between these symbols. For symbols based on area, the formulas are

$$Z_i = \frac{v_i - v_S}{v_L - v_S}$$

$$A_i = Z_i(A_L - A_S) + A_S$$

where v_S and v_L = smallest and largest data values
Z_i = proportion of the data range associated with the data value v_i
A_s and A_L = smallest and largest areas desired
A_i = area of a symbol associated with the data value v_i

This approach will produce an unclassed map, but the symbols will not be scaled proportional to the data (that is, a data value twice another will not have a symbol twice as large). An advantage of the approach, however, is that it can enhance the map pattern, just as do an arbitrary exponent and range grading, as is described in sections 7.3.2 and 7.3.3.

7.3.2 Perceptual Scaling

Numerous studies have shown that the perceived size of proportional symbols does not correspond to their mathematical size; rather, people tend to underestimate the size of larger symbols. For example, in viewing the larger mathematically scaled circle in Figure 7.6, most people would estimate it to be less than 20 times larger than the smaller circle.

If larger symbols are underestimated, it seems reasonable to suggest that formulas for mathematical scaling might be modified (or "corrected") to account for underestimation; this process is known as **perceptual** (or psychological) **scaling.**

Formulas for Perceptual Scaling

To develop formulas for perceptual scaling, it is necessary to consider how researchers have summarized the results of experiments dealing with perceived size. The relation between actual and perceived size typically has been stated as a *power function* of the form

$$R = cS^n$$

where R = response (or perceived size)
S = stimulus (or actual size)
c = a constant
n = an exponent

To differentiate this exponent from the one for symbol scaling introduced above, we will term it the **power function exponent.**

The power function exponent is the key to describing the results of experiments involving perceived size. If size is estimated correctly, the exponent will be close to 1.0. Underestimation and overestimation are represented by exponents appreciably below and above 1.0. For example, an oft-cited study by Flannery (1971) found that for circles $R = (0.98365)S^{0.8747}$; here the exponent of 0.8747 is indicative of underestimation.

The power function equation states what response arises from a certain stimulus. For constructing symbols, we need to know the reverse of that: what stimulus must be shown to get a certain response. Therefore, the power function must be transposed by dividing each side by c, and then raising each side to the $1/n$ power. The result is

$$S = c_1 R^{1/n}$$

where c_1 is a constant equal to $(1/c)^{1/n}$. In Flannery's study, the transpose was $S = (1.01902)R^{1.1432}$.* To simplify computations, the value of the constant (1.01902) in the transposed equation can be ignored because it is close to 1.0; thus we have $S = R^{1.1432}$. Since this equation expresses the relation between the areas of circles, and circles are constructed on the basis of radii, we need to take the square root of both sides, producing $S^{0.5} = R^{0.5716}$. Again for simplicity, the value 0.5716 has normally been rounded to 0.57. As a result, a perceptual scaling formula for circles is

$$r_i = \left(\frac{v_i}{v_L}\right)^{0.57} \times r_L$$

The result of using an exponent of 0.57 in the circle scaling formulas can be seen for the perceptually scaled circles in Figure 7.6; for most readers, the larger circle should now appear closer to 20 times larger than the smallest circle.

The magnitude of the power function exponent varies, depending on the symbol type. For squares, Crawford (1973) derived an exponent of 0.93 (which yields an exponent of 0.54 for symbol scaling), indicating that squares are estimated better than circles. For bars, Flannery found that underestimates were balanced by overestimates, and thus recommended no corrective formula. These results suggest that when precise estimates are desired, squares and bars should be used in preference to circles.

Power function exponents for "drawn" three-dimensional symbols have been appreciably lower than for two-dimensional symbols, indicating severe underestimation. For example, for spheres and cubes Ekman and Junge (1961) derived exponents of 0.75 and 0.74 (corresponding to exponents of 0.44 and 0.45 for symbol scaling). Interestingly, research by Ekman and Junge also indicated that when truly three-dimensional cubes were used (they were "made of steel with surfaces polished to a homogeneous, dull silvery appearance," p. 2), the exponent was 1.0. This result suggests that the manner in which three-dimensional symbols are portrayed in two-dimensional space might have an impact on the exponent. For example, if an interactive graphics program gives the impression of traveling through three-dimensional space, then we might expect an exponent closer to 1.0.

Problems in Applying the Formulas

Unfortunately, there are numerous problems in applying the above formulas for perceptual scaling: (1) the value of a power function exponent can be affected by various experimental factors, (2) using a single exponent may be unrealistic because of the variation that exists between subjects and within an individual subject's responses, (3) the formula fails to account for the spatial context within which symbols are estimated, and (4) experimental studies for deriving exponents have dealt only with acquiring specific map information (as opposed to

* Flannery failed to raise $1/k$ to $1/n$, so his constant differed slightly from the value reported here.

considering memory and general information). We will briefly consider each of these problems.

Chang (1980) provided a good summary of the various experimental factors that can affect a power function exponent. One is whether subjects are asked to complete a ratio or magnitude estimation task. **Ratio estimation** involves comparing two symbols and indicating how much larger or smaller one symbol is than another (noting that "this symbol appears to be five times larger than this one"); on a map, this involves comparing symbols without consulting the legend. Flannery's exponent of 0.87 for circles was developed on this basis.

Magnitude estimation involves assigning a value to a symbol on the basis of a value assigned to another symbol; for example, if a single circle is included in a map legend, values for other circles on the map can be estimated by comparing them with the legend circle. Using this approach, Chang (1977) found that the use of a moderately sized circle in the legend led to an exponent of only 0.73 for circles.

A second experimental factor affecting the exponent is the size of the standard symbol against which other symbols are compared. For example, in magnitude estimation studies, Chang (1977) and Cox (1976) showed that the value of the exponent gradually increased as the size of the standard increased. Other experimental factors noted by Chang that influenced the exponent included the wording of the instructions to subjects, the range of stimuli presented, and the order in which estimates were made.

Griffin (1985) is one of several researchers who have noted that reporting a single exponent for an experiment hides the variation between and within subjects. Although overall Griffin found a power function exponent of 0.88 for circles (nearly identical to Flannery's 0.87), he stressed that the exponents for individuals varied from approximately 0.4 to 1.3. He also noted that "perceptual rescaling was shown to be inadequate to correct the estimates of poor judges, while seriously impairing the results of those who were more consistent" (p. 35).

One solution to the problems noted by Chang and Griffin is to apply no correction and stress the importance of a well-designed legend. For example, Chang, in discussing the experimental factors affecting the exponent recommended, including "three standards [in the legend]—small, medium, [and] large"; he based this recommendation, in part, on a magnitude estimation study in which he found an exponent of 0.94 when three legend circles were used (Chang 1977). Chang (1980, 161) also recommended including the statement that "circle areas are made proportional to quantities," to encourage users to make estimates based on area.

Another solution to the problems noted by Chang and Griffin is to train users how to read proportional symbol maps. For example, Olson (1975a) found that when readers were asked to make estimates of circle size and then given feedback on the correct answers that the exponent for the power function moved closer to 1.0. She also found, however, that only when practice was combined with perceptual scaling did the dispersion of errors also decrease, suggesting that training and perceptual scaling might need to be used in concert.

The importance of spatial context on developing a perceptual scaling formula can be seen by considering the **Ebbinghaus illusion** shown in Figure 7.7: the two circles in the middle of the surrounding circles are identical in size, but the one surrounded by larger circles appears smaller. Applying this principle to a map, a small circle surrounded by larger circles should appear smaller than it really is, while a large circle surrounded by smaller circles should appear larger than it actually is. In an experimental study with maps, Gilmartin (1981) found that spatial context actually had these effects. On this basis, one could argue for a formula in which each circle is scaled to reflect its local context. Gilmartin argued against this, indicating it would "have the undesirable effect of weakening the overall pattern perception" (p. 162).

Another limitation of perceptual scaling is that experiments for deriving an appropriate exponent have focused solely on specific map information. In Chapter 1, it was argued that the portrayal of general information is a more important function of maps. Thus, it seems that general tasks should be considered in developing an exponent, or at least that the effect of the exponent should be considered on the overall look of

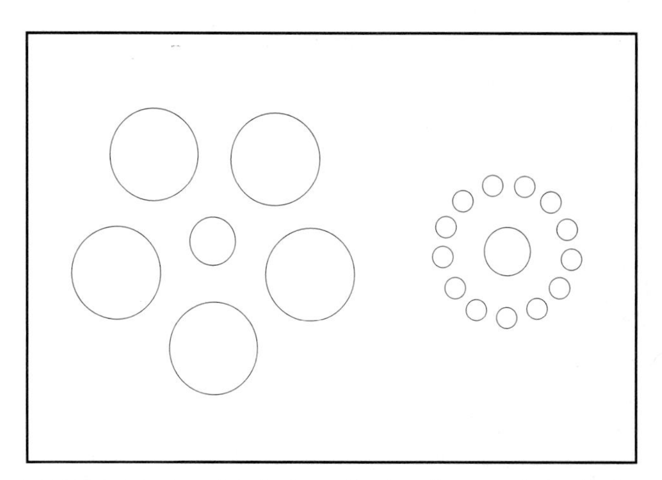

FIGURE 7.7. The Ebbinghaus illusion: the two circles in the middle of the surrounding circles are identical in size, but the one surrounded by larger circles appears smaller.

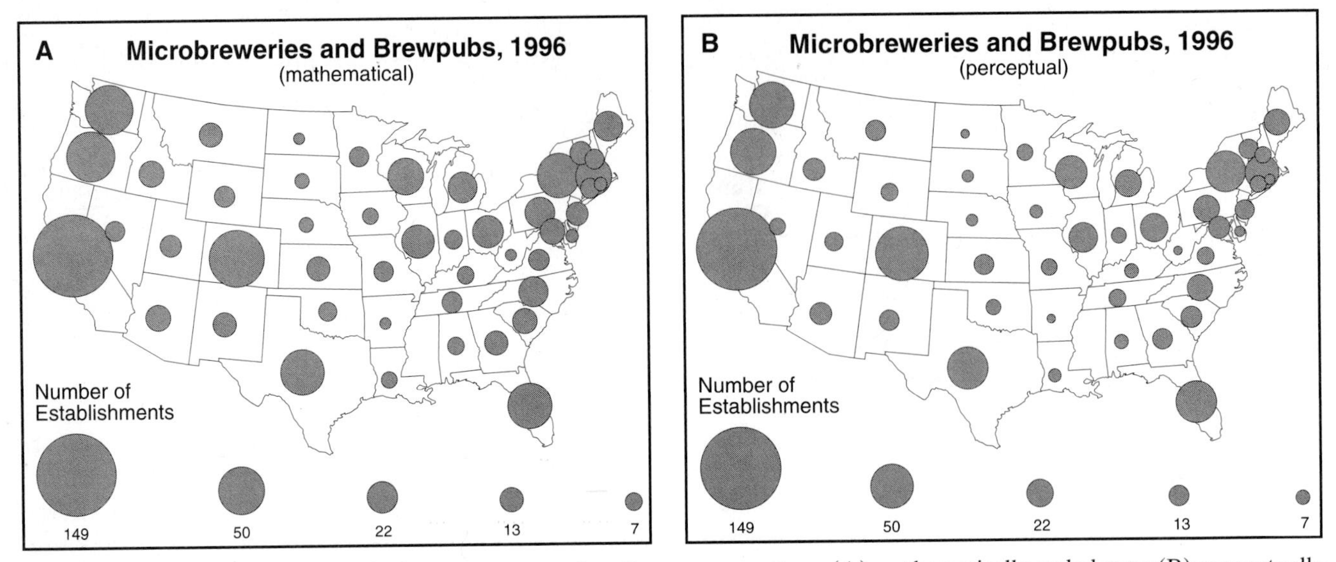

FIGURE 7.8. Effect of mathematical versus perceptual scaling on map pattern: (A) mathematically scaled map; (B) perceptually scaled map based on Flannery. Note that perceptual scaling appears to have little effect on the overall pattern when the largest circle on both maps is the same size.

the map. To illustrate this point, consider Figure 7.8, which compares mathematically and perceptually scaled maps of the microbrewery and brewpub data. Here the largest circle size on each map has been held constant to minimize the effects of circle overlap on the two maps. Although a larger range of circle sizes appears on the perceptually scaled map, the patterns on the two maps appear very similar, suggesting that perceptual scaling has little effect on general information.

Rather than basing the exponent on specific circle estimates, Olson (1976a, 155–156; 1978) has suggested that an arbitrary exponent might be used to enhance the recognition of patterns on the map. To illustrate, Figure 7.9 compares a mathematically scaled map (the exponent in the circle-scaling formula is set to 0.5) with one having an arbitrary exponent (the exponent is 1.0 in the circle-scaling formula) for the standardized birth data. The narrow range of data on the mathematically scaled map (Figure 7.9A) makes it very difficult to detect any patterns. On the arbitrarily scaled map (Figure 7.9B), it is somewhat easier to detect a pattern; for example, note the lower percentage of out-of-wedlock births in the extreme south-central part of the country (in Texas).

A final limitation of perceptual scaling is that most of the experiments dealing with the derivation of an exponent, and thus any correctional formulas, have been based only on *acquiring* information, as opposed to *memory* for that information. Studies by psychologists (for example, Kerst and Howard 1984), have revealed that underestimation of larger sizes occurs twice, once when acquiring information and once again when recalling that information. As a result, the exponent for memory tasks is appreciably lower than that for acquisition (values obtained are approximately the square of that for acquisition). Although the exponent in the perceptual-scaling formulas might be adjusted to account for this, I suspect that most cartographers would not make this adjustment because map readers are not expected to remember precise, specific information. It is, however, interesting to note that the lower exponent resulting from this approach would have an effect similar to that just suggested for enhancing spatial patterns.

7.3.3 Range-Graded Scaling

Range grading groups raw data into classes, and represents each class with a different sized symbol (as an example, if the data are grouped into five classes, then five symbol sizes are used). Three basic decisions must be made in range grading: the number of classes to be shown, the method of classification to be used, and the symbol sizes to be used for each class. The first two decisions are standard in any classification of numerical data and were discussed extensively in Chapter 4. (The reader may wish to review Figure 4.7, which summarizes the advantages and disadvantages of each classification method.)

The sizes of range-graded symbols are normally selected to enhance the visual discrimination among classes. Figure 7.10 portrays two sets of symbols that cartographers have developed for this purpose. The first (Figure 7.10A) was developed by Meihoefer (1969) in a visual experiment. Meihoefer indicated that people "were

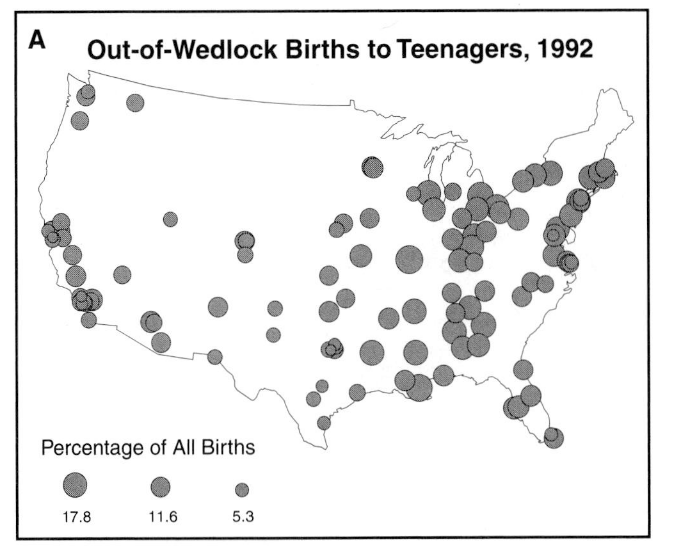

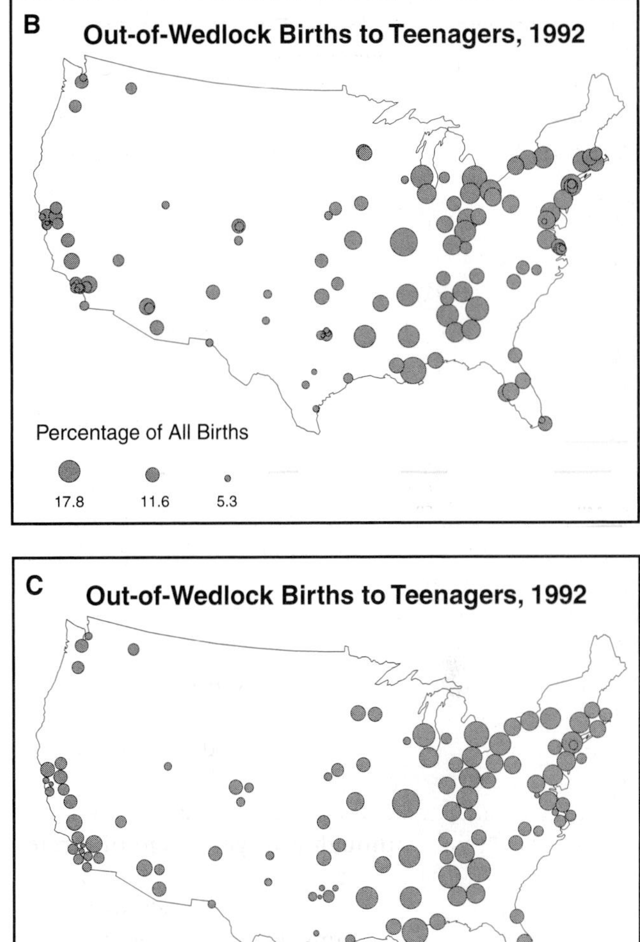

FIGURE 7.9. Manipulating symbol scaling and symbol position to enhance map pattern: (A) mathematically scaled map (the exponent in the circle scaling formula is set to .5); (B) an arbitrarily scaled map (the exponent in the circle scaling formula is set to 1.0); (C) circles are moved away from areas of concentration.

able generally to differentiate" (p. 112) among all of these circles in a map environment. Arguing that some of Meihoefer's circles were difficult to discriminate, Dent (1996, 174–175) developed the set shown in Figure 7.10B. (Dent's were based on his own experience in designing maps.) For both sets, the mapmaker simply selects n adjacent circles, where n is the number of classes to be depicted on the map.

Range grading is generally considered advantageous because readers can easily discriminate symbol sizes and thus can readily match map and legend symbols; another advantage is that the contrast in circle sizes may enhance the map pattern, in a fashion similar to the use of an arbitrary exponent described in the preceding section. To illustrate the latter, consider Figure 7.11, which compares range-graded and mathematically scaled maps for both the standardized birth data and the raw microbrewery and brewpub data.* In the case of the microbrewery and brewpub data, we see that range grading had relatively little effect on the spatial pattern because the range of circle sizes on the range-graded and mathematically scaled maps was similar. For the birth data, however, the results were rather dramatic, with range grading resulting in considerable overlap.

Since specific range-graded sizes have been recommended only for circles, individual mapmakers must de-

* In this case, Dent's five smallest circles were used and the data were classed using Jenks's optimal approach (see section 4.1.6).

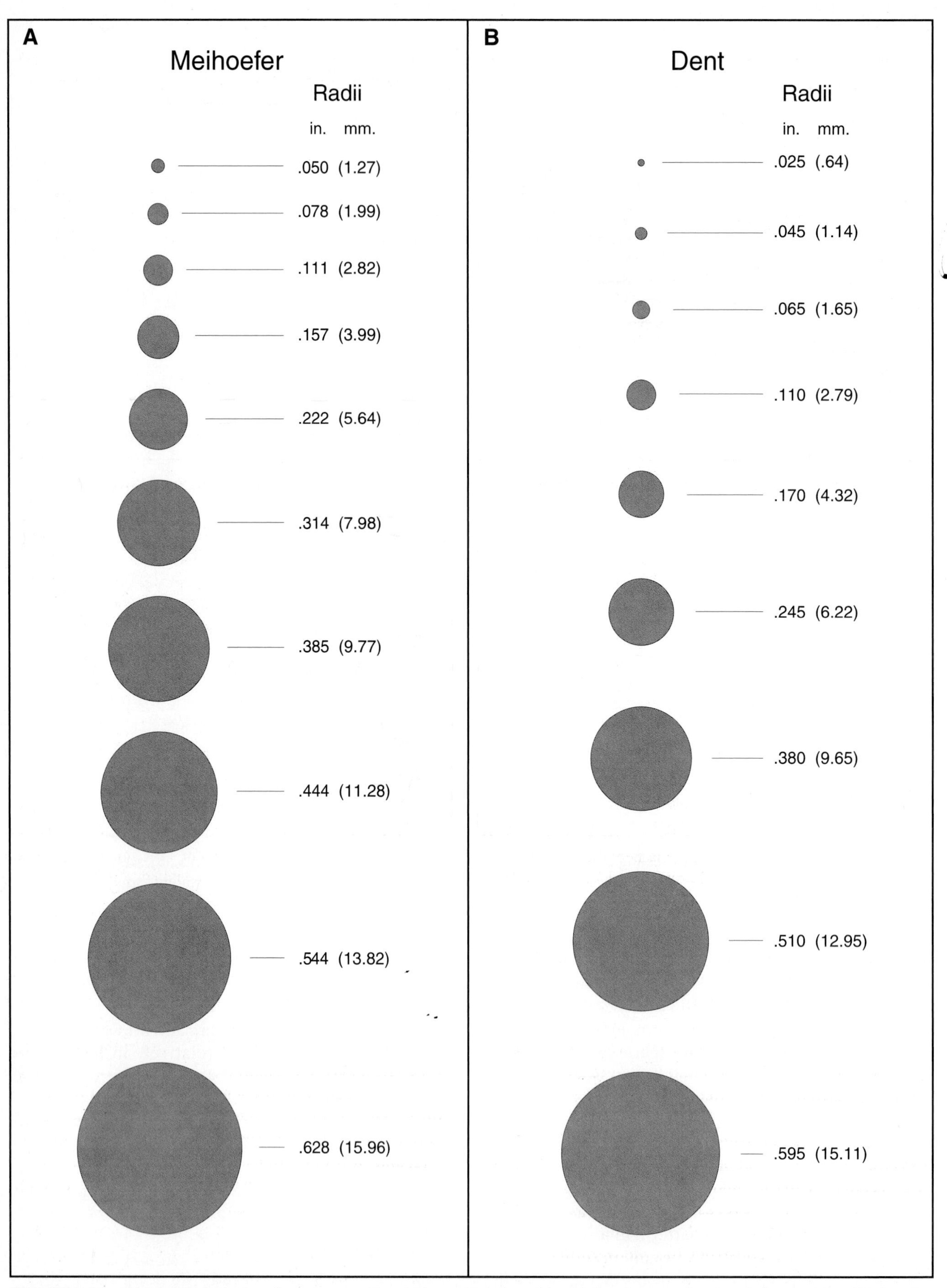

FIGURE 7.10. Potential circle sizes for range grading: (A) a set developed by Meihoefer in a visual experiment (after Meihoefer, 1969, figure 4); (B) a set developed by Dent based on practical experience (after Dent, 1996, 175).

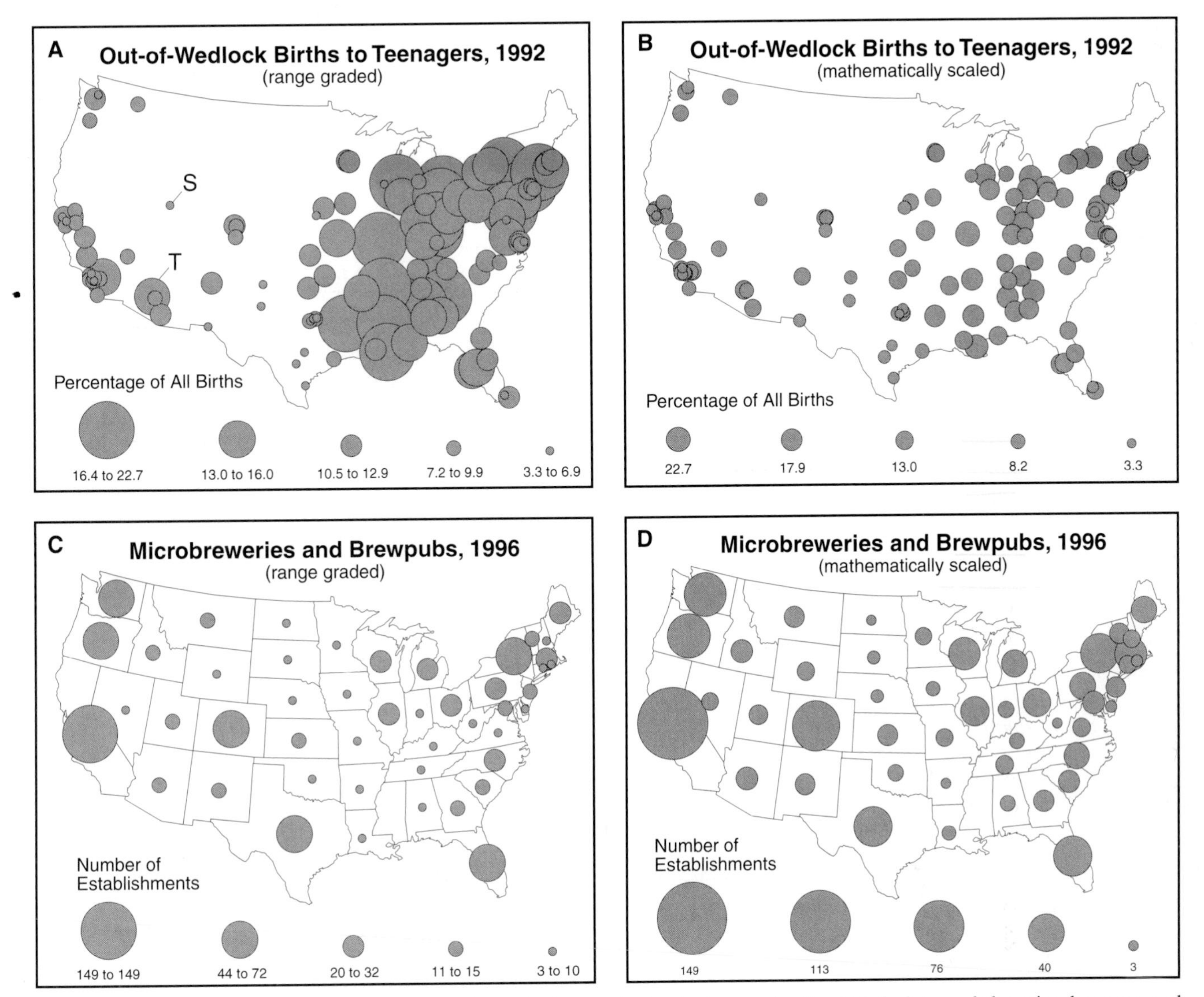

FIGURE 7.11. Range-graded and mathematically scaled maps for both the standardized birth data and the microbrewery and brewpub data. Maps A and C are range-graded based on Dent's five smallest circles. Maps B and D are mathematically scaled.

termine appropriate sizes for other symbol types. For example, to create the range-graded proportional-square map shown in Figure 7.12, distinctly different small and large squares were selected, and then three intermediate squares were specified so that their areas were evenly spaced between the smallest and largest square.

Range grading is particularly desirable for pictographic symbols because their unusual shape often makes precise relationships between symbols awkward to compute (for the mapmaker) and to estimate (for the map reader). Thus, it makes sense to select intermediate-sized symbols by eye, as opposed to spacing them regularly on the basis of area or volume. This was the approach taken to construct the pictographs of beer mugs shown in Figure 7.3.

A disadvantage of range grading is that readers may misinterpret specific information if they do not pay careful attention to the legend. For example, a reader failing to examine the legend might say that the value for circle T in Figure 7.11A is considerably larger than the value for circle S (say, approximately 30 times as large), but the numbers specified in the legend indicate that the values differ only by about a factor of 4. Another disadvantage is that range grading creates clear differences in circle size, and thus potentially creates a pattern, when there might not be a meaningful one. For example, for a data set with minimum and maximum values of 10 and 11 percent, respectively, range grading would create obvious differences among circles, even though a 1 percent difference might be of no interest or significance to the reader.

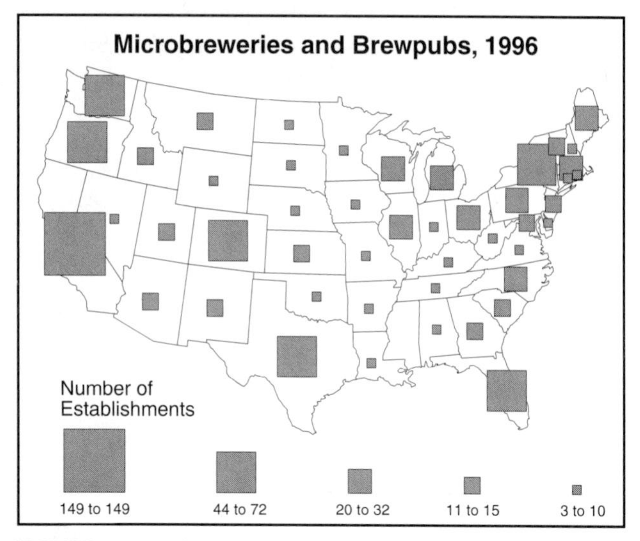

FIGURE 7.12. A range-graded map using squares as a point symbol.

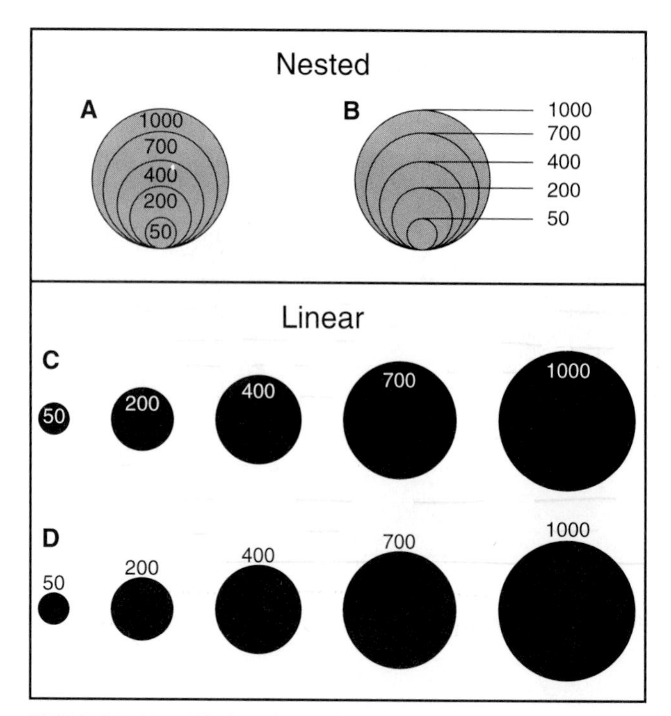

FIGURE 7.13. Various legend arrangements: (A) nested with numbers displayed inside symbols; (B) nested with numbers displayed outside symbols; (C) linear with numbers displayed inside symbols; (D) linear with numbers displayed outside symbols.

7.4 LEGEND DESIGN

There are two basic problems in designing legends for proportional symbol maps: deciding *how* the symbols should be arranged and determining *which* symbols should be included.

7.4.1 Arranging Symbols

Two basic legend arrangements are used on proportional symbol maps: nested and linear. In the **nested-legend arrangement,** smaller symbols are drawn within larger symbols, while in the **linear-legend arrangement,** symbols are placed adjacent to each other in either a horizontal or vertical orientation (Figure 7.13). An advantage of the nested arrangement is that it conserves map space. In contrast, the linear arrangement permits a solid fill; this is advantageous because a solid fill tends to enhance figure-ground contrast. Note that if all circles are sufficiently large, numeric values may be placed within them (Figure 7.13A and C), providing a rather attractive design alternative.

When a linear horizontal arrangement is used, the mapmaker must decide whether the symbols should be ordered with the smallest on the left and the largest on the right, or vice versa. Displaying larger symbols on the right seems most desirable given that the traditional number line progresses from left to right. One must, however, consider where space is available on the map; for example, the greater amount of open space in the lower left-hand corner of the U.S. map caused me to place larger symbols on the left in this case.

When a linear vertical arrangement is used, the mapmaker similarly must decide whether symbols should be ordered with the largest on the top or bottom. Using MacEachren's argument that "people associate 'up' with 'higher' and 'higher' with larger data values," it makes sense to place larger symbols at the top and smaller ones at the bottom. As with the horizontal arrangement, available map space may, however, alter this guideline.

7.4.2 Which Symbols to Include

With range grading, the symbols shown in the legend are determined by the classes displayed on the map (for example, a five-class map yields five symbols in the legend). For mathematical and perceptual scaling, there are two general methods for selecting the symbols shown in the legend. One is to include the smallest and largest symbol actually shown on the map and then interpolate several intermediate-sized symbols (Figure 7.14A). A second method is to select a set of symbols that are most representative of those appearing on the map, which should minimize estimation error. The latter method can be implemented by applying Jenks's optimal classification routine to the raw data and then constructing legend symbols based on the median (or mean) of each class (Figure 7.14B). This method might also be refined to include circles representing the extremes in the data because of the difficulty of extrapolating beyond symbols shown in the legend (Dobson 1974).

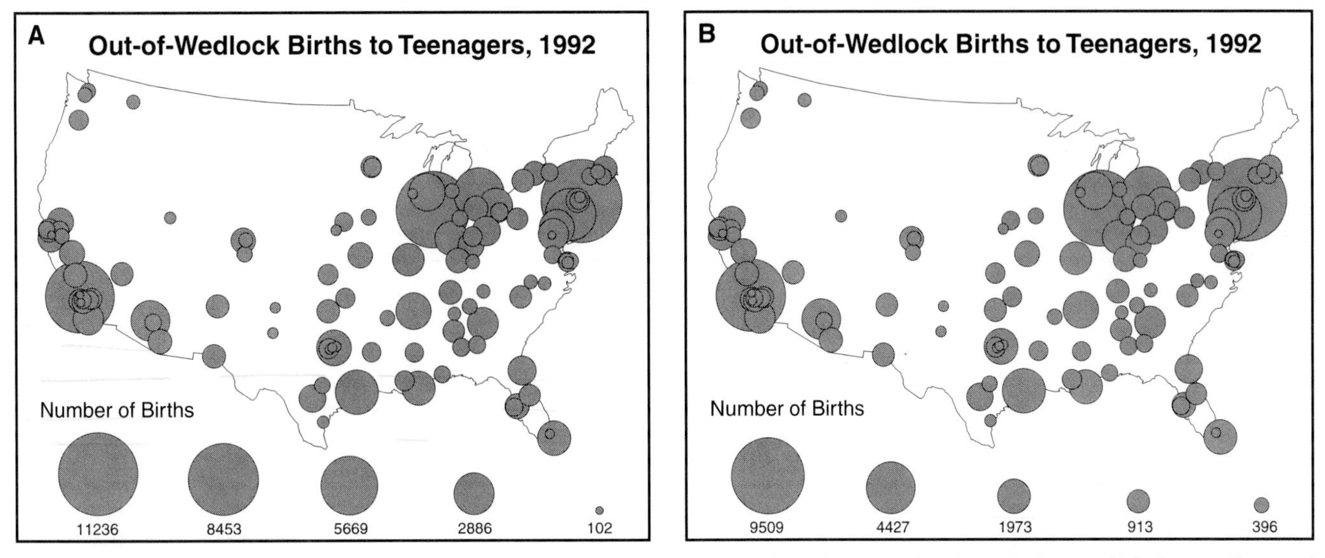

FIGURE 7.14. Approaches for selecting circles for the legend on an unclassed proportional symbol map: (A) the smallest and largest symbol actually shown on the map are used, along with several intermediate-sized symbols; (B) the raw data are classified, and a representative value of each class is used (in this case the median).

In addition to selecting one of these general methods for mathematical and perceptual scaling, the mapmaker must also decide *how many* symbols will be shown in the legend. One possibility would be to use three symbols, as recommended by Chang. Although three symbols may be sufficient for data with a small range, it makes sense to use more than three when the range is large. As a rule of thumb, I suggest using as many symbols as appear to be easily discriminated.

7.5 HANDLING SYMBOL OVERLAP

A major issue in proportional symbol mapping is deciding how large symbols should be, and consequently how much overlap there will be. Small symbols cause little or no overlap, and thus a map potentially devoid of spatial pattern; in contrast, large symbols can create a cluttered map, making it difficult to interpret individual symbols. This section considers two issues: deciding how much overlap there should be, and how the overlap should be symbolized.

7.5.1 How Much Overlap?

Unfortunately, there are no rules regarding the appropriate amount of overlap. Rather, cartographers have suggested subjective guidelines; for instance, Robinson et al. (1984) indicate that the map should appear "neither 'too full' nor 'too empty'" (p. 294). Examples of improper amounts of overlap are shown in Figures 7.15A and 7.15B. Although most cartographers would probably agree that such extreme cases should be avoided, there would likely be disagreement as to which map should be shown between the extremes. (Figures 7.15C and 7.15D are two possibilities.)

The role that overlap plays is determined to some extent by whether the map is to be used for communication or for data exploration. If the map is to be used for communication, then the amount of overlap will depend on the mapmaker's purpose. For example, if the mapmaker perceives a certain character of the distribution that she wishes readers to also note, then overlap could be manipulated to enhance that character. In the case of Figure 7.15, map C may be a more appropriate choice than map D if the mapmaker wishes readers to note the concentration of microbreweries in the northwest part of the United States.

If the map is to be used for data exploration, then mapping software should provide an option to easily change symbol sizes. See, for example, Using Visual Basic to Teach Programming for Geographers, Appendix G.

7.5.2 Symbolizing Overlap

Two basic solutions for symbolizing overlap are transparent and opaque symbols. **Transparent symbols** enable readers to see through overlapping symbols, while **opaque symbols** are stacked on top of one another (see Figure 7.16). On black-and-white maps, the fill within both transparent and opaque symbols can be varied from white to solid black. Color can, of course, provide considerable flexibility in overlap, as different colors can be used for the boundary and interior of symbols.

Transparent symbols allow background information (such as a road network) to be seen beneath the symbols (Figures 7.16A, 7.16C, and 7.16E), and make it eas-

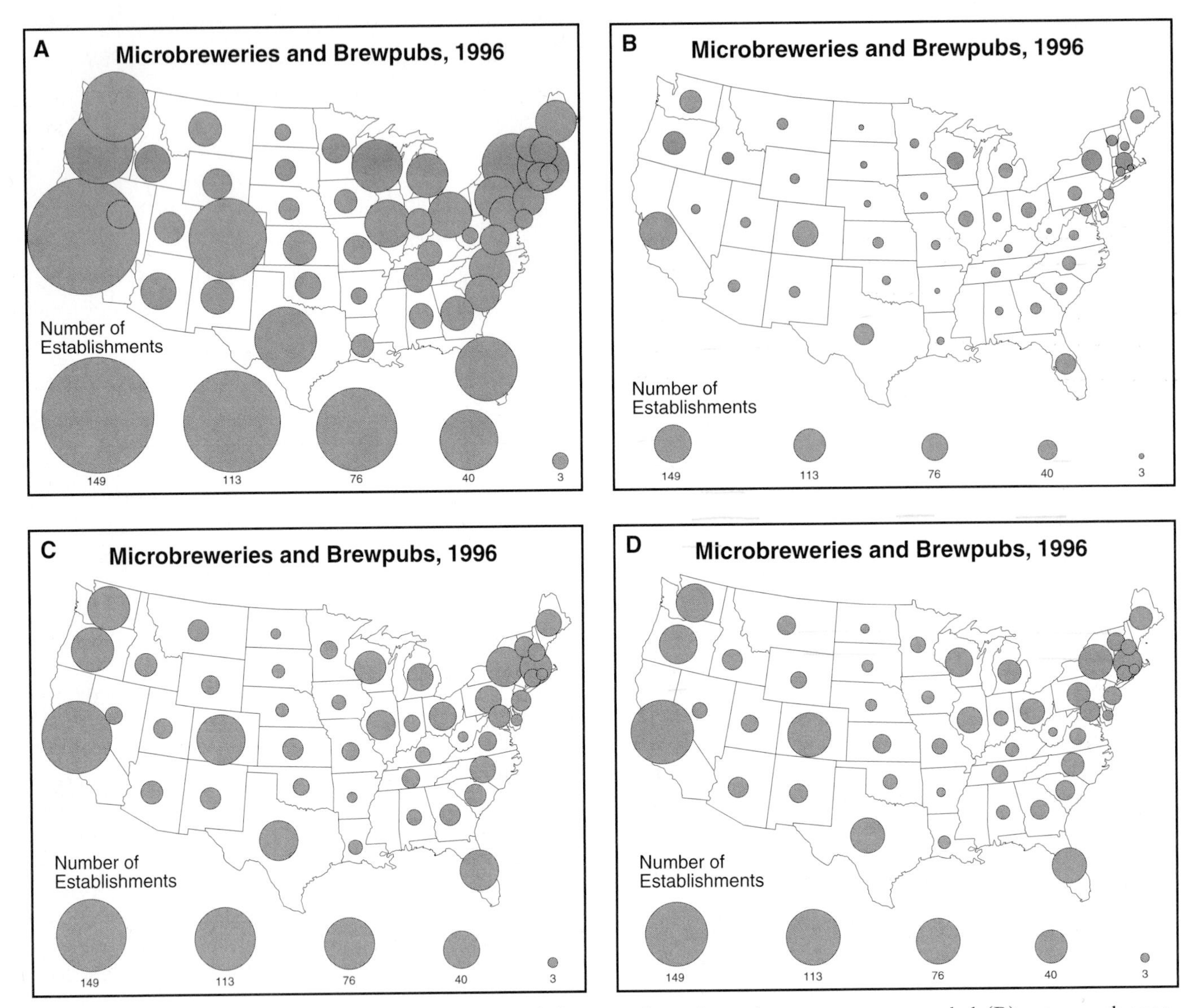

FIGURE 7.15. Effect of varying the amount of overlap: (A) too much overlap—the map appears crowded; (B) not enough overlap—the map appears empty; (C) and (D) examples of maps having an appropriate amount of overlap.

ier to determine the bounds of individual symbols, simplifying the estimation of symbol size. In contrast, the opaque approach enhances **figure-ground contrast** because the proportional symbols tend to appear as a figure against the background of the rest of the map; this is most apparent for the gray and black symbols in Figure 7.16. The opaque approach also promotes a visual hierarchy, as the circles appear "above" other map information (such as a road network).

Studies by Griffin (1990) and Groop and Cole (1978) confirm the above hypotheses. Griffin found that people were about equally split in their preference for transparent and opaque symbols; transparent symbols were liked because they provided "maximum information," while those favoring opaque symbols "stressed the quality of clarity" (p. 24). With respect to fill, Griffin found that for transparent circles, gray and black were about equally popular, while for opaque circles, black was most popular; almost no one liked a white fill. Groop and Cole found that transparent symbols were more accurately estimated than opaque symbols, and that for opaque symbols there was a strong correlation between the amount of overlap and the error of estimation.

Other solutions for handling overlap include **inset maps,** which portray a congested area at an enlarged scale; the use of a **zoom function** in an interactive graphics environment; and the possibility of moving symbols slightly away from the center of congested areas. Zooming is typically implemented by enclosing the area to be enlarged by a rectangular box, clicking a mouse button, and having the outlined area fill the screen. MapTime contains such a zoom function. Figure 7.9C illustrates the

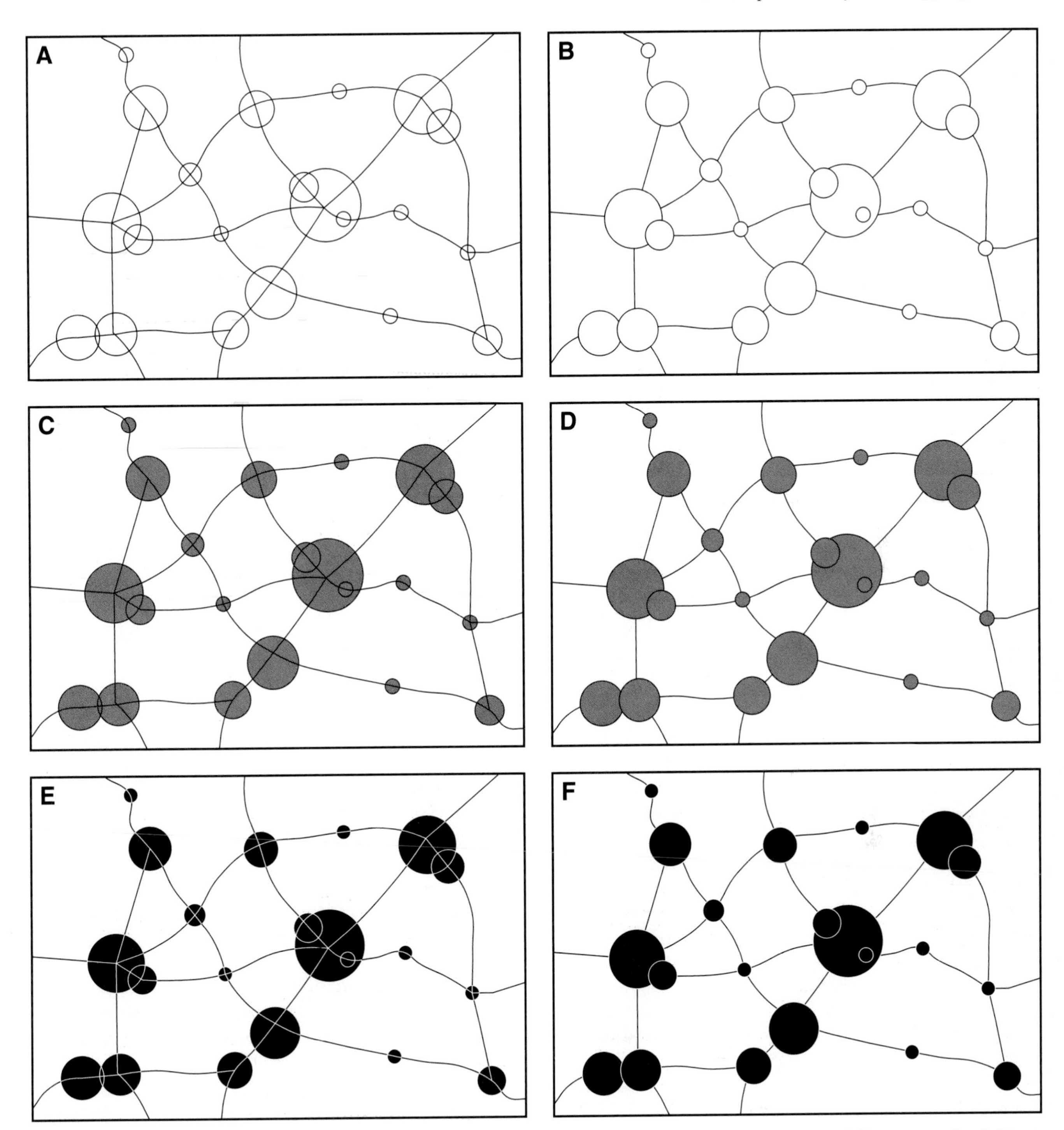

FIGURE 7.16. Transparent and opaque methods of handling overlap. Maps on the left are transparent, while maps on the right are opaque. From top to bottom different fills are shown, ranging from white to gray to black. (After Griffin 1990, 23.)

effect of moving symbols away from congested areas. In comparing 7.9C with Figure 7.9B, we can see that in 7.9C it is easier to compare the sizes of individual circles and the spatial pattern is more obvious. (Note the cluster of small circles in the south-central part of the United States; these circles are similar in size to those near the border to the south.)

7.6 REDUNDANT SYMBOLS

Redundant symbols portray a single attribute with two or more visual variables. For example, on proportional symbol maps, the visual variables size and value could be used (Figure 7.17). The logic of using redundant symbols

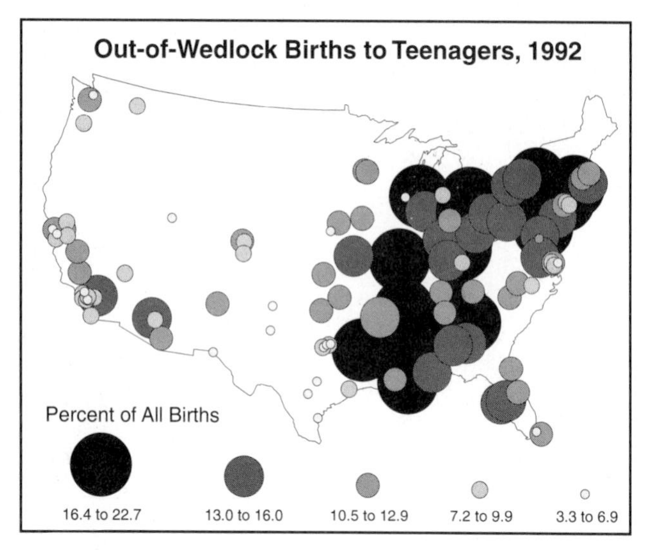

FIGURE 7.17. Use of redundant symbols (the visual variables size and value) on proportional symbol maps.

is that it is easier to discriminate symbols on the basis of two visual variables than one. As a result, users are able to perform specific map tasks more easily. For example, in Figure 7.17 it is possible to differentiate the large circles from others, even though there is considerable overlap (compare with Figure 7.11A).*

Although redundant symbols promote the communication of specific information, they are seldom used on printed maps. I suspect that redundant symbols are not used because mapmakers feel that they impede the communication of general information (it is difficult to combine symbols to form a region when the symbols differ in both size and value). Redundant symbols seem more appropriate for a data exploration environment in which multiple maps can easily be displayed (redundant symbols could be used for acquiring specific information, and nonredundant symbols for acquiring general information).

SUMMARY

In this chapter, we have covered a number of important principles involved in constructing **proportional symbol maps.** We have stressed that two kinds of data can be mapped using proportional symbols: **true point data,** which can actually be measured at a point location, and **conceptual point data,** which are collected over an area (or volume) but are conceived as being located at a point for the purpose of symbolization. Respective examples would be the water released by wells associated with center-pivot irrigation systems and the number of fatalities due to drunk driving in each state. It is often useful to standardize either true or conceptual point data by considering raw counts (or totals) for one variable relative to values of another variable. For example, we might divide the gallons of water pumped at a well by the area covered by the associated center-pivot irrigation system.

Two basic kinds of proportional symbols are possible: geometric and pictographic. **Geometric symbols** (for example, circles and spheres) generally do not mirror the phenomenon being mapped, while **pictographic symbols** (for example, drawings of oil derricks to represent oil production) do. Conventionally, geometric symbols have been more frequently used, but pictographic symbols are becoming common because of the ease with which they can be created with software. Three-dimensional symbols (such as spheres) have traditionally been frowned upon because of the difficulty readers have in estimating their size, but they can produce attractive eye-catching graphics and can be useful for representing a large data range.

There are three methods for scaling (or sizing) proportional symbols: mathematical, perceptual, and range grading. In **mathematical scaling,** symbols are sized in direct proportion to the data; thus, a data value 20 times another is represented by an area (or volume) 20 times as large. In **perceptual scaling,** a correction is introduced to account for visual underestimation of larger symbols; thus, larger symbols are made bigger than would normally be specified by mathematical scaling. For example, when constructing proportional circles, a **Flannery correction** is normally applied so that larger circles appear even larger (a symbol scaling exponent of 0.57 is used). We noted a number of problems with applying such a correction (for example, using a single exponent may be unrealistic because of the variation in perception that exists between subjects). In **range grading,** data are grouped into classes, and a single symbol size is used to represent all data falling in a class. Range grading is considered advantageous because readers can easily discriminate symbol sizes and thus readily match map and legend symbols, and because the contrast in circle sizes may enhance the map pattern.

Other important issues in proportional symbol mapping include legend design, symbol overlap, and symbol redundancy. Basic problems of legend design include the arrangement of symbols (nested or linear) and the number and size of symbols. A nested arrangement conserves map space, while a linear arrangement permits a solid fill, which enhances **figure-ground contrast** (symbols tend to appear as a figure against the background of the rest of the map). Regarding circle overlap, small symbols cause little or no overlap, and thus potentially a map devoid of spatial pattern; in contrast, large symbols can cre-

* For an experimental study dealing with redundancy on proportional symbol maps, see Dobson (1983).

ate a cluttered map, making it difficult to interpret individual symbols. **Redundant symbols** represent a single attribute (say, the magnitude of production at water wells) by two visual variables (for example, size and value), thus enabling users to perform specific tasks more easily and accurately. Since redundant symbols are arguably less effective for portraying general information, they are best used in a data exploration framework.

FURTHER READING

Chang, K. (1980) "Circle size judgment and map design." *American Cartographer* 7, no. 2:155–162.

Summarizes various experimental factors that can affect a power function exponent.

Dent, B. D. (1996) *Cartography: Thematic Map Design.* 4th ed. Dubuque, Iowa: William C. Brown Publishers.

Chapter 8 provides an overview of proportional symbol mapping.

Dobson, M. W. (1983) "Visual information processing and cartographic communication: The utility of redundant stimulus dimensions." In *Graphic Communication and Design in Contemporary Cartography,* ed. D. R. F. Taylor, pp. 149–175. Chichester, England: John Wiley & Sons.

An experimental study dealing with redundancy on proportional circle maps.

Flannery, J. J. (1971) "The relative effectiveness of some common graduated point symbols in the presentation of quantitative data." *Canadian Cartographer* 8, no. 2:96–109.

A frequently cited study for determining an appropriate circle scaling exponent.

Gilmartin, P. P. (1981) "Influences of map context on circle perception." *Annals, Association of American Geographers* 71, no. 2:253–258.

Examines spatial context as a factor in estimating proportional symbol sizes.

Griffin, T. L. C. (1985) "Group and individual variations in judgment and their relevance to the scaling of graduated circles." *Cartographica* 22, no. 1:21–37.

Considers the notion that power function exponents may vary as a function of the individual.

Griffin, T. L. C. (1990) "The importance of visual contrast for graduated circles." *Cartography* 19, no. 1:21–30.

Covers various methods for handling the overlap of proportional circles.

Kerst, S. M., and Howard, J. H. J. (1984) "Magnitude estimates of perceived and remembered length and area." *Bulletin of the Psychonomic Society* 22, no. 6:517–520.

Considers the notion that symbol scaling exponents should consider memory for information.

Lindenberg, R. E. (1986) "The effect of color on quantitative map symbol estimation." Unpublished Ph.D. dissertation, University of Kansas, Lawrence, Kansas.

Examines the role of color in the adjustment of circle scaling exponents and concludes that when all circles are the same color, Flannery's adjustment is appropriate.

Meihoefer, H.-J. (1969) "The utility of the circle as an effective cartographic symbol." *Canadian Cartographer* 6, no. 2:105–117.

Develops an appropriate set of range-graded circles.

Olson, J. M. (1975a) "Experience and the improvement of cartographic communication." *Cartographic Journal* 12, no. 2:94–108.

Considers the notion that readers might be trained to improve their estimates of symbol size.

Olson, J. M. (1976a) "A coordinated approach to map communication improvement." *American Cartographer* 3, no. 2:151–159.

Considers how changes in both map design and reader training might be used to enhance map communication. Of particular interest is the discussion of adjustments to the symbol scaling exponent to enhance map pattern.

Patton, J. C., and Slocum, T. A. (1985) "Spatial pattern recall: An analysis of the aesthetic use of color." *Cartographica* 22, no. 3:70–87.

Examines the effect of circle color on the ability to recall the pattern of proportional circles.

Rice, K. W. (1989) "The influence of verbal labels on the perception of graduated circle map regions." Unpublished Ph.D. dissertation, University of Kansas, Lawrence, Kansas.

An extensive study of the role that verbal labels play in the perception of regions on proportional circle maps.

Slocum, T. A. (1983) "Predicting visual clusters on graduated circle maps." *American Cartographer* 10, no. 1:59–72.

Describes a model for predicting the visual clusters that readers see on proportional circle maps.

8

Interpolation Methods for Smooth Continuous Phenomena

OVERVIEW

Smooth continuous phenomena, such as snowfall or the earth's topography, are commonly depicted as ***isarithmic maps*** *(or* ***contour maps****). Chapter 8 considers the problem of* interpolating *data between irregularly spaced control points of known value; for example, if we wish to create an isarithmic map of snowfall in Michigan, initially we would be provided data only at irregularly spaced weather stations and have to interpolate data for locations in between the stations. Chapter 9 considers the problem of* symbolizing *data representing smooth continuous phenomena. Such data can come from the interpolation methods described in this chapter or from an equally spaced grid (as is commonly done for digital elevation data).*

It is important to recognize that isarithmic maps can be created for both true and conceptual point data. True point data are measured at point locations, while conceptual point data are collected over an area but treated as points. ***Isometric*** *and* ***isopleth maps*** *are isarithmic maps produced using these two forms of data. For example, an isarithmic map of temperatures recorded at weather stations is an isometric map, while an isarithmic map of murder rates based on census tracts is an isopleth map.*

Sections 8.1 to 8.3 discuss three common interpolation methods appropriate for true point data: triangulation, inverse distance, and kriging. ***Triangulation*** *fits a set of triangles to control point locations and then interpolates along the edges of these triangles.* ***Inverse-distance*** *methods lay an equally spaced grid of points on top of the control points, estimate values at each grid point as a function of their distance from control points, and then interpolate between grid points.*

Kriging *is similar to inverse distance methods in that a grid is overlaid on control points and values are estimated at each grid point as a function of distance to control points. Rather than considering distances to control points independently of one another, however, kriging considers the spatial autocorrelation in the data, both between the grid point and the surrounding control points, and among the control points themselves. Although the mathematical basis underlying kriging is complex, learning about it is useful because kriged maps are often more accurate than other interpolation methods (kriging is sometimes said to produce an* ***optimal interpolation****).*

Section 8.4 discusses six criteria that can be considered in selecting an interpolation method for true point data: (1) correctness of estimated data at control points (does the method honor the raw data?), (2) correctness of estimated data at other points (how well does the method predict unknown points?), (3) ability to handle discontinuities (such as geologic faults), (4) execution time of the method, (5) time spent selecting interpolation parameters, and (6) ease of understanding the method.

Section 8.5 considers a study by Mulugeta (1996) that evaluated some of the limitations *of automated interpolation approaches for true point data (as compared with traditional manual procedures). Mulugeta performed a quantitative analysis of the accuracy of automated and traditional manual interpolation approaches and had experts (those with knowledge about the phenomenon being mapped) evaluate the quality of the automated approaches. Although his quantitative analysis revealed that the computer-based maps were at least as accurate as the manually produced ones, experts noted a number of limitations of the computer-based maps, such as "jagged lines" and "spurious details."*

The last section of the chapter covers ***pycnophylactic interpolation,*** *a technique appropriate for conceptual point data. Pycnophylactic interpolation begins by assuming that each enumeration unit is raised to a height proportional to the value of its associated control point. This three-dimensional surface is then gradually smoothed, keeping the volume within each individual enumeration unit constant; thus, volume added somewhere within an enumeration unit must be balanced by taking volume from somewhere else in that unit. This is accomplished using a cell-based smoothing process analogous to generalizing procedures used in image processing.*

8.1 TRIANGULATION

Triangulation is logical to address first because it emulates how contour maps are made manually. In manual contouring, we connect control points via straight lines and then linearly interpolate along these lines. For example, Figure 8.1A illustrates a set of straight lines connecting control points, along with the resulting manually contoured lines. Note that when the straight lines are connected, a set of triangles is formed.

Triangulation uses a set of triangles analogous to those employed in manual contouring. Early software for creating these triangles was problematic because "simply entering the [control] points in a different sequence could result [in a different set of triangles and] . . . conspicuously different-appearing contour lines" (Davis 1986, 356). Fortunately, this problem has disappeared as **Delaunay triangles** now provide a unique solution regardless of the order in which control points are specified.

Delaunay triangles are closely associated with **Thiessen polygons,** which are formed by drawing boundaries between control points such that all hypothetical points within a polygon are closer to that polygon's control point than to any other control point. For example,

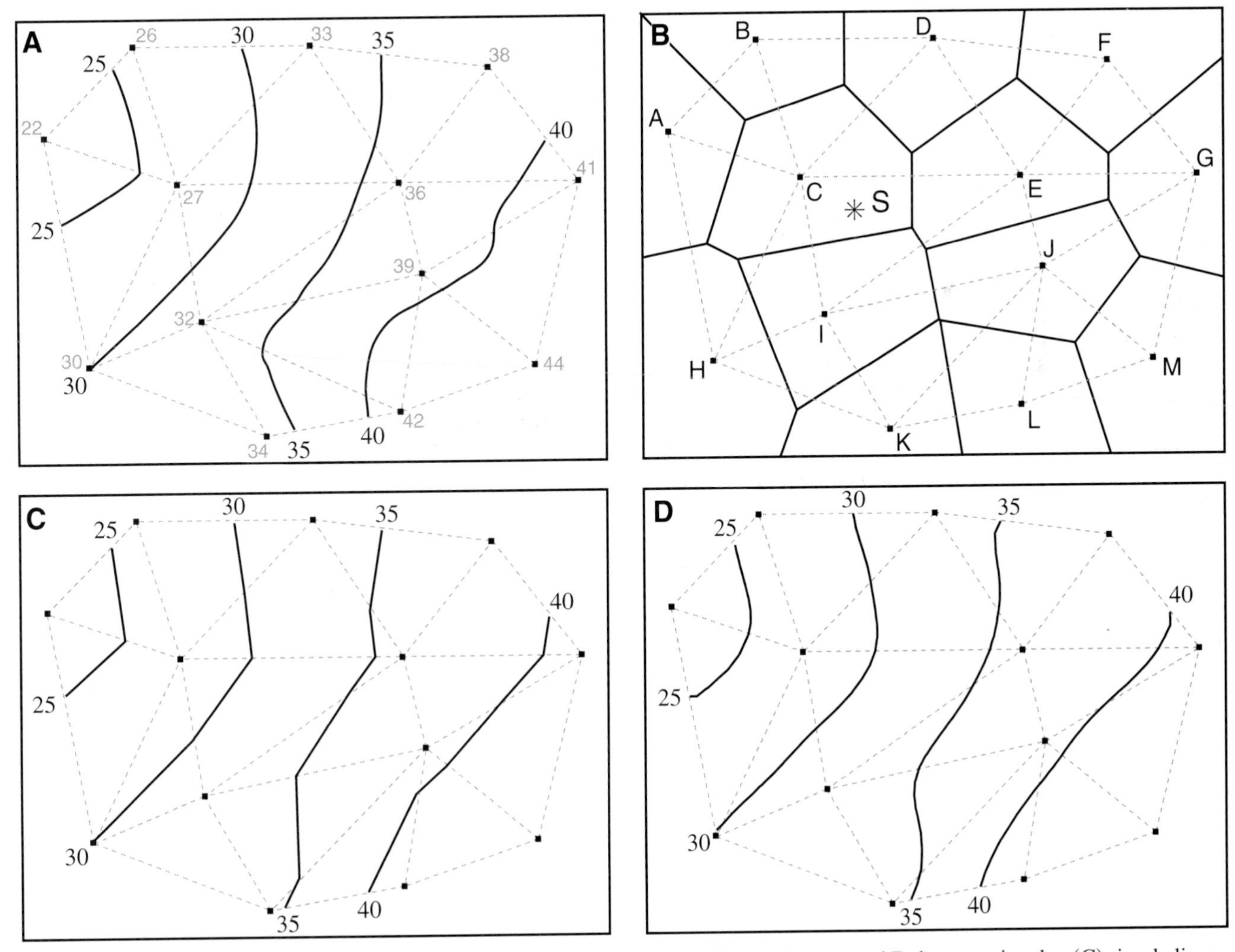

FIGURE 8.1. Hypothetical data showing (A) manual contouring, (B) Thiessen polygons and Delaunay triangles, (C) simple linear interpolation along Delaunay triangle edges, and (D) smoothed interpolation along Delaunay triangle edges.

in Figure 8.1B, hypothetical point S is closer to control point C than to any other control point, so it is part of the Thiessen polygon associated with C. Delaunay triangles are created by connecting control points of neighboring Thiessen polygons. For example, in Figure 8.1B, triangle ABC is formed because the Thiessen polygons associated with control points A, B, and C are all neighbors of one another.

Once Delaunay triangles are formed, contour lines are created by interpolating along the edges of triangles in a fashion similar to the manual process. Delaunay triangles are desirable for this purpose because the longest side of any triangle is minimized, and thus the distance over which interpolation must take place is minimized. A simple linear interpolation along triangle edges leads to angular contour lines, as shown in Figure 8.1C. Smoothing of these lines is obviously necessary if the smooth continuous phenomenon is to be properly represented. Although details of the smoothing procedure are beyond the scope of this text, the process is the three-dimensional equivalent of splining, which will be illustrated below for inverse-distance methods (Davis, 361). The result of smoothing is shown in Figure 8.1D.*

To assist in comparing triangulation with the inverse-distance and kriging approaches to be discussed, Figure 8.2 portrays contour maps of all three methods. These

* Interested readers with a mathematical background should consult McCullagh (1988, 757–761) for more details on smoothing methods for both triangulation and inverse-distance methods.

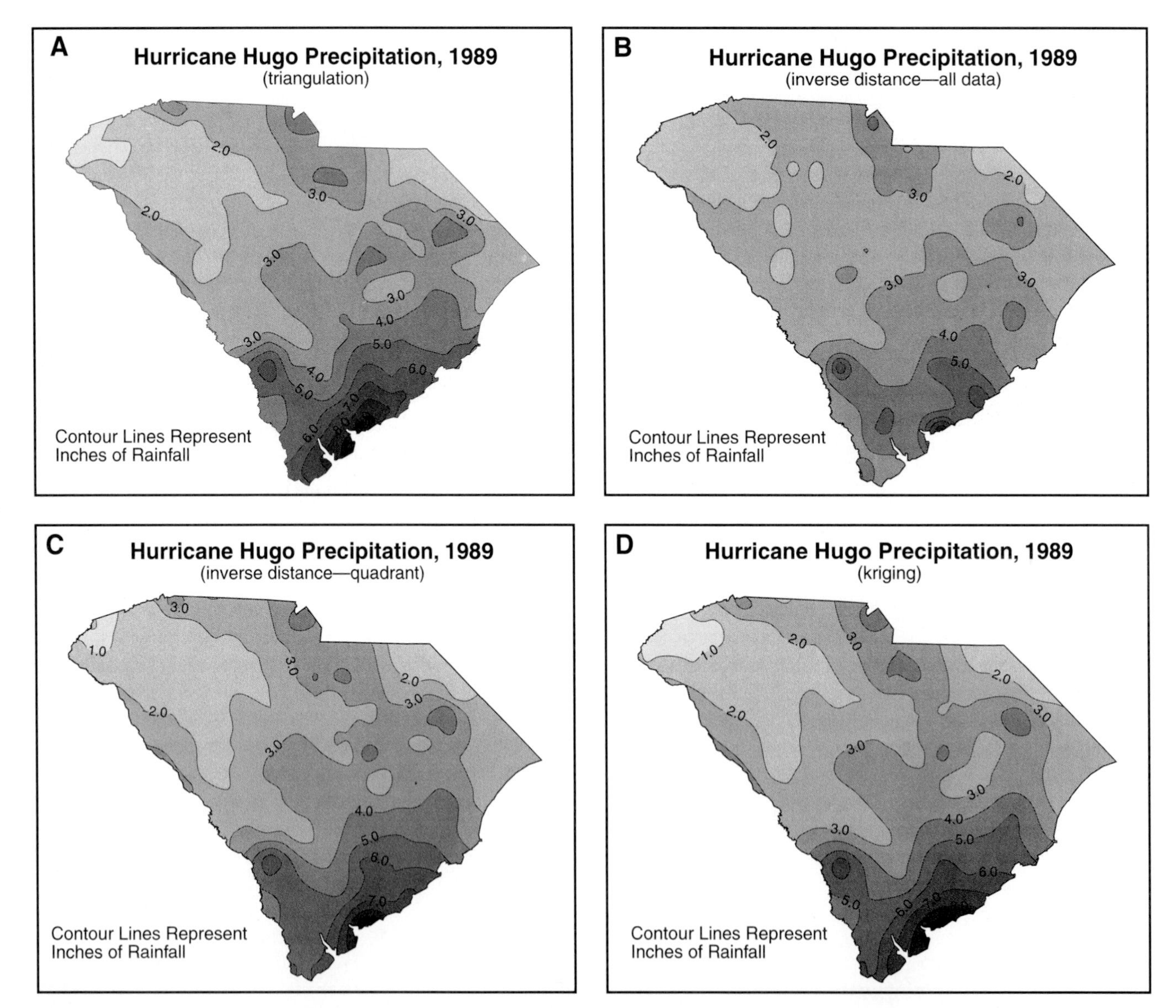

FIGURE 8.2. Contour maps of precipitation data associated with Hurricane Hugo: (A) triangulation, (B) inverse distance using all data, (C) inverse distance using a quadrant approach, and (D) kriging. (Data Source: Southeast Regional Climate Center; Telnet to 198.202.229.3.)

maps are based on precipitation values for 376 weather stations within South Carolina, North Carolina, and Georgia. More specifically, the data cover a four-day period associated with Hurricane Hugo in 1989. Although only South Carolina is shown in Figure 8.2, data for adjacent states were included so that the contour map would be accurate along the state boundary of South Carolina. The results for triangulation are shown in Figure 8.2A.

8.2 INVERSE DISTANCE

Inverse-distance interpolation methods lay a grid on top of the control points, estimate values at each grid point (or grid node) as a function of distance to control points, and then interpolate between the grid points. Because a grid is used, this method is also sometimes termed **gridding.** There is no analogy to this method in manual contouring; a grid is used because "it is much easier to draw contour lines through an array of regularly spaced grid points than it is to draw them through the irregular pattern of the original points" (Davis, 365).

The term *inverse distance* is appropriate for describing gridding methods because control points are weighted as an inverse function of their distance from grid points: control points near a grid point are weighted more than control points far away. The basic formula used for inverse distance is

$$\hat{Z} = \frac{\sum_{i=1}^{n} Z_i/d_i^k}{\sum_{i=1}^{n} 1/d_i^k}$$

where $\hat{Z}$ = estimated value at a grid point
Z_i = data values at control points
d_i = euclidean distances from each control point to a grid point
k = power to which distance is raised
n = number of control points used to estimate a grid point

To illustrate, Figure 8.3 depicts a portion of a hypothetical grid laid on top of four control points, along with calculations for the distances from each control point to the central grid point. If we assume $k = 1$, then the central grid point is calculated as follows:

$$\hat{Z} = \frac{(Z_1/d_1^1) + (Z_2/d_2^1) + (Z_3/d_3^1) + (Z_4/d_4^1)}{(1/d_1^1) + (1/d_2^1) + (1/d_3^1) + (1/d_4^1)}$$
$$= \frac{(40/2.24) + (60/1.00) + (50/1.00) + (40/1.41)}{(1/2.24) + (1/1.00) + (1/1.00) + (1/1.41)}$$
$$= 49.5$$

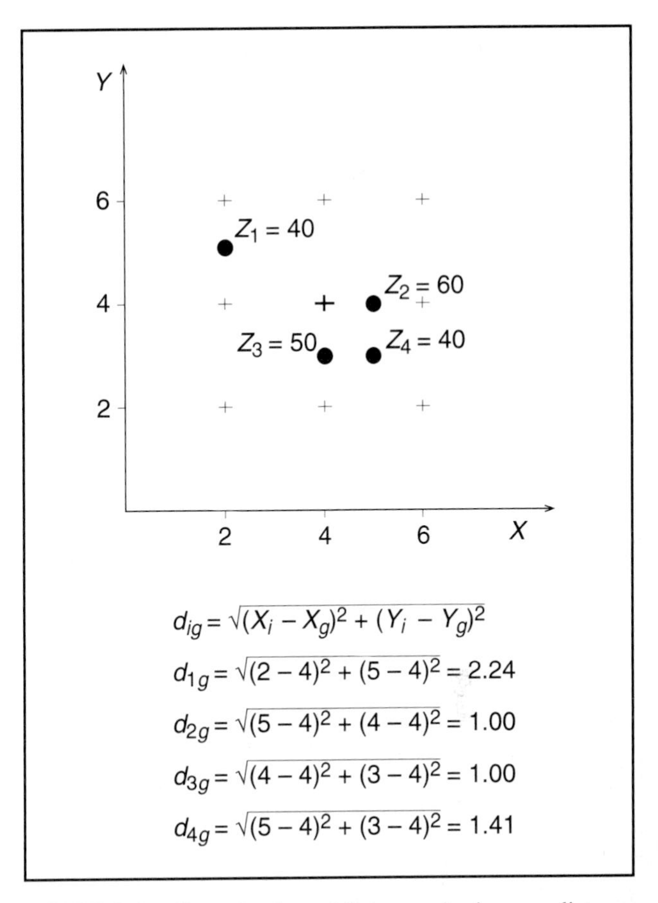

FIGURE 8.3. Computation of distances for inverse-distance contouring. Four control points are overlaid with a hypothetical grid; distance calculations are shown from each control point to the central grid point.

k is normally set to 2, which cancels the square root computation in distance calculations, and thus saves computer time (Isaaks and Srivastava 1989, 259).

With real-world data, obviously, the number of grid and control points will be greater than this simple hypothetical case. A larger number of control points raises the question, Which control points should be used to estimate each grid point? One approach is to use all control points; in this case, n in the formula for $\hat{Z}$ would simply be the number of control points. This approach seems unrealistic, however, because control points far away from a grid point are unlikely to be correlated with that grid point. This sort of thinking and the need to minimize computations led software developers to create strategies for selecting an appropriate subset of control points.

To illustrate the range of strategies, we will consider four available within Surfer, a common contouring package: All Data, Simple, Quadrant, and Octant. The All Data strategy uses *all* control points in the calculation of each grid point. The Simple strategy requires that control points fall within an ellipse (typically a circle) of fixed size; normally, only a subset of these points (say, the eight nearest) is used to make an estimate. Quadrant and Octant strategies also require that control points fall within an ellipse, but the ellipse is divided into four and eight sectors, respectively, with a specified number of control points used within each sector. Surfer also includes options for the minimum number of points that must fall within the ellipse and the maximum number of empty sectors permitted.

To better understand these strategies, the Simple, Quadrant, and Octant strategies are illustrated with hypothetical data in Figure 8.4. For sake of comparison, assume that each strategy requires a total of eight points to make an estimate. Note that the Simple strategy takes the nearest eight points regardless of the direction of the points; Quadrant, the nearest two points within each of four sectors; and Octant, the nearest point in each of eight sectors. For these hypothetical data, it appears that either the Quadrant or Octant strategy is preferable to the Simple strategy because the latter does not use any control points southwest of the grid point.

To further illustrate the effect of the strategies, consider the contour maps of the Hurricane Hugo data using the All Data and Quadrant strategies (Figure 8.2B and 8.2C). For the Quadrant strategy, I specified that two control points be used in each sector, permitted a grid point to be calculated using as few as six control points, and allowed a maximum of one sector to have no control points. I chose these parameters, in part, to demonstrate how the All Data and Quadrant strategies can lead to quite different looking maps. Since precipitation data were not available for the ocean surface, the latter two parameters also insured that grid points would be calculated near the land-ocean interface.

For the Quadrant strategy, I also specified differing ellipse radii (2 degrees latitude and 2.4 degrees longitude) because the control points were specified in degrees latitude and longitude, and at this latitude 1 degree of latitude is not equal to 1 degree of longitude.* One could also modify the radii to account for the possibility that the precipitation data might be more highly related (autocorrelated) along one axis than another. For example, the user's guide for Surfer (Golden Software 1995, 5–27) suggests that for temperature data in the upper Midwest of the United States a longer axis in the east-west direction might be appropriate because temperatures are more similar in that direction. For the precipitation data, one might consider orienting the ellipse along the direction of the storm track, assuming that it was constant.

Comparing the All Data and Quadrant maps in Figure 8.2, we see that they have similar overall patterns, but that the All Data map exhibits more gradual changes; for example, if you compare the north-central portion of the maps, you will note that the distance between the 2.0 and 3.0 contour lines is greater on the All Data map. This result is not surprising given

* Differing ellipse radii would not be necessary if the control points were specified in terms of an equidistant projection.

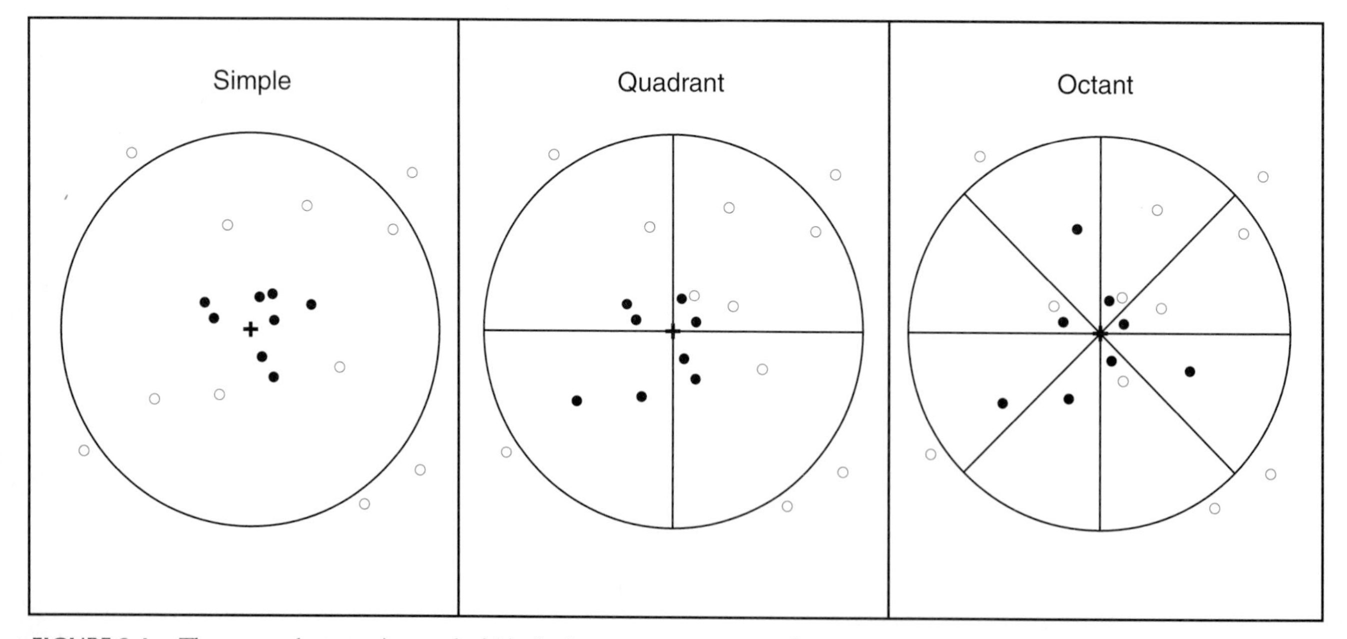

FIGURE 8.4. Three search strategies used within Surfer, a common contouring package. The cross in the middle represents a grid point location to be estimated, and the large circle represents the search radius within which control points must be located. Actual control points selected in each case are shown in black.

that all control points are included in grid calculations on the All Data map and thus we would expect estimated values for grid points closer to the average of control point values. It is also interesting to note that the Quadrant map is very similar to the triangulated map. One wonders whether this would be true with other forms of data; we will deal with this question in more detail in section 8.4.

In covering inverse distance methods, we have focused on grid point calculations because of the considerable number of options involved. Remember, however, that after grid point calculation, an interpolation must also take place between grid points. As with triangulation, a simple linear interpolation leads to angular contour lines; angularity can be avoided either by increasing the number of grid points (which increases computations and memory requirements) or by smoothing the contours. Smoothing is accomplished by **splining,** which involves fitting a mathematical function to points defining a contour line. Details of the mathematics are beyond the scope of this text (see Davis 1986, 204–211, for details), but the concept can be illustrated graphically, as shown in Figure 8.5.

One problem with the inverse-distance methods described thus far is that they cannot account for trends in the data. For example, a visual examination of a set of control points may suggest a value for a grid point higher than any of the control points, but the inverse-distance method cannot produce such a value. This problem exists because the formula for inverse distance is a weighted *average* of control points: an average of values cannot be lesser or greater than any input values. One solution to this problem is to fit a **trend surface** to a set of control points surrounding a grid point, and then insert X and Y coordinates for the grid point into the trend surface equation to estimate a value at the grid point.

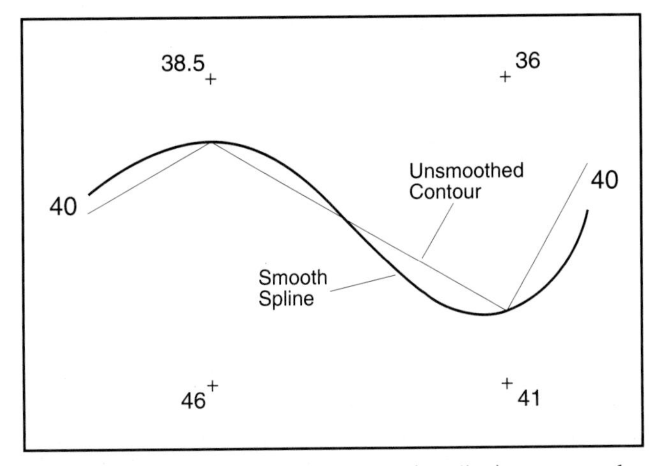

FIGURE 8.5. Fitting a smooth curve (a spline) to an angular contour line. (After Sampson 1978, 50; courtesy of Kansas Geological Survey.)

To illustrate the trend surface approach, consider a first-order surface (or a plane), which has an equation of the form

$$Z = b_0 + b_1X + b_2Y$$

where Z = attribute being mapped
X and Y = geographic coordinates of the data
b_0 = intercept of the plane with the Z axis
b_1 and b_2 = slopes in the X and Y directions

If X, Y, and Z values for control points are inserted in the equation, it is possible to solve for the b values. These b values and the X and Y values for a grid point can then be used to estimate a value at a grid point. Davis (368–371) provides a more detailed discussion of this concept, and the contouring package SURFACE III contains implementations of it.

8.3 KRIGING

The term **kriging** comes from Daniel Krige, who developed the method for geological mining applications. (For more on the origins of kriging, see Cressie 1990.) Kriging is similar to inverse-distance methods in that a grid is overlaid on control points, and values are estimated at each grid point as a function of distance to control points. Rather than simply considering distances to control points independently of one another, however, kriging considers the spatial autocorrelation in the data, both between the grid point and the surrounding control points and among the control points themselves. Understanding how spatial autocorrelation is used in kriging requires an understanding of semivariance and the semivariogram.

8.3.1 Semivariance and the Semivariogram

Consider the simplified graphic shown in Figure 8.6A, which portrays attribute data for five hypothetical equally spaced control points. Normally, of course, control points are not equally spaced, but it is easier to understand semivariance if initially we assume that they are. **Semivariance** is defined as

$$\gamma_h = \frac{\sum_{i=1}^{n-h} (Z_i - Z_{i+h})^2}{2(n-h)}$$

where Z_i = values of the attribute at control points
h = multiple of the distance between control points
n = number of sample points

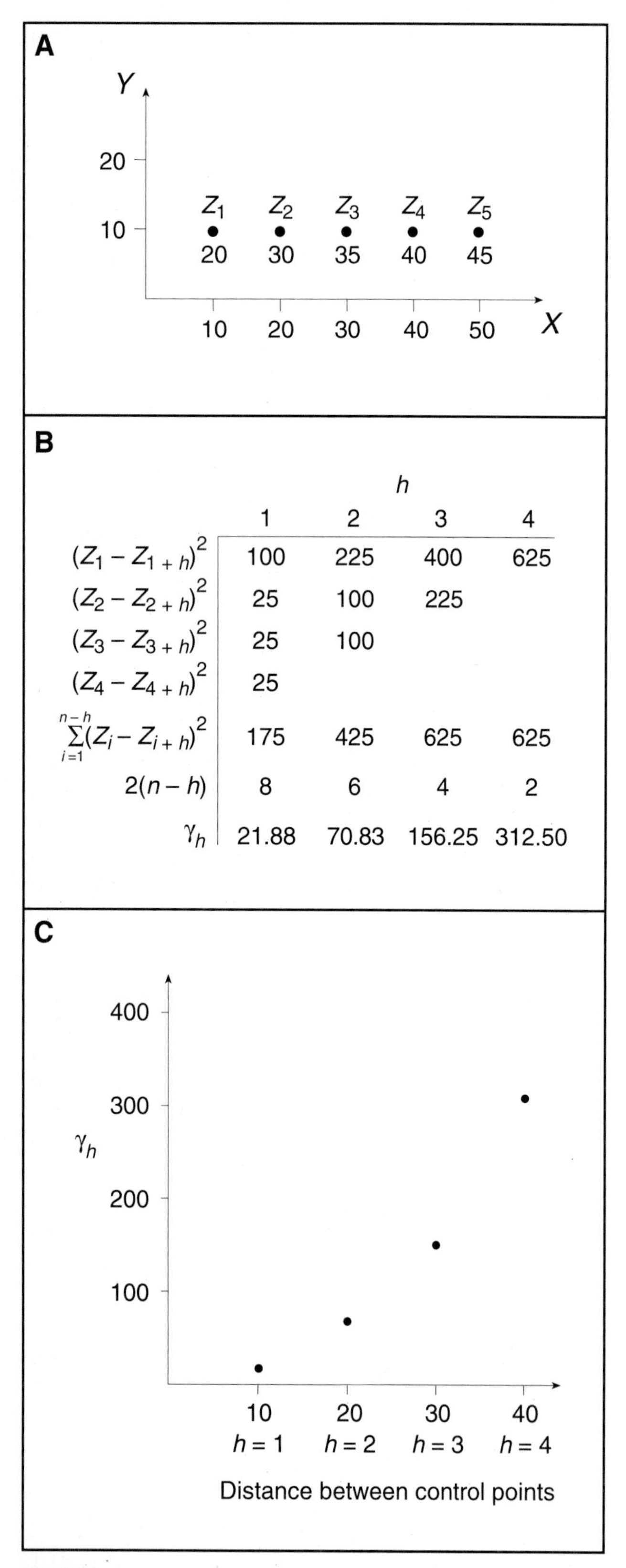

	h = 1	2	3	4
$(Z_1 - Z_{1+h})^2$	100	225	400	625
$(Z_2 - Z_{2+h})^2$	25	100	225	
$(Z_3 - Z_{3+h})^2$	25	100		
$(Z_4 - Z_{4+h})^2$	25			
$\sum_{i=1}^{n-h}(Z_i - Z_{i+h})^2$	175	425	625	625
$2(n-h)$	8	6	4	2
γ_h	21.88	70.83	156.25	312.50

FIGURE 8.6. Computation of semivariance using equally spaced hypothetical control points: (A) the equally-spaced points, (B) semivariance computations, and (C) semivariogram.

In Figure 8.6 the distance between control points is 10; h can take on values of 1, 2, 3, and 4, and n is 5.

To illustrate computations for semivariance, we can insert the data from Figure 8.6A in the above formula for $h = 1$.

$$\gamma_1 = \frac{\sum_{i=1}^{5-1} (Z_i - Z_{i+1})^2}{2(5-1)}$$

$$= \frac{(Z_1 - Z_2)^2 + (Z_2 - Z_3)^2 + (Z_3 - Z_4)^2 + (Z_4 - Z_5)^2}{8}$$

$$= \frac{(20 - 30)^2 + (30 - 35)^2 + (35 - 40)^2 + (40 - 45)^2}{8}$$

$$= 21.88$$

A summary of these calculations for all values of h is shown in Figure 8.6B. I encourage the reader to compute the column for $h = 2$ to insure that the method of computation is understood.

Real-world computation of semivariance requires two modifications to this approach. First, since control points are not normally equally spaced, a tolerance must be allowed in both distance and direction between control points. For example, consider the control points shown in Figure 8.7. If we assume the distance between control points is 5 meters, note that no control points occur exactly 5 meters east of point A. If, however, we permit a distance tolerance of 1 meter and an angular tolerance of 20 degrees, then four of the control points are "east"

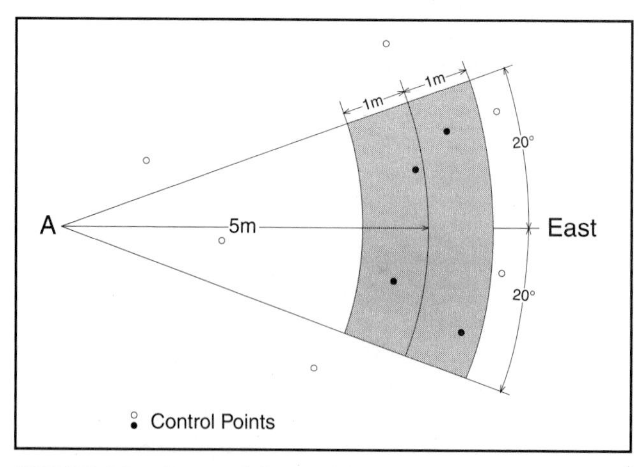

FIGURE 8.7. Determining which control points will be used in semivariance computations. A fixed distance yields no points directly 5 meters east of point A. A tolerance of 1 meter and 20 degrees, however, yields four control points (From *Applied Geostatistics* by E. H. Isaaks and R. Mohan Shrivastava. Copyright © 1989 by Oxford University Press, Inc. Used by permission of Oxford University Press, Inc.)

of point A. The second modification is to calculate semivariance in a variety of directions; thus, in addition to computing along the x axis (east-west direction), computations should also be made in north-south, northwest-southeast, and northeast-southwest directions. For now, we'll assume that such computations are combined to create a single semivariance value.

The **semivariogram** is a graphical expression of how semivariance changes as h increases. For example, Figure 8.6C is a semivariogram for the hypothetical data shown in Figure 8.6A. Clearly, as h increases (i.e., as control points become more distant), the semivariance also increases. This basic feature is characteristic of most data and should not surprise us: we expect nearby geographical data to be more similar than distant geographical data.

The behavior of semivariograms with larger data sets is characterized by the idealized curve shown in Figure 8.8. Note that the semivariance increases as it did in Figure 8.6C, but eventually reaches a plateau. The value for semivariance at which this occurs is known as the *sill*, and the distance at which it occurs is known as the *range*. The flattening normally indicates that the data are no longer similar to nearby values, but rather that the semivariance has approached the variance in the entire data set.

When using the semivariogram in kriging, we will see that it is necessary to make an estimate of the semivariance for some arbitrary distance between points. For example, in Figure 8.6C we might wish to estimate the semivariance for a distance of 26. This is normally accomplished by fitting a curve (or model) to the set of points comprising the semivariogram, a process known as *modeling the semivariogram*. Once an equation for the model is determined, a value for distance can be inserted in the equation, and a value for semivariance computed. The simplest model is a straight-line (linear) one; other common models include the spherical and exponential (Figure 8.9).

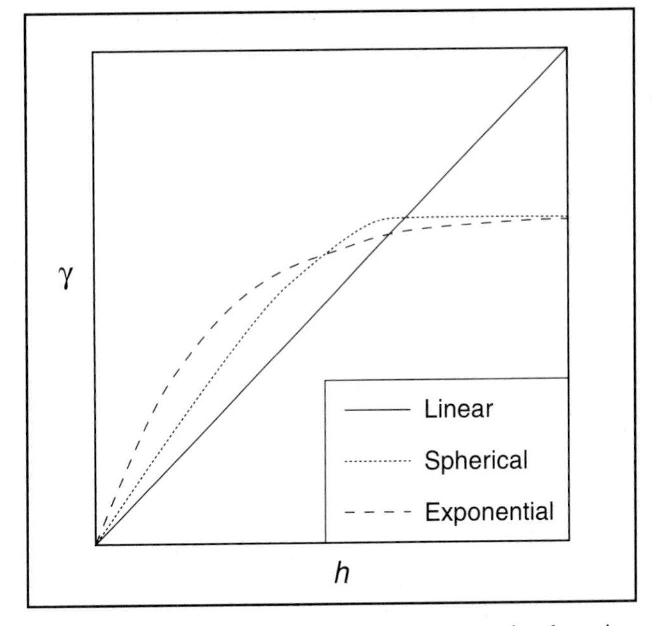

FIGURE 8.9. Models commonly used to summarize the points comprising the semivariogram: linear, spherical, and exponential. (After Olea 1994, 31.)

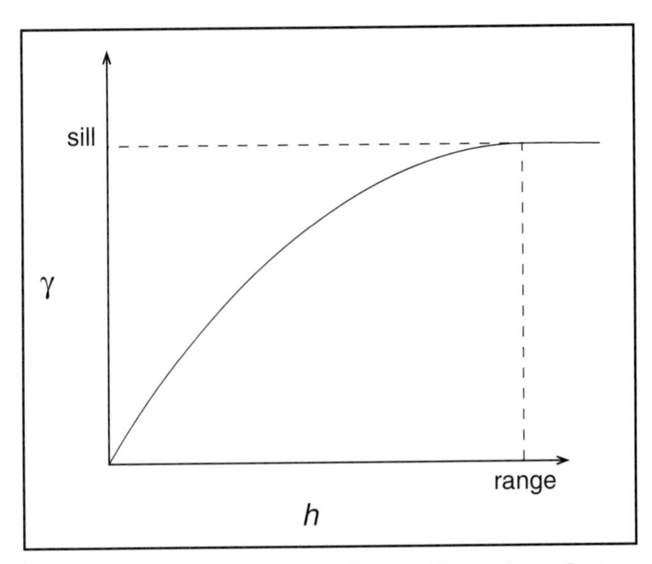

FIGURE 8.8. Idealized semivariogram illustrating a flattening in the semivariance values. The distance at which this occurs is known as the range and the associated semivariance value is the sill.

8.3.2 Kriging Computations

There are two major forms of kriging: punctual and universal. We will focus on **punctual** (or ordinary) **kriging** in which the mean of the data is assumed constant throughout geographic space (there is no trend or drift in the data). The reader is referred to Davis (1986, 393–405) for a discussion of the more complicated method of **universal kriging,** which does account for the trend in the data. To simplify the discussion, we will assume that only three control points are used to estimate a grid point; later we will relax this constraint.

In a fashion similar to inverse-distance methods, kriging uses a weighted average to compute a value at a grid point. For three control points, the equation is

$$\hat{Z} = w_1 Z_1 + w_2 Z_2 + w_3 Z_3$$

where $\hat{Z}$ = estimated value at a grid point
Z_1, Z_2, and Z_3 = data values at the control points
w_1, w_2, and w_3 = weights associated with each control point

The w_i are analogous to the $1/d$ values used in inverse-distance computations; the formula appears simpler for kriging because the w_i are constrained to sum to 1.0.

In kriging, the weights (w_i) are chosen to minimize the difference between the estimated value at a grid point and the true (or actual) value at that grid point. This is analogous to the situation in regression analysis in which we minimize the difference between the estimated value of a dependent variable and its true value (see section 3.3.2). In kriging, minimization is achieved by solving for the w_i in the following simultaneous equations:

$$\begin{aligned} w_1\gamma(h_{11}) + w_2\gamma(h_{12}) + w_3\gamma(h_{13}) &= \gamma(h_{1g}) \\ w_1\gamma(h_{12}) + w_2\gamma(h_{22}) + w_3\gamma(h_{23}) &= \gamma(h_{2g}) \\ w_1\gamma(h_{13}) + w_2\gamma(h_{23}) + w_3\gamma(h_{33}) &= \gamma(h_{3g}) \end{aligned}$$

where $\gamma(h_{ij})$ = the semivariance associated with the distance between control points i and j, and $\gamma(h_{ig})$ is the semivariance associated with the distance between the ith control point and a grid point (Davis 1986, 385). Note that these equations consider not only the distance from control points to the grid point (used in calculating $\gamma(h_{ig})$), but also the distances between the control points themselves (used in calculating $\gamma(h_{ij})$; contrast this approach with inverse-distance methods, which consider only distances between the grid point and control points.

To illustrate the nature of these equations, assume that we are given three hypothetical control point values and wish to estimate a value for a grid point (Figure 8.10). Figure 8.10B lists the X and Y coordinates for all points and 8.10C lists the distances between pairs of points. Also assume that Figure 8.10D is an appropriate linear model for a semivariogram associated with a larger data set from which the three points were taken: the model (equation) states that the semivariance is 10 times the distance (when h is 5, γ is 50).

Determining a semivariance for use in the simultaneous equations requires entering the semivariogram in 8.10D with the distance between two points, and then reading off the corresponding semivariance. For example, for control points 1 and 2, the distance is 3.16 and the semivariance is approximately 30. A more precise estimate can be obtained by using the equation for the linear semivariogram: 10 times 3.16 yields 31.6. Repeating this process for each pair of points produces the semivariance values shown in Figure 8.10E. Inserting these values into the simultaneous equations, we have

$$\begin{aligned} w_1 0.00 + w_2 31.6 + w_3 22.4 &= 22.4 \\ w_1 31.6 + w_2 0.00 + w_3 22.4 &= 10.0 \\ w_1 22.4 + w_2 22.4 + w_3 0.00 &= 14.1 \end{aligned}$$

Davis (pp. 385–388) provides details of how these equations can be solved using matrix algebra. I did so using the

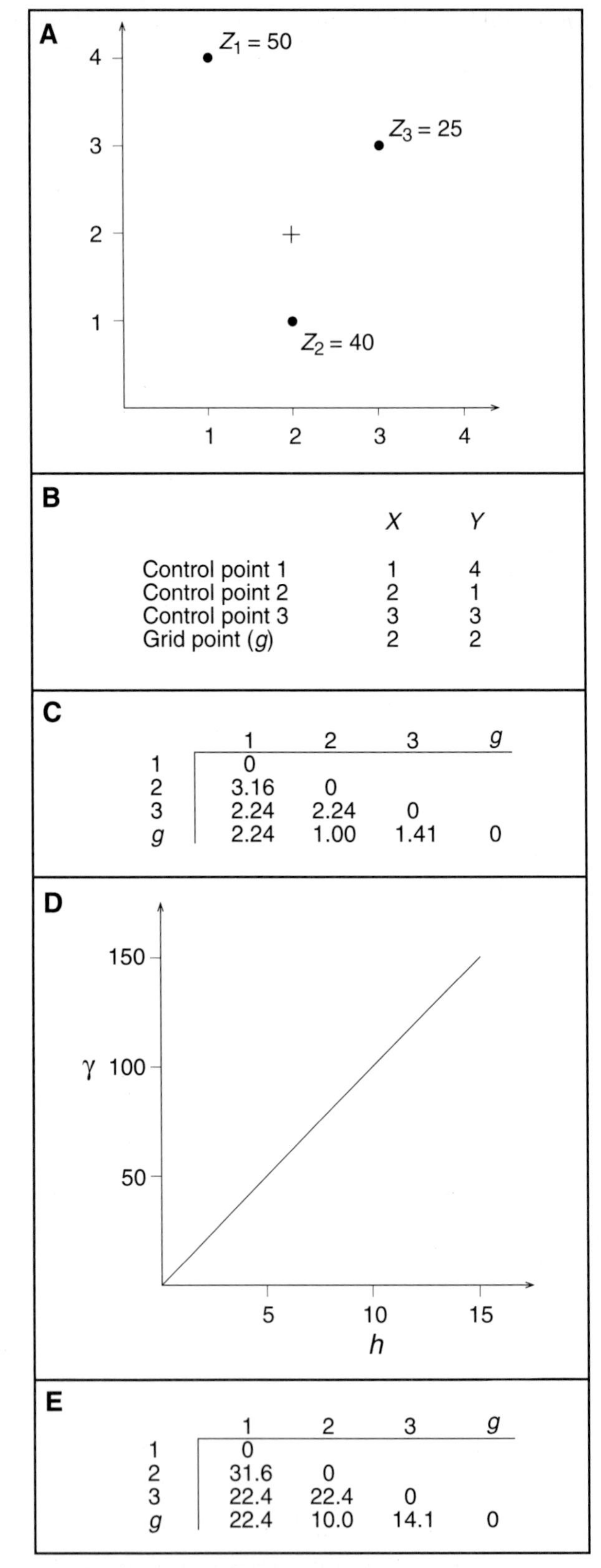

FIGURE 8.10. Semivariance computations for a grid point and three control points: (A) graph of the grid point and control points; (B) X and Y coordinates for the points; (C) distances between points; (D) linear model of a semivariogram for a data set from which the points were presumed to have been taken; (E) semivariances associated with distances between points. (After Davis 1986, 387.)

statistical package SPSS, and found the following values for w: $w_1 = 0.15$, $w_2 = 0.55$, and $w_3 = 0.30$. Inserting these and the values for the control points into the basic weighted-average formula for kriging, we have

$$\hat{Z} = 0.15(50) + 0.55(40) + 0.30(25) = 37$$

Using kriging in real-world situations (such as with the Hurricane Hugo data) differs from this simplified example in two important ways: (1) more than three control points normally are used to estimate each grid point, and (2) more than one semivariogram may be appropriate. With respect to the number of control points, one normally specifies a search strategy similar to that used for the inverse distance method. For example, Isaaks and Srivastava (1989, Chapter 14) suggest using 3 points in each quadrant, for a total of 12 points overall, and a search radius slightly larger than the average spacing between control points.

For the Hurricane Hugo data, I followed Isaaks and Srivastava's recommendations (using the same ellipse radii as for my Quadrant map). Additionally, I permitted a grid point to be calculated using as few as six control points and a maximum of one sector to have no control points; remember that these parameters were necessary to have grid points estimated along the land-water boundary. The resulting map (Figure 8.2D) appears very similar to both the Quadrant and triangulated maps. Again, one wonders whether the similarity among these maps is a function of this particular data set. Or do these various interpolation methods yield similar results for a wide variety of data?

More than one semivariogram may be appropriate for a particular data set because the nature of spatial autocorrelation may vary as a function of the direction of semivariance calculation (for example, east-west versus north-south), as in the upper Midwest temperature example discussed when considering inverse-distance methods. Ideally, this notion should be explored by creating separate semivariograms for each direction of semivariance computation; concomitantly, it may also be necessary to fit different models (equations) to each semivariogram. Since this procedure can become quite complex, I refer the reader to Isaaks and Srivastava (1989) for a more detailed discussion.

For the Hurricane Hugo precipitation data, two semivariograms might have been useful if we wished to consider the direction of hurricane movement. For example, if over the four-day period of rainfall, the hurricane moved in a constant direction, we might have used one semivariogram parallel to the movement and one perpendicular to it. Since I did not have data regarding hurricane movement, I chose a single semivariogram. I chose a linear model based on a visual examination of a sample semivariogram.

Although kriging is admittedly more complex than other methods of interpolation, it can produce a more accurate map; as such, it is often said to produce an **optimal interpolation.** It must be stressed that this is only true if one has made the proper specification of the semivariogram(s) and associated semivariogram models. This is important to recognize because software for kriging may provide simplified defaults (such as a single semivariogram and linear model) that will not produce an optimal kriged map.

In addition to being termed an optimal method, another advantage of kriging is that it provides a measure of the error associated with each estimate, known as the *standard error of the estimate.* This error measure can be used to establish a *confidence interval* for the true value at each grid location, assuming a normal distribution of errors. For example, if the kriged estimate at a grid point is 2.3 inches of rainfall and the corresponding standard error is 0.5 inches, then a 95 percent confidence interval for the true value at that grid point would be $2.3 \pm 2(0.5)$. This is equivalent to saying we are 95 percent certain that the true value is within the range 1.3 to 3.3.*

8.4 CRITERIA FOR SELECTING AN INTERPOLATION METHOD FOR TRUE POINT DATA

So far in the chapter, we have covered three methods for interpolating true point data: triangulation, inverse distance, and kriging. Given the similarity of the maps resulting from applying these three methods to the Hurricane Hugo data, how does one decide which method should be used? This section considers six criteria that can be considered in selecting an interpolation method: (1) correctness of estimated data at control points (does the method honor the raw data?), (2) correctness of estimated data at noncontrol points (how well does the method predict unknown points?), (3) ability to handle discontinuities (such as geologic faults), (4) execution time, (5) time spent selecting interpolation parameters, and (6) ease of understanding. It should be noted that it is assumed for the present discussion that the goal is to create an isarithmic map depicting *values* of a smooth continuous phenomenon. If the intention is to map various aspects related to elevation (such as slope and drainage), then triangulation is a natural choice (Clarke 1995, 144–148).

8.4.1 Correctness of Estimated Data at Control Points

One advantage often attributed to triangulation is that estimated values for control points will be identical to the raw values originally measured at those control

* See Olea (1992, 15–16) for approaches to handling nonnormal data.

points; this is known as **honoring the control point data.** For example, for the triangulated map shown in Figure 8.1D, the contour line for 30 passes directly through control point H, which has a raw value of 30. In contrast, inverse-distance and kriging methods do not, in general, honor control points. An exception would be when control and grid points coincide: in this case, the distance between control point and grid point will be zero, and the value for the grid point can simply be made equivalent to the control point value.

Although honoring control point data is worthwhile, using it as the key criterion for evaluating interpolation seems risky for two reasons. First, the process presumes that data are measured without error; this is, of course, incorrect. It is common knowledge that physical recording devices (such as a rain gauge) are imperfect, and their readings may also be subject to human error. Rather than exactly honoring control points, it makes sense that such estimates should be within a small tolerance of the raw value.

A second problem with honoring control points is that the process does not necessarily guarantee correctness at noncontrol point locations. For example, in an evaluation of inverse-distance methods, Davis (1975) found that methods providing good estimates at control point locations did poorly at noncontrol point locations.

8.4.2 Correctness of Estimated Data at Noncontrol Points

Two basic approaches have been used for evaluating correctness of data at noncontrol points: cross validation and data splitting. **Cross validation** involves removing a control point from the data to be interpolated, using other control points to estimate a value at the location of the removed point, and then computing the *residual,* the difference between the known and estimated control point values; this process is repeated for each control point in turn. If cross validation is done for a variety of interpolation methods, the resulting sets of residuals can be compared. Isaaks and Srivastava (1989, Chapter 15) provide a thorough discussion of this approach.

In **data splitting,** control points are split into two groups, one to create the contour map, and one to evaluate it. Residuals are computed for each of the control points used in the evaluation stage. One problem with data splitting is that it is impractical with small data sets because it makes sense to use as many control points as possible to create the contour map.

Ideally, mapmakers should use either cross validation or data splitting to evaluate the accuracy of contour maps. Since this can be time-consuming, it is useful to consider the results of those who have already done this sort of analysis. The results of two such studies are summarized below.*

Isaaks and Srivastava (1989, 249–277, 313–321, 338–349) evaluated triangulation, inverse distance, kriging, and other methods for a set of digital elevation values in the Walker Lake area near the California-Nevada border. One important conclusion of their study was that the " 'best' [interpolation method] depends on the yardstick we choose" (p. 272). For example, triangulation produced a smaller standard deviation of residuals than an inverse-distance method (using all control points within a specified search ellipse), but the inverse-distance method minimized the size of the largest residual. Overall, their study revealed that kriging was best (pp. 318–321), although it was only slightly more effective than an inverse-distance approach in which four control points were required in each quadrant surrounding a grid point.

Declercq (1996) evaluated triangulation, inverse distance, and kriging for two types of data: one relatively smooth and a second more abrupt in nature. Smooth data consisted of mean annual hours of sunshine for the European Union, while abrupt data consisted of soil erodability (*K* values) in northern Belgium. One of Declercq's major conclusions was that there was little difference in inverse-distance and kriging methods. He argued that the number of control points used to make an estimate is more important than the general interpolation method: he recommended few control points (4 to 8) for smooth data, and many (16 to 24) for more abrupt data. He also recommended using either a quadrant- or octant-sector approach. Lastly, he recommended not using triangulation approaches because they produce "highly inaccurate values and erratic images . . . in poorly informed regions or with abruptly changing data" (p. 143).

8.4.3 Handling Discontinuities

In addition to honoring the data, another advantage of triangulation is its ability to handle discontinuities, such as geologic faults or cliffs associated with a topographic surface. McCullagh (1988, 763) indicates that "If the locations of these special lines are entered as a logically connected set of points, the triangulation process . . . will automatically relate them to the rest of the data." He further indicates that although grids can handle discontinuities, recognizing a fault in gridded data is almost impossible (p. 766).

* For other studies that have compared interpolation methods, see Mulugeta (1996).

8.4.4 Execution Time

The amount of computer time it takes to create a contour map can be an important consideration for large data sets. In general, a simple linear interpolated triangulation is faster than either inverse distance or kriging because of the numerous computations that must be done at individual grid points with the latter methods. Kriging is especially time-consuming because of the simultaneous equations that must be solved ($n + 1$ simultaneous equations must be solved for each grid point, where n is the number of control points). *Smoothed* triangulation methods, however, can take substantially longer than either inverse-distance or kriging methods (McCullagh 1988).

The study by Declercq provides some comparative figures. Using a 50 megahertz 486-based microcomputer, Declercq computed grid point estimates for a 121×361 grid in less than 5 minutes using either linear triangulation or inverse distance, but almost an hour was required for some kriging computations. A smoothed triangulation approach (using a quadratic-spline interpolation) was clearly the slowest, requiring almost 12 hours.

Given the similarity of contour maps resulting from inverse distance and kriging, the faster execution time of inverse distance would seem to favor it when many grid points must be computed. One must bear in mind, however, that microcomputers are becoming substantially faster, and that for large data sets, one can simply let the computer run while doing something else.

8.4.5 Time Spent Selecting Parameters and Ease of Understanding

Time spent selecting parameters refers to how long it takes to make decisions such as the appropriate power for distance weighting, the number of control points to use in estimating a grid point, which model to use for the semivariogram, or which smoothing method to use for triangulation. One can avoid this time by simply taking program defaults, but such an approach is risky. Kriging is the most difficult in this regard because of the need to consider whether one or more semivariograms are appropriate and what sort of model should be used to fit each semivariogram.

The time spent selecting parameters will obviously become longer if one wishes to evaluate correctness using cross validation or data splitting. This time can be eliminated by using the evaluations of others, such as those described above, although more studies appear necessary to cover the broad range of data types that geographers are apt to deal with.

Ease of understanding refers to the ease of comprehending the conceptual basis underlying the interpolation method. Clearly, kriging scores poorly on this criterion, as the kriging literature (both published articles and software user guides) is often complex.

8.5 LIMITATIONS OF AUTOMATED INTERPOLATION APPROACHES

Although automated interpolation is desirable because it eliminates the effort involved in manual interpolation, a study by Mulugeta (1996) illustrated some of the potential limitations of automated interpolation. Two data sets were used in Mulugeta's study: precipitation totals for a single storm event (June 28, 1982) in a five-county area of Michigan, and bedrock-surface altitudes for a township in Michigan. These data were automatically interpolated using a kriging method and manually interpolated by experts (climatologists and meteorologists for the precipitation data and geomorphologists for the bedrock-surface altitude data).

A quantitative analysis using data splitting revealed that the computer-interpolated precipitation map was significantly more correct than the manually interpolated one, and that there was no significant difference in computer- and manually interpolated bedrock-surface maps. Together these results suggested that the computer-interpolated maps were at least as good as, if not better than, the manual ones. Comments elicited from experts, however, raised serious questions about the appropriateness of automated interpolation.

Criticisms noted by the experts can best be understood by comparing the computer-interpolated map of the storm event with a couple of the manually interpolated maps created by the experts (Figure 8.11). The most significant problem noted by experts was that "isolated peaks and depressions [were] overly smoothed and too circular." (Consider A1–A5 in Figure 8.11A.) The experts would have preferred "elongated bands . . . characteristic . . . of storm-precipitation surfaces" (p. 335).

A second problem was that the computer-interpolated map portrayed "features . . . not warranted by the data points" (p. 335). An example was the peak at B4 in the middle of the computer-drawn map, which "many experts showed . . . as a saddle between two elongated 2[-inch] isarithms [Figure 8.11B] or enclosed . . . in a 1.5[-inch] isarithm [Figure 8.11C]" (p. 335). Other problems noted by experts included "jagged lines" and "spurious details" (D1–D4 and F1–F4 in Figure 8.11A).

Mulugeta indicated that, ideally, mapmakers should edit automated contour maps to alleviate such problems. Unfortunately, if the mapmaker does not have expertise in the domain being mapped, this task will be difficult; for example, those unfamiliar with storm precipitation surfaces would be unlikely to see the need for

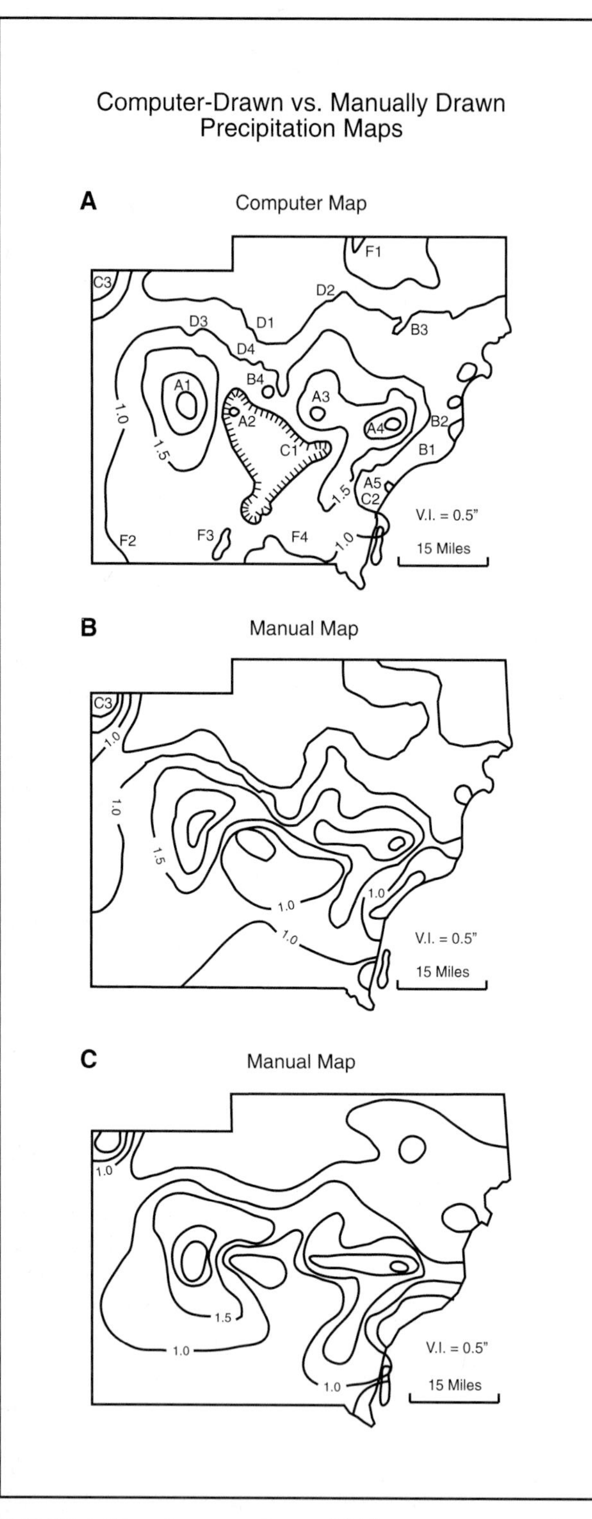

FIGURE 8.11. A comparison of (A) a computer-produced contour map and (B–C) manually produced contour maps. (Mulugeta, 1996. First published in *Annals, Association of American Geographers,* 86(2), p. 335. Reprinted with permission of Association of American Geographers.)

"elongated bands." Other problems, such as "jagged lines" and "spurious details" are capable of being edited by the unsophisticated mapmaker, and thus I encourage readers to correct such problems within common design programs such as Freehand.

8.6 TOBLER'S PYCNOPHYLACTIC APPROACH: AN INTERPOLATION METHOD FOR CONCEPTUAL POINT DATA

Although the methods discussed in sections 8.1 to 8.3 were developed to handle true point data, they also have been commonly used for conceptual point data. For example, I used kriging to create the contour map for the wheat harvested data for Kansas counties described in Chapter 2 (the map is repeated in Figure 8.13B). I accomplished this by assigning the percentage of wheat harvested data to the centroids of each county, which served as control points for interpolation.

Tobler's (1979) pycnophylactic (or volume-preserving) method is a more sophisticated approach for handling conceptual point data. To visualize this method, consider the standardized data associated with enumeration units as a clay model in which each enumeration unit is raised to a height proportional to the data (as in Figure 2.1B). The objective of the pycnophylactic method is to "sculpt this surface until it is perfectly smooth, but without allowing any clay to move from one [enumeration unit] to another and without removing or adding any clay" (p. 520). Relating this concept to the other interpolation approaches we have considered, we can think of volume preservation as a form of *honoring* the data associated with each *enumeration unit.*

To illustrate the pycnophylactic method in more detail, we will use a simplified algorithm developed by Lam (1983, 148–149).* For this purpose, presume that we are given raw counts (RC_i) for the three hypothetical enumeration units shown in Figure 8.12A. These raw counts might be the number of people or acres of wheat harvested in each enumeration unit.

In step 1 of Lam's algorithm, a set of square cells is overlaid on top of the enumeration units, and it is determined which cells fall in each enumeration unit (Figure 8.12B). A cell is considered part of an enumeration unit if its center falls within that unit; this can be determined by a so-called **point-in-polygon test** (Clarke 1995, 207–209). Note that four cells fall within enumeration unit E_1.

In step 2 of the algorithm, a raw count for each cell is determined by dividing the raw count for each enumeration unit by the number of cells in that unit; for example, cells within enumeration unit 1 receive a value of 25/4, or 6.25 (Figure 8.12C). Note that this step essentially standardizes the data by computing a density measure.

* See http://www.ncgia.ucsb.edu/pubs/gdp/pop/pycno.html for a visual representation of pycnophylactic interpolation.

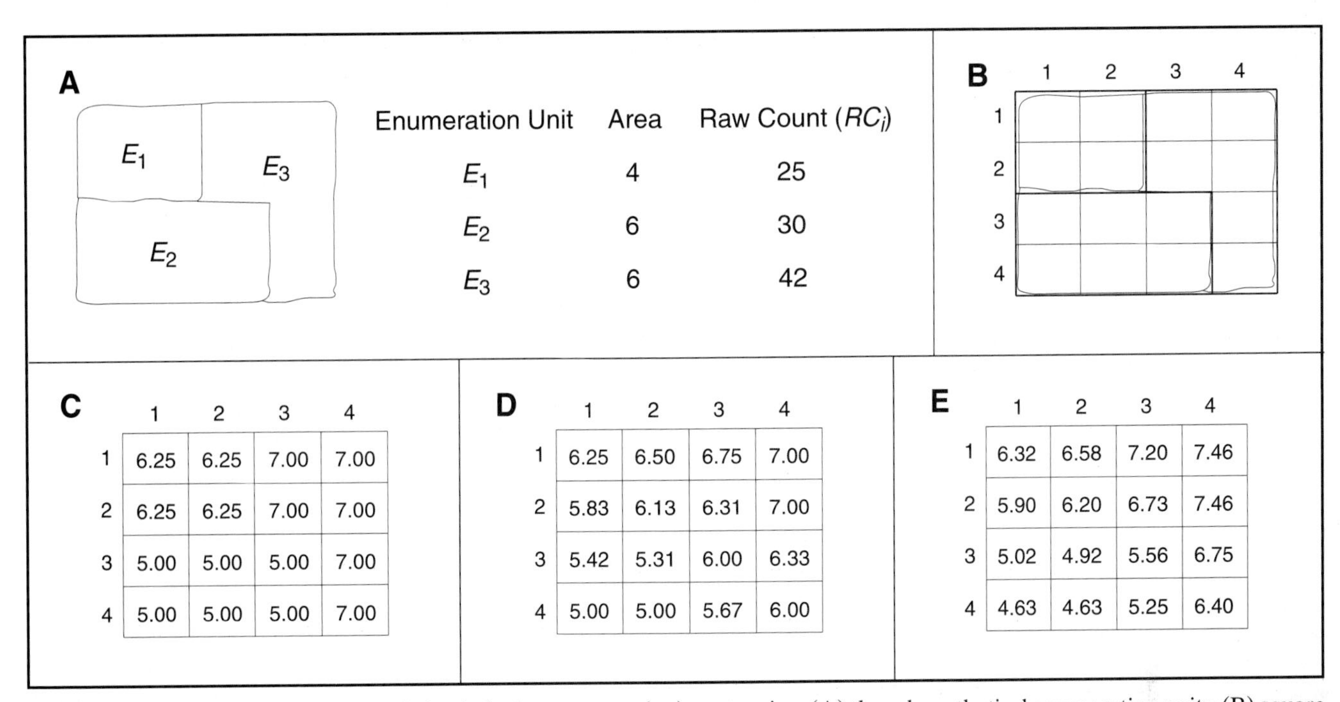

A

Enumeration Unit	Area	Raw Count (RC_i)
E_1	4	25
E_2	6	30
E_3	6	42

C

	1	2	3	4
1	6.25	6.25	7.00	7.00
2	6.25	6.25	7.00	7.00
3	5.00	5.00	5.00	7.00
4	5.00	5.00	5.00	7.00

D

	1	2	3	4
1	6.25	6.50	6.75	7.00
2	5.83	6.13	6.31	7.00
3	5.42	5.31	6.00	6.33
4	5.00	5.00	5.67	6.00

E

	1	2	3	4
1	6.32	6.58	7.20	7.46
2	5.90	6.20	6.73	7.46
3	5.02	4.92	5.56	6.75
4	4.63	4.63	5.25	6.40

FIGURE 8.12. Basic steps of pycnophylactic (volume-preserving) contouring: (A) three hypothetical enumeration units; (B) square cells overlaid on the enumeration units; (C) initial density values for each cell (computed by dividing the raw count for each enumeration unit by the number of cells in that unit); (D) smoothed cell values (achieved by averaging neighboring cells); (E) smoothed values adjusted so that the sum within an enumeration unit equates to the original total sum for that enumeration unit (the volume is preserved). (After Lam 1983, 148–149.)

Steps 3 to 5 of the algorithm are executed in an iterative fashion. The steps are as follows:

Step 3. Each cell is computed as the average of its nondiagonal neighbors. For example, cell (2,2) becomes

$$\frac{6.25 + 6.25 + 7.00 + 5.00}{4} = 6.13$$

The results for all cells for the first iteration are shown in Figure 8.12D. This step accomplishes the smoothing portion of the algorithm.

Step 4. The cell counts within each enumeration unit at the end of step 3 (Figure 8.12D) are added to obtain a total smoothed count value, SC_i for each enumeration unit. For example, the total for enumeration unit 1 is

$$SC_1 = 6.25 + 6.50 + 5.83 + 6.13 = 24.71$$

The results for enumeration units 2 and 3 are 32.40 and 39.39, respectively.

Step 5. All cell values are multiplied by the ratio RC_i/SC_i. For cell (2,2), the result is

$$6.13 \times (RC_1/SC_1) = 6.13 \times (25/24.71) = 6.20$$

The results for all cells for the first iteration are shown in Figure 8.12E.

Note that if the counts in each cell of an enumeration unit in Figure 8.12E are added, the resulting sum is equal to the original raw count for that unit; for example, for enumeration unit 1, we have

$$6.32 + 6.58 + 5.90 + 6.20 = 25.00$$

As a result, steps 4 and 5 enforce the pycnophylactic (volume-preserving) constraint.

Remember that steps 3 to 5 are executed in an iterative fashion. For the second iteration, the results shown in E would be placed where C currently is depicted in Figure 8.12, and steps 3 to 5 would be executed again. Iteration continues until there is no significant difference between the raw and smooth counts for each enumeration unit or there is no significant change in the cell values compared with the last iteration.

One issue stressed by Tobler and not dealt with in the above algorithm is how the boundary of the study area is handled. His computer program for pycnophylactic interpolation, PYCNO, provides two options: one in which zeros are presumed to occur outside the bounds of the region, and one in which a constant gradient of change is presumed across the boundary. The former would be appropriate if the region is surrounded by water, as when mapping population along the coast of the United States. The latter would be appropriate when mapping a

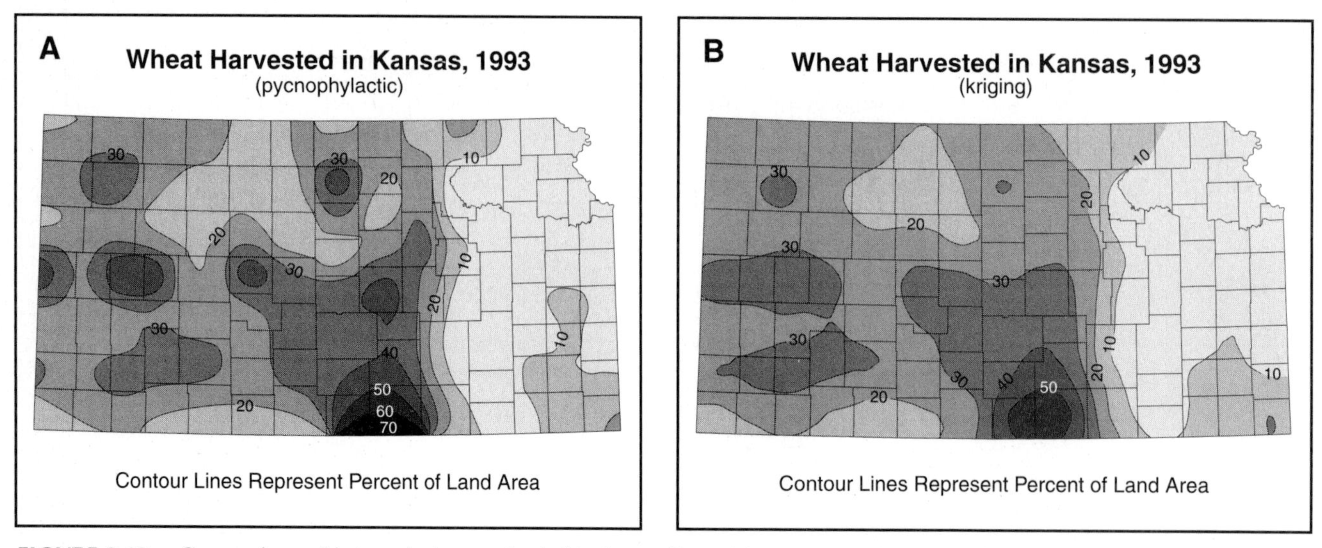

FIGURE 8.13. Comparison of interpolation methods for data collected for enumeration units: (A) the pycnophylactic approach, which expressly deals with the fact that the data were collected from enumeration units; (B) the kriging approach, which treats the areal data as conceptual point locations.

phenomenon that is presumed to have similar characteristics in surrounding enumeration units, as when contouring wheat harvested data for the state of Kansas.

To contrast pycnophylactic interpolation with methods intended for true point data, Figure 8.13 portrays contour maps of the wheat harvested data examined in Chapter 2, using both the pycnophylactic and kriging methods. Although the two maps appear similar, there are some notable differences. One is that the magnitude of the highest contour on the pycnophylactic map is 70, while the highest on the kriged map is only 50. A higher-valued contour appears on the pycnophylactic map because of its volume-preserving character. To understand this, imagine each of the counties in the form of the clay model described by Tobler above. In such a model, you would find that Sumner County is highest, with a raw standardized data value of 58.5 percent. In order to show a smooth transition to lower-valued surrounding counties, Sumner's edges must be beveled off, but to retain the same volume, its center portion must be built up, thus resulting in a higher value than 58.5 in its interior. In contrast, punctual kriging cannot produce a value higher than any control point because each grid point is a weighted average of surrounding control points.*

Although the pycnophylactic method is arguably more appropriate for conceptual point data than point-based interpolation methods, it must be emphasized that the method should only be used with continuous phenomenon. If the phenomenon is not continuous, nothing is gained by using the pycnophylactic approach. Instead, it may be more appropriate to use another mapping method, such as the dot map introduced in Chapter 2.

* Technically, this statement is true only of punctual kriging. Universal kriging can estimate values beyond the range of the data, but it would not honor the data associated with the enumeration unit.

SUMMARY

In this chapter, we have covered a variety of methods for interpolating data associated with smooth continuous phenomena. Two basic forms of data serve as input for interpolation: true point data and conceptual point data. True point data can actually be measured at point locations. An example would be the hours of sunshine received over the course of the year in the United States; such data are collected at irregularly spaced weather stations. Conceptual point data are collected over areas (or volumes), but conceived as located at points. An example of the latter would be birth rates for census tracts within an urban area.

We covered three interpolation methods appropriate for true point data: triangulation, inverse distance, and kriging. **Triangulation** fits a set of triangles to control points and then interpolates along the edges of these triangles. **Delaunay triangles** are commonly used because the longest side of any triangle is minimized, and thus the distance over which interpolation must take place is minimized. Triangulation is advantageous in that it honors the original control points, handles discontinuities well, and is an obvious choice for elevation data when information such as slope and drainage are also desired. A disadvantage of triangulation is that *smoothed* contours (a nonlinear interpolation) can require a lengthy execution time.

Inverse-distance methods lay an equally spaced grid of points on top of control points, estimate values at each grid point as a function of their distance from control

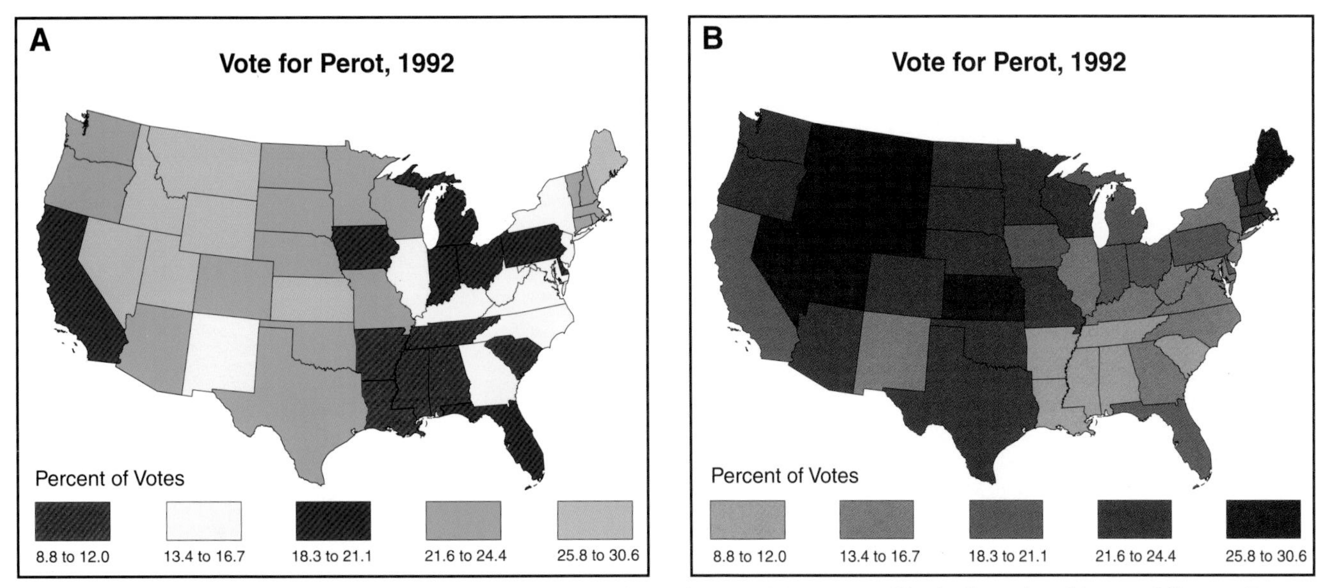

COLOR PLATE 1.1. The choropleth map: an example of a thematic map. Map A uses illogically ordered hues, while map B uses logically ordered shades of a single hue. Although map A may allow the reader to discriminate easily between individual states, it does not permit the reader to perceive the overall spatial pattern as readily as map B. (Data Source: Famighetti 1993, 583.)

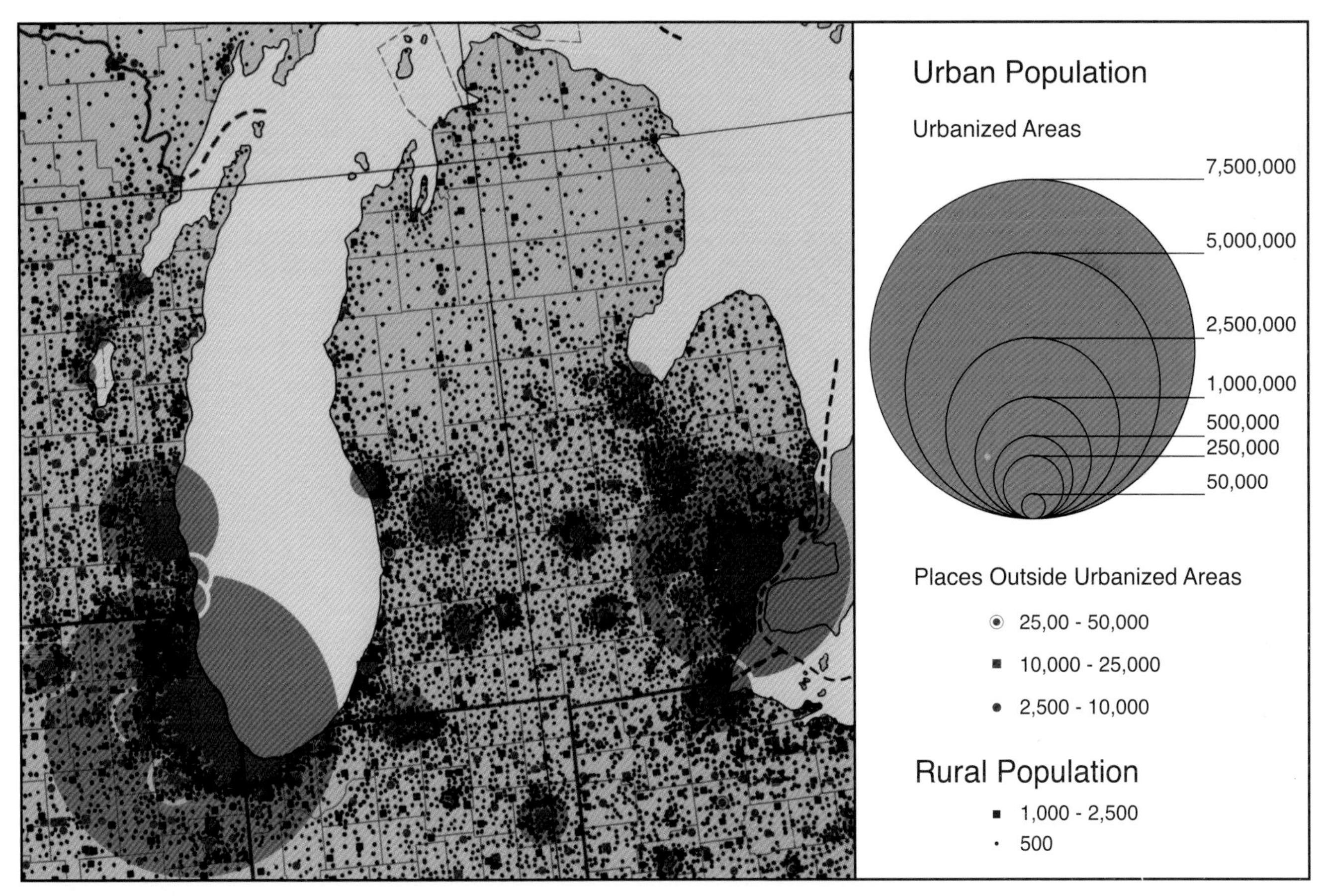

COLOR PLATE 1.2. A combined proportional symbol-dot map that attempts to represent what the population might look like in the "real world." (After U.S. Bureau of the Census 1970.)

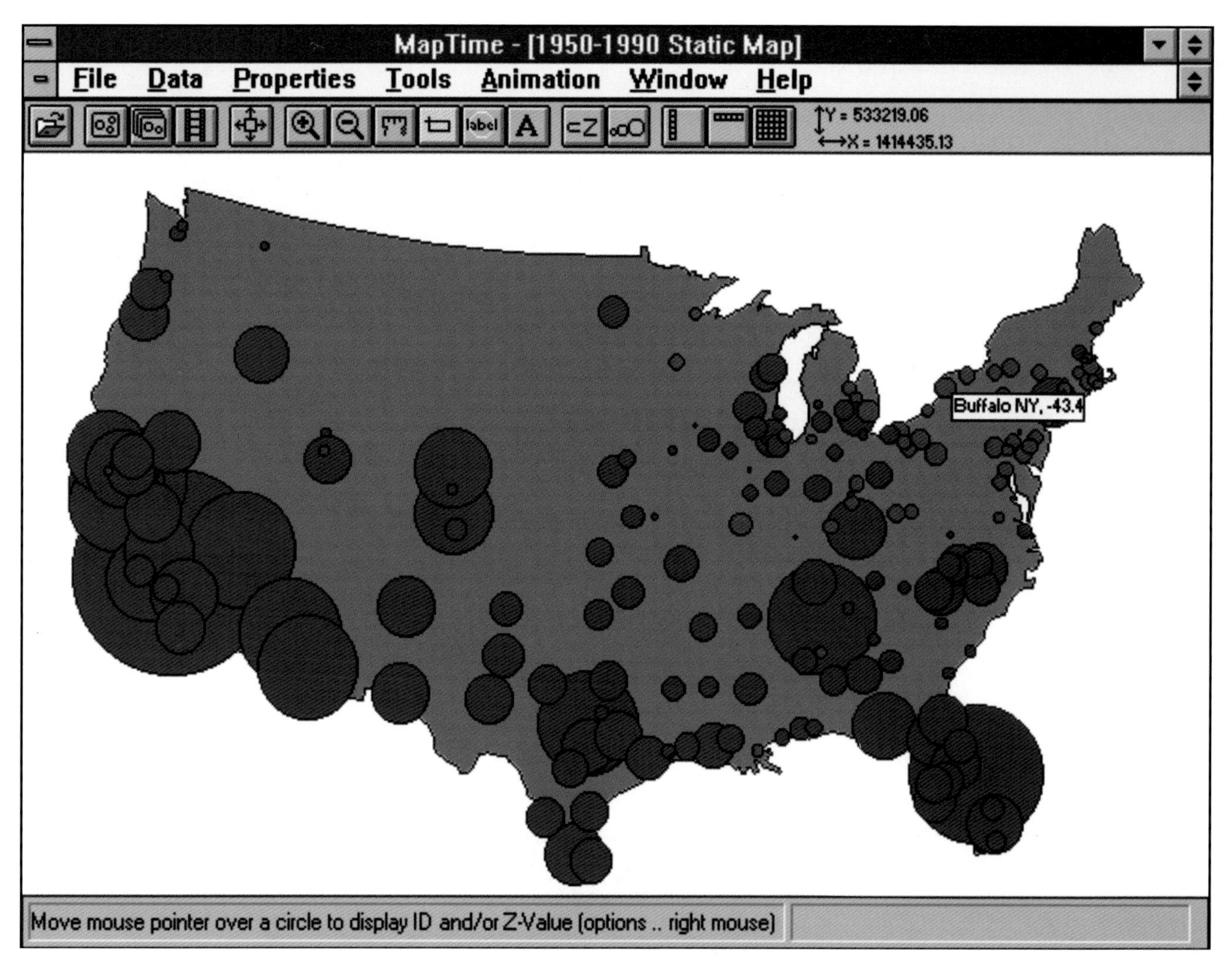

COLOR PLATE 1.3. A change map created using MapTime. Red and blue circles represent population decreases and increases between 1950 and 1990. Note the distinctive region of population losses in the northeast; this pattern was not revealed in an animation of the data over this time period. (Courtesy of Stephen C. Yoder.)

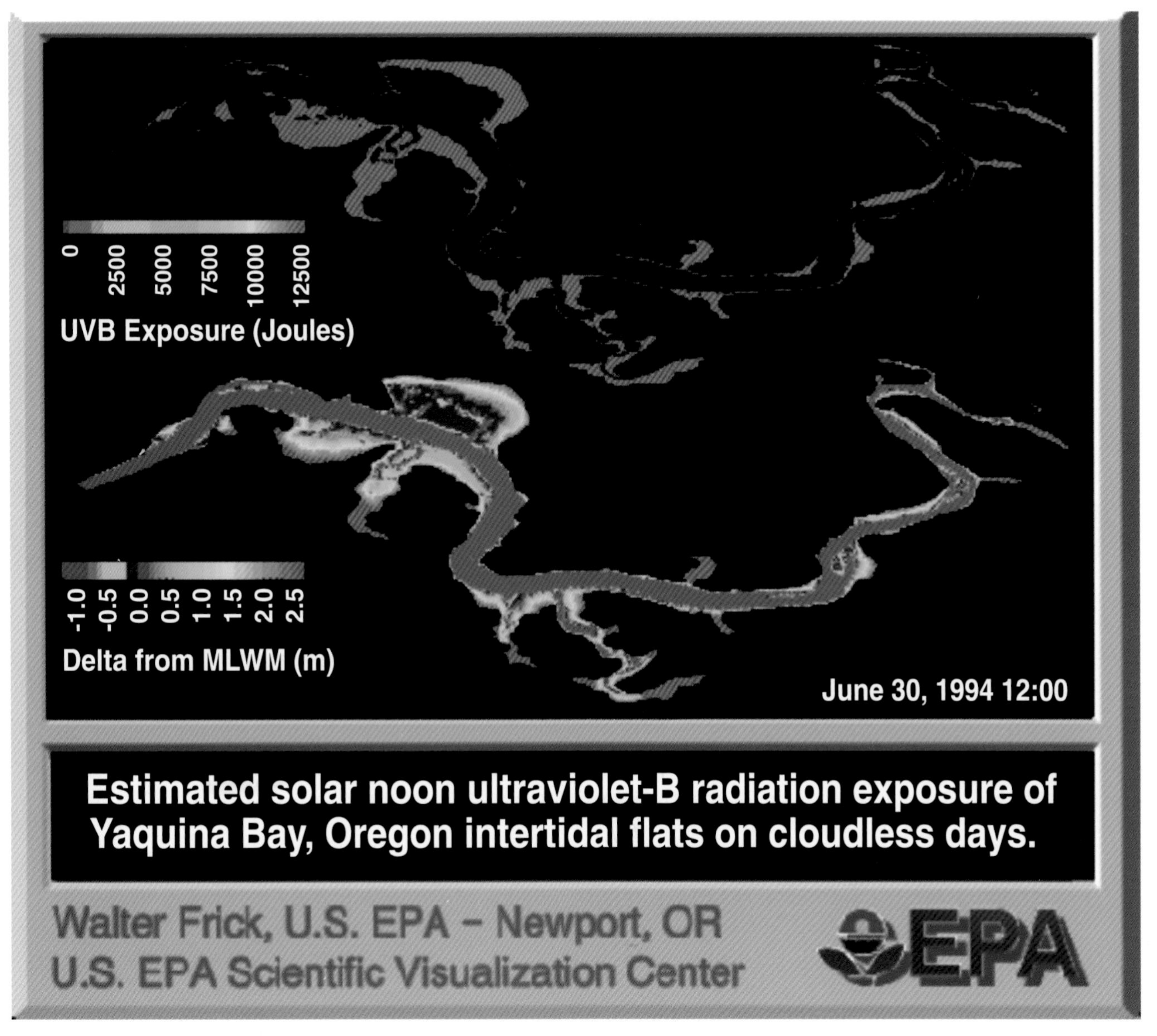

COLOR PLATE 1.4. An example of scientific visualization. Here the relationship between depth of water and ultraviolet B radiation is examined. The bottom image represents the deviation of a daily low tide June 30, 1994 from the yearly average, while the upper image depicts the predicted ultraviolet B exposure at noon of the same day. Areas well above the average daily low tide (red, orange, and yellow) are associated with high ultraviolet B radiation (red). For information on this and other visualizations, see http://www.epa.gov/nescweb0/15_services/ar_svc.html.

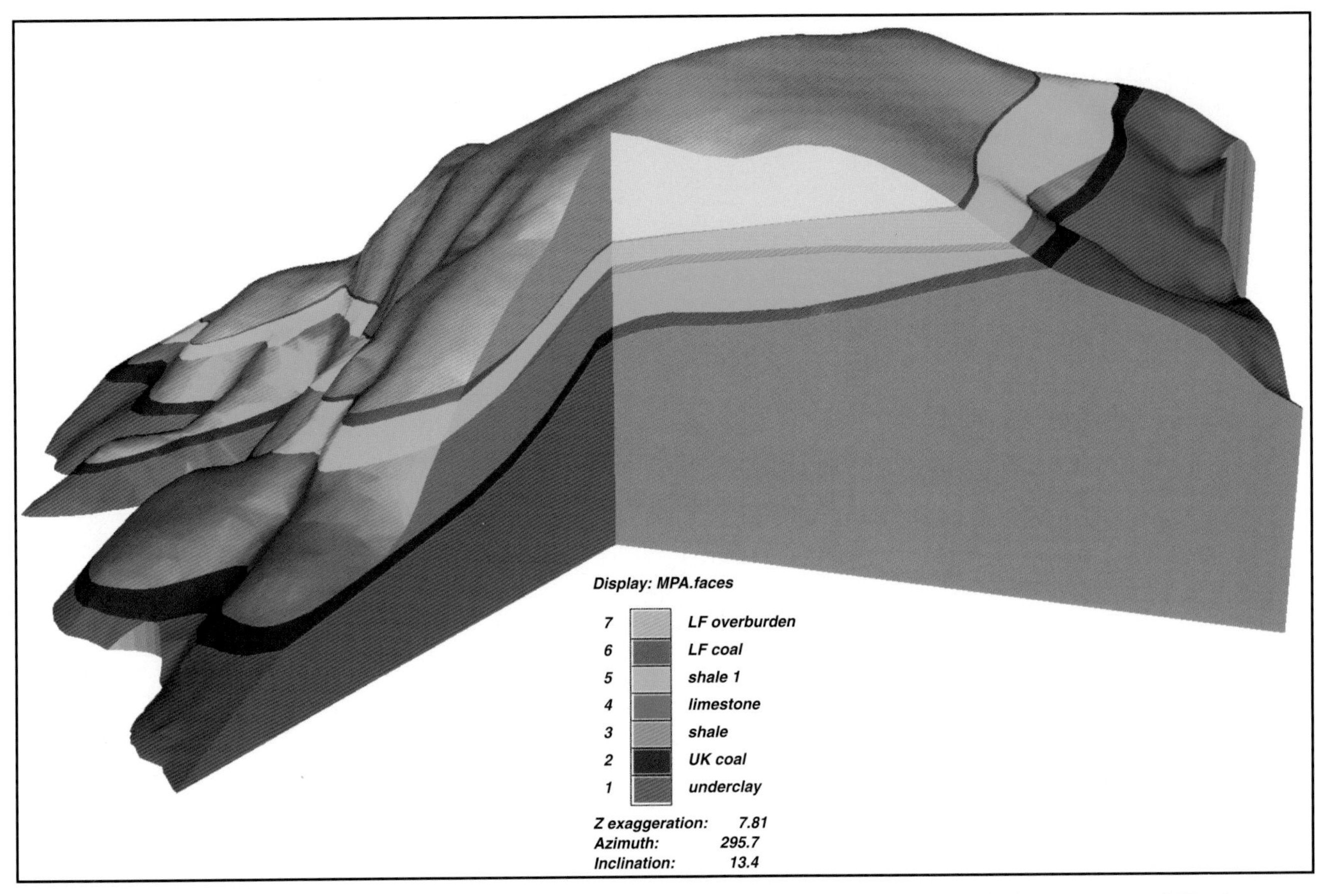

COLOR PLATE 2.1. 3D model of an open-pit coal mining site. (Data source: Pennsylvania Department of Environment Protection; courtesy of Dynamic Graphics, Inc., Alameda, CA.)

	Point	Linear	Areal	2½-D	3-D
Hue					
Lightness					
Saturation					

COLOR PLATE 2.2. Visual variables for colored maps. For visual variables for black-and-white maps, see Figure 2.2.

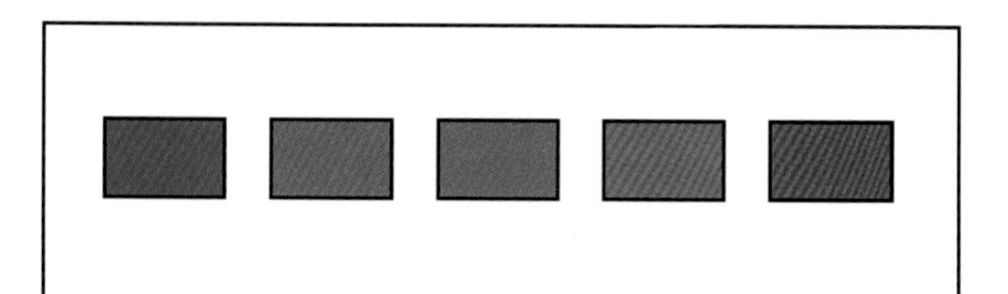

COLOR PLATE 2.3. An illustration of saturation, holding hue and value constant. The area symbols shown for saturation in Color Plate 2.2 are arranged from a desaturated red (gray) to a fully saturated red.

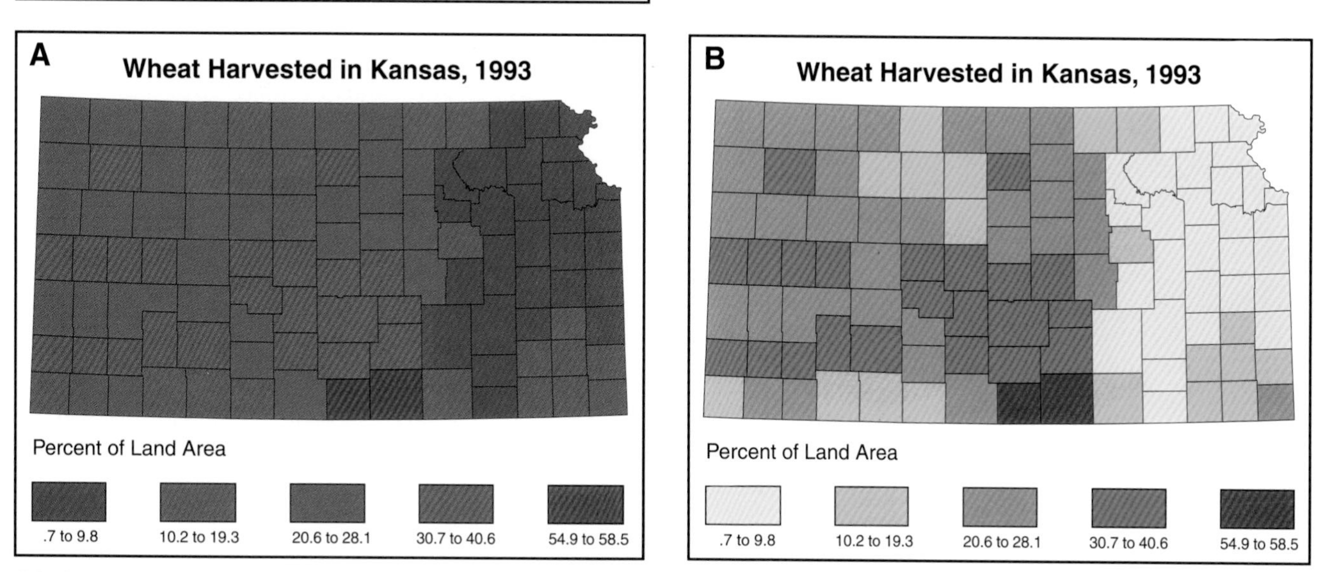

COLOR PLATE 2.4. Representing the percentage of wheat harvested in Kansas counties using different visual variables: (A) saturation, (B) hue. For black-and-white visual variables, see Figure 2.9.

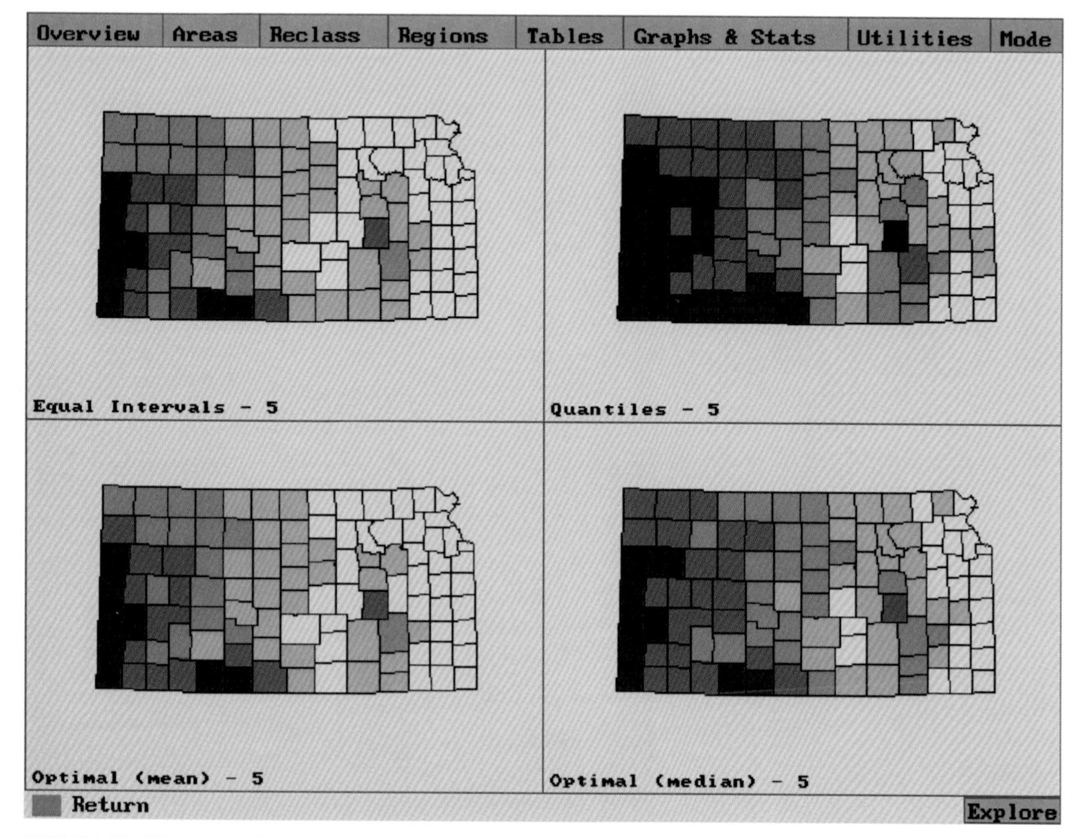

COLOR PLATE 4.1. Using the program ExploreMap to simultaneously display a distribution utilizing four methods of classification.

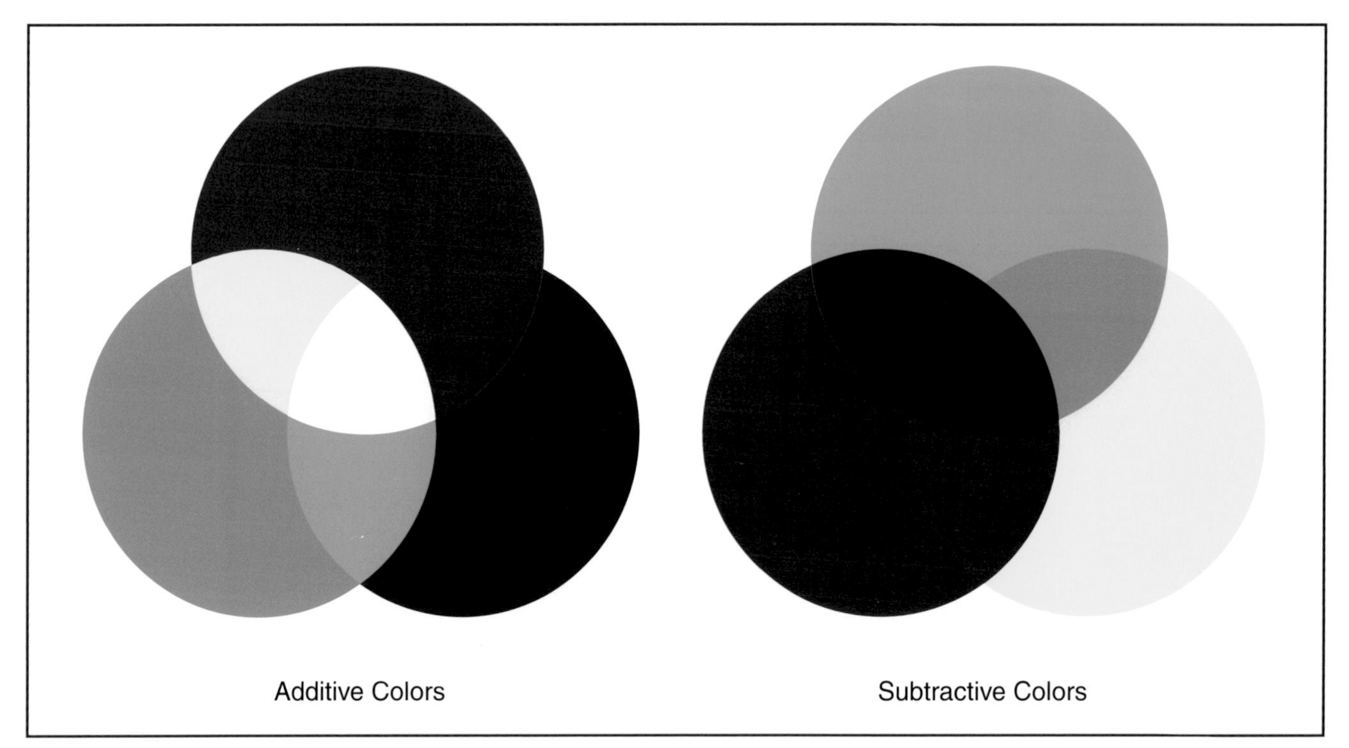

COLOR PLATE 5.1. Principles of additive and subtractive color. For additive color, overlapping red, green, and blue lights reveal how cyan, magenta, yellow, and white can be created. For subtractive color, the reverse is the case: cyan, magenta, and yellow combine to produce red, green, blue, and black. To obtain a true black with subtractive colors, it is often necessary to add a black layer.

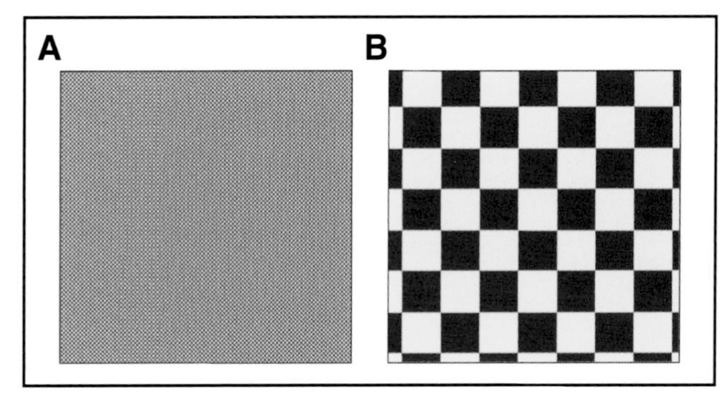

COLOR PLATE 5.2. The dithering process. (A) shade created using dithering; (B) a blow-up of a portion of (A) showing the individual colors used to create the dithered color.

COLOR PLATE 5.3. The Munsell color model. (Courtesy of GretagMacbeth.)

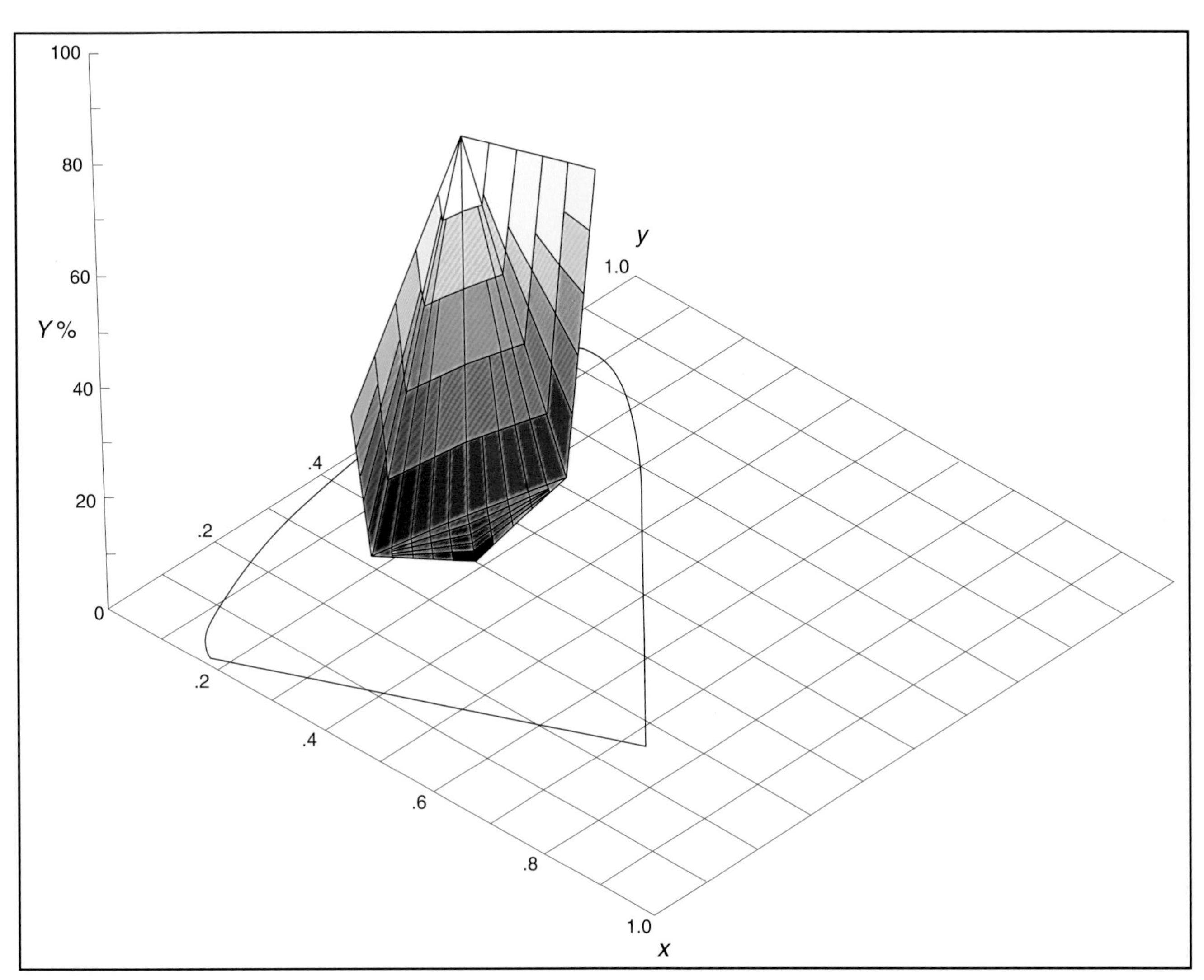

COLOR PLATE 5.4. A three-dimensional view of the Yxy CIE system. (Courtesy of A. Jon Kimerling.)

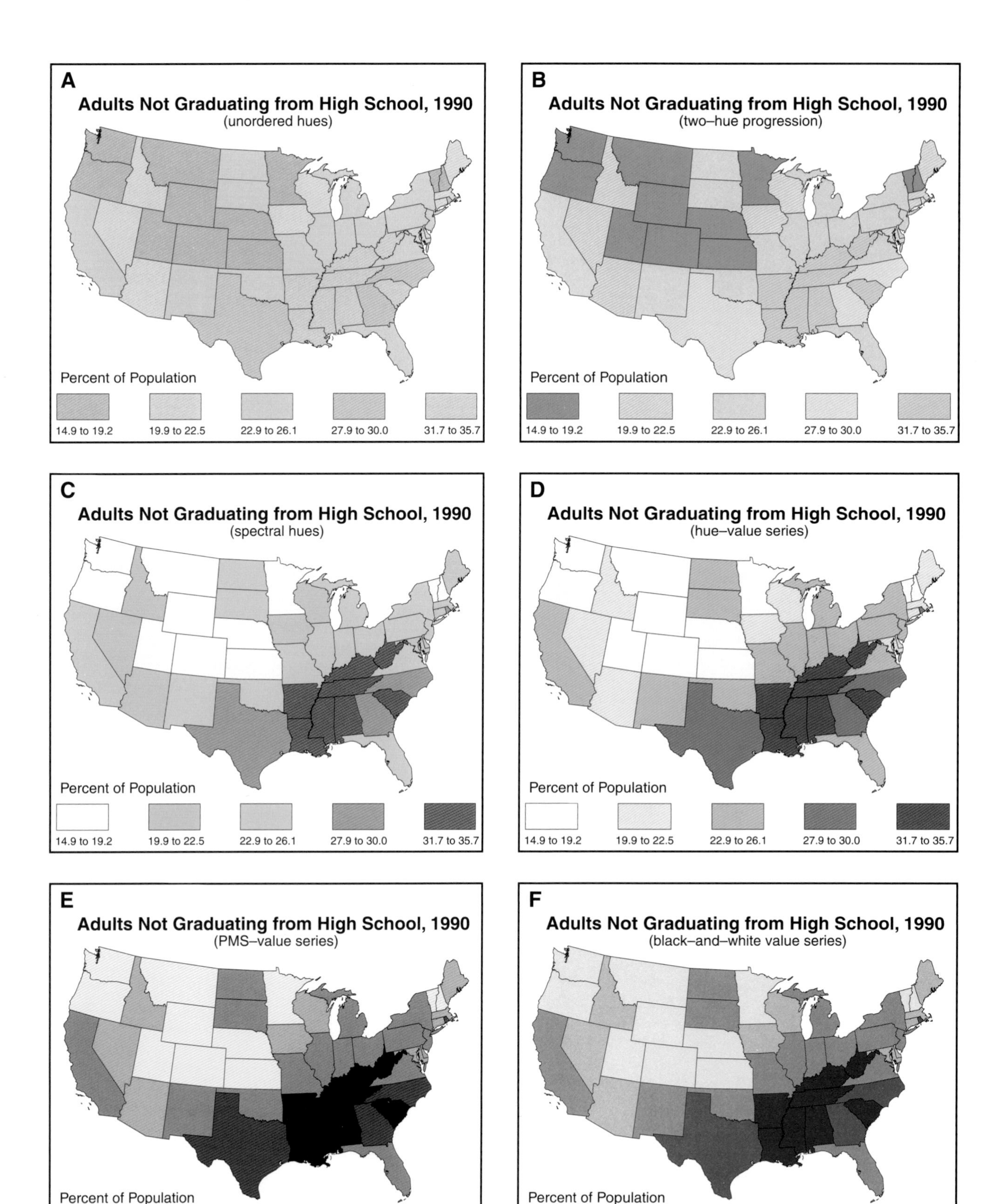

COLOR PLATE 6.1. Color schemes used in a study by Mersey (1990): (A) unordered hues, (B) a two-hue progression, (C) spectral hues, (D) a hue-value series, (E) a PMS-value series, and (F) a black-and-white value series. Colors shown are based on Mersey's specification of CMY and PMS colors. The PMS value series involved overprinting an orange-brown color with a black ink in order to increase the value range of the orange-brown color. Mersey found that the hue-value series worked best overall. (After Mersey 1990.)

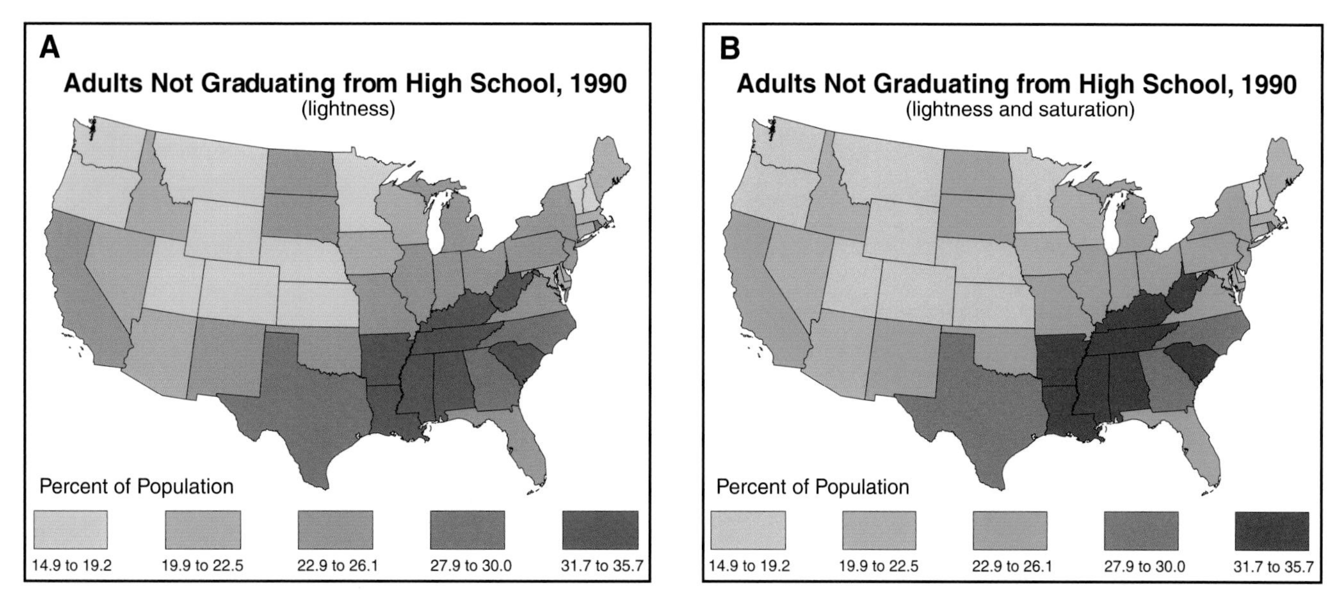

COLOR PLATE 6.2. A comparison of two sequential color schemes: (A) only lightness varies; (B) both lightness and saturation vary, with saturation increasing from the first to the third class and then decreasing for the latter two classes. (Colors based on CMYK values shown in Brewer 1989.)

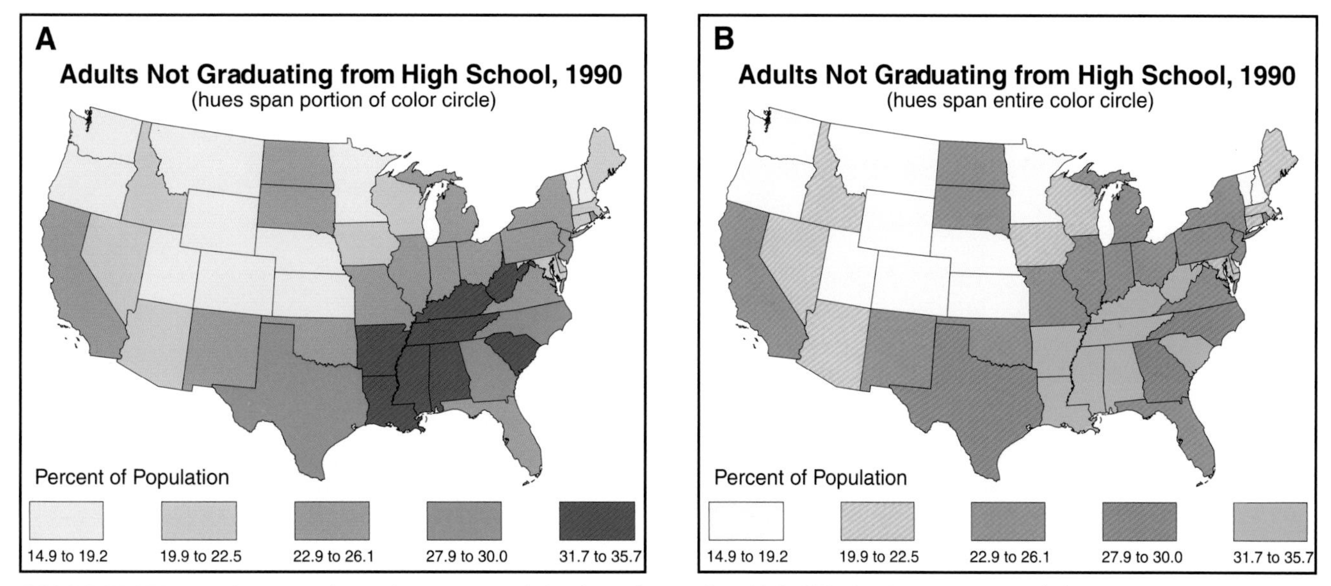

COLOR PLATE 6.3. A comparison of two sequential color schemes in which differing hues are used: (A) the hues span a portion of the color circle; (B) the hues span the entire color circle. (CMYK values for maps provided by Cynthia Brewer.)

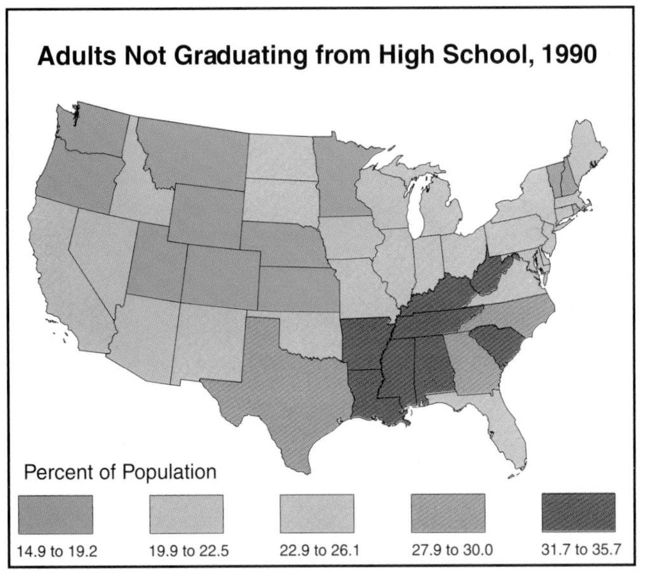

COLOR PLATE 6.4. An example of a diverging scheme in which two sequential schemes converge on a neutral gray. This scheme would arguably be most appropriate for bipolar data, but is used here (with unipolar data) for comparison with the schemes shown in the other color plates.

COLOR PLATE 6.5. A combination of contour lines and continuous spectral color scheme developed by Eyton (1990) are applied to the Grand Canyon. When viewed with special glasses (available with Jumping Colours), the color stereoscopic effect is exaggerated, producing a striking three-dimensional image. (Courtesy of J. Ronald Eyton.)

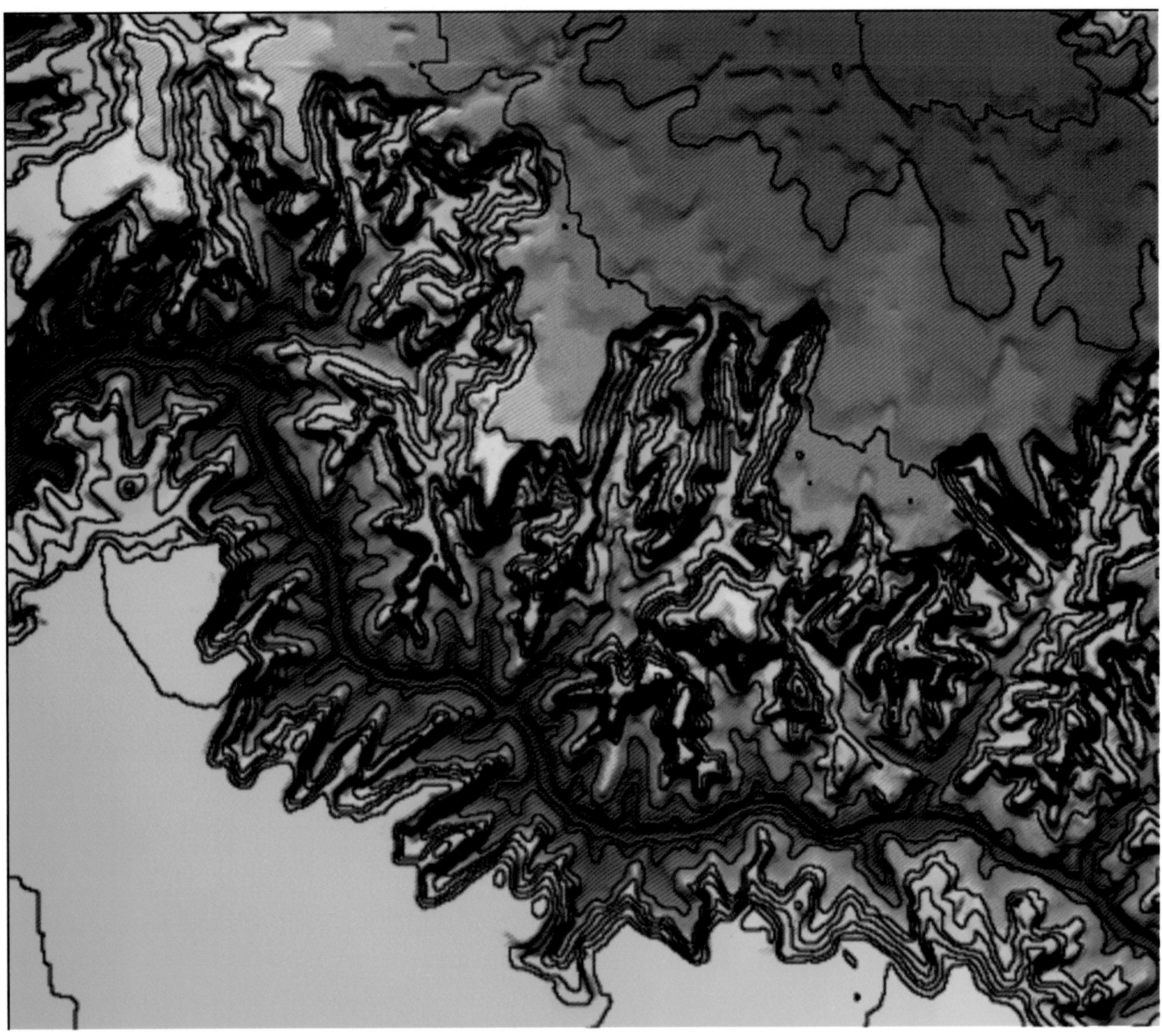

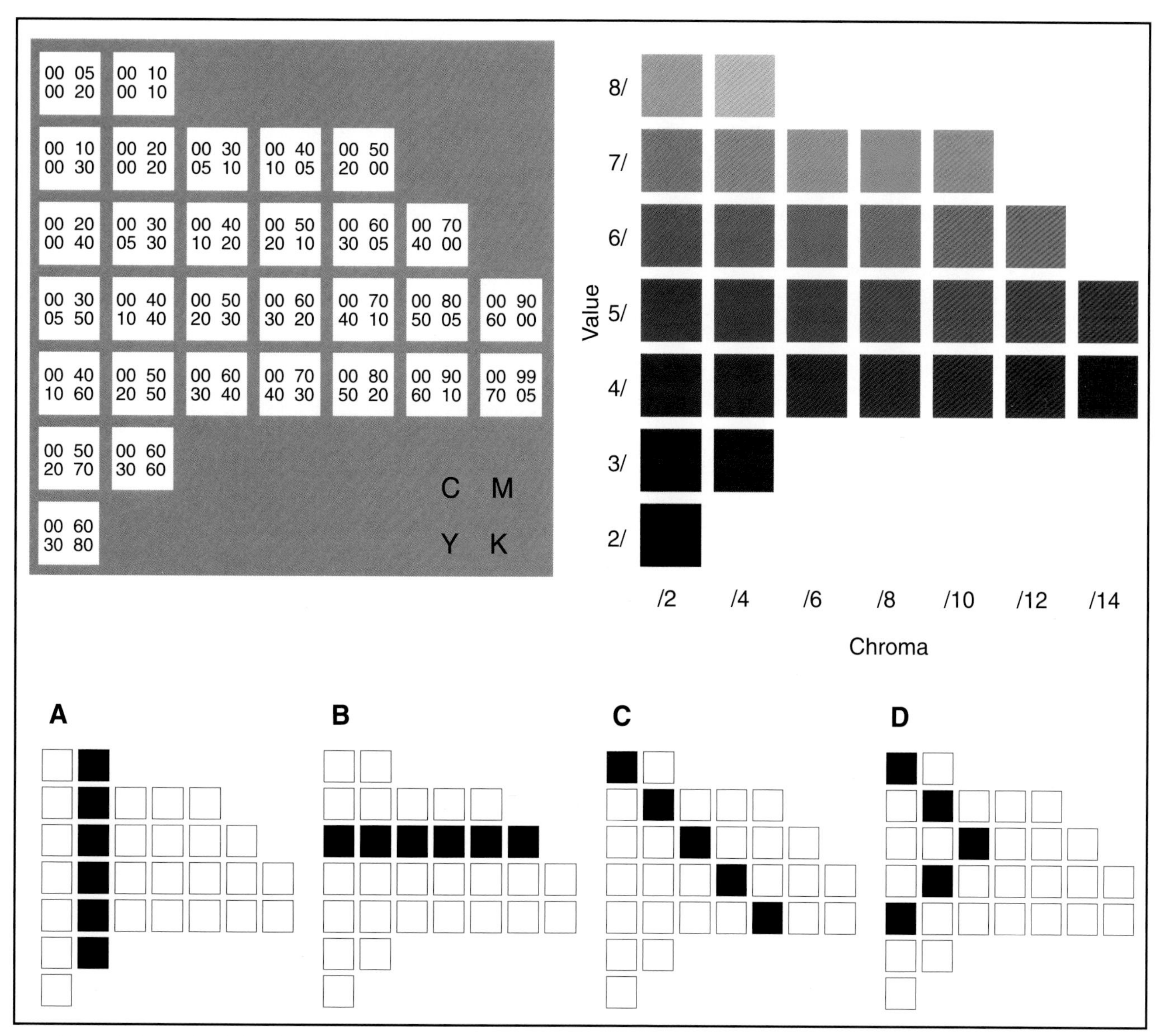

COLOR PLATE 6.6. Relation between CMYK and Munsell colors for the Munsell hue 5R. A to D represent a series of color progressions that might be appropriate for numerical data. (After Brewer, C.A., 1989. First published in the American Cartographer 16(4) pp. 274, 278. Reprinted with permission from the American Congress on Surveying and Mapping.)

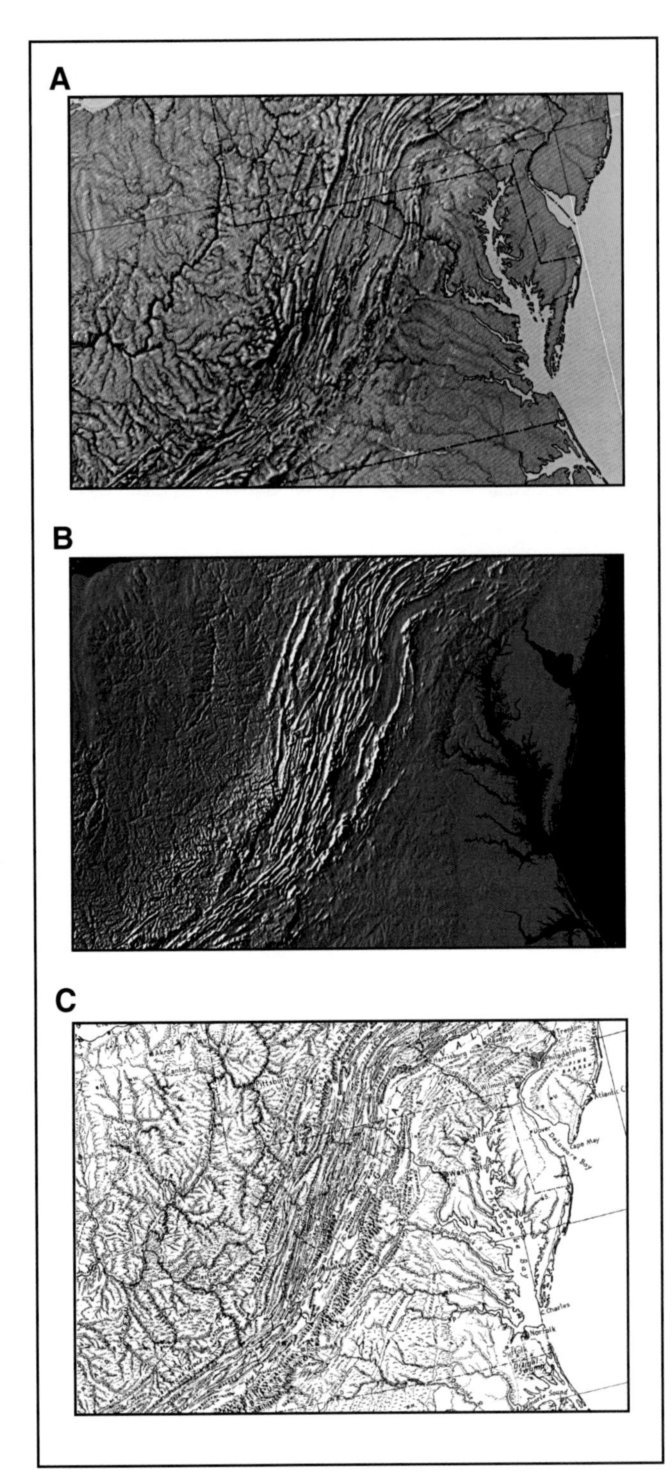

COLOR PLATE 9.1. A comparison of three methods of depicting topography: (A) Harrison's manual shaded-relief method; (B) a USGS digital shaded-relief portrayal developed by Thelin and Pike; (C) Raisz's physiographic method. (Map C copyright by Erwin Raisz, 1967. Reprinted with permission by GEOPLUS, Danvers, Massachusetts.)

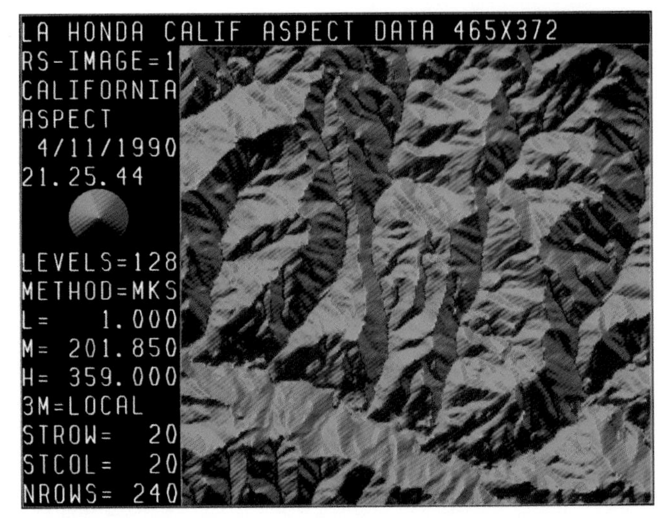

COLOR PLATE 9.2. Moellering and Kimerling's MKS-ASPECT™ approach for symbolizing aspect based on the opponent-process theory of color. The image was created using 128 colors and was scanned from a slide provided by Moellering. (Courtesy of Harold Moellering and A. Jon Moellering.)

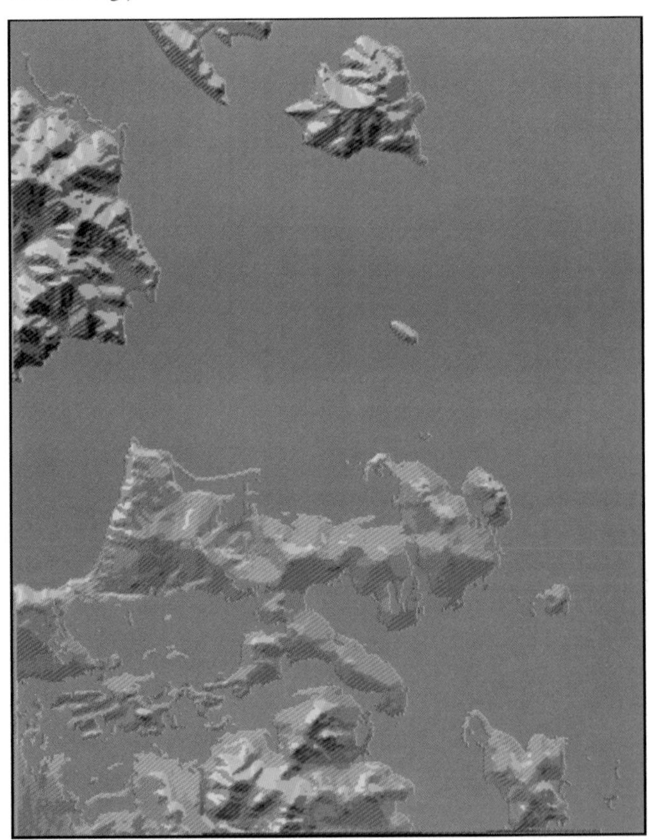

COLOR PLATE 9.3. Brewer and Marlow's method for portraying aspect and slope. The image is for the San Francisco North 7 and one-half minute USGS quad. Caution should be taken in interpreting these colors because the RGB values recommended by Brewer and Marlow have been converted to CMYK values for printing.

COLOR PLATE 9.4. A portrayal of the earth's topography developed by Treinish (1994) using the IBM Visualization Data Explorer (DX). The technique uses Gouraud shading to represent aspect and slope and hypsometric tints to represent elevation above or below sea level. (Courtesy of Lloyd Treinish, IBM T.J. Watson Research Center.)

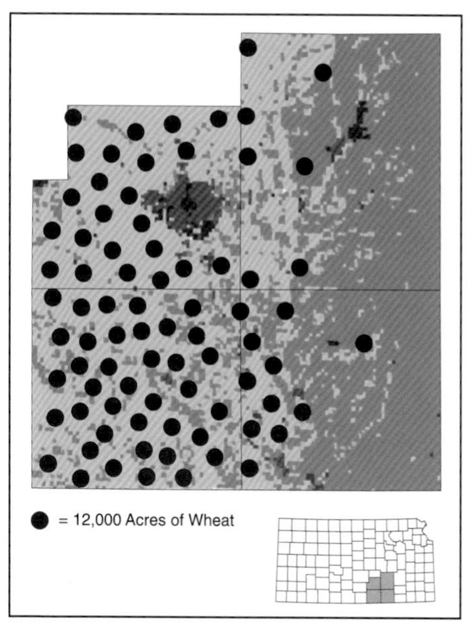

COLOR PLATE 10.1. A portion of the Kansas land use/land cover map illustrating the variability in shape and size of subregions within which dots were placed. (Cropland appears in yellow.) (Courtesy of Kansas Applied Remote Sensing Program, University of Kansas.)

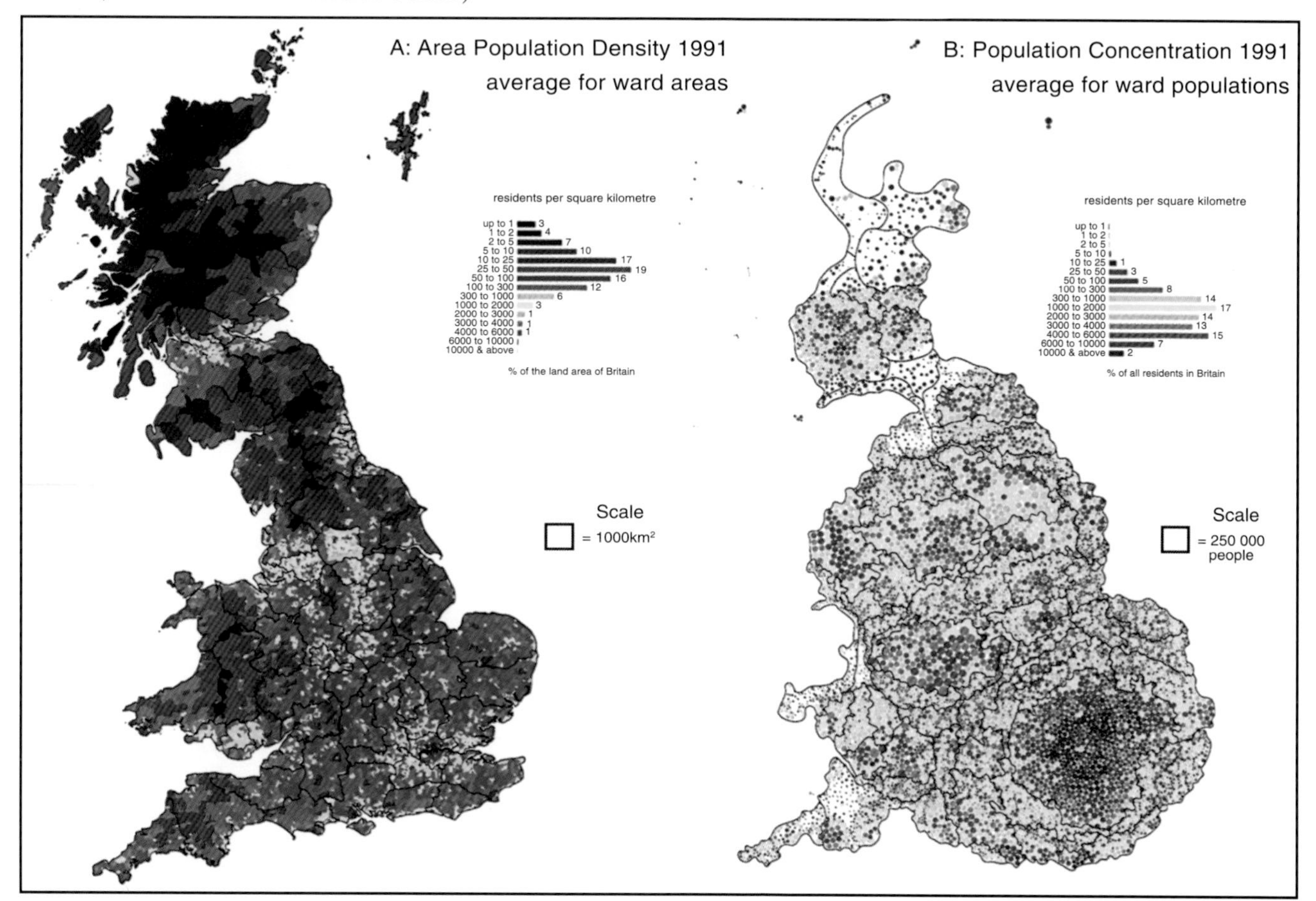

COLOR PLATE 11.1. Maps of population density in Britain using a traditional (A) equal-area projection and (B) a cartogram. The equal-area projection suggests that most of Britain is dominated by relatively lower population densities (the blues and greens), while the cartogram provides a detailed picture of the variation in population densities within urban areas. (From Dorling 1995a, p. xxxiii. Courtesy of Daniel Dorling.)

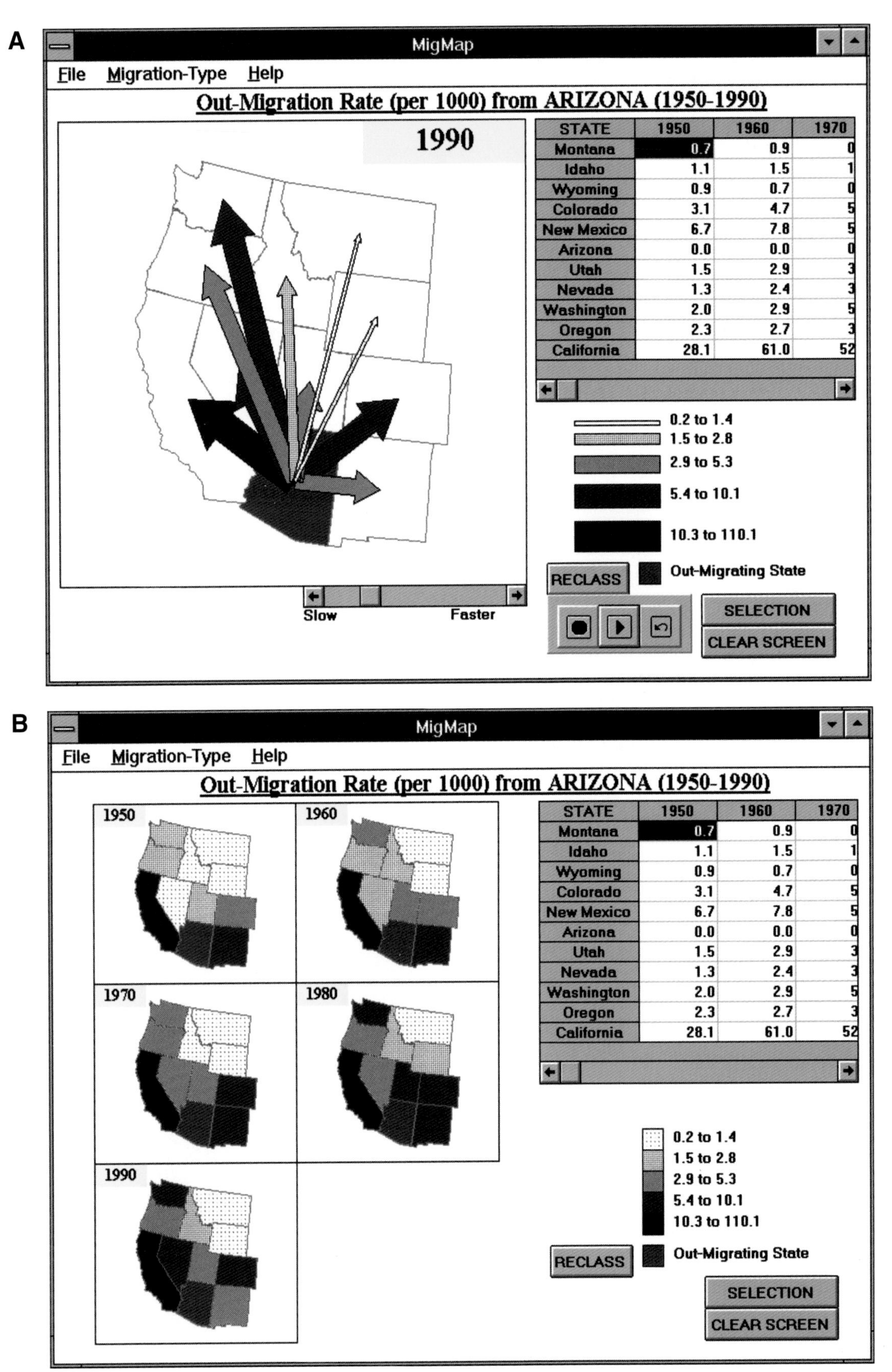

COLOR PLATE 11.2. Screens from Yadav's MigMap, a program for depicting migration flows: (A) a frame from an animation; (B) small multiples using choropleth symbols. Note the use of redundancy in (A) (migration is a function of both the width and color of the symbol). (Courtesy of Sunita Yadav-Pauletti.)

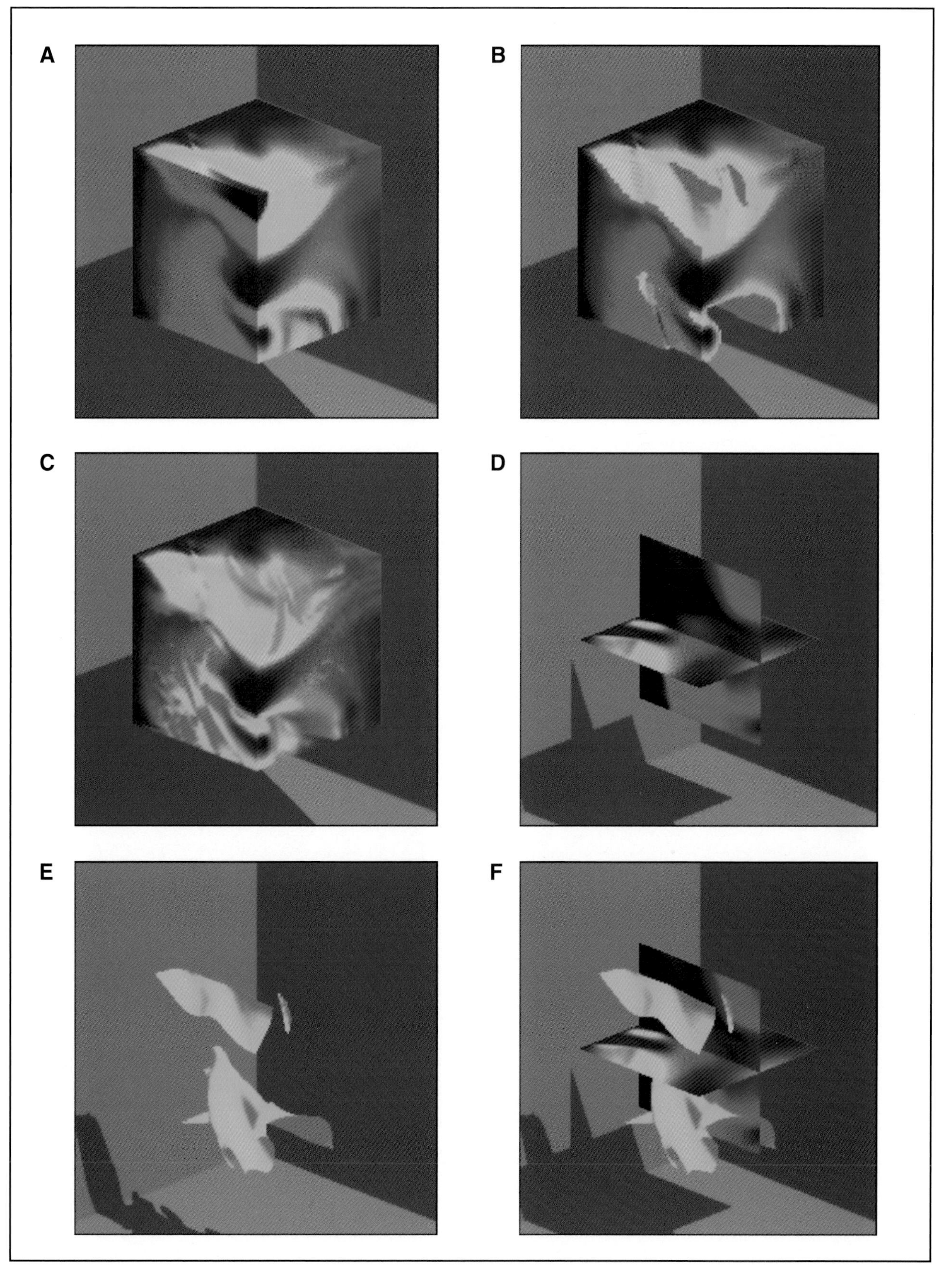

COLOR PLATE 11.3. Using T3D to visualize a true 3-D data set, upflows and downflows within a thunderstorm: (A) an opaque default Rainbow color scheme is used; (B) a portion of the color scheme is made transparent; (C) the transparency of the color scheme is varied continuously; (D) examples of slices taken through the 3-D surface; (E) an example of an isosurface, the three-dimensional equivalent of a contour line; (F) combining isosurfaces and slices. (Images created with Noesys Visualization Pro, courtesy of Fortner Software LLC.)

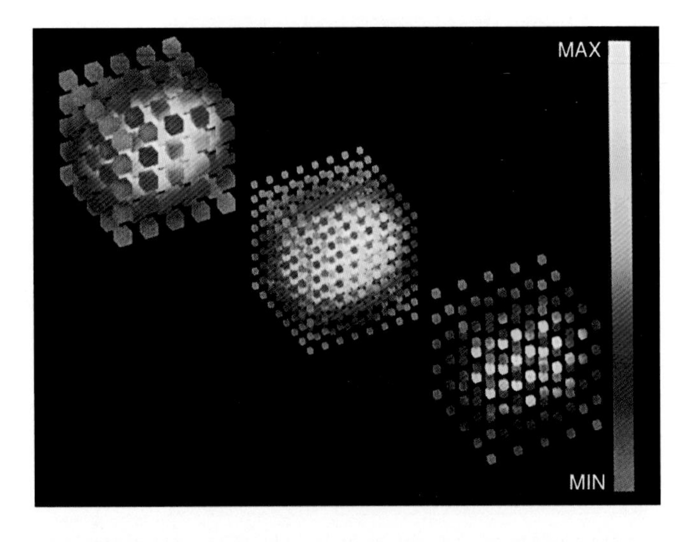

COLOR PLATE 11.4. Nielson and Hamann's tiny cubes method for visualizing true 3-D phenomena. (From Keller and Keller 1993, Visual Cues: Practical Data Visualization, p. 148; © 1993 IEEE.)

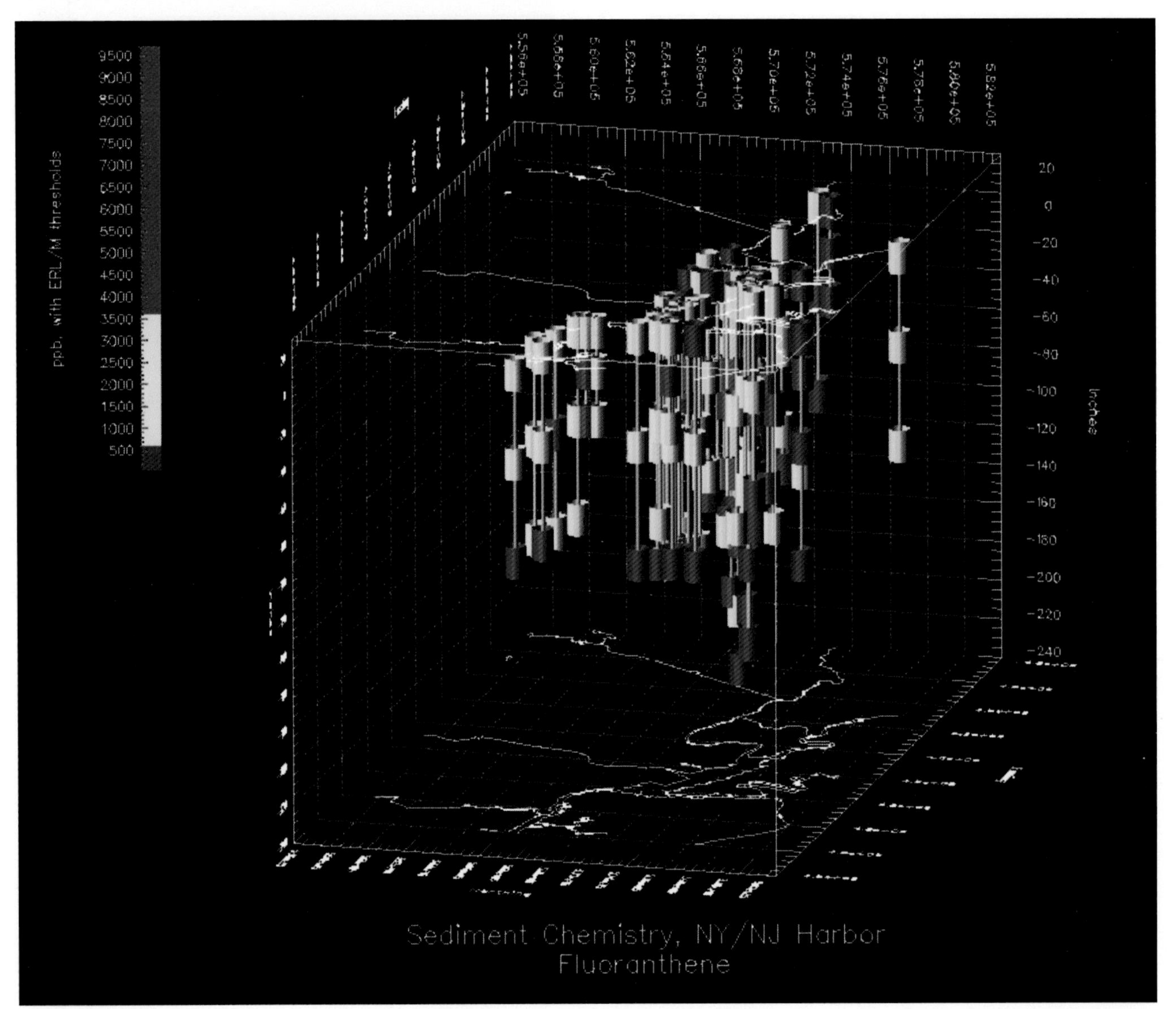

COLOR PLATE 11.5. Waisel's use of DX to visualize the concentration of fluoranthene in sediment in New York Harbor. Each cylinder represents a homogeneous sample within a core. (Courtesy of Laurie Waisel.)

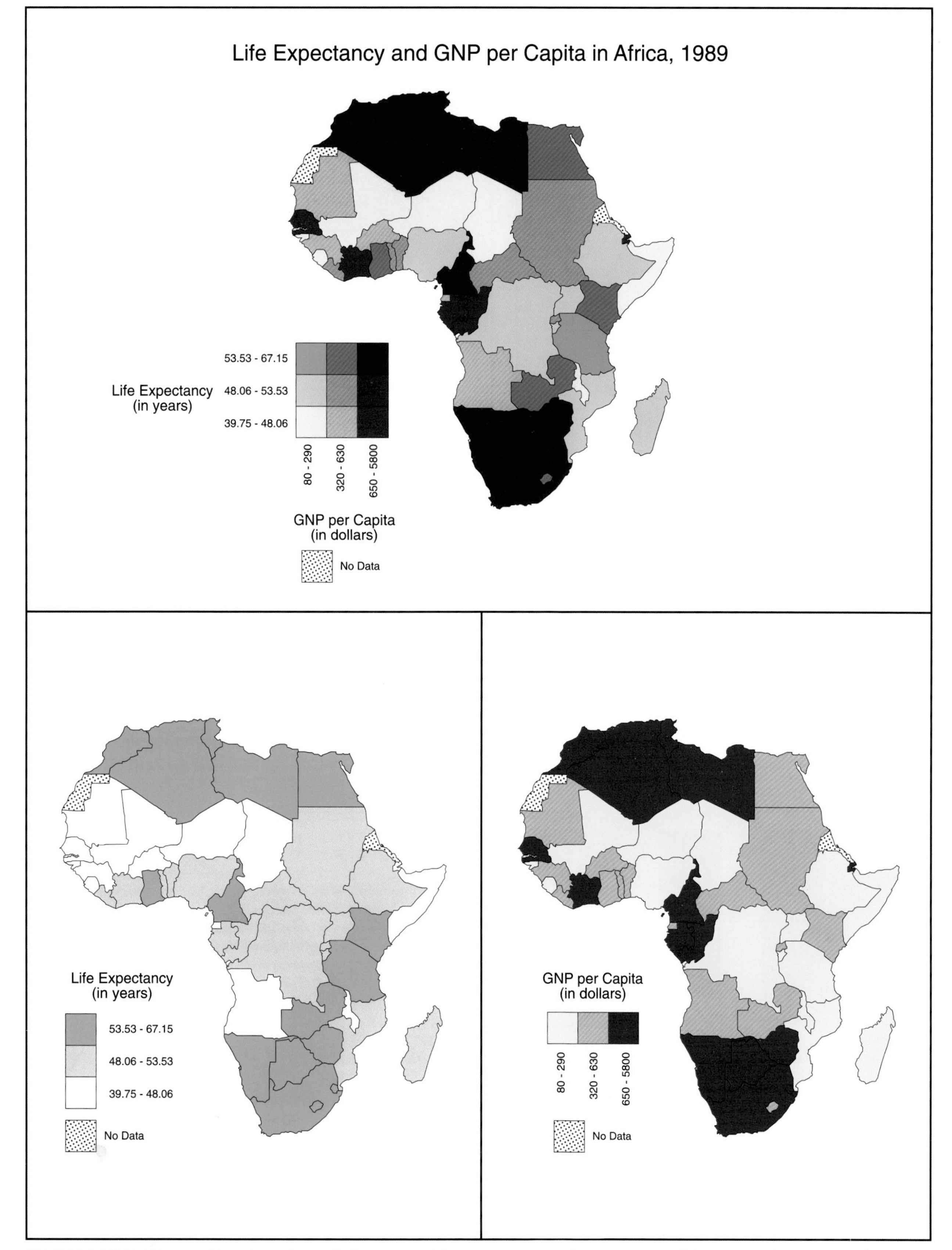

COLOR PLATE 12.1. A bivariate choropleth map and its component univariate maps. The color scheme was taken from Olson (1981, 269) and is similar to those popular on U.S. Bureau of the Census maps in the 1970s. A quantiles classification was used for each map because the GNP data were positively skewed. (Data Source: ArcView 3.)

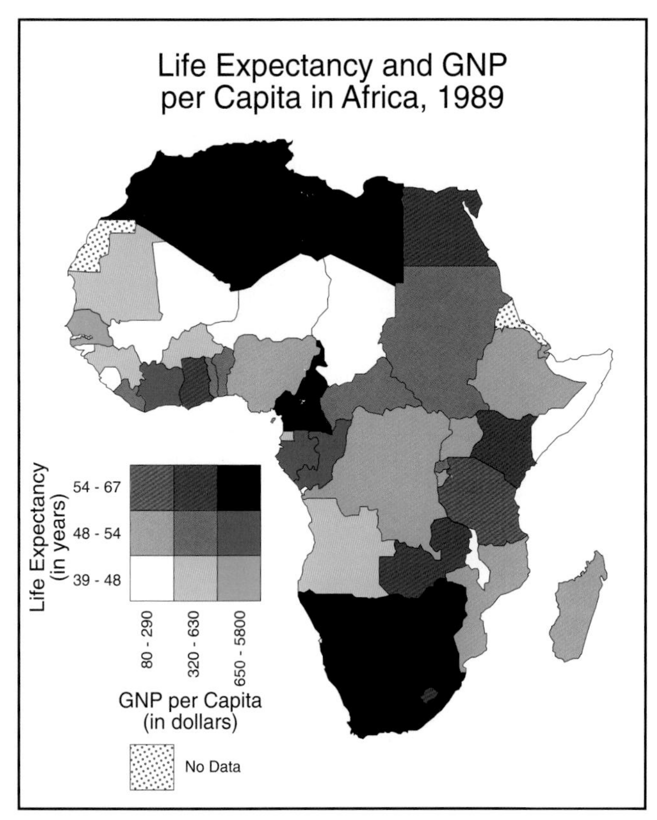

COLOR PLATE 12.2. A bivariate choropleth map based on the complementary colors red and cyan. (After Eyton 1984a.)

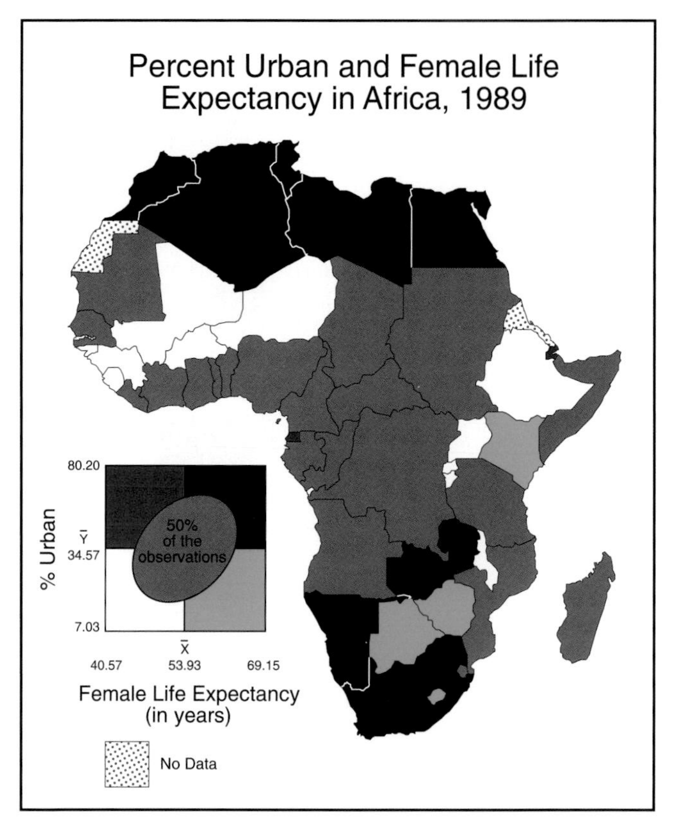

COLOR PLATE 12.3. A bivariate choropleth map using complementary colors in which legend classes are based on the means of the variables and an equiprobability ellipse enclosing 50 percent of the data. Since the ellipse is based on a bivariate normal distribution, only normally distributed data should be mapped using this method. (After Eyton 1984a; data source: ArcView 3.)

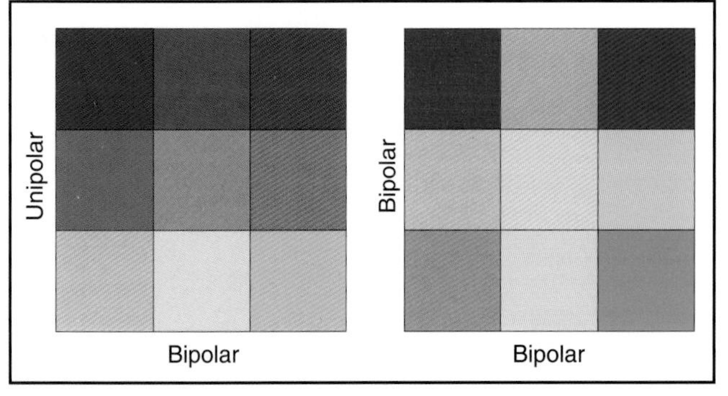

COLOR PLATE 12.4. Color schemes suggested by Brewer for bivariate maps in which the numerical data are not both unipolar.

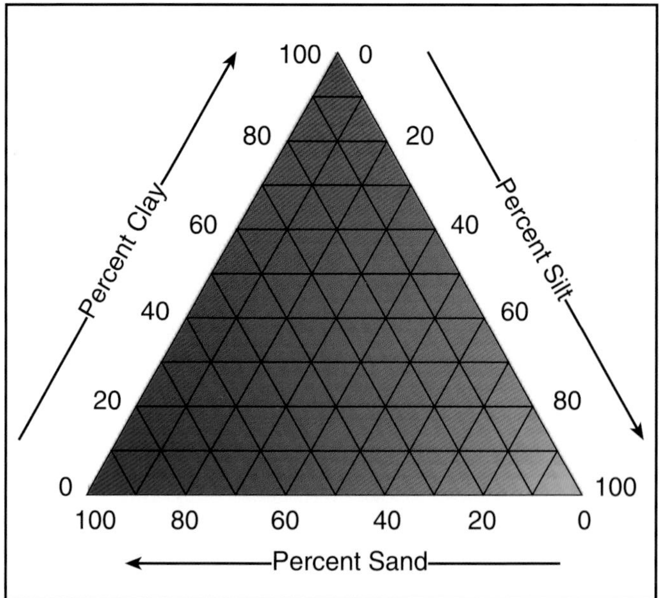

COLOR PLATE 12.5. An RGB color scheme for creating a trivariate choropleth map. (After Byron 1994, p. 126.)

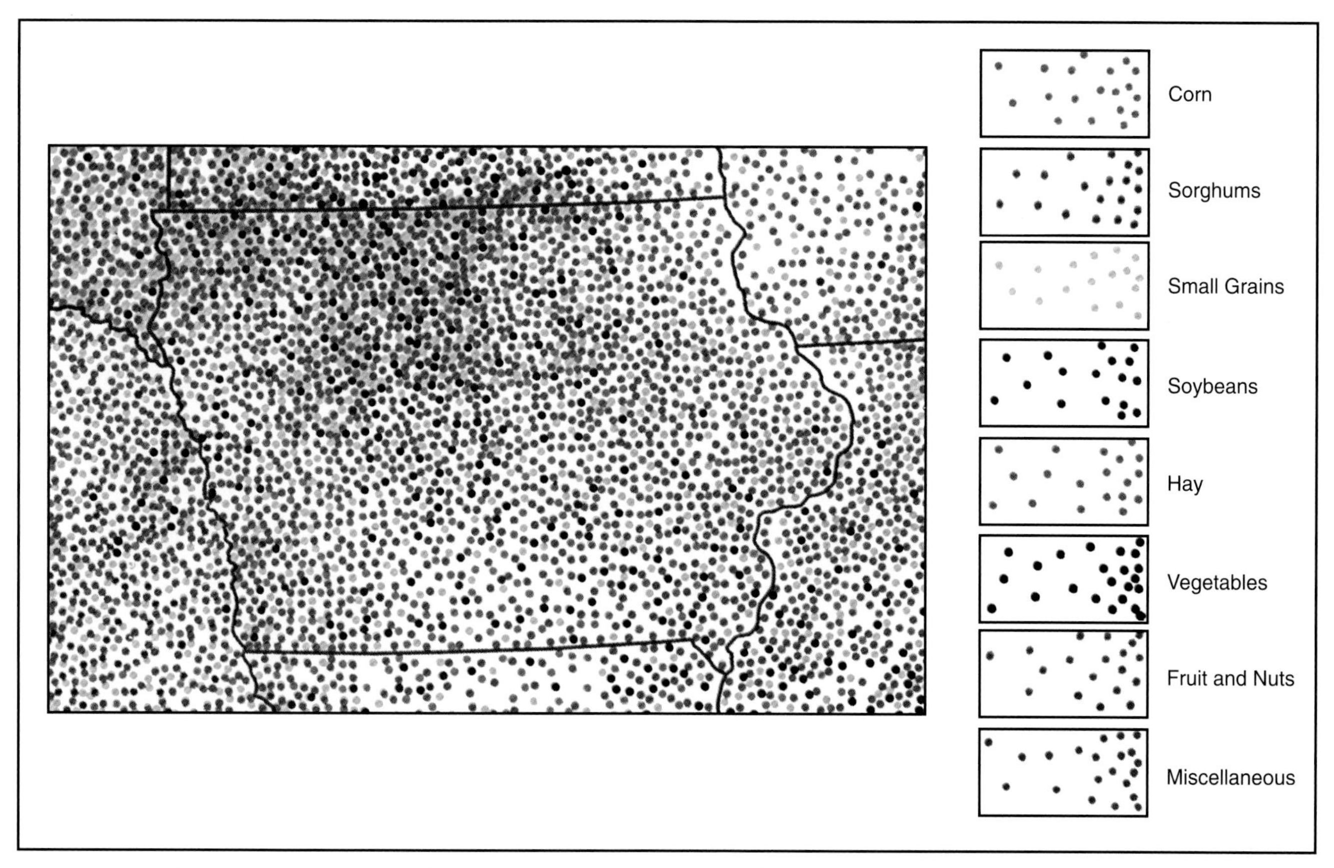

COLOR PLATE 12.6. A portion of a multivariate dot map constructed on the basis of pointillism. Since this map was scanned from the original map, colors shown must be considered approximations of the actual colors. (After Jenks, 1961.)

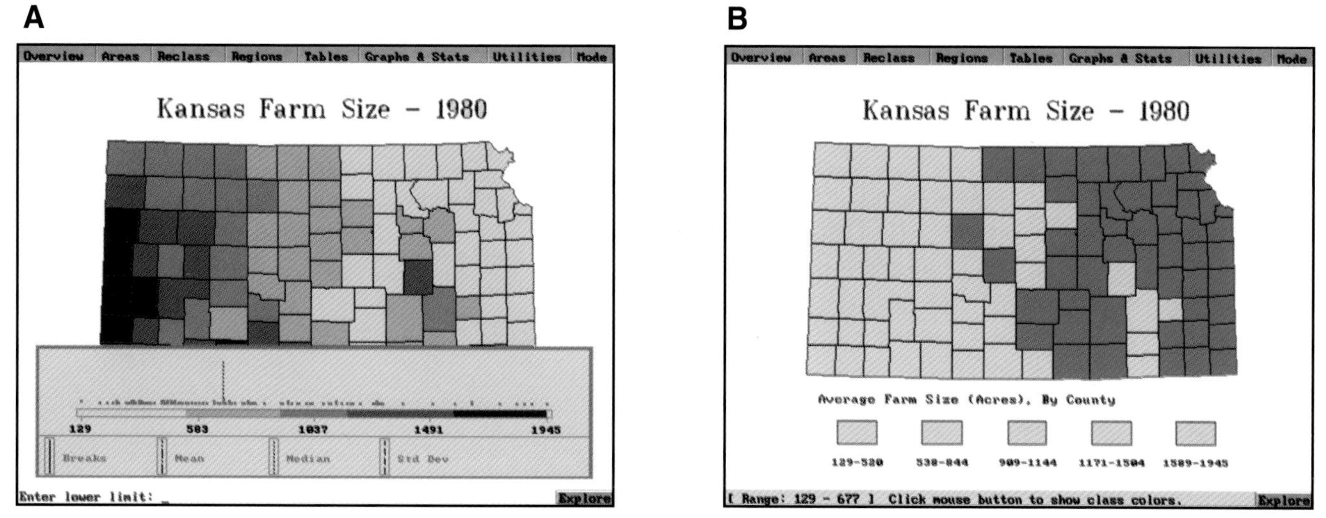

COLOR PLATE 13.1. The Subset option for ExploreMap: (A) A dispersion graph appears illustrating the distribution of the data. Note that the current class breaks, mean, median, and standard deviation can be toggled on; in this case, only the median is shown; (B) The user selects all values less than the median, and these are displayed in blue.

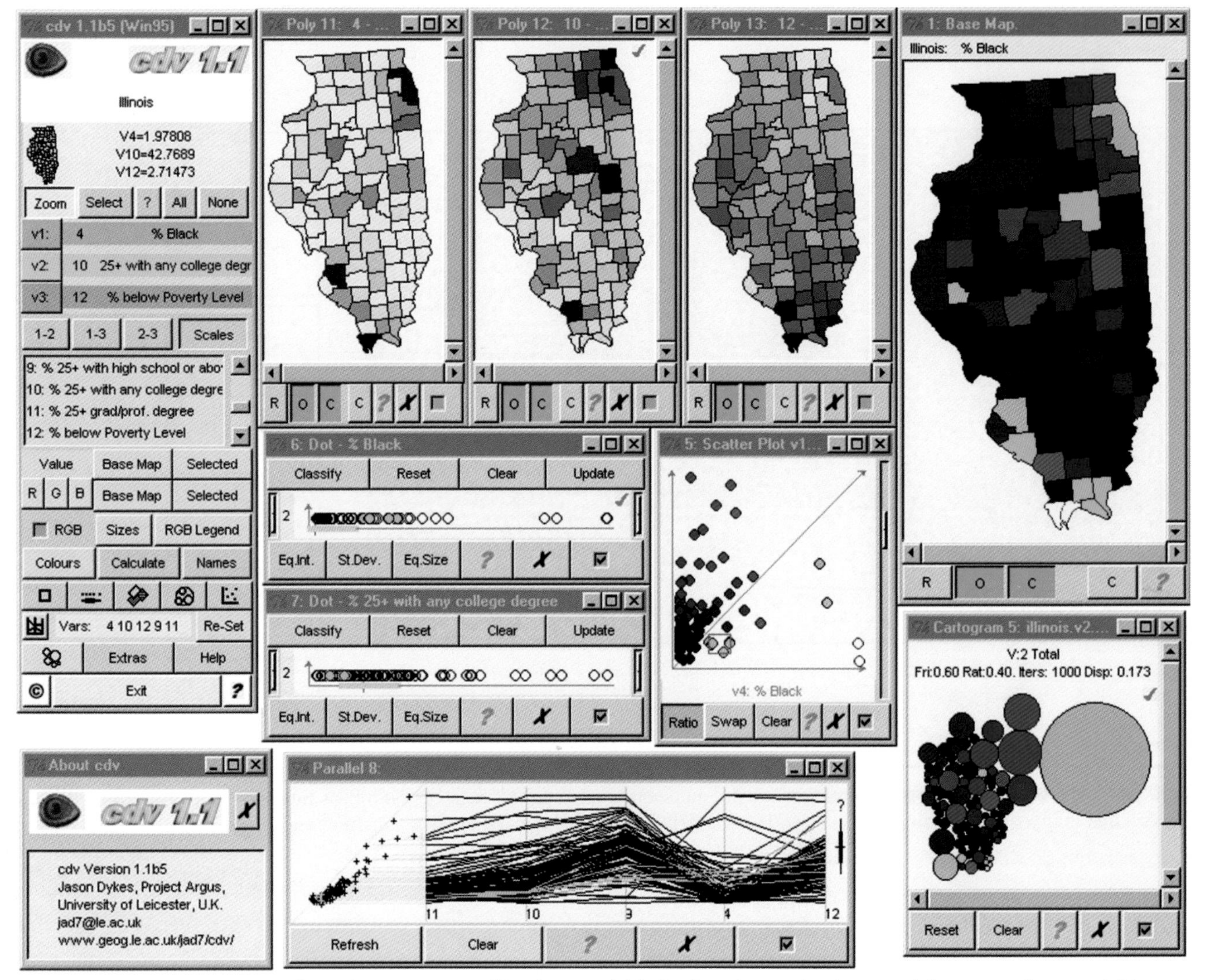

COLOR PLATE 13.2. A screen from the "enumerated" software within Project Argus. At the top of the screen three variables are shown: percent black, percent of population 25 years and older with any college education, and percent below the poverty level. In the top right is a bivariate choropleth map of the first two variables; high values on the first two variables are represented by increasing amounts of yellow and blue respectively. In the central portion are point graphs and a scatterplot of the first two variables. The lower and lower right views show a parallel coordinate plot and Dorling's cartogram method. (Jason Dykes, Department of Geography, University of Leicester.)

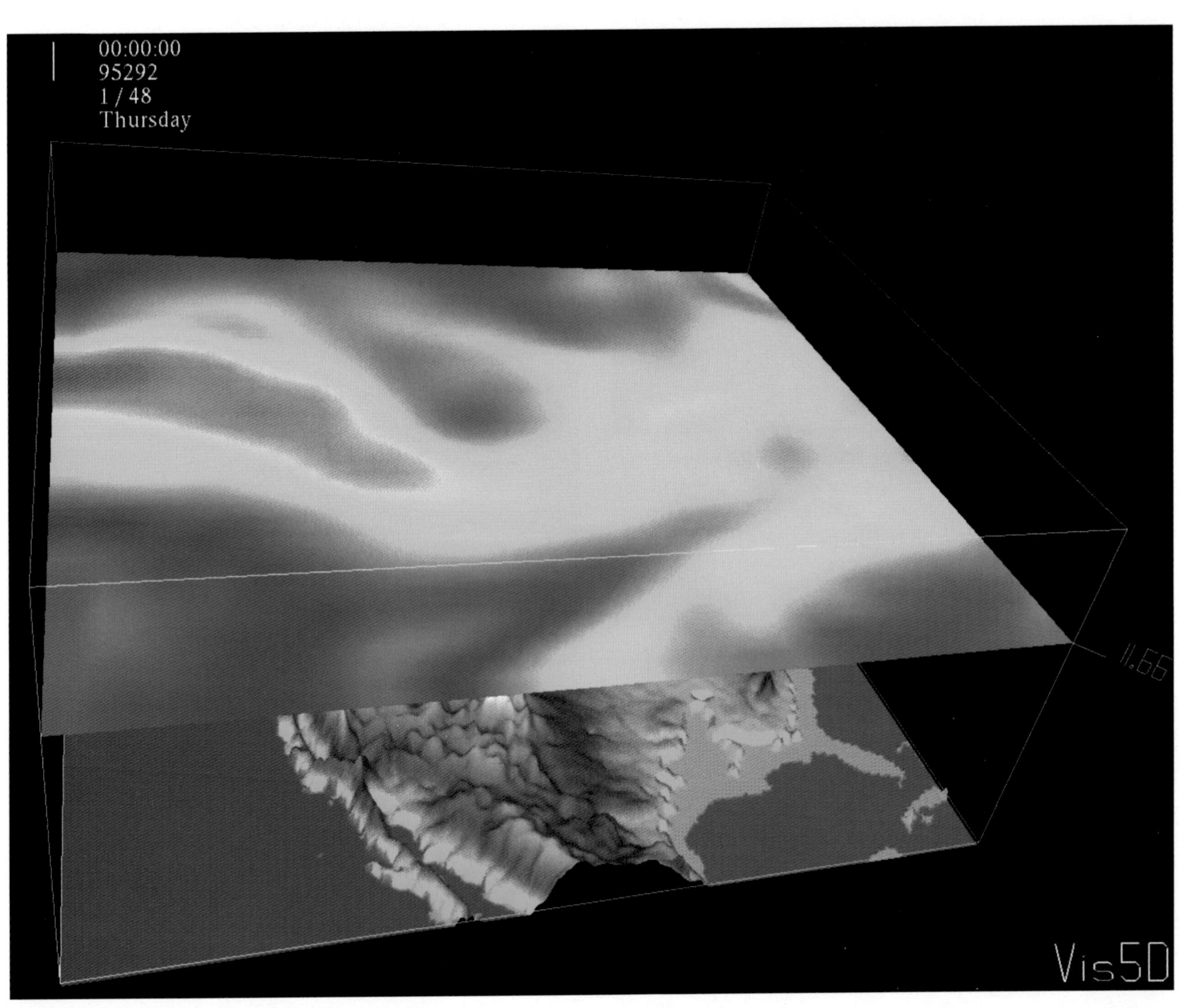

COLOR PLATE 13.3. In Vis5D, hyposometric tints are used to symbolize a horizontal slice through true 3-D wind speed data. Note the height of the slice is 11.66 kilometers, and thus the high-valued reddish-orange region represents a portion of the jet stream. (Courtesy of William L. Hibbard.)

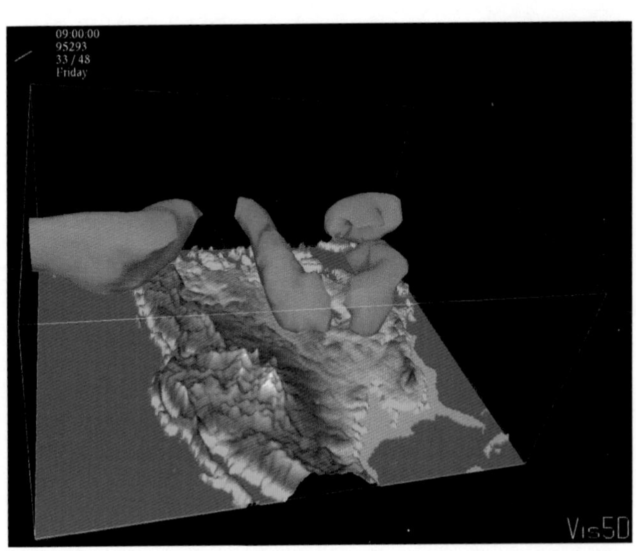

COLOR PLATE 13.4. An isosurface of wind speed created using Vis5D. An isosurface is a surface bounded by a particular value of an attribute; in this case, the 45 meter per second value for wind speed is used. (Courtesy of William L. Hibbard.)

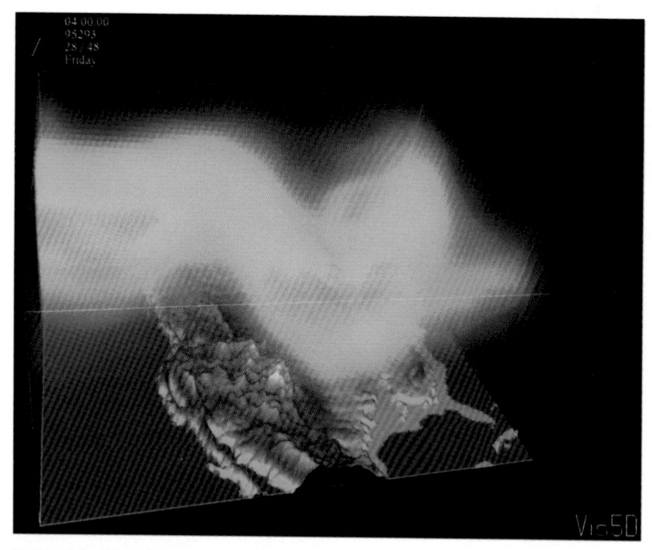

COLOR PLATE 13.5. Volume (or transparent fog) symbolization of wind speed created using Vis5D. Each point in the three-dimensional data is symbolized using a spectral color scheme. (Courtesy of William L. Hibbard.)

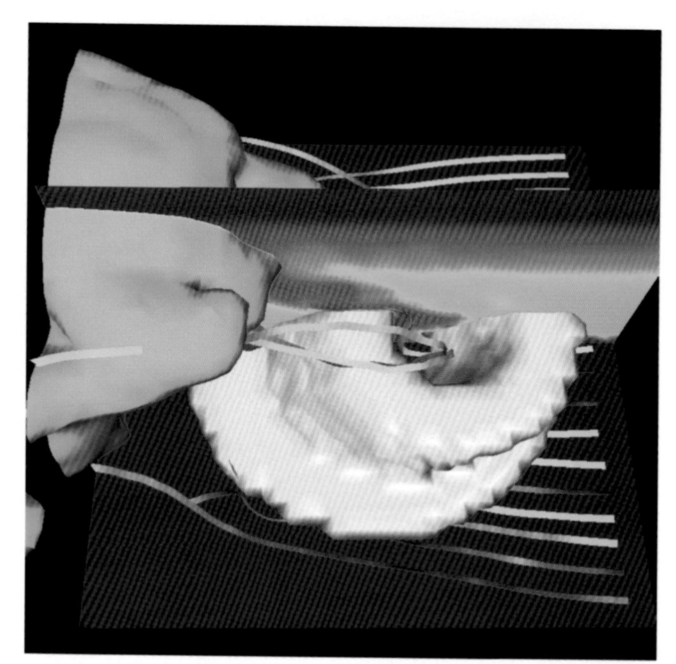

COLOR PLATE 13.6. Multiple variables depicted using Vis5D; see text for explanation. (From Hibbard, W.L., Paul, B.E., Santek, D.A., Dyer, C.R., Battaiola, A.L., and Voidrot-Martinez, Marie-F. 1994, Interactive Visualization of Earth and Space Science Computations, p. 67; © 1994 IEEE.)

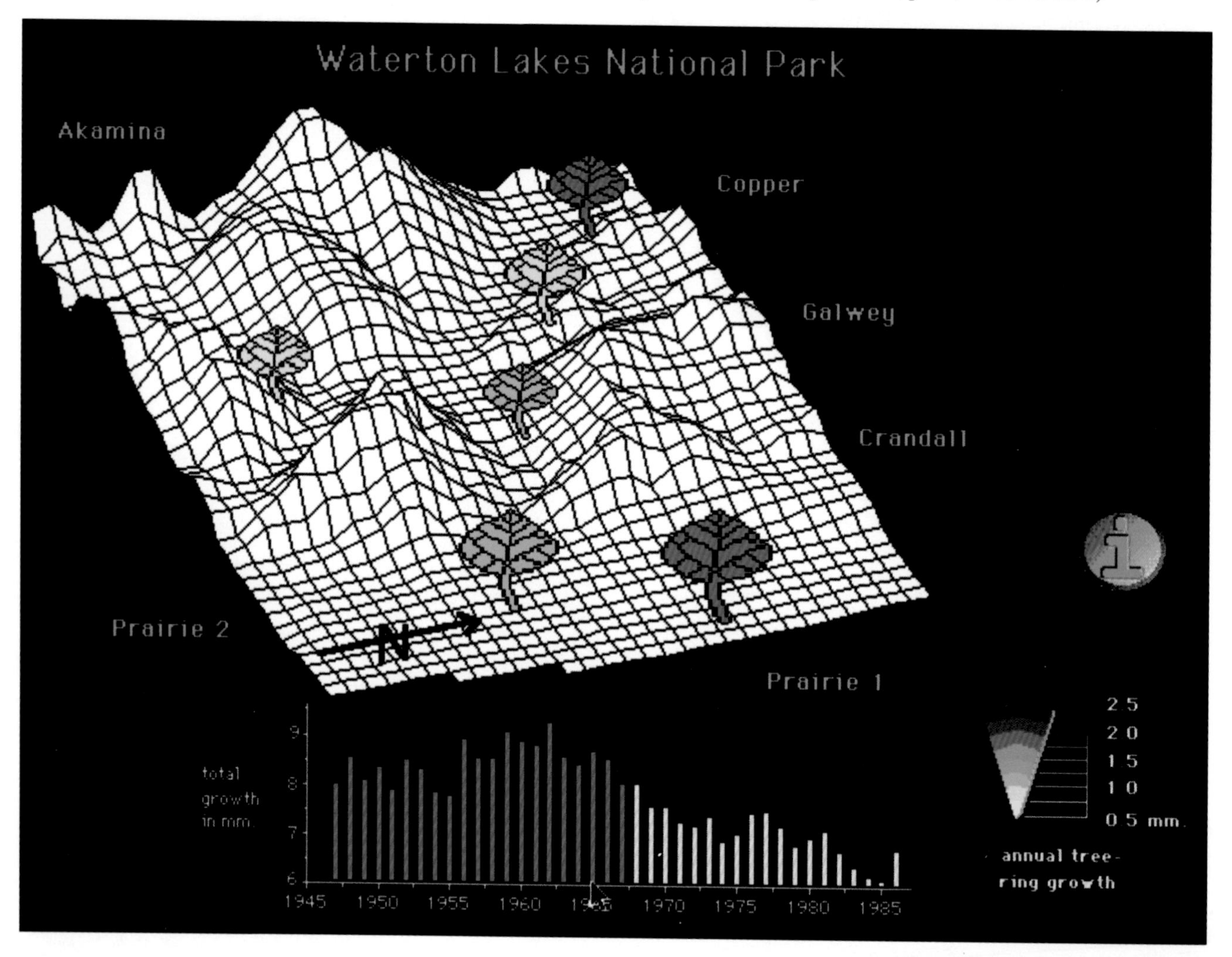

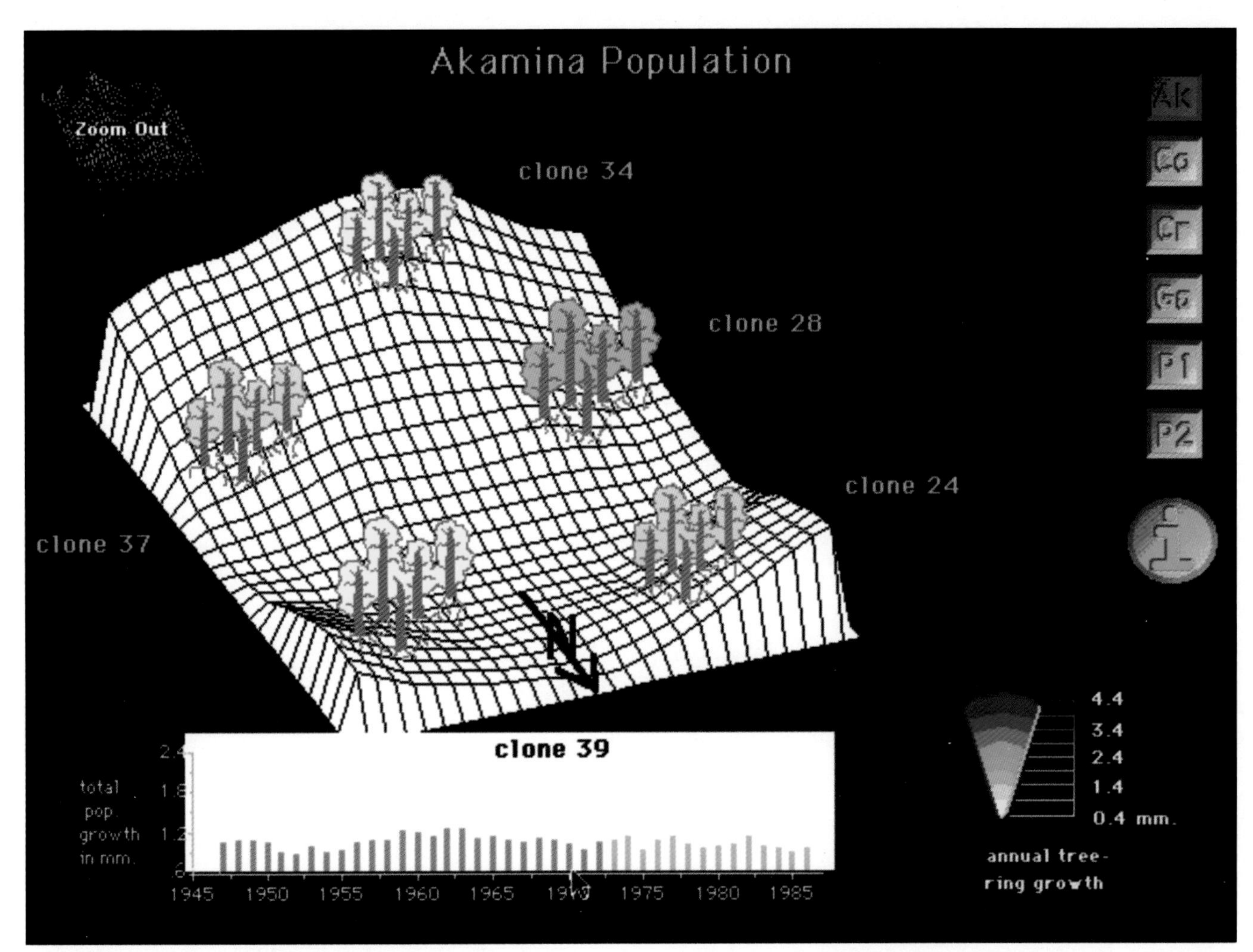

COLOR PLATE 13.8. A frame of an animation obtained when a leaf in Color Plate 13.7 is clicked on. Size of trees represents total cumulative growth, while tree color represents incremental growth. Clicking on an icon in the upper right can lead to animations of clones for other study sites. (Courtesy of Barbara P. Buttenfield and Christopher R. Weber.)

←

COLOR PLATE 13.7. A frame of an animation from Aspens. Size of leaf represents total cumulative tree growth for a study site, while leaf color represents incremental growth for each year. Clicking on a leaf reveals an animation for clones associated with that study site (Color Plate 13.8). (Courtesy of Barbara P. Buttenfield and Christopher R. Weber.)

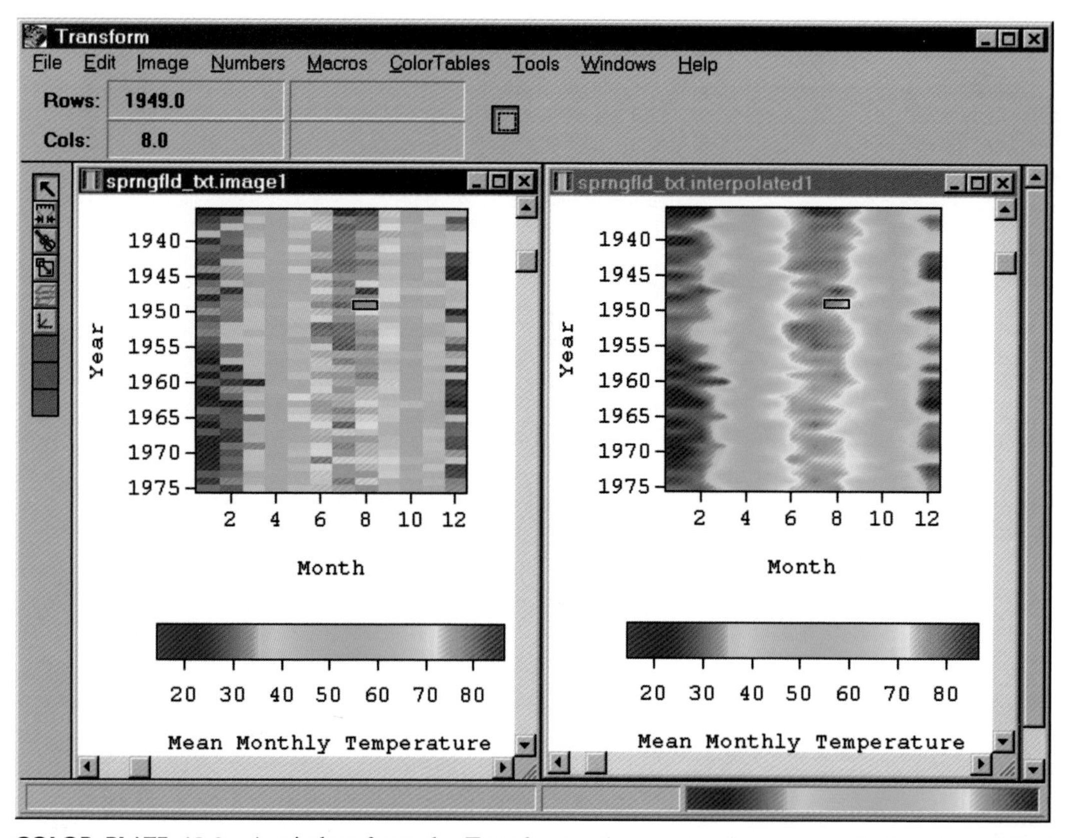

COLOR PLATE 13.9. A window from the Transform software showing unsmoothed and smoothed images (the left and right respectively) of mean monthly temperatures for Springfield, Illinois. Years are shown along the y axis, while months of the year are shown along the x axis. Compare with the table of these data shown in Figure 13.6. (Image created with Noesys Visualization Pro, courtesy of Fortner Software LLC.)

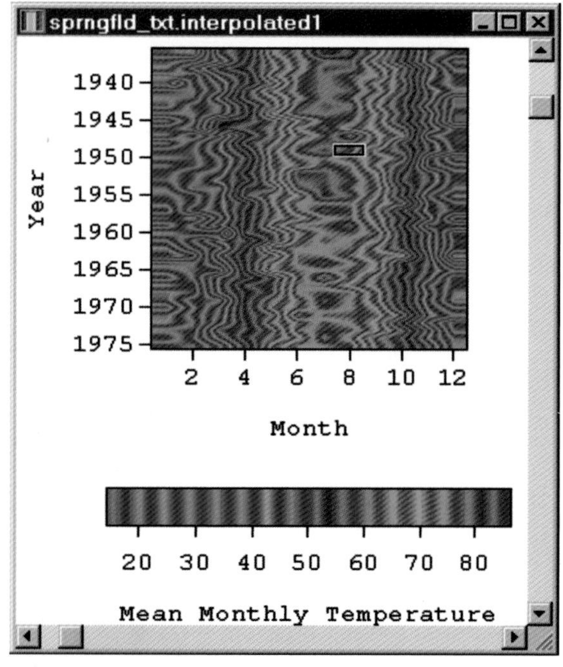

COLOR PLATE 13.10. A window from the Transform software showing a smoothed "Lava Waves" color scheme. Compare with the smoothed "Rainbow" scheme shown in Color Plate 13.9. (Image created with Noesys Visualization Pro, courtesy of Fortner Software LLC.)

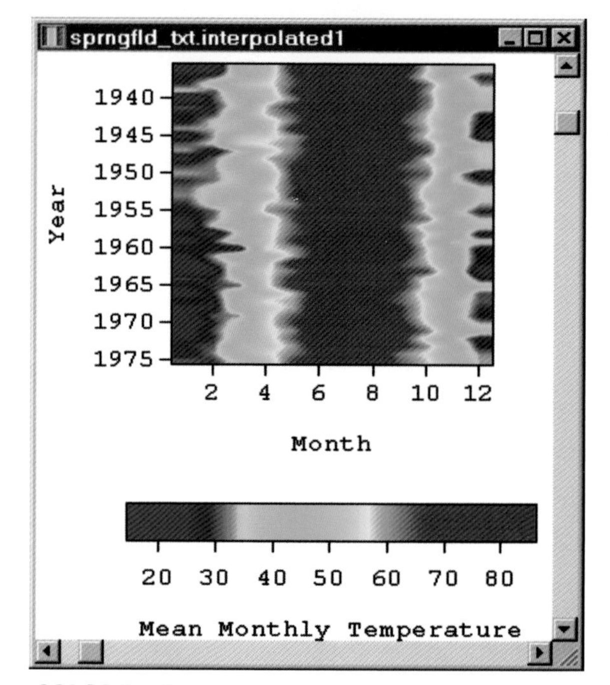

COLOR PLATE 13.11. A window from the Transform software showing a smoothed "Rainbow" color scheme shifted left and compressed (compare with Color Plate 13.9). (Image created with Noesys Visualization Pro, courtesy of Fortner Software LLC.)

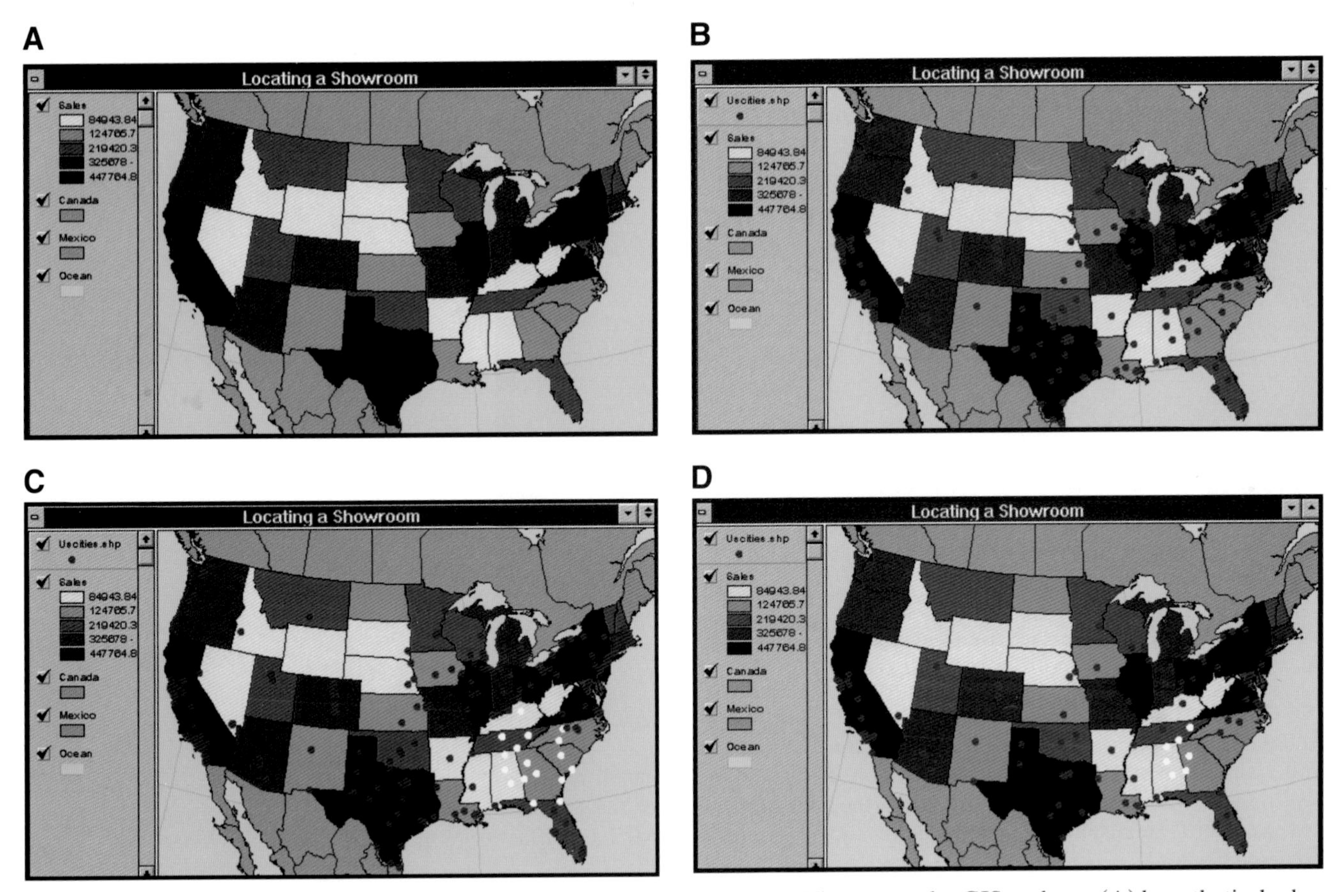

COLOR PLATE 13.12. The integration of analysis and display functions in ArcView, a popular GIS package: (A) hypothetical sales information by state is displayed as a choropleth map; (B) cities with a population over 80,000 are determined using a query builder and automatically displayed as blue circles; (C) cities within 300 miles of Atlanta are automatically displayed as yellow circles; (D) different spatial constraints are used (150,000 for population and 200 miles for distance from Atlanta). (Graphic image created using ArcView® GIS software. Source data: ESRI.)

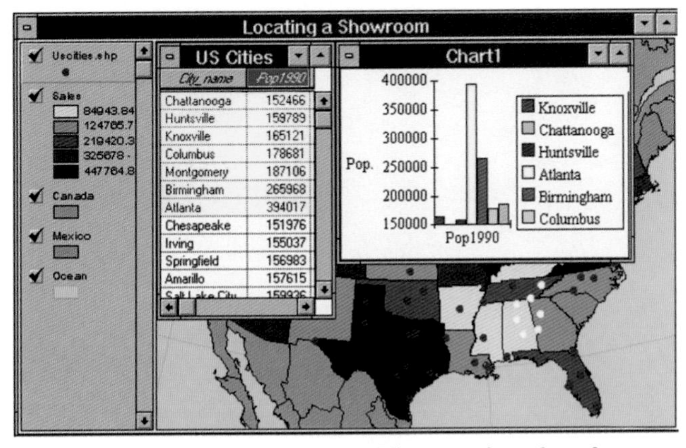

COLOR PLATE 13.13. Using ArcView to view data from a variety of perspectives: as a map, table, or chart. Note that cities highlighted on the map are highlighted in the table and also displayed in the chart. (Graphic image created using ArcView® GIS software. Source data: ESRI.)

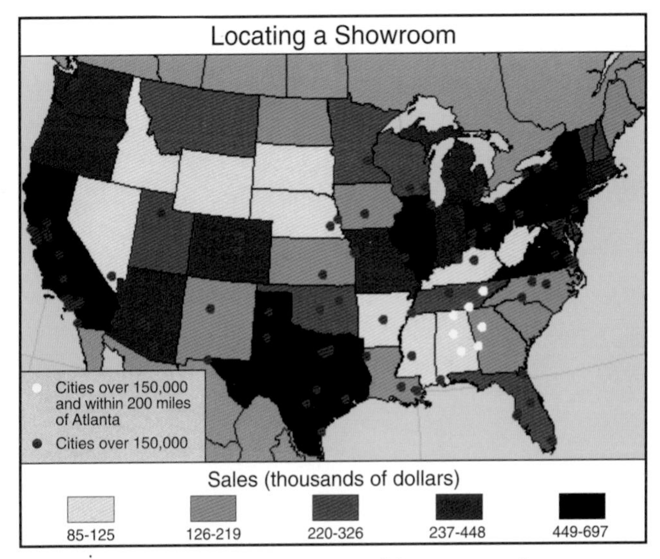

COLOR PLATE 13.14. A map created for presentation purposes. Cities in yellow that are located in states falling in the two lowest-valued classes meet the three criteria specified for locating the showroom. (Graphic image created using ArcView® GIS software. Source data: ESRI.)

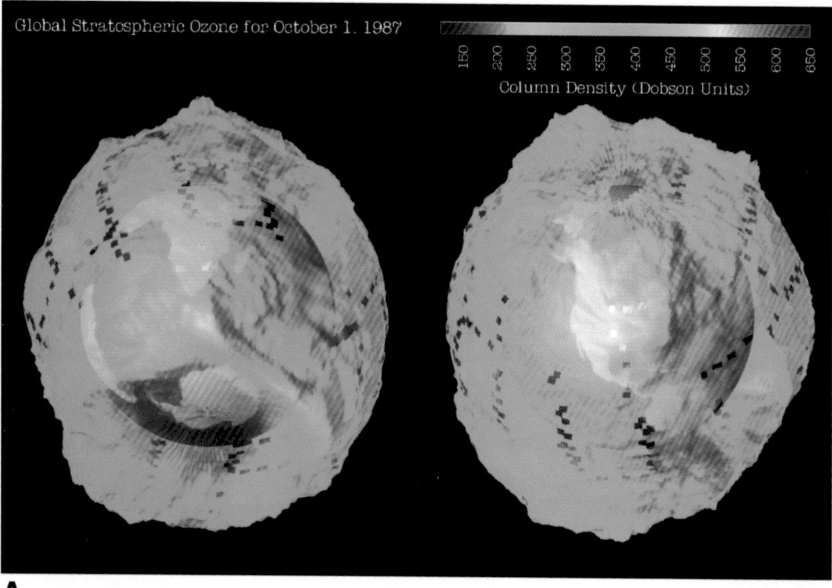

A

COLOR PLATE 14.1. Treinish's use of DX to portray the ozone hole over Antarctica: (A) the ozone symbolization is shown as an orthographic projection (this symbolization was used in an animation of the data); (B) small multiples depicting monthly averages of ozone during the spring warming for the Southern and Northern Hemispheres. (Courtesy of Lloyd Treinish, IBM T.J. Watson Research Center.)

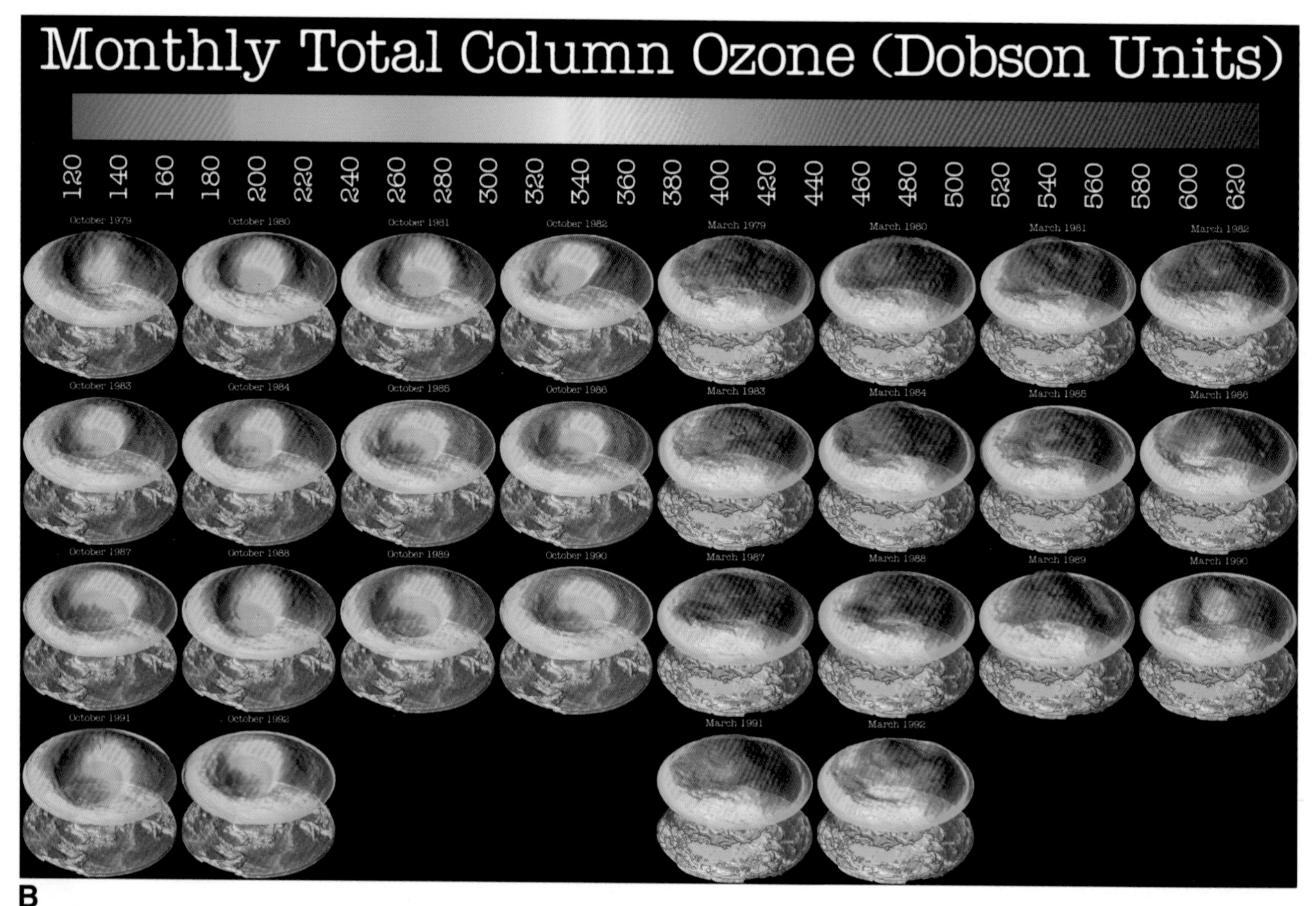

B

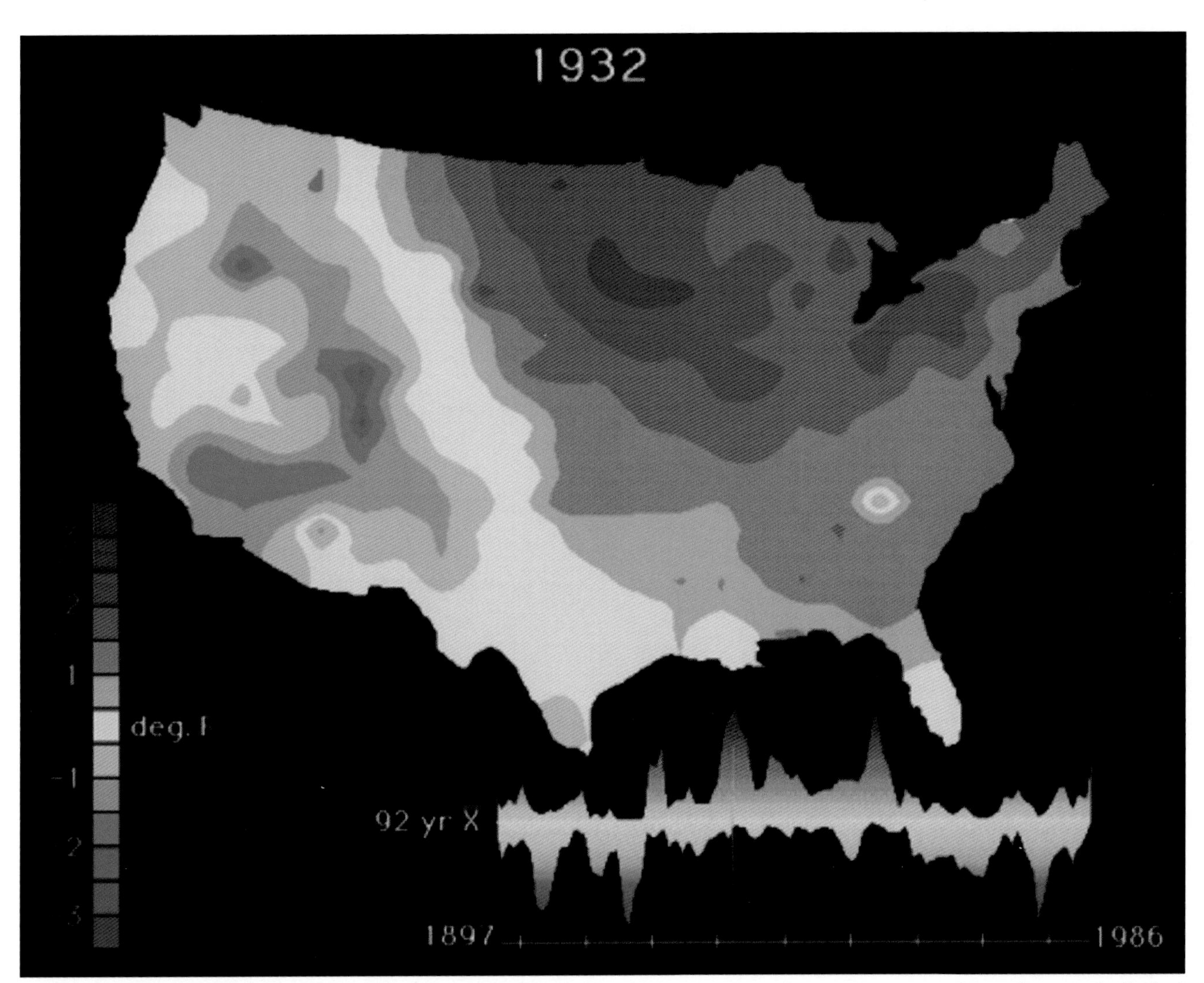

COLOR PLATE 14.2. A frame from Weber and Buttenfield's animation of U.S. surface temperatures. Note the use of a logical color progression, with reds for positive deviations and blue for negative deviations. (Courtesy of Christopher R. Weber and Barbara P. Buttenfield.)

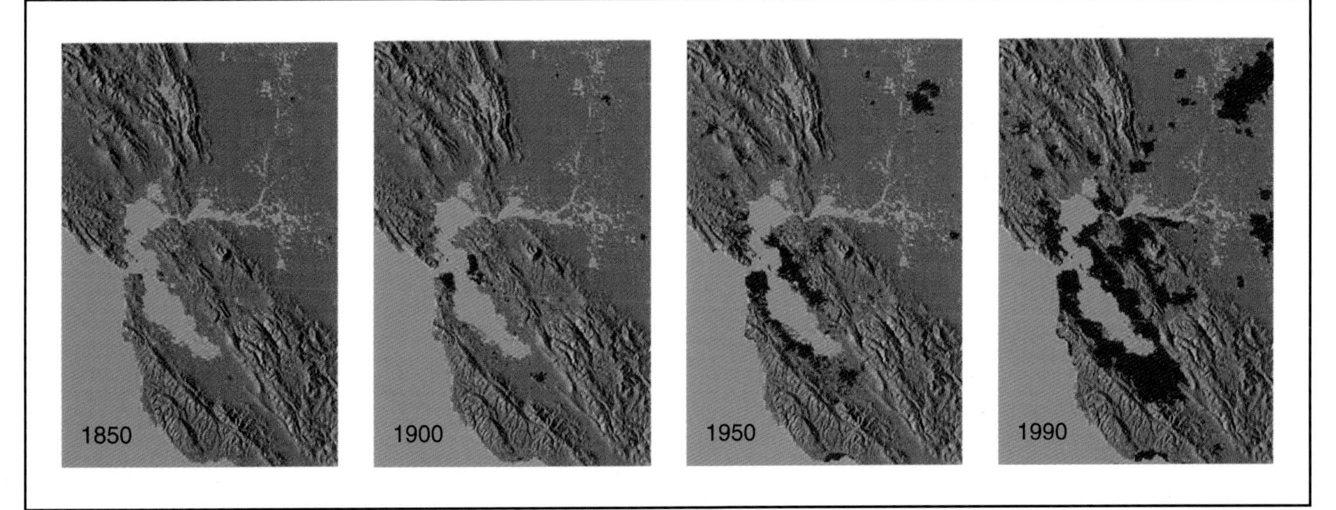

COLOR PLATE 14.3. Four frames from the USGS's animation of urban sprawl in the San Francisco–Sacramento region described in Buchanan and Acevedo (1996). Note the relatively smooth transitions between the frames. (Courtesy of William Acevedo.)

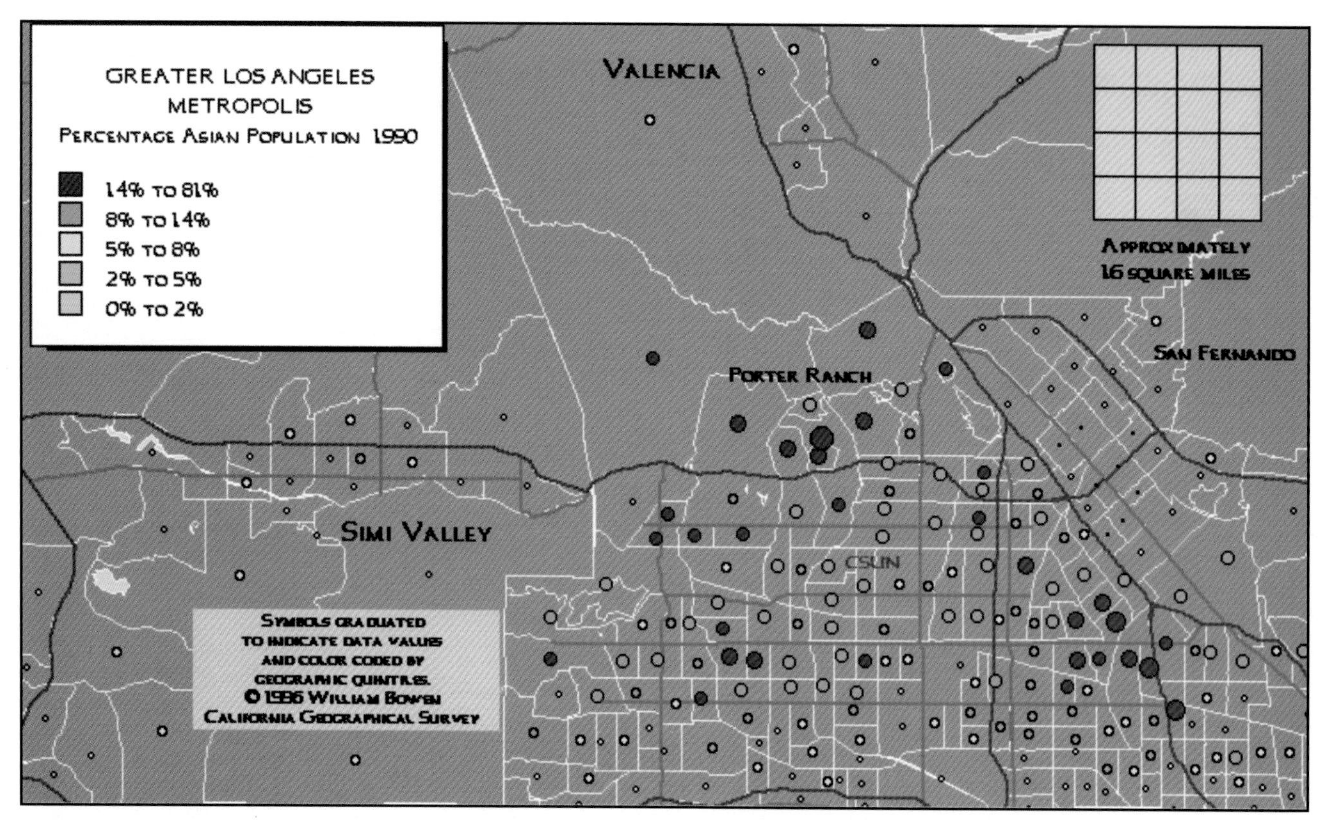

COLOR PLATE 15.1. A portion of a sample map from the Digital Atlas of California. This atlas emulates traditional paper atlases in the sense that no interaction with the maps is possible. (Courtesy of William Bowen.)

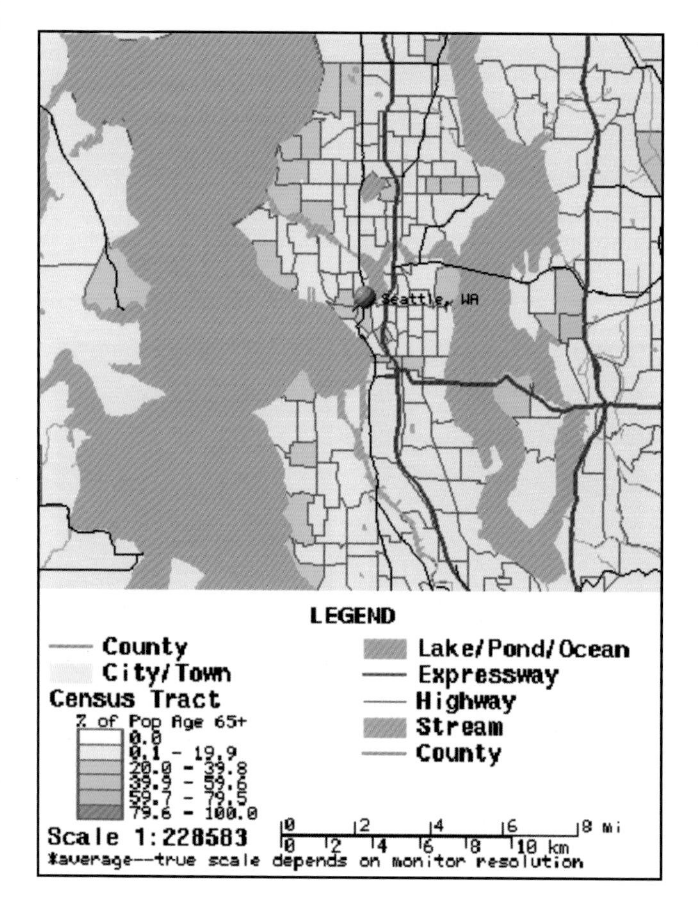

COLOR PLATE 15.2. A sample map created using the TIGER Mapping Service. Base data include water bodies and interstate highways: the theme is the percent of the population 65 years and older within census tracts.

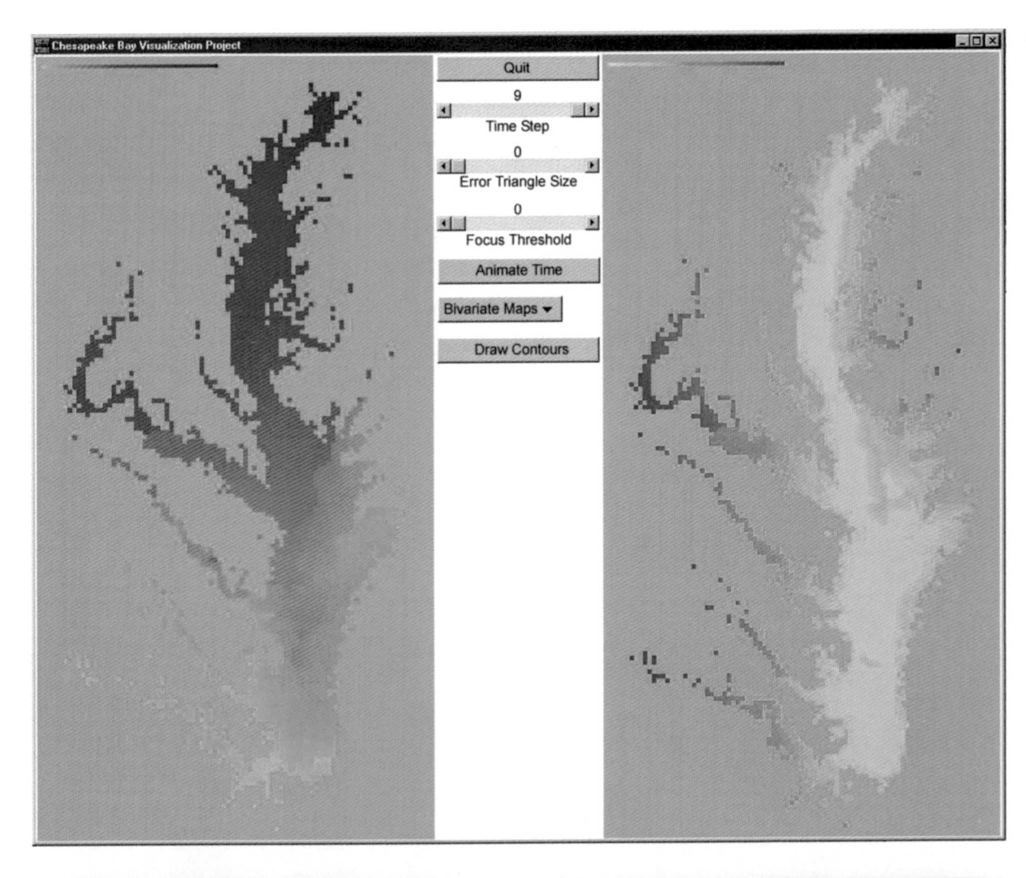

COLOR PLATE 16.1. The default display for R-VIS. On the left is a contour map resulting from interpolating between the 49 point locations for which dissolved inorganic nitrogen (DIN) values were collected. On the right is a reliability map based on a 95 percent confidence interval computed using kriging. Note that different hues are used for each map, and that lightness and saturation varies within each hue. (Courtesy of David Howard and Alan MacEachren.)

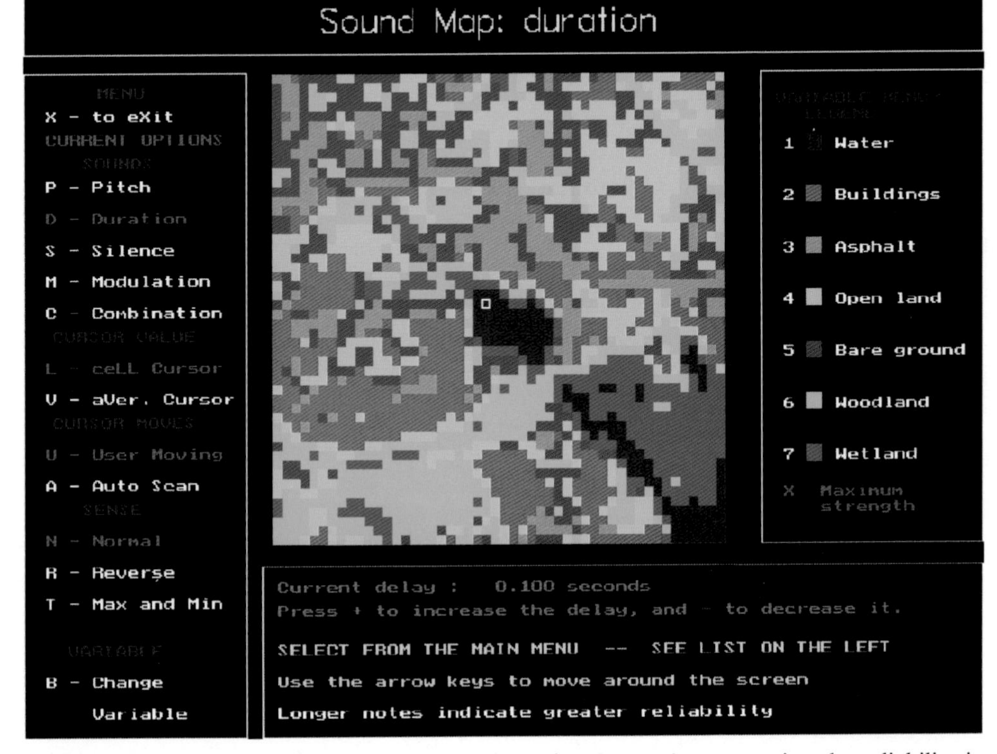

COLOR PLATE 16.2. A screen from Fisher's (1994a) software for portraying the reliability in remotely sensed images via sound. As the cursor moves across the image, the user hears a sound representing the reliability associated with the current pixel location. In the case depicted here, a long duration would indicate a pixel with a high reliability. (Courtesy of Peter F. Fisher.)

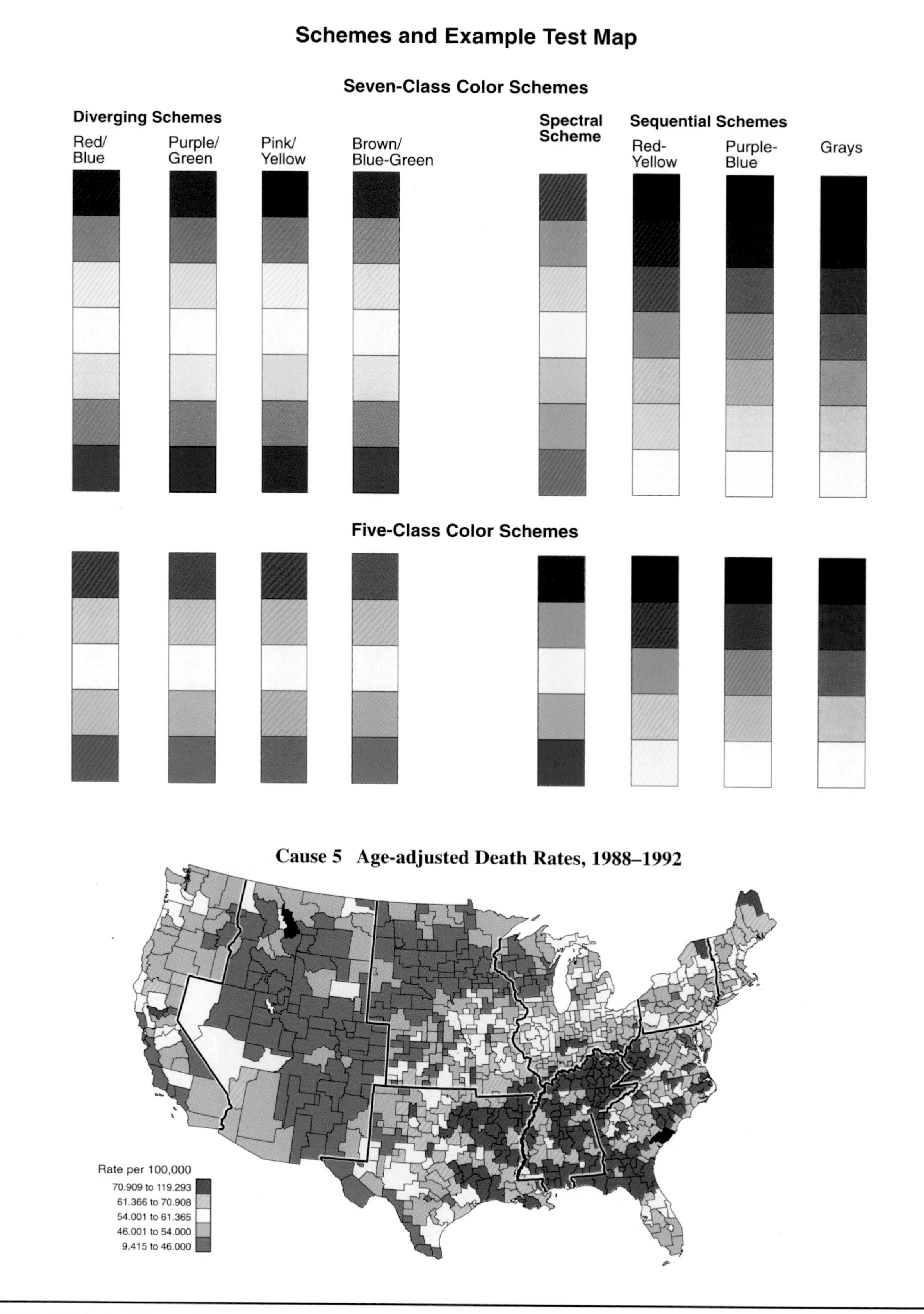

COLOR PLATE 16.3. Color schemes and a sample test map evaluated by Brewer et al. (1997). Color schemes for seven- and five-class maps are shown above, while a sample map using the purple-green five-class scheme is shown below. (From Brewer, C.A., MacEachren, A.M., Pickle, L.W., and Herman, D., 1997. First published in Annals, Association of American Geographers, 87. Reprinted with permission of Association of American Geographers.)

COLOR PLATE 16.4. An example of the visual realism that can be achieved in Virtual Los Angeles (Courtesy of Bill Jepson, Director UCLA Urban Simulation Laboratory).

COLOR PLATE 16.5. An image produced by CLRview, a package for creating visually realistic views. The image is part of an attempt to depict development changes in central Switzerland, a project that is described in Hoinkes and Lange (1995) (from http://www.clr.utoronto.ca:1080/LINKS/GISW/origarticle.html; courtesy of Eckert Lange and the Swiss Federal Office of Topography (Digital terrain model DHM25)).

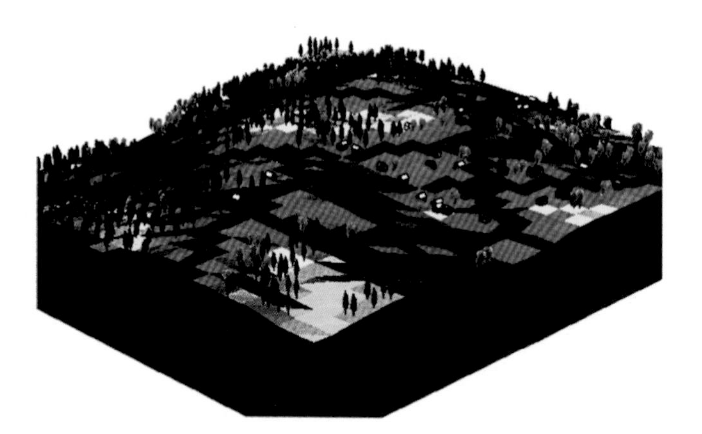

COLOR PLATE 16.6. An example of a view created using Ervin's Emaps. Visually realistic trees and houses have been added to a traditional hue-based color scheme using a traffic light analogy: undesirable areas close to a proposed road appear in red, while more desirable areas far away appear in green, and transitional areas appear in yellow. (Courtesy of Stephen Ervin.)

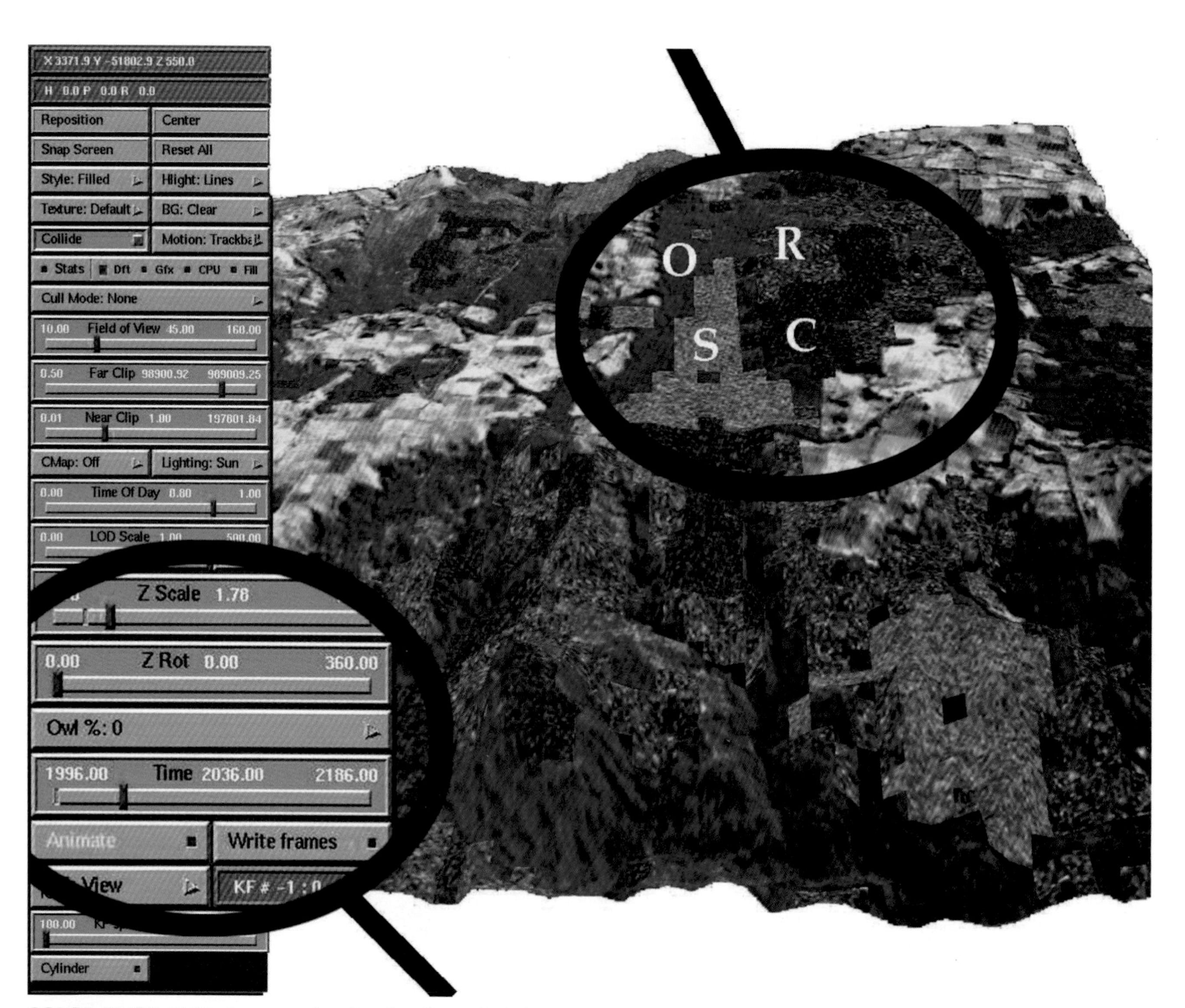

COLOR PLATE 16.7. An example of a visually realistic image created by Bishop and his colleagues. They term this a form of image-based visualization because aerial photographs or remotely sensed images are used to represent different vegetation types found in a forested region. (Courtesy of Ian Bishop.)

points, and then interpolate between grid points. The term **inverse distance** comes from the fact that nearby control points are weighted more than distant points in calculating a grid point. Since points far away from a grid point are apt to be uncorrelated with the grid point, it makes sense to use a limited set of control points in calculating a grid point; for example, a study by Declercq suggested using few control points (4 to 8) for smooth data, and many (16 to 24) for more abrupt data. Rather than simply considering the nearest control points, it also makes sense to consider the direction of the control points from the grid point (for example, an octant search can be used). An advantage of inverse distance is that large data sets can be efficiently interpolated.

Kriging is similar to inverse-distance methods in that a grid is overlaid on control points and values are estimated at each grid point as a function of distance to control points. Rather than considering distances to control points independently of one another, however, kriging considers the spatial autocorrelation in the data, both between the grid point and the surrounding control points, and among the control points themselves. The spatial autocorrelation in the data is evaluated by creating a **semivariogram,** which illustrates changes in the **semivariance** (variation in the data associated with control points) with increasing distance between control points.

Like inverse-distance methods, kriging calculates an estimate at a grid point by using a weighted average of nearby control points. The weights, however, are not simply a function of the distance of the control points from the grid point, but rather consider the nature of the semivariogram computed for the control points. An advantage of kriging is that it can produce an **optimal interpolation** (if the semivariogram(s) and associated **semivariogram models** are properly specified). Disadvantages of kriging are that execution times can be lengthy for large data sets, time is required to select the appropriate semivariogram(s) and associated models, and the method is difficult to understand.

Ideally, mapmakers should evaluate the accuracy of interpolation methods by evaluating **residuals** (the difference between known and estimated control point values). This can be accomplished using either cross validation or data splitting. In **cross validation,** a control point is removed from the data, and other control points are used to interpolate a value at the location of the eliminated control point. If this process is repeated for each control point, then a set of residuals can be computed for each control point. In **data splitting,** control points are split into two groups, one to create the contour map and one to evaluate it. Residuals are computed for each of the control points used in the evaluation stage. Since the process of evaluating residuals can be time-consuming, an alternative for evaluating the accuracy of interpolation methods is to use the results of past studies such as Declercq's.

Although automated interpolation methods for true point data clearly avoid the effort of manual interpolation, they are not without problems. Experts note that automated interpolation methods do not consider the nature of the underlying phenomenon; for example, an automated interpolation of precipitation data is unlikely to consider the "elongated bands . . . characteristic . . . of storm-precipitation surfaces." Automated interpolation methods also tend to produce "jagged lines" and "spurious details" not warranted by the data. One solution to these problems is to graphically edit contour maps using design software such as Freehand, but such editing often requires knowledge of the phenomenon being mapped.

This chapter also considered one method for handling conceptual point data: Tobler's **pycnophylactic interpolation.** The pycnophylactic method begins by assuming that each enumeration unit is raised to a height proportional to the value of its associated control point. This three-dimensional surface is then gradually smoothed, keeping the volume within each individual enumeration unit constant; thus, volume added somewhere within an enumeration unit must be balanced by taking volume from somewhere else in that unit. The smoothing is accomplished using a cell-based generalization process analogous to procedures used in image processing. Although pycnophylactic interpolation is appropriate for conceptual point data, it should only be used if the assumption of a smooth continuous phenomenon seems reasonable.

This chapter has focused on methods of interpolating data associated with smooth continuous phenomena. In the subsequent chapter, we will consider methods for symbolizing data representing smooth continuous phenomena.

FURTHER READING

Cressie, N. (1990) "The origins of kriging." *Mathematical Geology* 22, no. 3:239–252.

Discusses the origins of kriging within a variety of disciplines.

Cressie, N. A. C. (1993) *Statistics for Spatial Data.* rev. ed. New York: John Wiley & Sons.

An advanced treatment of statistical methods for handling spatial data; includes an extensive section on kriging.

Davis, J. C. (1986) *Statistics and Data Analysis in Geology.* 2d ed. New York: John Wiley & Sons.

Triangulation, inverse distance, and kriging methods are discussed on pages 353–377, 239–248, and 383–405. The section on kriging covers both punctual and universal kriging.

Declercq, F. A. N. (1996) "Interpolation methods for scattered sample data: Accuracy, spatial patterns, processing time."

Cartography and Geographic Information Systems 23, no. 3:128–144.

Evaluates the accuracy of various interpolation methods using a variety of approaches.

Isaaks, E. H., and Srivastava, R. M. (1989) *Applied Geostatistics.* New York: Oxford University Press.

A widely referenced text on interpolation methods. The focus is on kriging.

Kumler, M. P. (1994) "An intensive comparison of triangulated irregular networks (TINs) and digital elevation models (DEMs)." *Cartographica* 31, no. 2:1–99.

Triangulation and an equally spaced grid are two common approaches for representing elevation (topographic) data. This study compares the effectiveness of these two approaches.

Lam, N. S. (1983) "Spatial interpolation methods: A review." *Cartography and Geographic Information Systems* 10, no. 2:129–149.

An overview of a variety of interpolation methods, many of which are not covered in the present chapter.

McCullagh, M. J. (1988) "Terrain and surface modelling systems: Theory and practice." *Photogrammetric Record* 12, no. 72:747–779.

Describes and compares grid-based (using inverse distance or kriging) and triangulation interpolation methods.

Miller, E. J. (1997) "Towards a 4D GIS: Four-dimensional interpolation utilizing kriging." In *Innovations in GIS 4,* ed. Z. Kemp, pp. 181–197. Bristol, Penn.: Taylor & Francis.

The present chapter deals with interpolation methods appropriate for 2½-D phenomena in which one z value is associated with each x and y location. Miller's work considers interpolation methods for true 3-D phenomena in which multiple z values are associated with each x and y location. Miller also considers spatial data with a temporal component.

Mulugeta, G. (1996) "Manual and automated interpolation of climatic and geomorphic statistical surfaces: An evaluation." *Annals, Association of American Geographers* 86, no. 2:324–342.

Compares the accuracy of automated and manual methods for interpolation.

Oliver, M. A., and Webster, R. (1990) "Kriging: A method of interpolation for geographical information systems." *International Journal of Geographical Information Systems* 4, no. 3:313–332.

Reviews kriging and provides examples of how it can be used.

Robeson, S. M. (1997) "Spherical methods for spatial interpolation: Review and evaluation." *Cartography and Geographic Information Systems* 24, no. 1:3–20.

Reviews and evaluates spherical methods for interpolation, which account for curvature of the earth and are essential when interpolating global-scale phenomena.

Tobler, W. R. (1979) "Smooth pycnophylactic interpolation for geographical regions." *Journal of the American Statistical Association* 74, no. 367:519–536.

A detailed discussion of the pycnophylactic interpolation method. Understanding this paper requires a solid background in calculus.

Tobler, W., Deichmann, U., Gottsegen, J., and Maloy, K. (1997) "World population in a grid of spherical quadrilaterals." *International Journal of Population Geography* 3:203–225.

Describes the development of a subnational database of population for the world based on pycnophylactic interpolation.

Veve, T. D. (1994) "An assessment of interpolation techniques for the estimation of precipitation in a tropical mountainous region." Unpublished M.A. thesis, Pennsylvania State University, University Park, Pennsylvania.

Compares several interpolation methods, including inverse distance, kriging, and co-kriging. The latter permits the inclusion of ancillary variables in the kriging process (for example, elevation in the case of interpolating precipitation in a mountainous area).

Ware, J. M., and Jones, C. B. (1997) "A multiresolution data storage scheme for 3-D GIS." In *Innovations in GIS 4,* ed. Z. Kemp, pp. 9–24. Bristol, Penn.: Taylor & Francis.

Describes a sophisticated triangulation-based approach for representing topographic and geologic data.

9

Symbolizing Smooth Continuous Phenomena

OVERVIEW

This chapter presents a variety of methods for symbolizing smooth continuous phenomena (or 2½-D phenomena). Remember that such phenomena can be thought of as a 3-D surface for which each x *and* y *value has a single associated* z *value. For our purposes, it is useful to distinguish between two types of smooth continuous phenomena: the earth's topography and nontopographic phenomena. The earth's topography (both above and below sea level) can actually be seen in the real world, while nontopographic phenomena can only be conceptualized. An example of a nontopographic phenomenon would be the precipitation associated with Hurricane Hugo, which was mapped in Chapter 8.*

We should be aware of the different kinds of data commonly used as a basis for mapping smooth continuous phenomena. Topographic data are commonly available in the form of an equally spaced gridded (or possibly triangulated) network known as a ***digital elevation model (DEM).*** *These data can be acquired from mapping agencies such as the USGS. Nontopographic data are usually available as a set of irregularly spaced points (such as weather station locations). A set of interpolated values are calculated between these locations using methods such as triangulation, inverse distance, and kriging.*

The methods described in section 9.1 can be used to symbolize either topographic or nontopographic data. Techniques covered include ***contour lines, hypsometric tints*** *(shading the areas between contour lines),* ***continuous-tone maps*** *(shades analogous to those used on unclassed choropleth maps), and* ***fishnet*** *symbolization (a netlike structure that simulates the three-dimensional character of a smooth continuous surface). With the exception of the fishnet approach, a limitation of these methods is that they do not readily depict the three-dimensional character of smooth continuous phenomena (that is, the reader may have difficulty "seeing" a surface).*

Section 9.2 covers methods that do enable viewers to "see" a three-dimensional surface; normally these methods are restricted to portraying topography. Methods covered include hachures, contour-based methods, shaded relief, and the use of color to depict aspect and slope. These methods assume that the topographic surface is illuminated, typically obliquely from the northwest; the various methods use different approaches to distinguish areas that appear illuminated from those that appear in the shadows. With ***hachures,*** *illumination is accomplished by changing the width of parallel lines drawn perpendicular to contours; lines facing the light source are narrow, while those facing away are wide. Contour-based methods involve manipulating traditional contour lines in some fashion; for example,* ***shadowed contours*** *are created by thickening contour lines appearing in the shadow.*

To create ***shaded relief,*** *areas facing away from a light source are shaded, while areas directly illuminated are not shaded; slope may also be a factor, with steeper slopes shaded darker. Since the effects of shaded relief can be rather dramatic, and it has become a popular computer-based method for depicting topography, we will cover some of the details of the computational formulas required to create shaded relief.*

Aspect and slope are two important elements of topography that often need to be visualized. ***Aspect*** *deals with the direction the land faces (for example, north or south), while* ***slope*** *considers the steepness of the land surface. For example, a developer of a ski resort might want to visualize north-facing aspects (to minimize snowmelt) and a* *diversity of slopes (to handle skiers' varied abilities). In this chapter, we will consider two approaches for using color to symbolize aspect and slope. The first of these, MKS-ASPECT™ developed by Moellering and Kimerling (1990) considers only aspect, while the second developed by Brewer and Marlow (1993) combines aspect and slope.*

Section 9.2 also covers two topics that illustrate the capability of modern visualization and animation software. First, we consider Treinish's (1994) rendition of global topography, which was created using the IBM Visualization Data Explorer (DX). This rendition provides an attractive spherical representation of both the land surface and undersea topography. Second, we consider the ability of animation software to create fly-bys of terrain. A ***fly-by*** *is a form of map animation in which the viewer is given the feeling of flying over a landscape. Until recently, fly-bys were possible only in a sophisticated workstation environment; today they can be developed in a PC environment that most of us have access to.*

9.1 METHODS FOR SYMBOLIZING TOPOGRAPHY OR NONTOPOGRAPHIC PHENOMENA

Contour lines (Figure 9.1A) represent the intersection of horizontal planes with the three-dimensional surface of a smooth continuous phenomenon. They have been one of the most frequent methods used to depict smooth continuous phenomena and were particularly common when maps were produced manually because little production effort was involved. An obvious problem with contours, however, is that visualizing the surface requires careful examination of the position and values of individual contour lines.

The addition of **hypsometric tints** (or shaded areas) between contour lines (Figure 9.1B) enhances the ability to visualize a three-dimensional surface because light and dark tints can be associated with low and high values, respectively. This visualization can be further enhanced by using one of the color schemes introduced in Chapter 6 in place of gray tones.

One problem with hypsometric tints is that the limited number of tones suggest a stepped surface, as opposed to a smooth one. This problem can be ameliorated by creating a **continuous-tone map** (Kumler and Groop 1990), in which each point on the surface is shaded with a gray tone (or color) proportional to the value of the surface at that point (Figure 9.1D). This approach is analogous to unclassed choropleth mapping (see section 4.2), in which enumeration units are shaded with a gray tone proportional to the value of that unit. One problem with interpreting a continuous-tone map is that it is difficult to associate numbers in the legend with particular locations. This problem can be solved by overlaying continuous tones with traditional contour lines (Figure 9.1E).

Another approach that assists in interpreting smooth continuous phenomena is the **fishnet** symbolization (Figure 9.1C). With this approach, not only does the surface change gradually, but we can actually "see" that certain points are higher or lower than others. Although fishnet symbolization is useful, it has the same disadvantages as a prism map (discussed in section 2.5), including the blockage of low points by high ones, and that rotation may produce a view unfamiliar to readers who normally see maps with north at the top.

The three-dimensional look of fishnet symbolization can also be achieved by using stereo pairs and anaglyphs, both of which permit stereo views (Clarke 1995, 273–274). With **stereo pairs,** two maps of the surface are viewed with a stereoscope, which is normally used to create three-dimensional views of aerial photographs. With **anaglyphs,** two images are created, one in green or blue, and one in red; these images are viewed with anaglyphic glasses, which use colored lenses to produce a three-dimensional view.

When choosing among the various symbolization methods shown in Figure 9.1, it is useful to consider a study by Kumler and Groop (1990) that evaluated the methods using map tasks and readers' preference. Their study included black-and-white maps similar to those shown in Figure 9.1, plus several continuous-tone color maps. Subjects were asked to complete the following tasks while looking at the maps: "locat[e] surface extrema, interpret . . . slope directions between points on the surface, estimat[e] relative surface values, and estimat[e] absolute values at points on the surface" (p. 282). Based on the accuracy with which subjects completed these tasks, Kumler and Groop found that continuous-tone maps were the most effective approach. Also, when asked to pick their favorite approach, the majority of subjects selected one of the continuous-tone maps (a full-spectral scheme). Although their study suggests continuous-tone maps are particularly effective, the study was limited because it did not include noncontinuous-tone color maps. An interesting follow-up to their study would be a comparison of continuous-tone and noncontinuous-tone color maps.

Bear in mind that in a data exploration environment it is possible to show the map user more than one of the symbolization methods presented in Figure 9.1. For example, a user might be shown a continuous-tone map and be allowed to toggle the contour lines off and on. At the same time, a fishnet map could be shown, and the user could be allowed to manipulate it to provide different views of the surface.

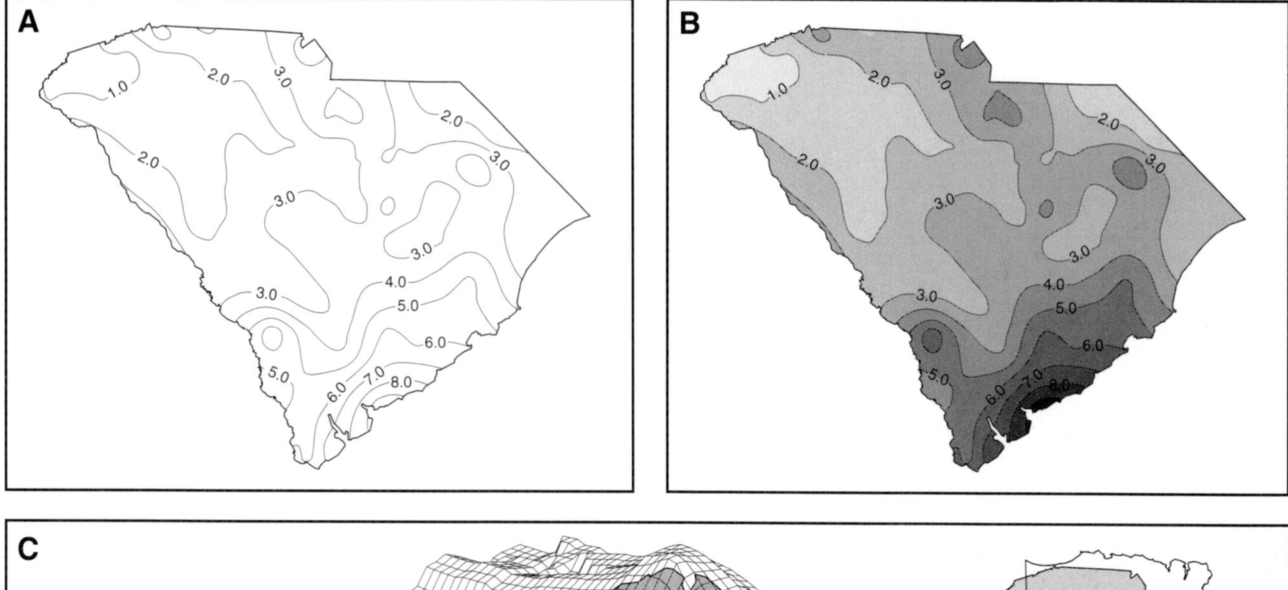

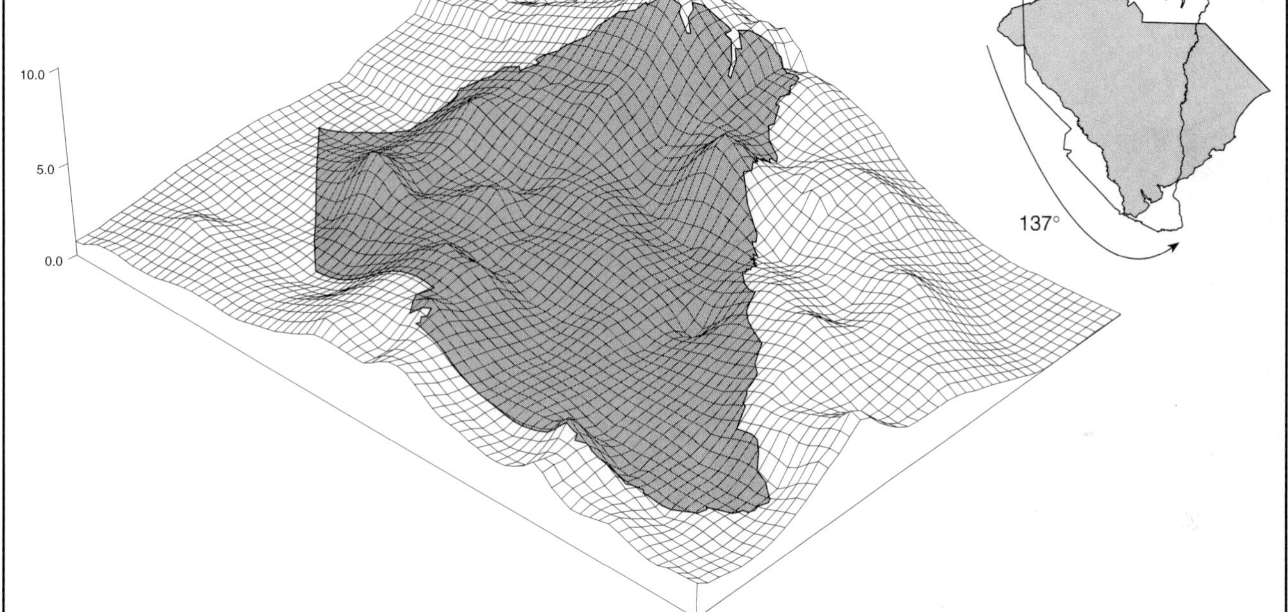

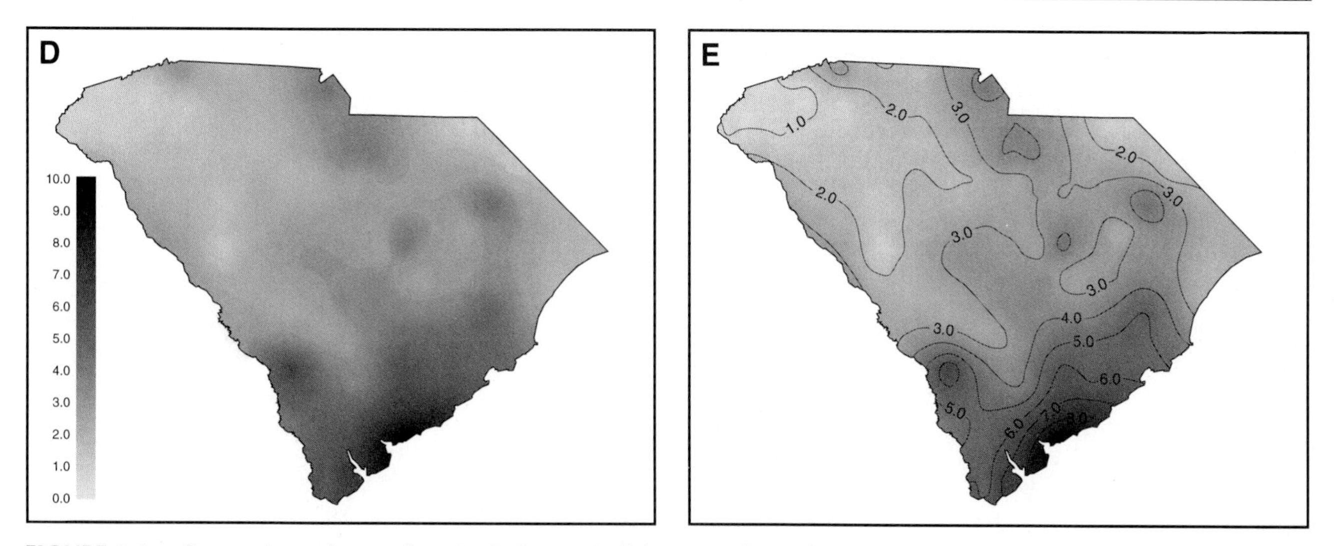

FIGURE 9.1. Comparison of several methods for symbolizing smooth continuous phenomena: (A) using only contour lines; (B) combining contour lines and hypsometric tints; (C) perspective block diagram; (D) continuous tones analogous to those used on unclassed choropleth maps; (E) continuous tones combined with contour lines.

9.2 METHODS FOR SYMBOLIZING TOPOGRAPHY

There are a broad range of methods that enable viewers to readily conceive of a topographic surface as three-dimensional in nature. Normally, these methods presume that the topographic surface is lit obliquely from the northwest. According to Imhof (1982, 178), a northwest light source is commonly used because of the preponderance of people who write with their right hand. When writing with the right hand, lighting is placed above and to the left to eliminate the possibility that a shadow will be cast over the written material. Our experience in using a light in this situation causes us to expect a light in a similar orientation in other situations such as relief depiction.

FIGURE 9.2. Hachures created using Yoeli's computer-based procedure. (From Yoeli 1985, (page 123); courtesy of The British Cartographic Society.)

9.2.1 Hachures

Hachures are constructed by drawing a series of parallel lines perpendicular to the direction of contours. Two forms of hachures were traditionally used in manual cartography (Yoeli 1985). In the first, known as *slope hachures,* the width of hachures was made proportional to the steepness of the slope; thus steeper slopes were darker. Since slope hachures generally failed to create the impression of a third dimension, *shadow hachures* were developed, in which an oblique light source was presumed: in this approach, the width of hachures was a function of whether the hachure was illuminated or in the shadow.

Although hachures normally are associated with the bygone era of manual cartography, Imhof (1982, 229) argued that they should not be forgotten.

> Hachures alone, without contours, are more capable of depicting the terrain than is [relief] shading alone. . . . Hachures also possess their own special, attractive graphic style. They have a more abstract effect than [relief] shading, and perhaps for this reason are more expressive.

Additionally, Yoeli (1985, 112) argued that hachures could be used for a structural analysis of the landscape. Figures 9.2 and 9.3 depict hachures produced from a computer program developed by Yoeli.

9.2.2 Contour-Based Methods

Several methods for portraying topography have been developed that involve manipulating contour lines; these include Tanaka's method, Eyton's illuminated contours, and shadowed contours. **Tanaka's method** (Tanaka 1950) involves varying the width of contour lines as a function of their angular relationship with a presumed light source; lines perpendicular to the light source have the greatest width, while those parallel to the light source are narrowest. Furthermore, contour lines facing the light source are drawn in white, while those in shadow are drawn in black (Figure 9.4). Although this technique was used as early as 1870 (Imhof 1982, 153), Tanaka is generally credited with originating it because of the attractive maps he produced (Figure 9.5) and the mathematical basis he provided.

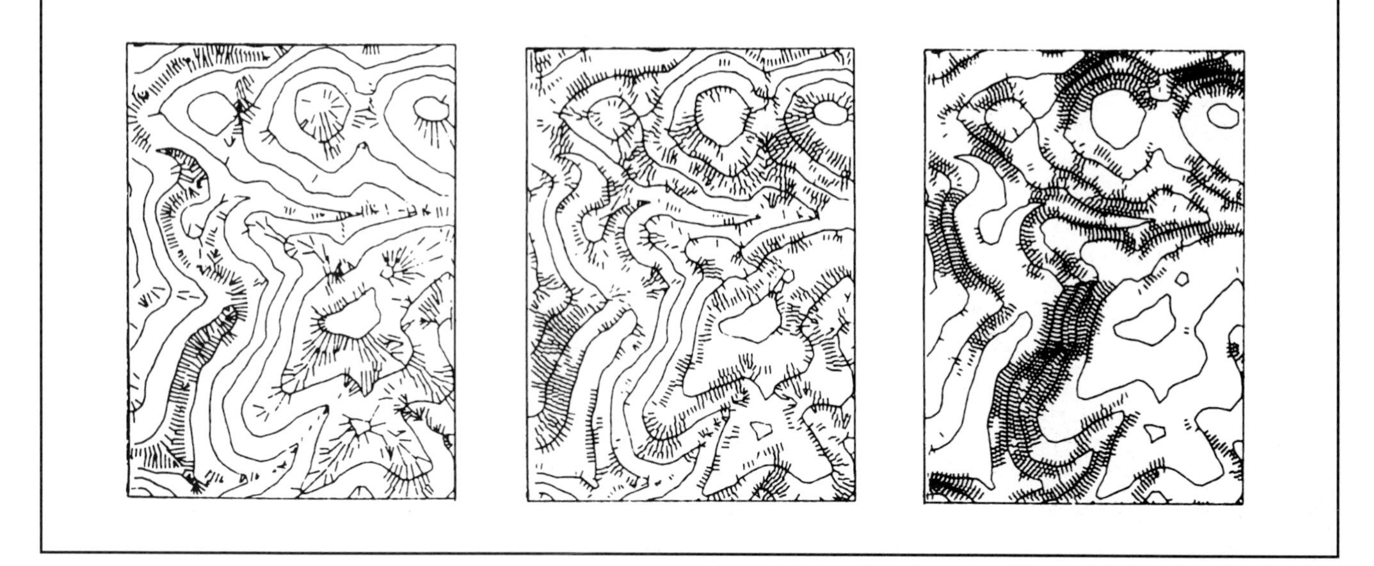

FIGURE 9.3. Hachures used for a structural analysis of the landscape, in this case depicting differing degrees of slope. (From Yoeli 1985, (page 124); courtesy of The British Cartographic Society.)

Like hachures, Tanaka's method was originally produced manually. Horn (1982), however, described a computer-based approach for implementing Tanaka's method. An analysis of maps Horn created suggests that Tanaka's method compares favorably with other computer-assisted methods for representing relief (see pp. 113–123 of Horn; method H is Tanaka's method).

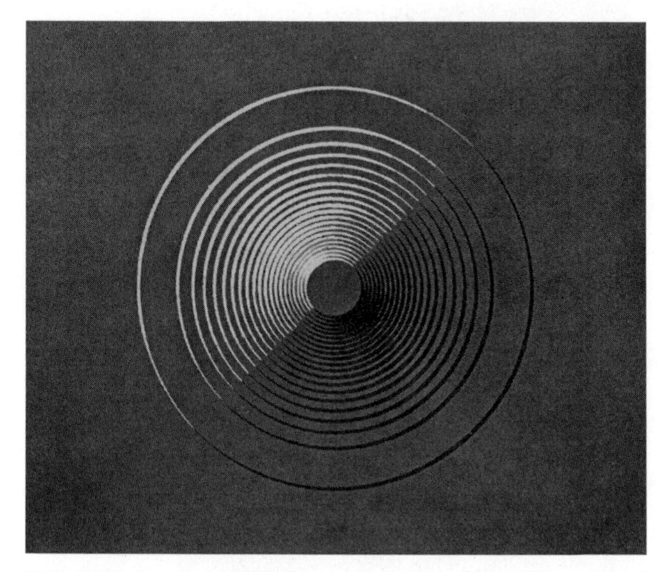

FIGURE 9.4. A simplified representation of Tanaka's method for depicting topography. The width of contour lines is a function of their angular relationship with a light source; contour lines facing the light source are drawn in white, while those in the shadow are drawn in black. (After Tanaka 1950. First published in *The Geographical Review* 40(3), p. 446. Reprinted with permission of the American Geographical Society.)

Eyton's illuminated contours, a raster (pixel-based) approach for computer-based contouring, was a simplification of Tanaka's method. In a fashion similar to Tanaka, Eyton (1984b) brightened contours facing the light source and darkened contours in the shadow. Eyton's approach, however, differed in that the width of contour lines was not varied (to speed computer processing). The topography resulting from Eyton's approach is not as distinctive as Tanaka's (compare Figures 9.5 and 9.6), but still gives the impression of three dimensions, especially when viewed from a distance.

Shadowed contours is still another approach for representing topography using modified contour lines. In this approach, contour lines facing the light source are drawn in a normal line weight, while those in the shadow are thickened (Figure 9.7). This technique differs from Tanaka's principally in that the contours facing the light source are drawn in black rather than white (compare Figures 9.4 and 9.7). Figure 9.8 depicts a computer-assisted implementation of shadowed contours developed by Yoeli (1983).

9.2.3 Shaded Relief

Shaded relief (also known as **hill shading** or **chiaroscuro**) has long been considered one of the most effective methods for representing topography. Swiss cartographers (Eduard Imhof in particular) have been widely recognized for their superb work. Like the preceding methods, this technique presumes an oblique light source, which creates shadows. Areas facing away

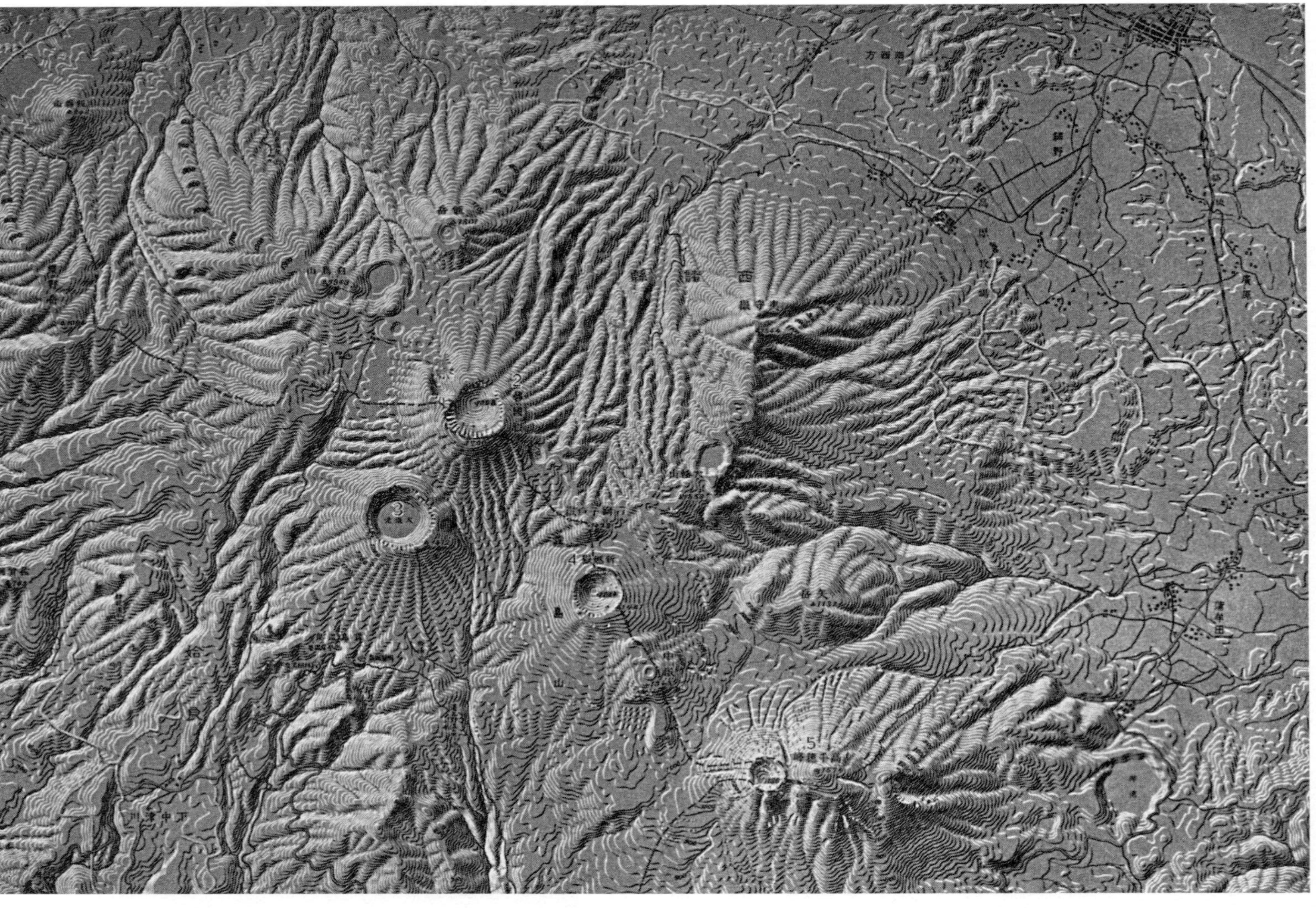

FIGURE 9.5. A map created using Tanaka's method for depicting topography. (After Tanaka 1950. First published in *The Geographical Review* 40(3), p. 451. Reprinted with permission of the American Geographical Society.)

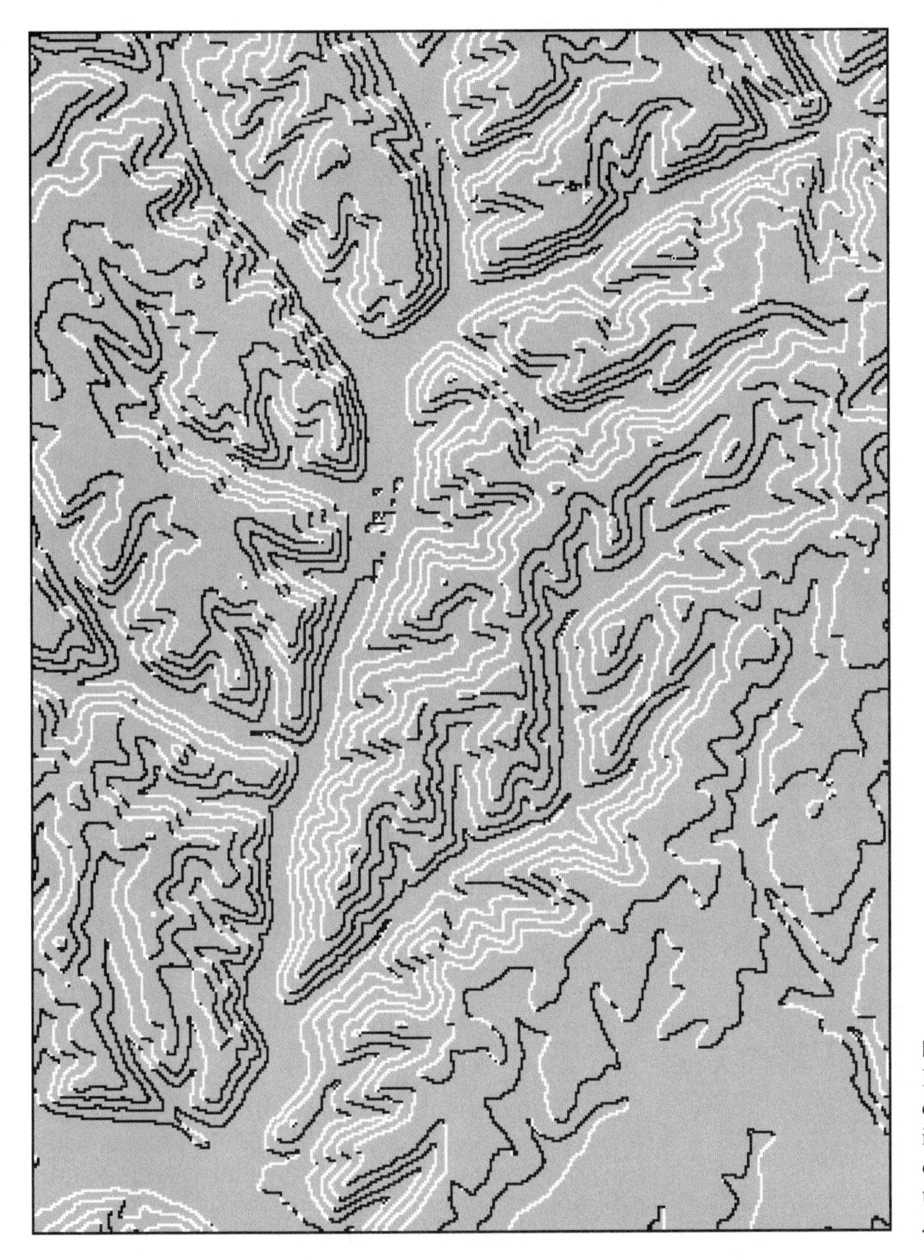

FIGURE 9.6. A map created using Eyton's illuminated contours in which contours facing the light source are drawn in white, while contours in the shadow are drawn in black; note that the width of contour lines is not varied. (Courtesy of J. Ronald Eyton.)

from the light source are shaded, while areas directly illuminated are not shaded; slope may also be a factor, with steeper slopes shaded darker. When produced manually, shading was accomplished using pencil and paper, and more recently with an airbrush. A classic example of a manually produced shaded relief map is the one Harrison (1969) created for the 1970 *U.S. National Atlas,* a portion of which is shown in Color Plate 9.1A.

Map I-2206, "Landforms of the Conterminous United States: A Digital Shaded-Relief Portrayal," developed by Thelin and Pike (1991) for the USGS, is a dramatic illustration of the capability of computer-based approaches for producing shaded relief (a small-scale version of the map is shown in Figure 1.8). Lewis (1992) describes the map as follows:

> Let us not mince words. The United States Geological Survey has produced a cartographic masterpiece: a relief map of the coterminous United States which, in accuracy, elegance, and drama, is the most stunning thing since Erwin Raisz published his classic "Map of the Landforms of the United States." (p. 289)

Portions of the USGS and Raisz (1967) maps are shown in Color Plate 9.1 for comparison with Harrison's map. In contrast to the USGS and Harrison maps, Raisz's was created using the **physiographic method** in which the earth's geomorphological features are represented by a

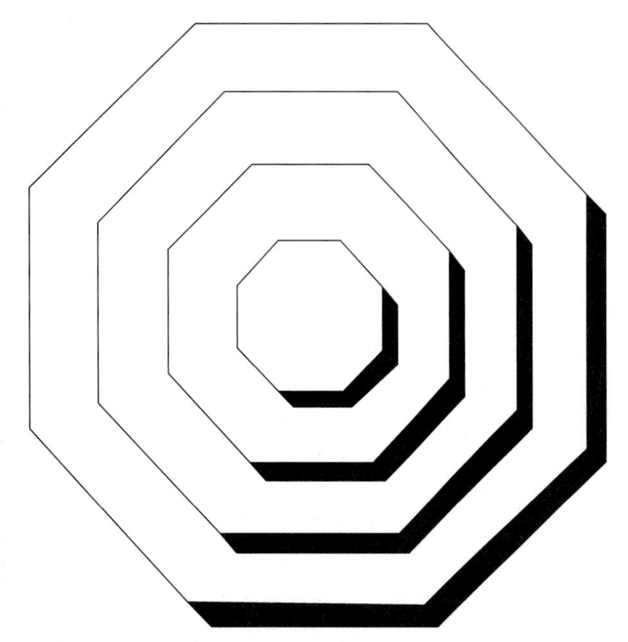

FIGURE 9.7. A simplified representation of shadowed contours in which contour lines facing a light source are drawn in black, while contour lines facing away from the light source produce a black shadow. (After Yoeli 1983. First published in *The American Cartographer* 10(2), p. 106. Reprinted with permission from the American Congress on Surveying and Mapping.)

FIGURE 9.8. A map created using shadowed contours. (After Yoeli 1983. First published in *The American Cartographer* 10(2), p. 109. Reprinted with permission from the American Congress on Surveying and Mapping.)

set of standard, easily recognized symbols (Figure 9.9 shows a portion of a set taken from Raisz, 1931).

Since the Raisz map has been so widely used by geographers, it is natural to ask which is better, the USGS map or the Raisz map? Lewis argues that both maps have merit. He states that the USGS map is "more like an aerial photograph" (1992, 298), allowing us to see the structure of the landscape as it actually exists (or at least within the constraints of the computer program). In contrast, the Raisz map represents one knowledgeable geographer's view of the landscape, with certain features emphasized as Raisz deemed appropriate.

Computation of Shaded Relief

Basic input for creating a computer-assisted shaded relief map is a **digital elevation model (DEM),** in which sample elevation values are provided on an equally spaced square gridded network.* Fundamental calculations involve manipulating 3 × 3 "windows" of elevation values within this gridded network. To illustrate this concept, we will use the window and hypothetical data shown in Figure 9.10. We will presume that our objective is to determine an appropriate gray tone for the central cell of the 3 × 3 window. The following steps are based on Eyton's (1991) formulas. For a discussion of alternative approaches for computing slope and aspect, see Chrisman (1997, 162–167).

* Digital elevation values may also be available in a triangulated form.

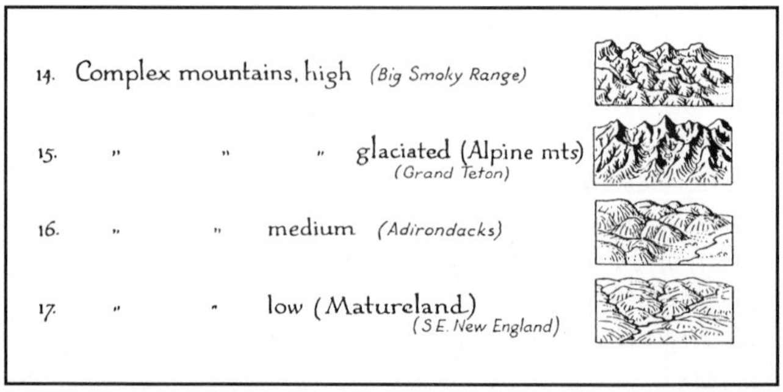

14. Complex mountains, high (Big Smoky Range)
15. " " " glaciated (Alpine mts) (Grand Teton)
16. " " medium (Adirondacks)
17. " " low (Matureland) (S.E. New England)

FIGURE 9.9. Examples of standard symbols that Raisz used to create his physiographic diagrams. (After Raisz 1931. First published in *The Geographical Review* 21(2), p. 301. Reprinted with permission of the American Geographical Society.)

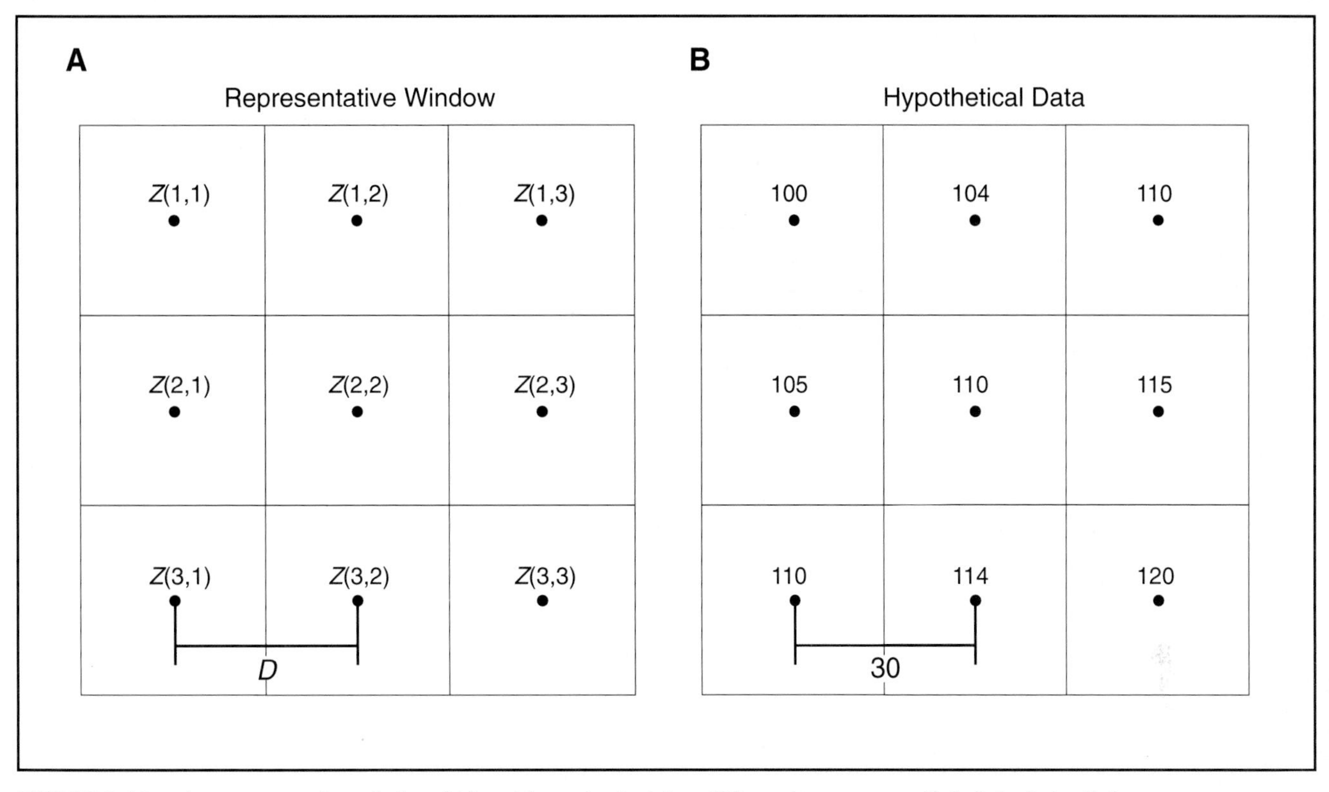

FIGURE 9.10. A representative window (A) and hypothetical data (B) used to compute digital shaded relief.

Step 1. Compute the slope of the window. The slopes along the x and y axes, sl_x and sl_y are

$$sl_x = \frac{Z(2,3) - Z(2,1)}{2D}$$

$$sl_y = \frac{Z(3,2) - Z(1,2)}{2D}$$

where the Z values represent digital elevation values as shown in Figure 9.10A, and D is the distance between grid points. For the sample data the results are

$$sl_x = \frac{115 - 105}{60} = 0.167$$

$$sl_y = \frac{114 - 104}{60} = 0.167$$

The overall slope, sl_o, is then defined as

$$\sqrt{(sl_x)^2 + (sl_y)^2}$$

For the hypothetical data, we have

$$\sqrt{(0.167)^2 + (0.167)^2} = 0.236$$

Since the resulting slope is expressed as a tangent (the ratio of rise over run), it can be converted to an angular value by computing the arc tangent ($\tan^{-1}$). The arc tangent for the hypothetical data is $\tan^{-1} 0.471 = 13.3°$. Thus, the hypothetical data have an overall slope of 13.3°.

Step 2. Compute the down-slope direction or azimuth for the 3 × 3 window. This is accomplished by calculating the local angle (ϕ), and converting this to an azimuth (an angle measured clockwise from a north base of 0°). ϕ is defined as

$$\phi = \cos^{-1}\left[\frac{sl_x}{sl_o}\right]$$

For the hypothetical data, we have

$$\phi = \cos^{-1}\left(\frac{0.167}{0.236}\right) = 45$$

To convert ϕ to an azimuth (θ), we use the signs shown in Table 9.1. For the hypothetical data, we compute $\theta = 270° + 45° = 315°$. Note that this result confirms a visual examination of the 3 × 3 window in Figure 9.10: the window appears to face northwest (an azimuth of 315°).

Step 3. Estimate the reflectance for the central cell of the 3 x 3 window, the cell associated with Z(2, 2). A simple, but reasonably effective approach, is to presume a *Lambertian reflector,* which reflects all incident light

TABLE 9.1. Conversion of local angle (ϕ) to azimuth (θ) for calculations associated with computer-based shaded relief

Sign of slope in x (sl_x)	*Sign of slope in y (sl_y)*	*Azimuth*
positive	positive	$270° + \phi$
positive	negative	$270° - \phi$
negative	positive	$90° - \phi$
negative	negative	$90° + \phi$

equally in all directions. The formula for Lambertian reflectance is

$$R_r = \cos(A_f - A_s)\cos E_f \cos E_s + \sin E_f \sin E_s$$

where R_r = relative radiance (scaled 0.0–1.0)
A_f = azimuth of a slope facet (0° to 360°)
A_s = azimuth of the sun (0° to 360°)
E_f = elevation of a normal to the slope facet (90° minus the slope magnitude in degrees)
E_s = elevation of the sun (0° to 90°)

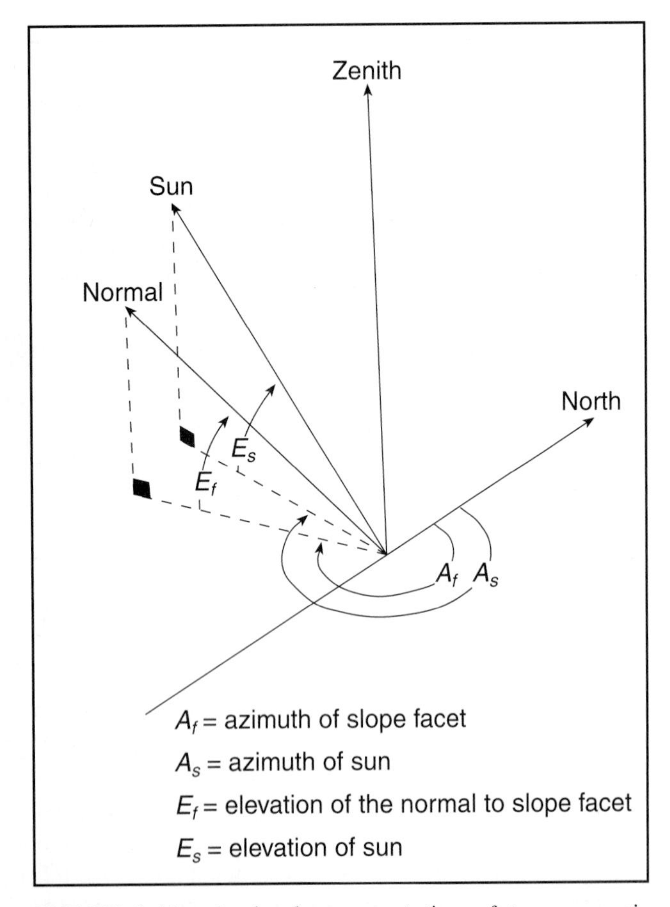

FIGURE 9.11. A visual representation of terms associated with computing Lambertian reflection. (From Eyton, Cartographica 27. Reprinted by permission of University of Toronto Press Incorporated. © University of Toronto Press 1990.)

A visual representation of these terms is shown in Figure 9.11. If we presume solar azimuth and solar elevations of 315° and 45°, respectively, then the Lambertian reflectance for the hypothetical data is

$$\begin{aligned} L_r &= \cos(315 - 315)\cos(90 - 13.3)\cos(45) \\ &\quad + \sin(90 - 13.3)\sin(45) \\ &= 0.850 \end{aligned}$$

Step 4. Convert the Lambertian reflectance values to either RGB or CMYK values appropriate for display or printing. Presuming a CRT display with intensities ranging from 0 to 255 for each color gun, RGB values would be calculated by multiplying the Lambertian reflectance by 255. For our hypothetical data the result would be $255 \times 0.850 = 217$ (the same value would be used for each color gun in order to produce a gray tone). Thus, the central cell of the 3×3 window would have a relatively bright reflectance value.

9.2.4 Using Color to Symbolize Aspect and Slope

Symbolizing Aspect: Moellering and Kimerling's MKS-ASPECT™ Approach

Moellering and Kimerling (1990) developed an approach for symbolizing aspect based on the opponent-process theory of color (see section 5.1.3). Their approach is known as MKS-ASPECT™ and is protected by two patents, No: 5,067,098 and No: 5,283,858, held by the Ohio State University Research Foundation in the name of Harold Moellering and A. Jon Kimerling. The approach

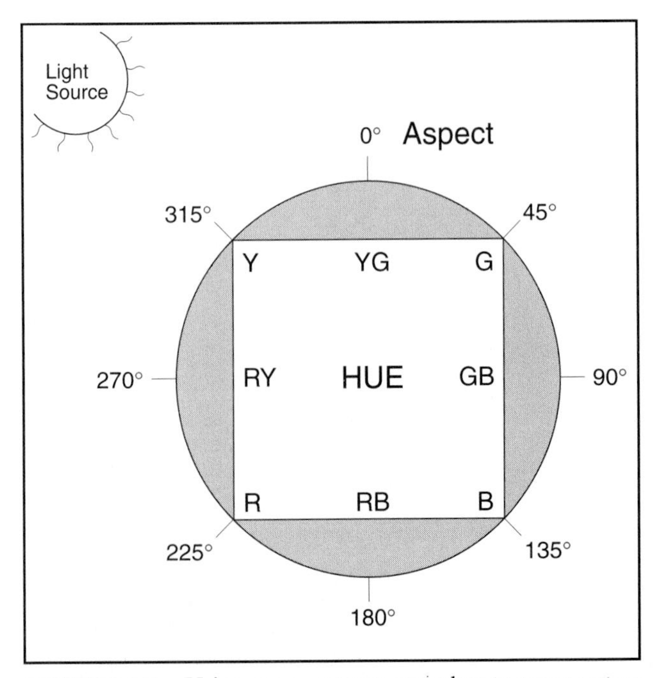

FIGURE 9.12. Using opponent process colors to represent aspect. (After Moellering and Kimerling 1990, 154.)

treats aspect as a nominal phenomenon (a northwest aspect is *different* than a southeast aspect) and assigns colors to opposing aspects based on opposing color channels; thus red and green, and blue and yellow are placed opposite one another. Mixtures of nonopposing hues can then be used for intermediate aspects (Figure 9.12).

Moellering and Kimerling argue that the **luminance** (or brightness) of a color should ideally be highest for a northwest aspect (assuming the usual light source from the northwest) and decrease to a minimum for a southeast aspect, thus providing the logic of using yellow for the northwest aspect. More specifically, they argue that colors should ideally fall on a curve represented by the following equation:*

$$L = 50 \times [\cos(315 - \phi) + 1]$$

where ϕ is the azimuth of the aspect and L is the relative luminance as measured by a **colorimeter,** a device for measuring the brightness of a color (Figure 9.13).

Moellering and Kimerling selected their colors using the HLS (hue-lightness-saturation) color specification model. The HLS model is similar to the HSV model described in section 5.3.3, except that it is a double rather than a single hexcone (compare Figure 9.14 with Figure 5.16). Initially, Moellering and Kimerling created colors by holding the lightness and saturation of the HLS model constant at 50 (one-half the maximum) and 100 (the maximum), respectively, and simply varying hue. Unfortunately, because HLS is not a perceptually based model, they found this approach produced colors relatively far from the ideal curve shown in Figure 9.13. Only after considerable experimentation, did they arrive at the set of eight colors plotted in Figure 9.13 and shown in Table 9.2. A map

* Their equation was a simplification of one described by Horn (1982).

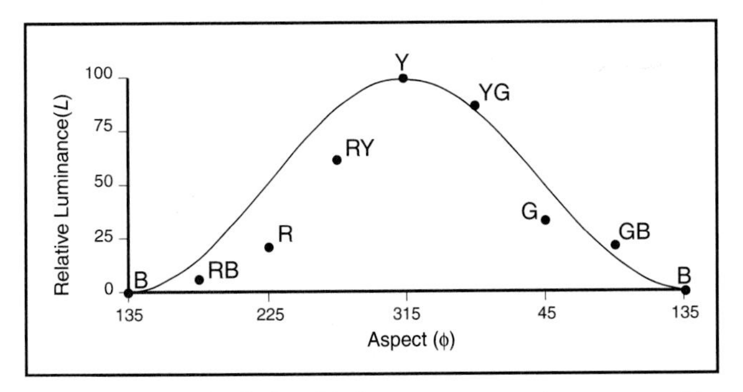

FIGURE 9.13. Relationship between aspect (as measured in degrees clockwise from a north value of 0) and relative luminance in Moellering and Kimerling's method for symbolizing aspect. The curve shows the ideal relationship, while dots represent the actual relationship. (After Moellering and Kimerling 1990, 158.)

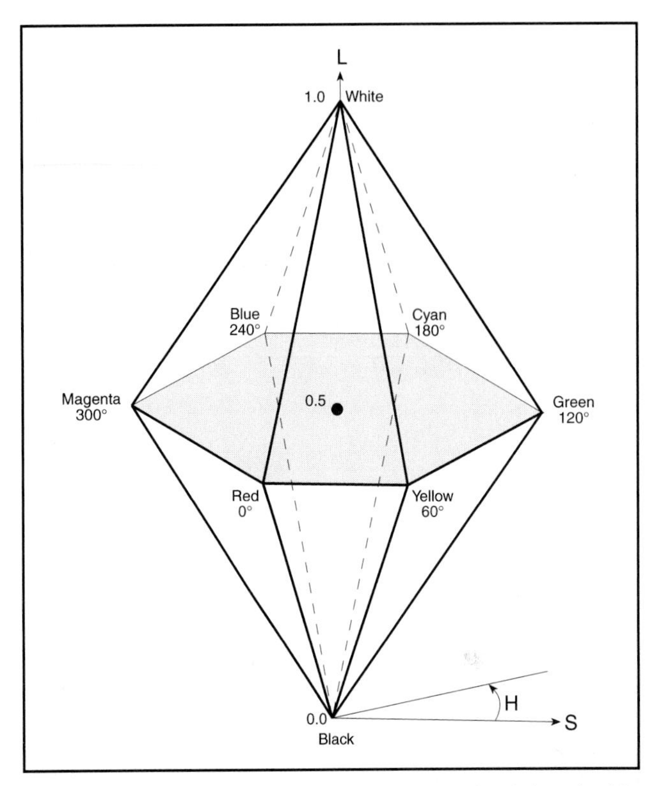

FIGURE 9.14. The HLS color model. Note that it is a double hexcone, as opposed to HSV, which is a single hexcone. (Compare with Figure 5.16.)

created using Moellering and Kimerling's color scheme is shown in Color Plate 9.2; in this case 128 colors are used (rather than 8) to provide greater color fidelity.

Symbolizing Aspect and Slope: Brewer and Marlow's Approach

Brewer and Marlow (1993) attempted to rectify Moellering and Kimerling's difficulty in working with HLS by employing a perceptually based model (the HVC model), while at the same time incorporating slope in addition to aspect. Like Moellering and Kimerling, Brewer and Marlow chose eight aspect hues. For each aspect, however, they also depicted three slope classes (5% to 20 percent, 20% to 40 percent, and >40 percent) as differing levels of saturation; they represented the <5 percent slope category as gray for all aspects. In total, they showed 25 classes of information (8 aspects × 3 slope classes, plus the <5 percent slope category).

Recall from section 5.3.5 that the HVC model (Figure 5.19) has three parameters: hue, measured in a circular fashion from 0 to 360 degrees; value, measured vertically from 0 to 100; and chroma, measured horizontally from 0 to 100. The colors selected by Brewer and Marlow are shown graphically in HVC coordinates in Figure 9.15. This graphic is a 2-D representation of the 3-D HSV space, so it requires some explanation.

TABLE 9.2. HLS, *Yxy*, and RGB specifications for Moellering and Kimerling's method for representing aspect.*

Aspect	*H*	*L*	*S*	*Y*	*x*	*y*	*R*	*G*	*B*
0	70	50	100	37.5	.354	.544	213	255	0
45	120	40	100	15.1	.288	.584	0	204	0
90	190	40	100	10.4	.200	.257	0	170	204
135	240	40	100	2.3	.160	.087	0	0	204
180	290	40	100	4.7	.250	.139	170	0	204
225	0	50	100	10.6	.594	.357	255	0	0
270	50	50	100	27.1	.446	.476	255	213	0
315	60	50	100	42.4	.396	.514	255	255	0

* Aspect is measured in degrees clockwise from north, which is 0 degrees. HLS values are based on the color model shown in Figure 9.14. *Yxy* values are taken from Moellering and Kimerling (1990, 158).

Each black dot in the graph represents one of the 25 colors Brewer and Marlow used. The eight lines extending from the interior to the edge of the circle represent the hues selected for the eight aspects. Brewer and Marlow did not select these hues on the basis of an equation (as Moellering and Kimerling did); rather they attempted to maximize the lightness differences among hues. The hues do not appear to be equally spaced (as one might suspect given the perceptual basis of HVC) because a wide range of lightness and saturation values are not possible for all hues; for example, it is not possible to create a variety of lightnesses and saturations between green and blue. Brewer and Marlow used three separate hues (135, 145, and 155) for the northeast direction to enhance the separation of slope-aspect classes in this region.

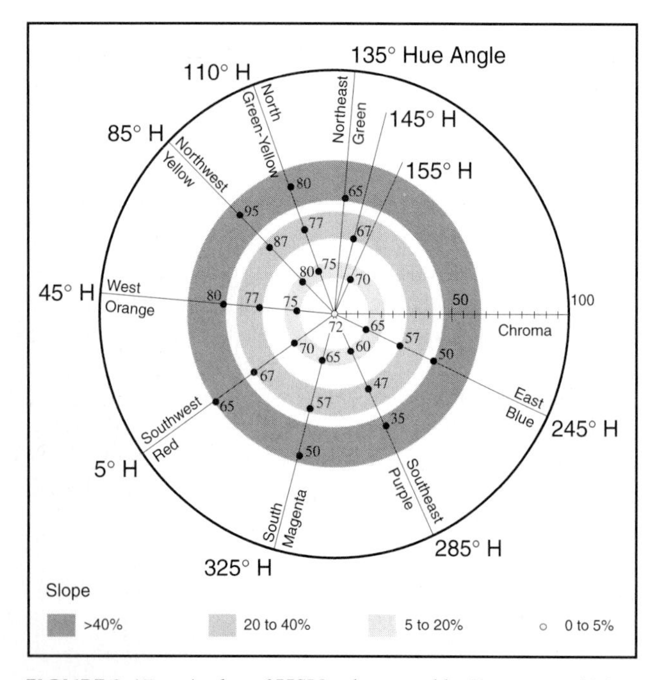

FIGURE 9.15. A plot of HSV colors used by Brewer and Marlow to depict aspect and slope; see text for explanation. (After Brewer and Marlow 1993, 331.)

The saturation (or chroma) of a slope class within an aspect is shown by the distance of a black dot from the center of the diagram; this distance can be determined using the scale labeled "Chroma" in the diagram. For example, for the blue hue, the third dot from the center of the diagram appears to have a saturation of approximately 45. The numbers printed alongside each dot represent the value (lightness) of each color. Note that higher lightness values tend to be found facing the direction of the light source (the northwest), and that lightness also varies with slope.

HVC, *Yxy*, and RGB specifications for the 25 slope-aspect classes developed by Brewer and Marlow are shown in Table 9.3, and a map created using their color scheme is shown in Color Plate 9.3.

9.2.5 Treinish's Depiction of Global Topography

Color Plate 9.4 illustrates a dramatic rendition of the earth's topography that was developed by Treinish (1994) using the IBM Visualization Data Explorer (DX). The attractiveness of this image is illustrated by the fact that a variant of it was chosen for the cover of the December 1992 issue of *GIS World*, a popular magazine for those interested in Geographic Information Systems. The image was created by combining shaded-relief techniques with hypsometric tints representing elevation values above and below sea level. The shaded-relief technique was based on **Gouraud shading,** a method that interpolates shades between vertices on a polygon (or cell) boundary, and thus produces a smoother appearance than one in which the same shade is produced throughout the polygon (Foley et al. 1996, 734–737). The color scheme for elevation values (see the top of Color Plate 9.4) was chosen so that the portion above sea level would have "the appearance of a topographic map" (Treinish 1994, 666); the portion below sea level used darker shades of blue to represent greater depth. DX is one of several specialized

TABLE 9.3. HVC, *Yxy*, and RGB specifications for Brewer and Marlow's method for representing aspect and slope.* (After Brewer and Marlow 1993. First published in *Auto-Carto II* proceedings. Reprinted with permission from the American Congress on Surveying and Mapping.)

Aspect	*H*	*V*	*C*	*Y*	*x*	*y*	*R*	*G*	*B*
Maximum Slopes (>40 percent slope)									
0	110	80	56	55.3	.339	.546	132	214	0
45	135	65	48	35.2	.252	.486	0	171	68
90	245	50	46	19.8	.180	.191	0	104	192
135	285	35	51	10.3	.240	.135	108	0	163
180	325	50	60	19.6	.350	.198	202	0	156
225	5	65	63	33.7	.442	.319	255	85	104
270	45	80	48	53.8	.421	.421	255	171	71
315	85	95	58	81.1	.395	.512	244	250	0
Moderate slopes (20 to 40 percent slope)									
0	110	77	37	50.6	.319	.451	141	196	88
45	145	67	33	37.0	.253	.400	61	171	113
90	245	57	31	25.7	.221	.238	80	120	182
135	285	47	34	17.4	.262	.209	119	71	157
180	325	57	40	25.1	.332	.240	192	77	156
225	5	67	42	34.3	.390	.316	231	111	122
270	45	77	32	47.3	.377	.384	226	166	108
315	85	87	39	63.7	.359	.445	214	219	94
Low slopes (5 to 20 percent slope)									
0	110	75	19	45.4	.305	.381	152	181	129
45	155	70	16	39.4	.270	.348	114	168	144
90	245	65	15	32.7	.260	.284	124	142	173
135	285	60	17	27.5	.280	.272	140	117	160
180	325	65	20	32.6	.311	.282	180	123	161
225	5	70	21	37.2	.339	.317	203	139	143
270	45	75	16	41.8	.328	.348	197	165	138
315	85	80	19	49.7	.321	.375	189	191	137
Near-flat slopes (0 to 5 percent slope)									
none	0	72	0	39.0	.290	.317	161	161	161

* Aspect is measured in degrees clockwise from north, which is 0 degrees.

visualization software packages that have been developed in recent years; we will consider these in more detail in Chapter 15.

9.2.6 Creating Fly-Bys of Terrain

A **fly-by** is a form of map animation in which the viewer is given the feeling of flying over a landscape (also see section 14.2.2). Until recently, fly-bys were possible only in a sophisticated workstation environment, but today they can be developed in a PC environment. Creating fly-bys typically involves combining digital elevation data and thematic information (such as land use). The digital elevation data are used to create a shaded-relief image, and the thematic information is then overlaid on top of this image. To create an animation, the resulting image is rendered from a variety of viewpoints (Figure 9.16 illustrates what a frame of such an animation might look like). Appendix G lists several software packages for creating fly-bys.

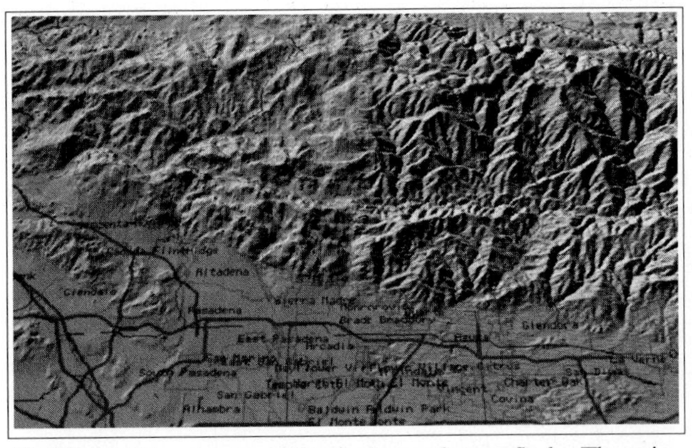

FIGURE 9.16. An example of a frame from a fly-by. The original illustration was in color, created using TruFlite software. (Courtesy of Keith Clarke.)

SUMMARY

This chapter has examined a broad range of methods for symbolizing smooth continuous phenomena. We noted that the data representing smooth continuous phenomena are usually available in one of two forms. Topographic data (representing the earth's topography) are generally provided in an equally spaced gridded format (as a **digital elevation model, or DEM**), while nontopographic data (such as snowfall amounts) are available as irregularly spaced points. A set of interpolated values must be calculated for nontopographic data using methods such as triangulation, inverse distance, and kriging, as was described in the preceding chapter.

A number of methods can be used to symbolize either topographic or nontopographic data, including **contour lines, hypsometric tints** (shading the areas between contour lines), **continuous-tone maps** (shades analogous to those used on unclassed choropleth maps), and **fishnet** symbolization (a netlike structure that simulates the three-dimensional character of a smooth continuous surface). The first three of these methods are limited in that they do not permit the map reader to actually "see" a three-dimensional surface. In contrast, the fishnet symbolization provides a three-dimensional perspective, but low points may be blocked by high ones, and rotation may produce a view unfamiliar to readers who normally see maps with north at the top. Data exploration software may enhance the visualization of the three-dimensional surface by permitting users to readily manipulate the view of the fishnet, while also permitting the examination of the data in other forms (such as a continuous-tone map).

Methods normally restricted to portraying topography include **hachures,** contour-based methods (for example, **Tanaka's method**), **shaded relief,** and the use of color to depict aspect and slope. These methods assume that the topographic surface is illuminated, typically obliquely from the northwest. The various methods use different approaches to distinguish areas that appear illuminated from those that appear in the shadows. For example, in Tanaka's method, contour lines perpendicular to the light source have the greatest width, while those parallel to the light source are narrowest; furthermore, contour lines facing the light source are drawn in white, while those in shadow are drawn in black. We presented detailed formulas for computing shaded relief because of its common use in computer-based systems and the dramatic maps that have been created using the technique. (The USGS map "Landforms of the Conterminous United States: A Digital Shaded-Relief Portrayal," created by Thelin and Pike, is a notable example.)

We also covered two topics that illustrate the capability of modern visualization and animation software. First, we considered Treinish's attractive rendition of global topography, which was created using the IBM Visualization Data Explorer (DX). Second, we examined the ability of animation software to create **fly-bys** of terrain. The development of such fly-bys was traditionally restricted to those with sophisticated workstation environments, but today fly-bys can be developed in a PC environment. In Chapter 14, we will consider fly-bys as one of several forms of map animation.

FURTHER READING

Brewer, C. A., and Marlow, K. A. (1993) "Color representation of aspect and slope simultaneously." *Auto-Carto 11 Proceedings,* Minneapolis, Minnesota, pp. 328–337.

Provides a more detailed discussion of Brewer and Marlow's approach for symbolizing aspect and slope.

Ding, Y., and Densham, P. J. (1994) "A loosely synchronous, parallel algorithm for hill shading digital elevation models." *Cartography and Geographic Information Systems* 21, no. 1:5–14.

Describes a sophisticated method for creating shaded relief that incorporates both shadows cast by other terrain features and atmospheric scattering; the method was implemented on a parallel-processing computer.

Eyton, J. R. (1984b) "Raster contouring." *Geo-Processing* 2:221–242.

Describes a variety of raster-based algorithms for symbolizing smooth continuous phenomena.

Eyton, J. R. (1990) "Color stereoscopic effect cartography." *Cartographica* 27, no. 1:20–29.

Combines shaded relief with the color stereoscopic effect (see section 6.3 of the present book for a detailed discussion).

Eyton, J. R. (1991) "Rate-of-change maps." *Cartography and Geographic Information Systems* 18, no. 2:87–103.

Includes basic calculations for creating a shaded relief map, but also several more advanced concepts such as computing and mapping the convexity and concavity of a three-dimensional surface.

Foley, J. D., van Dam, A., Feiner, S. K., and Hughes, J. F. (1996) *Computer Graphics: Principles and Practice.* 2d ed. Reading, Mass.: Addison-Wesley.

The notion of trying to simulate the appearance of topography by using a presumed light source (for example, shaded relief) is analogous to attempts to simulate the appearance of other phenomena, such as how a room looks when lit from a certain direction. Chapter 16 deals with simulating the appearance of these other phenomena.

Harrison, R. E. (1969) "Shaded Relief." In *The National Atlas of the United States of America, 1970.* U.S. Department of the Interior, Geological Survey, pp. 56–57.

A classic shaded relief map that was created manually. Scale 1:7,500,000.

Hobbs, F. (1995) "The rendering of relief images from digital contour data." *The Cartographic Journal* 32, no. 2:111–116.

Presents an algorithm for shaded relief that is appropriate for elevation data stored as contour lines. Also illustrates how multiple light sources can be used to enhance the visualization of terrain.

Horn, B. K. P. (1982) "Hill shading and the reflectance map." *Geo-Processing* 2, no. 1:65–144.

A dated, but thorough review of automated approaches for shaded relief.

Imhof, E. (1982) *Cartographic Relief Presentation.* Berlin: Walter de Gruyter.

A classic text on methods for symbolizing terrain.

Kumler, M. P., and Groop, R. E. (1990) "Continuous-tone mapping of smooth surfaces." *Cartography and Geographic Information Systems* 17, no. 4:279–289.

Introduces continuous-tone mapping and describes an experiment designed to test its effectiveness.

Lewis, P. (1992) "Introducing a cartographic masterpiece: A review of the U.S. Geological Survey's digital terrain map of the United States." *Annals, Association of American Geographers* 82, no. 2:289–304.

Discusses a digital shaded-relief map for the entire United States developed by Thelin and Pike (1991).

Moellering, H., and Kimerling, A. J. (1990) "A new digital slope-aspect display process." *Cartography and Geographic Information Systems* 17, no. 2:151–159.

Provides a more detailed discussion of Moellering and Kimerling's approach for symbolizing aspect.

Raisz, E. (1931) "The physiographic method of representing scenery on maps." *Geographical Review* 21, no. 2:297–304.

Describes the physiographic method for depicting geomorphological features which uses a set of standard, easily recognized symbols.

Raisz, E. (1967): Landforms of the United States

A classic map created using the physiographic method. Scale approximately 1:4,500,000. Available from GEDPLUS, Post Office Box 27, Danvers MA 01923; phone: 800-292-2102.

Rowe, J. (1997) "Is it real or data visualization? Windows-based packages bring terrain data to life." *GIS World* 10, no. 11:46–52.

Describes several Windows-based software packages for creating fly-bys of terrain.

Treinish, L. A. (1994) "Visualizations illuminate disparate data sets in the earth sciences." *Computers in Physics* 8, no. 6:664–671.

Discusses the creation of the dramatic rendition of the earth's topography shown in Color Plate 9.3, along with several other visualizations of spatial data.

White, D. (1985) "Relief modulated thematic mapping by computer." *The American Cartographer* 12, no. 1:62–68.

Describes a computer-based method for overlaying thematic information (such as land use) on top of shaded relief.

10

Dot and Dasymetric Mapping

OVERVIEW

Although data are frequently collected on the basis of enumeration units and mapped at that level (for example, as a choropleth map), such maps can be misleading because the distribution of an underlying phenomenon often varies within enumeration units. For instance, imagine that you had data on the approximate number of elephants living in each country of Africa. You could make a choropleth map of such data, assuming that you standardized the data appropriately; for example, you might divide the number of elephants by the area of each country to create a density figure. The resulting map would be misleading, however, because each country would have a uniform tone, suggesting no variation in the density of elephants within a country. Dot and dasymetric maps solve this problem by using ***ancillary variables*** *(or* ***ancillary information****) to create a more detailed map. In the case of elephants, appropriate ancillary variables might be the location of heavily vegetated areas, especially forests and wooded savannas, and protected areas, such as national parks.*

Recall from Chapter 2 that a ***dot map*** *is constructed by letting one dot equal a certain amount of some phenomenon and then locating dots where that phenomenon is most likely to occur. Raw-count data (such as the number of elephants or the number of acres of wheat harvested) are normally the basis for dot mapping. In contrast,* ***dasymetric maps*** *are constructed by using area symbols to represent zones of uniformity. Dasymetric maps are similar to choropleth maps, but the bounds of zones are based on ancillary variables, and thus need not match enumeration unit boundaries (Figure 10.1). As with choropleth maps, only standardized data should be mapped with the dasymetric method. Conventionally, cartographic texts have covered dot and dasymetric mapping in separate sections, reflecting the different kinds of data on which they are based and the different symbology used. We consider them together here because ancillary variables are a critical element of both.*

Traditionally, cartographers incorporated ancillary variables for dot and dasymetric maps through a tedious manual process. Today, this process can be automated through the use of GIS and remote-sensing techniques. In this chapter, we will consider two applications of this process: (1) the procedures that might be used to create a dot map of wheat harvested in Kansas (such a dot map was initially introduced in Chapter 2), and (2) an approach that Langford and Unwin (1994) have developed for mapping population density using dasymetric symbolization and a generalization (smoothing) operation.

It is important to note that this chapter focuses on the use of ancillary variables *to create detailed maps of data collected for enumeration units. A more detailed map can also be constructed through a* mathematical manipulation *of the data. Tobler's pycnophylactic method (section 8.6) is an example of this approach. Recall that Tobler's method began with a set of prisms (each raised to a height proportional to the data), which were sculpted to create a smooth surface. Although Tobler's method was introduced as a form of interpolation, it may be more proper to think of it as a form of reallocation (personal communication, Tobler 1997) in the sense that population is reallocated to different portions of an enumeration unit as a function of surrounding units. A similar mathematical approach has been developed by Martin (1989, 1996) that considers not only the population of each enumeration unit but also the centroid of the population within that unit.*

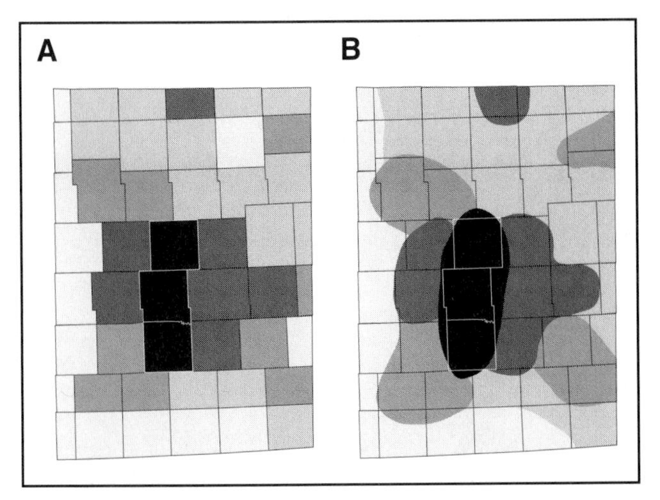

FIGURE 10.1. A hypothetical comparison of choropleth (A) and dasymetric (B) mapping.

10.1 TYPES OF ANCILLARY VARIABLES

Ancillary variables used to locate dots (for dot mapping) or to define zones (for dasymetric mapping) are commonly split into limiting and related variables. *Limiting variables* place absolute limits on where dots or zonal boundaries can be placed. For example, generally, it would not make sense to place dots representing population within water bodies. *Related variables* are those that are correlated with the phenomenon being mapped, but that do not place absolute limits on the location of dots or zonal boundaries. For example, certain types of soil in a region may favor a particular crop, but the crop may be found on all types of soil within the region.

10.2 MANUAL VERSUS AUTOMATED PRODUCTION

When maps were produced manually, creating dot and dasymetric maps was a tedious process. Initially, the cartographer had to collect a set of paper maps that provided an appropriate set of limiting and related variables. Next, the limiting and related variables had to be synthesized onto a single base map. This process was generally accomplished using an enlarging-reducing machine that allowed one to trace information from maps of varying scale onto the same base. For large, detailed maps, this could be a very time-consuming process. The cartographer then had to decide how the combined information would be used to place dots or zonal boundaries (for example, that more dots would be placed in one land slope category than another, or that no dots would be placed in wetlands). Finally, actual symbols would be placed on the map; in the case of the dot map, this was particularly time-consuming.

In theory, the techniques of GIS and remote sensing can alleviate some of the burden of the manual process. First, through its layering and associated analysis capability, GIS enables the automated processing of limiting and related variables. Second, remote sensing, as an integral part of GIS, can be used both to define limiting and related variables or, alternatively, to pinpoint the location of a particular phenomenon.

10.3 CREATING A DOT MAP

A dot map of wheat harvested in Kansas was initially introduced in Chapter 2 (Figure 2.8D is repeated in the lower right-hand portion of Figure 10.2). This section considers three issues relevant to creating such a dot map: (1) determining regions within which dots should be placed, (2) selecting dot size and unit value (the count represented by each dot), and (3) placing dots within selected regions. Although we will consider these issues in the context of the wheat harvested example, they are generic to most dot-mapping problems.

10.3.1 Determining Regions within Which Dots Should Be Placed

In constructing the dot map of wheat, I first considered several variables that might limit the distribution of wheat, and crops in general. Obvious limiting variables included the location of water bodies (in the case of Kansas, there are a number of reservoirs that would be relevant on a small-scale map), the location of urban areas (e.g., Wichita and Kansas City), and slope (tractors cannot be used on very steep slopes). Such limiting variables might be found on paper maps, and entered into a GIS as layers via *scanning* or *digitizing* (or they might already be available in digital form). The resulting layers could then be overlaid, and dots placed only in areas not constrained by the limiting variables.

Rather than employ an overlay approach, I used a Kansas land use/land cover map developed by Whistler et al. (1995) because it provided a convenient synthesis of limiting variables. The land use/land cover map, which was created from Landsat Thematic Mapper remote-sensing data, split the landscape into 10 classes: 5 urban (residential, commercial-industrial, grassland, woodland, and water) and 5 rural (cropland, grassland, woodland, water, and other). The map defined directly two of the limiting variables I had considered (urban areas and water bodies), and included two limiting variables that I had not immediately thought about: woodland and grassland. Most important, however, was the cropland class, which indirectly handled the slope variable and any other limiting and related variables likely associated with cropland.

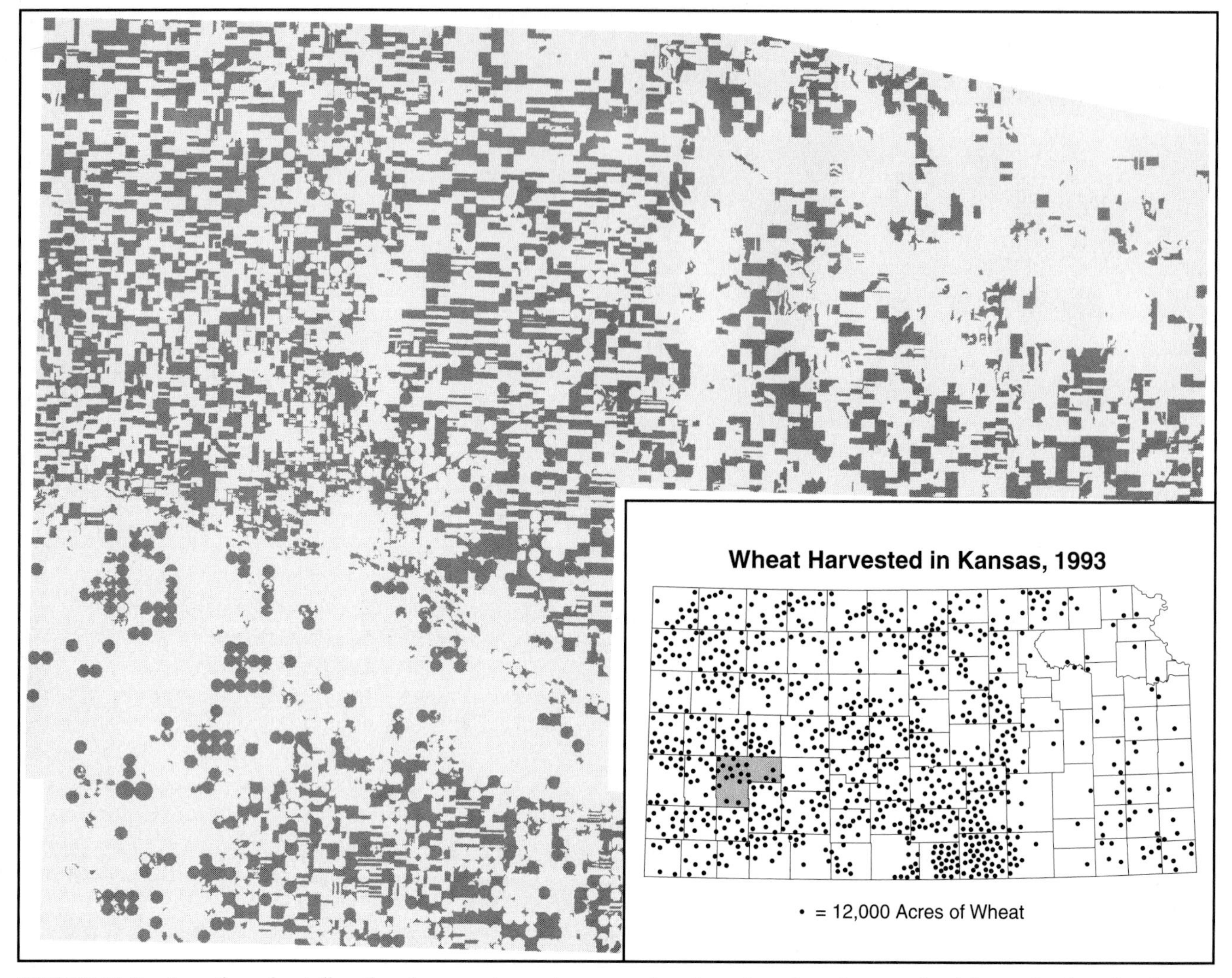

FIGURE 10.2. Locating wheat directly using remote sensing versus locating wheat based on cropland from a land use/land cover map. The upper left map depicts the distribution of wheat within Finney County based on an analysis of 30-meter resolution Landsat imagery, while the lower right map portrays the same distribution (the gray-shaded area) based on a 1-kilometer resolution land use/land cover map. (Courtesy of Kansas Applied Remote Sensing Program, University of Kansas.)

Although a mapmaker might be tempted to place all dots representing wheat within a cropland class, it is important to recognize that land use/land cover maps based on remote-sensing techniques are not without error. Whistler et al. (p. 773) estimated the overall accuracy of the Kansas land use/land cover map to be 85 to 90 percent or better, but individual categories within particular counties deviated considerably from this. For example, using data provided by Whistler et al., I found that only 53 percent of the cropland in Butler County was estimated to be correctly classified. To handle this sort of error, I placed some dots representing cropland within grassland areas (the category typically confused with cropland on imagery). The number of dots so placed was a function of the estimated percentage of incorrectly classified cropland in a county: if 47 percent of the cropland was estimated to be misclassified as grassland, then 47 percent of the dots were placed in the grassland category within that county. Although these percentage figures suggest large numbers of dots were affected, they were not because there was generally little wheat harvested in such counties; less than 5 percent of all dots were affected by this placement procedure.

The map shown in the lower right of Figure 10.2 is the result of using the above dot placement procedure. I considered this map satisfactory for this book because my primary purpose was to show readers that a more detailed map could be developed using the dot-mapping method than the choropleth method (see Figure 2.8); there was no need to create the "perfect" map. In other situations, however, a mapmaker might wish to create a more detailed map. Let's now consider how this might be done.

One obvious limitation of the Kansas land use/land cover map is that it does not distinguish the location of individual crops. This limitation might be handled by considering ancillary variables that could assist in locating wheat within the cropland areas. One potential variable would be precipitation, as it varies from a high of about 35 inches in eastern Kansas to only about 15 inches in western Kansas. Presuming an ideal amount of precipitation for wheat, a mathematical function could be developed that places a higher probability on locating dots near the ideal precipitation area. Note that such an approach would also have to consider other complicating variables, such as the timing of the precipitation.

Another ancillary variable that could be considered would be irrigated cropland. The *1994 Kansas Farm Facts* specifies the acres of wheat harvested from irrigated land, so if the location of irrigated cropland could be determined, a percentage of dots could be placed in such areas equivalent to the percentage of wheat irrigated in that area. Remote sensing could be used for this purpose (Eckhardt et al. 1990), particularly in the case of center-pivot irrigation, which traces a distinctive circle on the landscape (Astroth et al. 1990).

Still another possible ancillary variable would be the distribution of soils within each county. One might suspect that certain soils would be more conducive to wheat production than other crops. In discussing this possibility with those knowledgeable about Kansas agriculture (a farmer and a county extension agent), I found that the selection of crops at the county level was more likely a function of farm policy and associated programs rather than soil type. For example, wheat would continue to be planted at a particular location because a farm program specified that the same crop must be planted in order for financial support to be retained.

Although ancillary variables could assist in creating a more detailed dot map, remote sensing perhaps provides the capability to create the most precise map of wheat because of its ability to pinpoint the location of individual crops. For example, Egbert et al. (1995) used remote sensing to differentiate wheat, corn, grain sorghum (milo), alfalfa, fallow land, shortgrass prairie, and sand-sage prairie in Finney County, Kansas, and found classification accuracies as high as 99 percent for wheat. They accomplished this by using imagery for three different time periods, as opposed to the single-date imagery employed for the Kansas land use/land cover map. The map shown in the upper left of Figure 10.2 is illustrative of the kind of detail they were able to attain. Although using remote sensing in this fashion can provide considerable detail, the greater amount of imagery incurs a greater expense, and the necessary verification (ground truth) data is also costly (Egbert, personal communication, 1996). Thus, mapmakers wishing to use remote sensing must carefully consider its cost and benefits for a particular application.

10.3.2 Selecting Dot Size and Unit Value

Dot size (how large a dot is used) and **unit value** (the count represented by each dot) are important parameters in determining the look of a dot map. Traditionally, cartographers have argued that very small dots produce a "sparse and insignificant" distribution, while very large dots "give an impression of excessive density." Similarly, a small unit value produces a map that "gives an unwarranted impression of accuracy," while a large unit value results in a map that "lacks pattern or character" Robinson et al. (1995, 498). Generally, it has been argued that dots in the densest area should just begin to coalesce (merge with one another).

Although a graphical device known as a **nomograph** can assist in selecting dot size and unit value (Robinson et al., 499–500), some experimentation is generally required to determine appropriate values. For the wheat map, I selected several counties that had both a high percentage of cropland in wheat and a large value for acres of wheat harvested. Selecting counties on the basis of percentage of cropland in wheat insured that dense areas of wheat harvested would be located, while having a large value for acres harvested insured that the dense area would consist of a relatively large number of dots. I placed all necessary dots within these counties for several dot sizes and unit values and then subjectively evaluated the resulting maps for coalescence. Once I found an acceptable dot size and unit value, I used a graphic design program (Freehand) to locate all dots for that size and unit value.

The simplest way of determining the number of dots to place within a region is to divide the count figure (in our case acres of wheat harvested) associated with that region by the unit value. One problem with this approach is that it fails to consider the inability of readers to correctly estimate the number and density of dots within subregions of a map. In a fashion analogous to the underestimation of proportional symbol size (see section 7.3.2), readers also underestimate the number and density of dots. Olson (1977) indicated that it is possible to correct for this underestimation, but that the effect on the overall pattern is small, and thus of questionable utility.

It is interesting to note that mapmakers can ignore the issue of dot size and unit value when the phenomenon is located explicitly via remote sensing (as in the Finney County example shown in Figure 10.2); in this case, darkened or colored pixels tell us precisely where the phenomenon is located. One disadvantage of this approach, however, is that it may be difficult to make a visual estimate of the magnitude of production, as individual pixels merge together. Of course, since the map is automatically generated, the computer can be used to make an estimate of the acres of wheat grown in any area (albeit with the estimate a function of the ability to correctly interpret the phenomenon via remotely sensed imagery).

10.3.3 Placing Dots within Regions

In the above section, the reader may have been surprised to find that I manually located dots, as opposed to using an automated procedure. I did so because there is no readily accessible software for creating a good dot map. Software vendors have developed dot-mapping techniques (commonly referred to as "dot-density" routines), but these techniques generally place dots in a purely random fashion, which can lead to unrealistic clusters and gaps in the dot pattern. Another problem with such software is that it presumes mapmakers have already formed polygons within which dots are to be placed, and that these polygons are of sufficient size to place dots within them. Oftentimes, this is not the case; for example, the Kansas land use/land cover map was in raster format, and the cropland areas were quite variable in size (Color Plate 10.1).

The problem of random dot placement might be solved by modifying a **dot-density shading** technique that Lavin (1986) has developed for mapping continuous phenomena (such as rainfall). Below, we summarize the dot-density method and then consider how it might be applied to dot maps.

Step 1: Read an equally spaced gridded network of data values into the computer (Figure 10.3A). Such a grid would normally be the end result of the gridding approach for automated contouring (see section 8.2). Conceptually, each grid intersection is presumed to be surrounded by a square cell within which dots will be placed (Figure 10.3B).

Step 2: Compute the proportion of dot coverage, P_i *for each of the square cells.* The formulas are

$$z_i = \frac{v_i - v_s}{v_L - v_s}$$

$$P_i = z_i(P_{\max} - P_{\min}) + P_{\min}$$

where v_s and v_L = smallest and largest values in the grid
z_i = proportion of the data range associated with the grid value v_i
$P_{\min}$ and $P_{\max}$ = minimum and maximum desired proportions of dot coverage
P_i = proportion of dot coverage for grid value v_i

Assuming that the minimum and maximum desired proportions of dot coverage are 0.08 and 0.50, respectively, the following would be computed for the lower left-hand grid point in Figure 10.3:

$$z_i = \frac{27 - 18}{108 - 18} = 0.10$$

$$P_i = 0.10(0.50 - 0.08) + 0.08 = 0.1220$$

Step 3: Compute the number of dots, N_d, *to be placed within a square cell.* The formula is

$$N_d = \frac{P_i \times A_c}{A_d}$$

where A_c is the area of the cell, and A_d is the area of a dot. For the lower left-hand grid point, we have

$$N_d = \frac{0.1220 \times 0.1296}{0.0018} = 8.784 \approx 9$$

Step 4: Partition the square cells surrounding each grid point into n *equally sized subcells, where* n *is as close as possible to* N_d. In the case of the lower left-hand grid point, the number of cells (9) matches N_d exactly (Figure 10.3C). The subcells serve as plotting boundaries for each dot: a dot is plotted randomly within each cell and cannot extend outside the cell (Figure 10.3D). A map re-

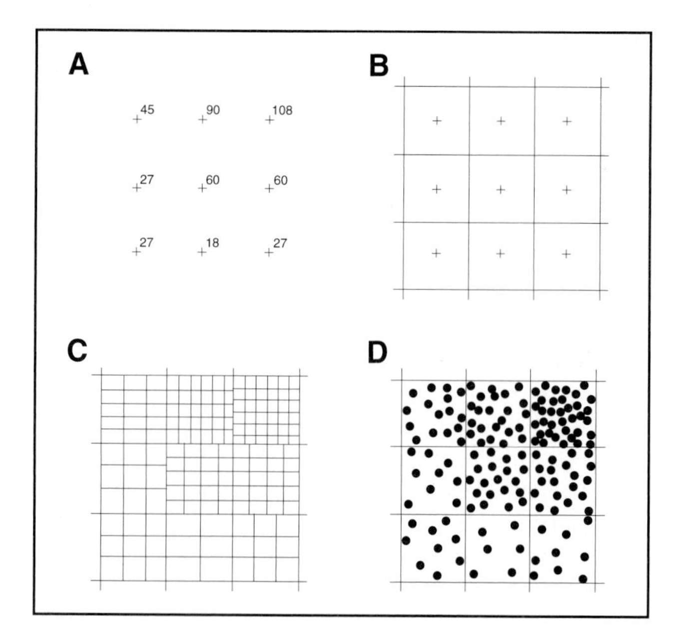

FIGURE 10.3. A depiction of Lavin's dot-density shading approach: (A) a matrix of grid points serves as input; (B) square cells are presumed to surround each grid point; (C) square cells are split into equally sized subcells, with the number of subcells a function of the number of dots to be placed; (D) a dot is placed within each subcell. (After Lavin 1986. First published in *The American Cartographer* 13(2), p. 145. Reprinted with permission of Stephen Lavin and the American Congress on Surveying and Mapping.)

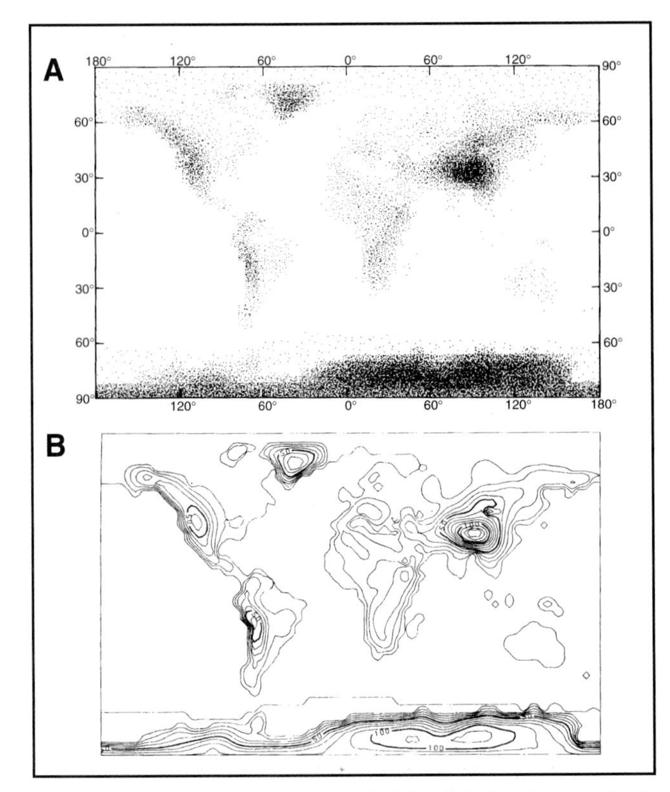

FIGURE 10.4. A comparison of (A) Lavin's dot-density shading approach and (B) the traditional isoline method. Maps are based on data for land elevations, and contour lines are in hundreds of feet. (After Lavin 1986. First published in *The American Cartographer* 13(2), p. 142. Reprinted with permission of Stephen Lavin and the American Congress on Surveying and Mapping.)

sulting from this process is shown in Figure 10.4, along with a traditional isoline map.

Applying dot-density shading to a dot map would involve the following steps. First, standardized values would have to be computed for each subregion within which dots are to be placed (subregions would be determined by limiting and related variables). This would be accomplished by dividing a count value (say, population) associated with a subregion by the area of that subregion. Second, a grid would be laid over the map. All grid points falling within a particular subregion would be assigned the standardized value for that subregion. Finally, steps 2 to 4 described above would be executed.

One problem with applying Lavin's approach to dot mapping is that subregions within which dots are to be placed are often in raster (pixel-based) form, and are quite variable in size and location, as in the Kansas land use/land cover map (Color Plate 10.1). In this case, note that individual subregions are frequently smaller than the dots that would be placed within them. In this case it might make more sense to develop a plotting algorithm based on pixels, such as the following:

Determine the proportion of cropland within an enumeration unit to be symbolized as wheat. (Divide acres of wheat harvested by acres of cropland.)
For each pixel within the enumeration unit:
  If the pixel is cropland,
    generate a random number between 0 and 1.
    If the random number is ≤ the proportion of cropland in wheat, set the pixel to wheat.

The resulting map would appear more like the detailed map of Finney County shown in the upper left-hand portion of Figure 10.2 than a traditional dot map. But it would not exhibit some of the regularities of the Finney County map, such as the circular features representing center-pivot irrigation.

10.4 MAPPING POPULATION DENSITY: LANGFORD AND UNWIN'S APPROACH

Traditionally, population density has been mapped using choropleth symbology because population data are readily available for enumeration units, and computer software frequently has the choropleth map as its only option. Langford and Unwin (1994, 22) noted three major problems with this approach: (1) larger enumeration units tend to have lower population densities (and conversely smaller enumeration units tend to have higher population densities); (2) enumeration units hide the variation that exists within them; and (3) enumeration unit boundaries are arbitrary, and thus unlikely to be associated with actual discontinuities in population density.

In response to these problems, Langford and Unwin developed a novel mapping method based on remote sensing, dasymetric mapping, and generalization techniques. First, they used Landsat Thematic Mapper imagery to split pixels within a region (northern Leicestershire, United Kingdom) into two groups: "Residential Housing" and "All Other Categories." A comparison of the resulting map (Figure 10.5B) with the traditional choropleth map (Figure 10.5A) makes clear the variation in population density that can exist within enumeration units, and the fact that enumeration unit boundaries generally do not coincide with discontinuities in population density.

Next, Langford and Unwin created a dasymetric map by combining their remote-sensing–based map with the population data associated with enumeration units. Standardized data for the dasymetric map were computed by dividing the population for each enumeration unit by the area of residential housing falling in that unit. They noted that the resulting density values varied considerably from those depicted on the choropleth map: the dasymetric map ranged from 2,820 to 14,900 people per square kilometer, while the choropleth map ranged from 50 to 10,000 people per square kilometer. The dasymetric map

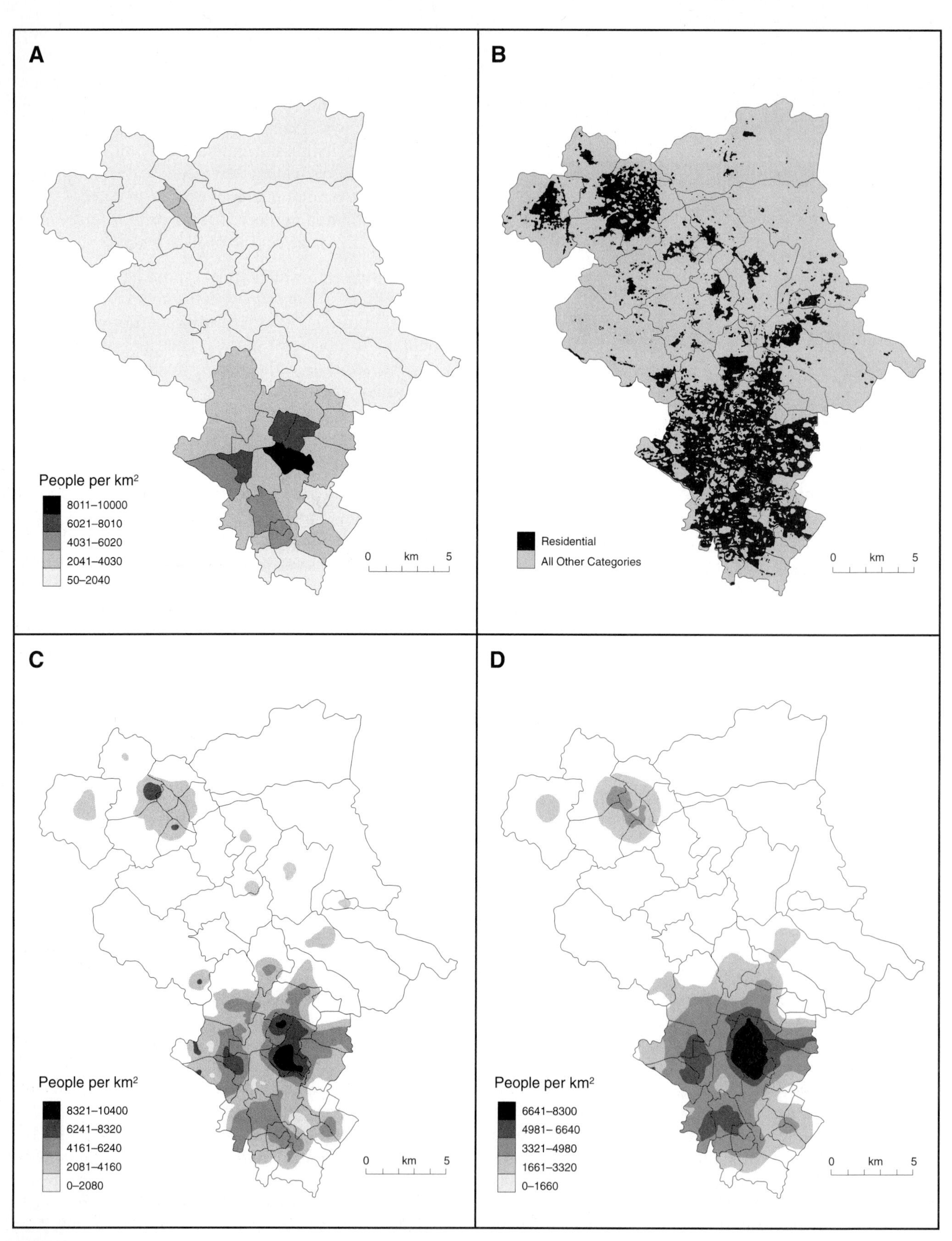

FIGURE 10.5. Langford and Unwin's approach for mapping population density: (A) a traditional choropleth map; (B) a binary map of "residential housing" and "all other categories" derived from remote sensing; a dasymetric map was created from this map by dividing the population for each enumeration unit by the area of residential housing falling in that unit; (C and D) generalized versions of the dasymetric map based on circular "windows" of 0.5 and 1 km centered on each pixel. (From Langford and Unwin 1994; courtesy of The British Cartographic Society.)

was symbolized by shading the residential housing pixels shown in Figure 10.5B with an intensity proportional to the standardized data associated with the enumeration unit within which the pixel fell. Unfortunately, Langford and Unwin found the resulting dasymetric map a "poor cartographic product . . . because the reader [could] see individual pixels and too much fine spatial detail" (p. 24).

To improve the visual quality of the dasymetric map, Langford and Unwin next applied a generalization operator. This was accomplished in three steps. First, they computed the population associated with a pixel by dividing the population of the enumeration unit by the number of residential housing pixels falling in that unit. Next they floated a circular "window" over each pixel and calculated the population falling within the window. Finally, a population density for each pixel was computed by dividing the population for the associated window by its area. When the resulting pixel values were symbolized as gray tones (Figure 10.5C), the map depicted relatively smooth changes in population density (as opposed to the fine irregularities of the dasymetric map).

Although Langford and Unwin's resulting map clearly provides a much more detailed depiction than a conventional choropleth map, their approach is not without problems. One is that remotely sensed categories have some error, as was indicated for the Kansas land use/land cover map. In more recent research, Fisher and Langford (1996, 308) suggest that this is not a serious problem, as "errors at the pixel level . . . can be large . . . without impacting the accuracy of estimates of regional amounts."

Another problem is that remote sensing may not provide information about the type of housing unit (for example, a single-family dwelling versus high-rise apartment complex). Langford and Unwin argue that this is not a serious problem if "housing type is relatively uniform within the study region" (p. 24). A third problem is that the resulting map is a function of the search radius used (compare Figures 10.5C and 10.5D). This "problem" can also be viewed as an advantage, as it provides alternative views of the data (a form of data exploration).

SUMMARY

This chapter has examined a number of principles involved in dot and dasymetric mapping, techniques which are used to more precisely map data commonly collected for enumeration units. **Dot maps** are used to portray *raw-count data* (for example, the number of people of Swedish descent living in counties in Wisconsin), and are constructed by letting each dot equal a specified value (e.g., one dot per 100 people of Swedish descent). In contrast, **dasymetric maps** are used to portray *standardized data* (for example, the proportion of the population of Swedish descent), and use area symbols to represent zones of uniformity.

Both dot and dasymetric mapping use **ancillary variables** (or **ancillary information**) to determine appropriate locations for mapped phenomena. For example, you might have information available on where Swedes tended to settle in Wisconsin, and thus place dots on this basis. Of course, if you had data available for very fine enumeration units (say, census tracts) and you were constructing a small-scale map, the use of ancillary variables would be unnecessary.

Remote sensing can be particularly useful in determining suitable ancillary variables. For example, we saw that a land use/land cover map for Kansas created via remote-sensing imagery could be used to determine locations for dots representing wheat. We also saw that remote sensing could be used to determine precisely where wheat was grown (in the case of Finney County, Kansas, an accuracy of 99 percent was obtained). Finally, in the case of Langford and Unwin's mapping of population density, we saw how remote sensing could be used to determine locations of "residential housing," which served as the basis for dasymetric mapping.

An important issue in dot mapping is selecting the **dot size** and **unit value** (the count represented by each dot). Normally, dot size and unit value are selected so that dots in the densest area just begin to coalesce (merge with one another). The number of dots to place within a region is computed by dividing a count figure (for example, acres of wheat harvested) associated with a region by the unit value. Although it is possible to adjust such numbers to account for the *perceived* number and density of dots, this is normally not done because the effect on the overall dot pattern is small.

Another issue in dot mapping is the actual placement of dots on a map. Although so-called dot-density routines have been developed for this purpose, these methods are undesirable because they (1) produce unrealistic clusters and gaps in the dot pattern and (2) presume that mapmakers have already formed polygons within which dots are to be placed. A potential solution to the first problem is to modify Lavin's **dot-density shading** developed for smooth continuous phenomena and apply it to the abrupt discrete phenomena characteristic of dot maps. The second problem might be handled by developing a raster-based plotting algorithm: the net result is that small square pixels would be plotted, as opposed to traditional round dots.

This chapter focused on how *ancillary variables* can be used to create detailed maps of data collected for enumeration units. Bear in mind that a more detailed map can also be constructed through a *mathematical manipulation* of the data. Tobler's pycnophylactic method (described in section 8.6) is an example of one such approach.

FURTHER READING

Dent, B. D. (1996) *Cartography: Thematic Map Design.* 4th ed. Dubuque, Iowa: William C. Brown Publishers.

Chapter 7 covers basic principles of dot mapping. For more on basic principles, see pages 497–502 of Robinson et al. (1995).

Egbert, S. L., Price, K. P., Nellis, M. D., and Lee, R.-Y. (1995) "Developing a land cover modelling protocol for the high plains using multi-seasonal thematic mapper imagery." ACSM/ASPRS Annual Convention & Exposition, *Technical Papers,* Vol. 3, pp. 836–845.

Describes how remote sensing can be used to differentiate a variety of crops.

Fisher, P. F., and Langford, M. (1996) "Modeling sensitivity to accuracy in classified imagery: A study of areal interpolation by dasymetric mapping." *Professional Geographer* 48, no. 3:299–309.

Examines to what extent the accuracy of a dasymetric map is a function of the error in remotely sensed imagery.

Gerth, J. D. (1993) "Towards improved spatial analysis with areal units: The use of GIS to facilitate the creation of dasymetric maps." Research paper for an M.A. degree, Ohio State University, Columbus, Ohio.

Describes how ARC/INFO software can be used to create a dasymetric map. To acquire the paper, contact the Department of Geography at Ohio State University (e-mail: geography@osu.edu).

Langford, M., and Unwin, D. J. (1994) "Generating and mapping population density surfaces within a geographical information system." *Cartographic Journal* 31, no. 1:21–26.

Provides detail on Langford and Unwin's dasymetric method.

Lavin, S. (1986) "Mapping continuous geographical distributions using dot-density shading." *American Cartographer* 13, no. 2:140–150.

Describes a method for using dot symbols to represent smooth continuous phenomena. In the present chapter, Lavin's equations are used to create dot maps, which are normally used for phenomena that are not smooth and continuous.

Martin, D. (1996) "An assessment of surface and zonal models of population." *International Journal of Geographical Information Systems* 10, no. 8:973–989.

Describes mathematical approaches for reallocating population within enumeration units. The method presumes that population density peaks at the centroid of population within an enumeration unit.

McCleary, G. F. J. (1969) "The dasymetric method in thematic cartography." Unpublished Ph.D. dissertation, University of Wisconsin, Madison, Wisconsin.

Covers the historical development of the dasymetric map and describes several forms of dasymetric mapping.

Monmonier, M. S., and Schnell, G. A. (1984) "Land use and land cover data and the mapping of population density." *International Yearbook of Cartography* 24:115–121.

Indicates how the denominator in a ratio for choropleth mapping might be adjusted to produce a more meaningful map. For example, rather than dividing population by the total area of an enumeration unit, it could be divided by the area within which people were likely to live.

Olson, J. M. (1977) "Rescaling dot maps for pattern enhancement." *International Yearbook of Cartography* 17:125–136.

Describes how the pattern on dot maps may be enhanced by increasing the contrast between areas of similar dot density.

Ryden, K. (1987) "Environmental Systems Research Institute Mapping." *American Cartographer* 14, no. 3:261–263.

Briefly describes how ARC/INFO was used to create a population map of various ethnic groups in California.

Wright, J. K. (1936) "A method of mapping densities of population: With Cape Cod as an example." *Geographical Review* 26, no. 1:103–110.

A classic article on dasymetric mapping. Introduces a formula for "density of parts" that enables densities to be calculated for portions of an enumeration unit based on ancillary information.

11

Developments in Univariate Mapping Methods

OVERVIEW

The purpose of this chapter is to introduce some novel univariate mapping methods that have been developed in recent years. The focus is on static maps, although we will see that some methods are most effective in the exploration and animation realms. Methods covered include Cleveland and McGill's (1984) framed-rectangle symbols, MacEachren and DiBiase's notion of modeling geographic phenomena (and the associated chorodot map), Dorling's (1993) cartograms, techniques for mapping flows, and techniques for mapping true 3-D phenomena.

***Framed-rectangle symbols** are created by changing the proportion of area filled within a rectangular frame of constant size. They have been proposed as an alternative to the choropleth map because readers can acquire specific information from them more accurately than from a choropleth map. A problem with framed-rectangle symbols, however, is that as point symbols, they are difficult to associate with the areal extent of enumeration units.*

*MacEachren and DiBiase have developed a set of models for representing geographic phenomena, along with a corresponding set of symbolization methods. The models form a matrix that spans the discrete-continuous and abruptness-smoothness continua introduced in Chapter 2. If the cartographer can determine the appropriate model for a geographic phenomenon being mapped, then an associated symbolization method can be chosen. Many of MacEachren and DiBiase's symbolization methods are conventional, but one, the chorodot map, is innovative. The **chorodot map** combines the features of dot and choropleth maps by shading small squares ("dots") with different intensities. MacEachren and DiBiase argue that a chorodot map is appropriate when a phenomenon falls near the middle of the discrete-continuous and abruptness-smoothness range.*

*A **cartogram** is a map in which spatial geometry is purposely distorted to reflect a theme; for example, sizes of countries might be made proportional to the population of each country. Conventionally, cartographers attempted to preserve the shape of enumeration units comprising a cartogram. More recently, Dorling has developed an algorithm in which uniformly shaped symbols (typically, circles) are used to represent each enumeration unit. A circle's size is a function of the magnitude of a phenomenon being mapped, which in Dorling's case is normally population. Within the circles, another theme (or themes) can be depicted, such as the percent of the adult population employed in manufacturing. Dorling contends that the resulting map more properly reflects the human geography of a region; this is in contrast to traditional choropleth maps, which he argues reflect the physical extent of the region.*

***Flow maps** are used to depict the movement of phenomena between geographic locations; probably most familiar to readers is the portrayal of flows between countries. Although general-purpose digital methods for depicting flows have not been developed, some specialized software has been devised for portraying migration and continuous vector–based flows (such as wind speed and direction). With respect to migration, we will consider Tobler's (1987) software, Tang's (1992) Visda and Yadav-Pauletti's (1996) MigMap, while for continuous vector–based flows, we will examine an approach developed by Lavin and Cerveny (1987).*

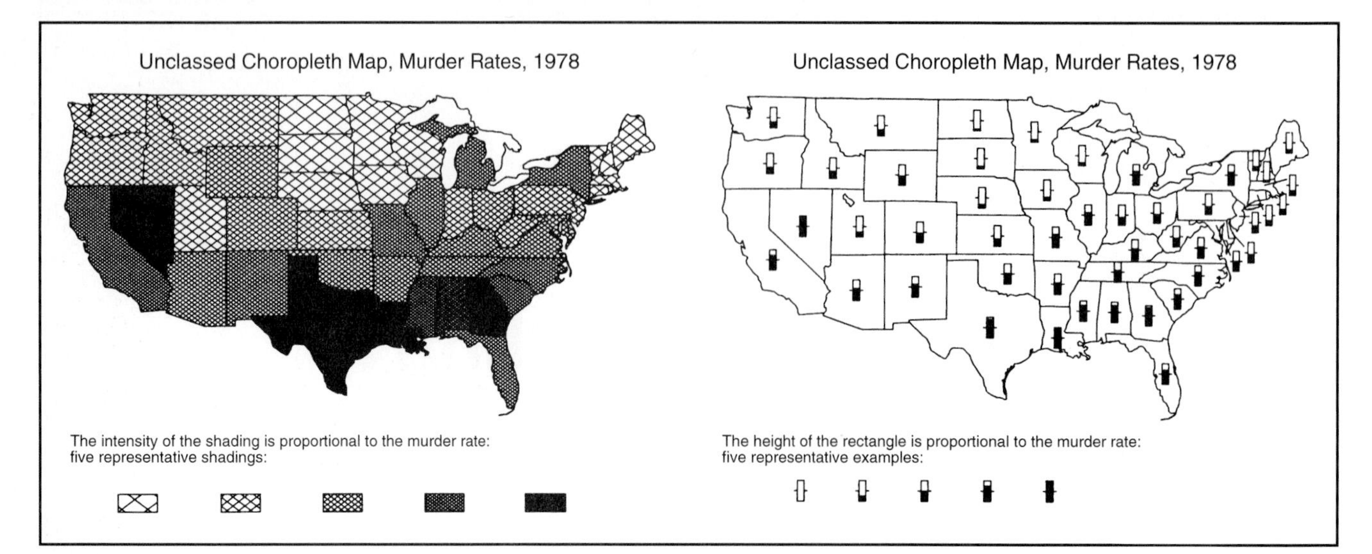

FIGURE 11.1. A comparison of (A) choropleth and (B) framed-rectangle maps. (From Dunn 1988, p. 124; courtesy of American Statistical Association.)

True 3-D phenomena *(such as level of carbon dioxide in the earth's atmosphere) are a challenge for mapping because they vary continuously in three-dimensional space: thus, a symbol at one location can hide a symbol at another location. In this chapter we will consider two approaches for symbolizing true 3-D phenomena: the software package T3D, which is intended for handling a broad range of true 3-D phenomena, and Waisel's (1996) use of the IBM Visualization Data Explorer (DX) to symbolize a particular phenomenon, the chemical composition of sediments deposited in New York Harbor. DX is an example of* ***visualization software,*** *which we will consider in greater depth in Chapter 15.*

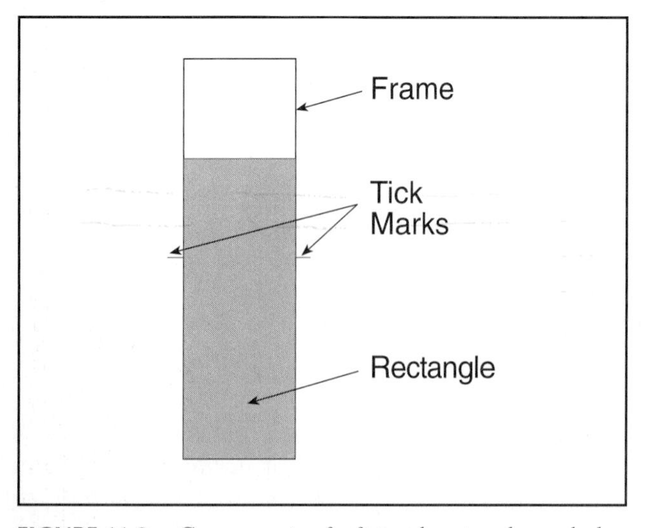

FIGURE 11.2. Components of a framed-rectangle symbol.

11.1 FRAMED-RECTANGLE SYMBOLS

The **framed-rectangle symbol** was developed by Cleveland and McGill (1984) as an alternative to the areal shading conventionally used on choropleth maps (Figure 11.1). Each framed-rectangle consists of a "frame" of constant size, within which a solid "rectangle" is placed (Figure 11.2). The tick marks on the side of the framed-rectangle were added by Cleveland (1994, 209) to enhance the reader's ability to estimate correct values. Note that framed-rectangle symbols allow similar values to be compared easily when those values would be difficult to distinguish with simple bars (Figure 11.3).

Cleveland and McGill argued that framed-rectangle symbols should be used in preference to choropleth shading because the estimation of values is a simpler perceptual task (assigning values to the height of a rectangle is easier than assigning values to a shade); and because larger enumeration units will not have a larger visual impact. We can add to this the fact that esti-

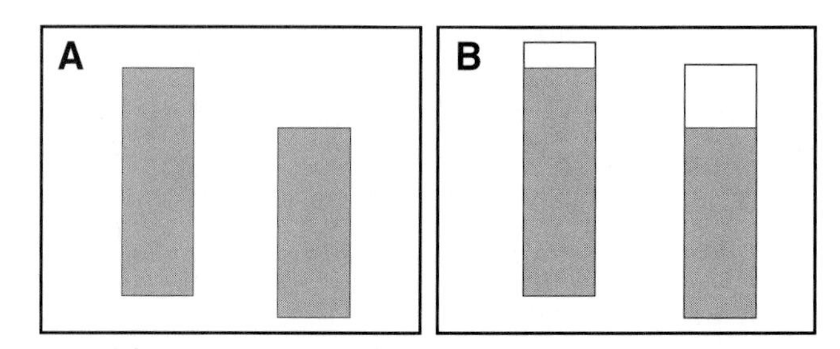

FIGURE 11.3. A comparison of (A) simple bars and (B) framed-rectangle symbols. Note the ease with which the frame-rectangle symbols can be differentiated. (After Cleveland and McGill 1984, p. 538.)

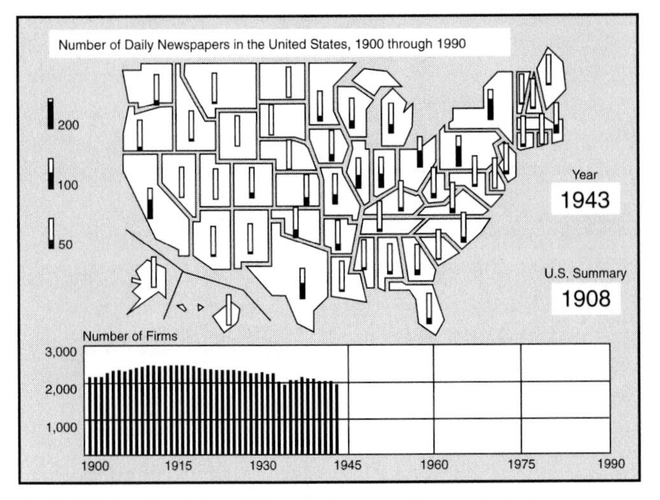

FIGURE 11.4. Use of framed-rectangle symbols to map unstandardized (count) data for enumeration units. The map is a frame from an animation of the number of daily newspapers subscribed to from 1900 to 1990. (After Monmonier 1992. First published in *Cartography and Geographic Information Systems* 19(4), p. 249. Reprinted with permission from the American Congress on Surveying and Mapping.)

mation of values on choropleth maps is adversely affected by simultaneous contrast (see Chapter 5). In an experimental study, Dunn (1988) found that framed-rectangle symbols had a smaller average level of error and a smaller variation in error, and that on choropleth maps variation in error was a function of the size of enumeration units.

A limitation of Dunn's study was that it dealt only with specific (as opposed to general) information. This is critical because a primary reason for creating thematic maps is to portray general information. It is arguably difficult to extract general information from a framed-rectangle map because the point symbols do not easily associate with the areal extent of the enumeration units. Imagine attempting to form regions on the framed-rectangle map shown in Figure 11.1 and then having to recall those regions at a later time. Extracting general information would be especially problematic when framed-rectangles must be drawn outside of small enumeration units to avoid overlap (as in the northeast portion of Figure 11.1).

The framed-rectangle symbol, however, is appropriate for mapping unstandardized (count) data associated with enumeration units because point symbols normally are used for that purpose (see Chapter 7). As an example, Monmonier (1992a) used the framed-rectangle symbol in an animation of the number of daily newspapers subscribed to over a 90-year period (a frame from his animation appears in Figure 11.4). Monmonier handled the problem of symbol overlap by adjusting the sizes of the states so that smaller ones would be more noticeable; however, this creates shape distortion, which may be disconcerting to some readers.

11.2 MODELING GEOGRAPHIC PHENOMENA AND THE CHORODOT MAP

Section 2.4 introduced the problem of selecting appropriate symbology by examining a variety of mapping techniques (choropleth, proportional symbol, isarithmic, and dot). In this process, we found that two major factors should be considered: (1) the need to accurately depict the underlying phenomenon, and (2) the kind of information that the mapmaker wishes the map user to acquire. This section delves into this problem in greater depth by considering models that MacEachren and DiBiase have developed for representing spatial phenomena.

Figure 11.5 is a graphic portrayal of MacEachren and DiBiase's models. Note that geographic phenomena are arranged on two axes: discrete-continuous and abrupt-smooth. As discussed in section 2.1.2, discrete phenomena occur at distinct locations (with space between elements of the phenomenon), while continuous phenomena are defined throughout the geographic region of interest. Smooth phenomena change in a gradual fashion, while abrupt phenomena change suddenly across geographic space. Figure 11.6 lists representative phenomena that might correspond to each of the nine models shown in Figure 11.5. We will now consider these phenomena in detail so that a clearer understanding of the two axes is obtained.

First, consider the continuous phenomena shown in the bottom row of Figure 11.6. Percent sales tax is an ob-

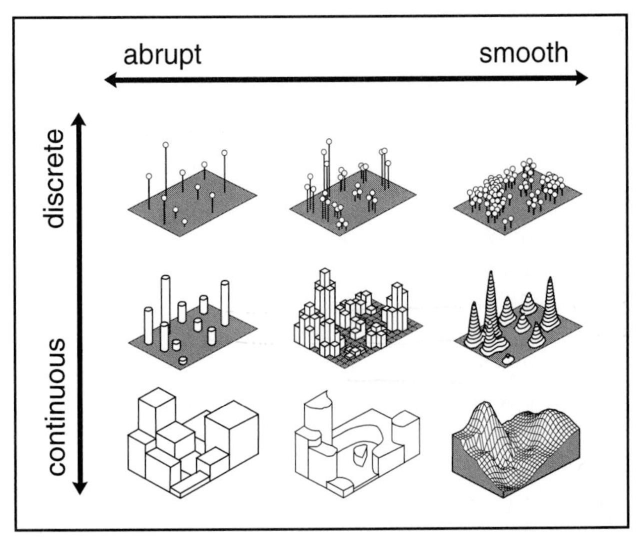

FIGURE 11.5. MacEachren and DiBiase's models for representing geographic phenomena. (From MacEachren 1992, p. 16; courtesy of North American Cartographic Information Society and Alan MacEachren.)

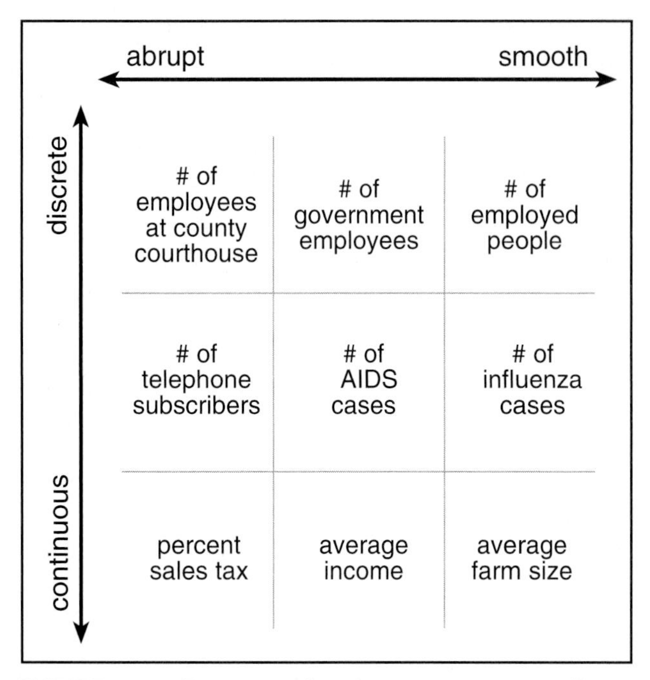

FIGURE 11.6. Representative phenomena corresponding to each of the nine models shown in Figure 11.5. (After MacEachren and DiBiase 1991, p. 223.)

vious *abrupt* continuous phenomenon, as it changes suddenly at the boundary between enumeration units (for example, one state's sales tax is different from another). In contrast, average farm size is an example of a *smooth* continuous phenomenon because we would expect it to vary in a relatively gradual fashion (as the climate becomes drier, we would expect the average farm size to increase). Average income falls somewhere between percent sales tax and average farm size on the abruptness-smoothness continuum. In some cases, average income would exhibit the abrupt changes of percent sales tax (as at the boundary between urban neighborhoods), while in others it would exhibit a more gradual change (as one moves up a hill toward a region of more attractive views, average income should increase).

In contrast to the bottom row, the top row represents a range of discrete phenomena. The number of employees located at county courthouses is clearly an *abrupt* discrete phenomenon, as there can be only one value for a county and it occurs at an isolated location. In contrast, the number of employed people (based on where they live, as opposed to where they work) is a *smooth* discrete phenomenon, since it gradually changes over geographic space. The number of government employees (again, based on where they live) falls somewhere between these; it may exhibit an abrupt character in the sense that government employees may live near government offices, but it will not exhibit the extreme abruptness of the courthouse example.

The middle row represents phenomena that can be classified as not clearly continuous or discrete, and that also span the abruptness-smoothness continuum. This row is probably most easily understood by considering the influenza case first. Since influenza is an infectious disease, it should exhibit a smooth character. Although individual influenza cases could be represented at discrete locations, it makes sense to suggest some degree of continuity if we wish to stress the potential of infection. At the other end of the row is the number of subscribers to a particular telephone company. Competition between telephone companies could lead to a distribution that changes abruptly but that exhibits continuity between the lines of abrupt change. Finally, the number of people with AIDS (acquired immune deficiency syndrome) is in the middle of the diagram. AIDS occupies a more abrupt position than influenza because of its mode of transmission (sexual intercourse, sharing of needles, and blood transfusions).

Once a cartographer has defined the nature of the underlying phenomena, a decision can be made regarding appropriate symbology. Figure 11.7 illustrates a set of symbolization methods corresponding to the models shown in Figure 11.5.* The bulk of these methods are conventional and have already been covered in this text. Others, notably the middle and middle right-hand ones, are novel. MacEachren and DiBiase focused their efforts on the middle one, which they termed the chorodot map.

As the name implies, the **chorodot map** is a combination of choropleth and dot maps (Figure 11.8). The "dot"

* Enumeration unit boundaries are shown on a number of these maps because MacEachren and DiBiase focused on data collected for enumeration units.

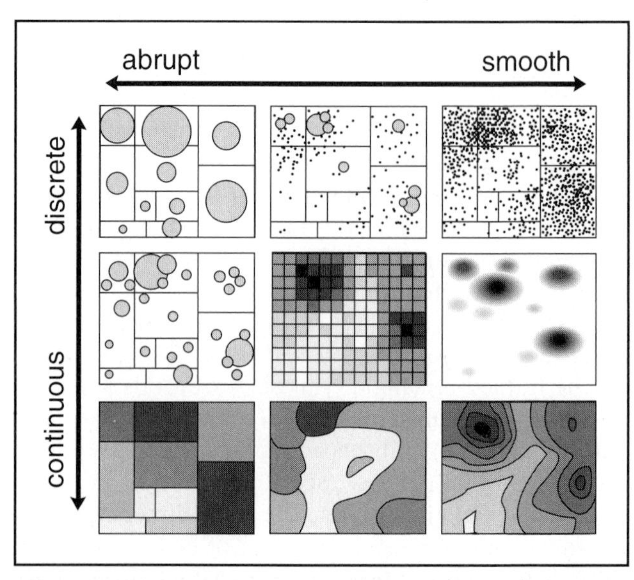

FIGURE 11.7. A set of symbolization methods that MacEachren and DiBiase argued would be appropriate for the models shown in Figure 11.5. (From MacEachren 1992, p. 16; courtesy of North American Cartographic Information Society and Alan MacEachren.)

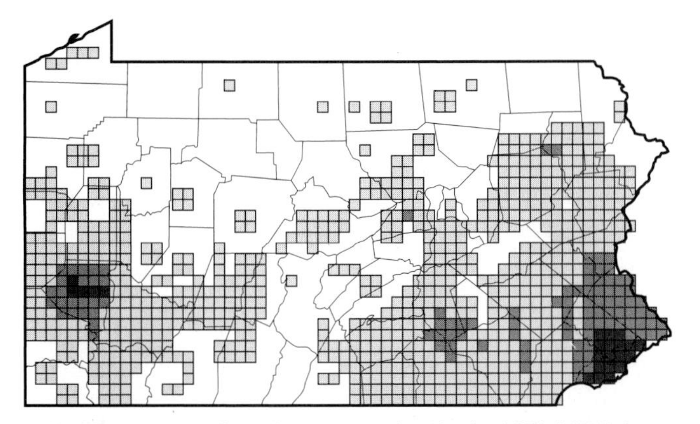

FIGURE 11.8. A chorodot map used to depict 1988 AIDS data for Pennsylvanian counties. The number of cases represented by each cell are as follows: light gray (1 case), medium gray (8 cases), and dark gray (64 cases). (From MacEachren 1994a, p. 63; courtesy of Association of American Geographers and Alan MacEachren.)

portion comes from the use of small squares within each enumeration unit, while the "choropleth" portion derives from the shading assigned to each square. For shading, MacEachren and DiBiase use a geometric classification: a light gray dot represents 1 AIDS case; a medium gray dot, 8 cases; and a dark gray dot, 64 cases. Darker dots were placed only after lighter ones completely fill a county; thus no medium gray dots can be placed until the entire county has been covered by light gray. The resulting map suggests the middle position of the AIDS phenomena in the models shown in Figure 11.5.

Although MacEachren and DiBiase intended the graphical portrayals in Figure 11.5 primarily to assist in *visualizing* different models, it is apparent that these portrayals could also be used for *symbolizing* phenomena. For example, MacEachren and DiBiase used the middle right-hand portrayal as another symbolization method for depicting AIDS (see also Gould 1989, 76).

In addition to providing mapmakers with models that may assist in selecting symbology, MacEachren and DiBiase's work also provides an illustration of how a geographer's intended purpose may affect symbol selection. In MacEachren and DiBiase's case, they were developing maps of AIDS for Peter Gould, who had two basic goals: to portray the large number of people affected by AIDS (to alarm the map viewer) and to illustrate a "classic example of hierarchical diffusion (modeled as a smooth, continuous function)" (MacEachren and DiBiase, 225). As a result, MacEachren and DiBiase created an animation of the raw number of AIDS cases through time via a series of isopleth maps (the animation can be found in DiBiase et al. 1991). MacEachren and DiBiase noted that the isopleth maps were technically improper because they required standardized data (see sections 2.4.1 and 2.4.3 for more discussion of this issue), but the maps did serve Gould's goals well.

11.3 DORLING'S CARTOGRAMS

When creating thematic maps, cartographers generally avoid distorting spatial reality. For example, an **equal-area projection** is normally used for a dot map so that the density of dots is solely a function of the underlying phenomenon (and not the map projection). Sometimes, however, cartographers purposefully distort space based on values of a theme: the resulting maps are known as **cartograms.**

Probably the most common cartograms used in everyday life are **distance cartograms** in which real-world distances are distorted to reflect some attribute, such as the time between stops on subway routes: here cartograms are appropriate because the time between (and order of) stops is more important than the actual distance between stops. In the geographic literature, **area cartograms** are more common.* Area cartograms are created by scaling (sizing) enumeration units on the basis of an attribute. For example, in Figure 11.9B the states and territories of Australia have been scaled in direct proportion to the population of each state or territory. Although population is the most typical attribute portrayed on area cartograms, any ordinal or higher-level attribute (unstandardized or standardized) could be used.

Two forms of area cartograms have traditionally been recognized: contiguous and noncontiguous. In contiguous cartograms, an attempt is made to retain the contiguity of enumeration units; as a result, the shape of units must be greatly distorted (as in Figure 11.9B). In noncontiguous cartograms, units' shapes are retained, but gaps are introduced between units (Figure 11.10).

Area cartograms are commonly used for their dramatic impact; in fact, it can be argued that an area cartogram should not be used if the relations between individual values of an attribute are similar to the relations between the sizes of enumeration units because the resulting map will not appear distorted.

Readers may have difficulty interpreting area cartograms because they look so different from other maps, which are frequently based on equal-area projections. To alleviate this problem, mapmakers should retain characteristic features (for example, the distinctive meanders of a river) along boundaries of enumeration units; include an inset map depicting actual areal relations of enumeration units (or alternatively the units should be labeled); and not use area cartograms if the anticipated readership is unfamiliar with the region.

Dorling (1993, 1994, 1995b, 1995c) has developed a new algorithm for creating cartograms based on a uni-

* Dent (1996) terms these *value-by-area cartograms;* I have used the term *area cartogram* for simplicity.

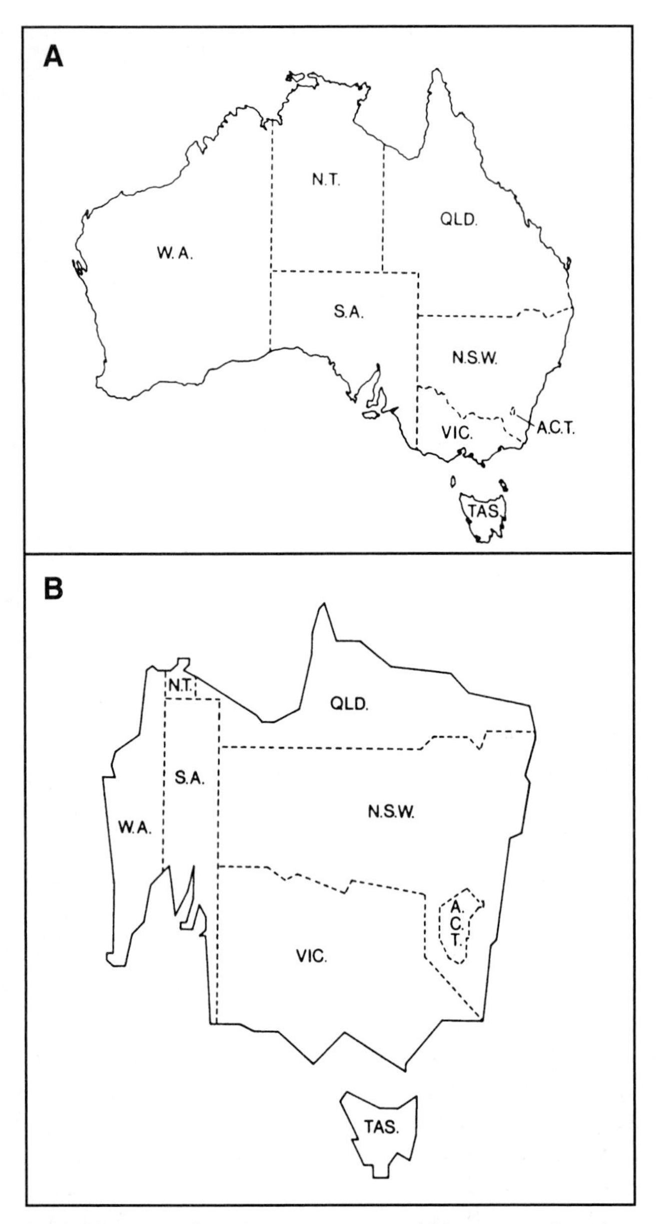

FIGURE 11.9. Creating a cartogram: (A) states and territories of Australia as they might appear on an equal-area projection; (B) a contiguous-area cartogram in which states and territories are scaled on the basis of 1976 population. (From Griffin 1983, p. 18; courtesy of T. L. C. Griffin.)

formly shaped symbol (typically, a circle). Dorling's algorithm begins by placing the uniformly shaped symbol in the center of each enumeration unit, with the size of the symbol scaled on the basis of population. Initially, symbols overlap one another because small enumeration units may have relatively large populations (Figure 11.11A). To eliminate overlap, an iterative procedure is executed in which symbols are gradually moved apart from one another (Figure 11.11B to 11.11E). Wherever possible, points of contact between symbols reflect the points of contact between actual enumeration units, but sometimes it is not possible to meet this constraint. As a result, Dorling terms this a "noncontiguous form of cartogram."*

The result of applying Dorling's algorithm to the 9,289 wards of England and Wales is shown in Figure 11.12. Figure 11.12A depicts the boundaries of wards as they would appear on a traditional equal-area projection, while Figure 11.12B displays the cartogram resulting from applying Dorling's algorithm. (The upper right-hand portion of each figure portrays wards grouped into counties.) The basic difference between the maps is that small land areas with large populations are much more apparent on the cartogram than on the traditional map. This is particularly apparent on the map of county boundaries: note that London is a mere dot on the traditional map, but encompasses a substantial portion of the corresponding cartogram. Dorling contends that his method is essential if we are to represent the *human geography* of a region (as opposed to focusing on the *physical extent*).

The major purpose of Dorling's algorithm is not to map population, but rather to serve as a base on which other variables can be displayed. For example, Color Plate 11.1 portrays population density in Britain on both a traditional equal-area projection and on his cartogram. Note the drastic difference in appearance between these two maps. The equal-area projection suggests that most of the country is dominated by relatively lower population densities (the blues and greens), while the cartogram provides a detailed picture of the variation in population densities within urban areas.

One obvious difference between Dorling's cartograms and traditional ones is that his provide no shape information for individual enumeration units. As a result, a conventional equal-area base map would seem to be essential when examining Dorling's cartograms. Another difference is that his cartograms typically show a large number of enumeration units, and thus considerable detail. As a result, readers should expect to study these maps meticulously.

The complexity of Dorling's cartograms suggests that they might be examined most effectively in a data exploration environment. For example, imagine having both the traditional map and the cartogram displayed on a computer screen at the same time. As the cursor is moved over an enumeration unit on the cartogram, this same enumeration unit could be highlighted and named on the traditional map (or vice versa). Alternatively, the user could zoom in to examine subregions in greater detail. We will consider some related work that Dorling has done in this context in Chapter 14.

It should be noted that Dorling is not the only one to focus on the human geography of a region by

* A Pascal implementation of Dorling's algorithm can be found in Dorling (1995c, 373–378).

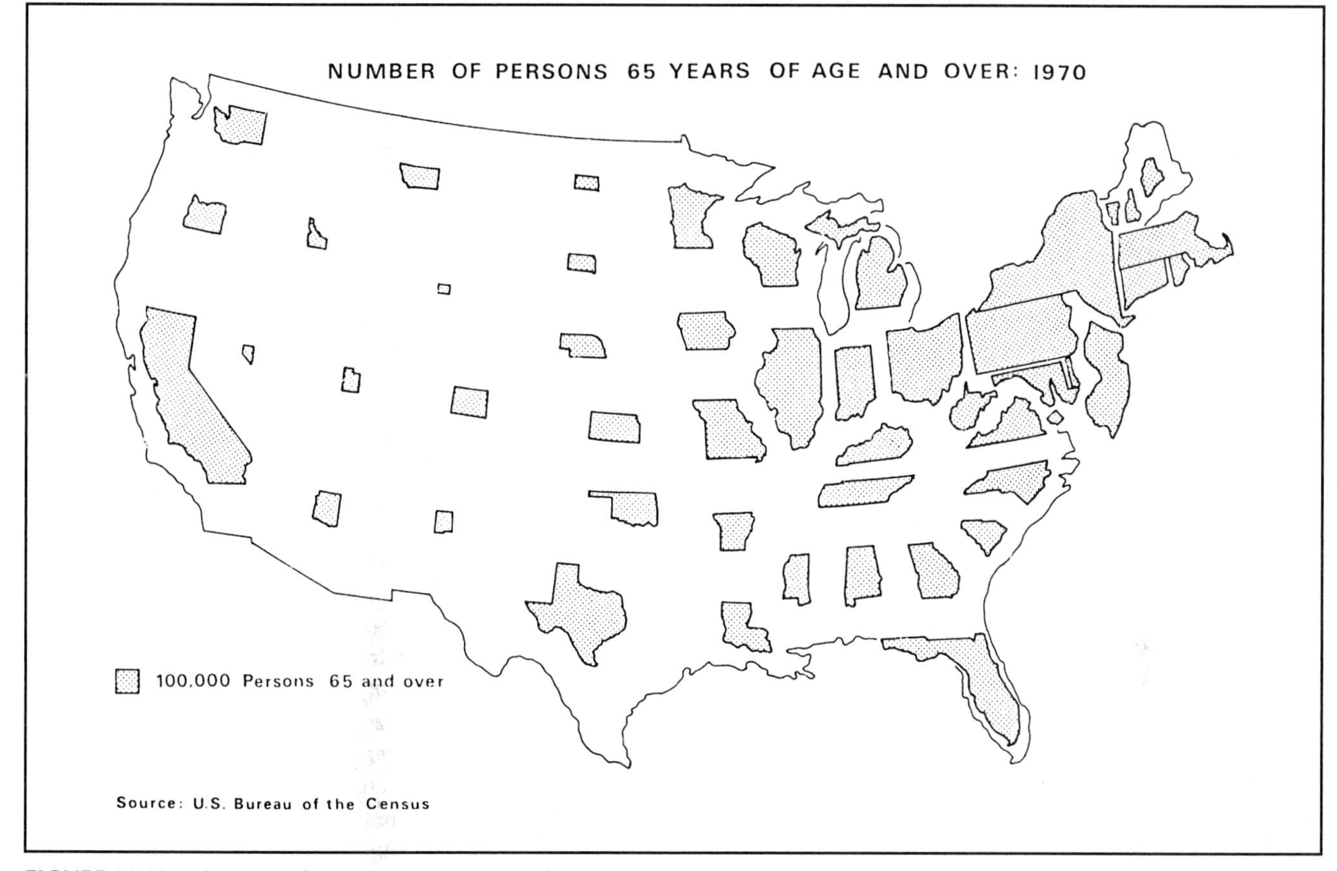

FIGURE 11.10. A noncontiguous area cartogram; shape of enumeration units is retained by introducing gaps between units. (From Olson 1976b, p. 372; courtesy of Association of American Geographers.)

mapping thematic variables onto a cartogram base. The *Historical Atlas of Massachusetts* (Wilkie and Tager 1991) contains several examples of this approach, but in this case an attempt is made to maintain the shape of enumeration units. For example, page 91 of the atlas illustrates "Irish Ancestry in 1980" using a cartogram to show the population of each city and town in Massachusetts and a green choropleth shading to represent the percentage of Irish ancestry in each city and town.

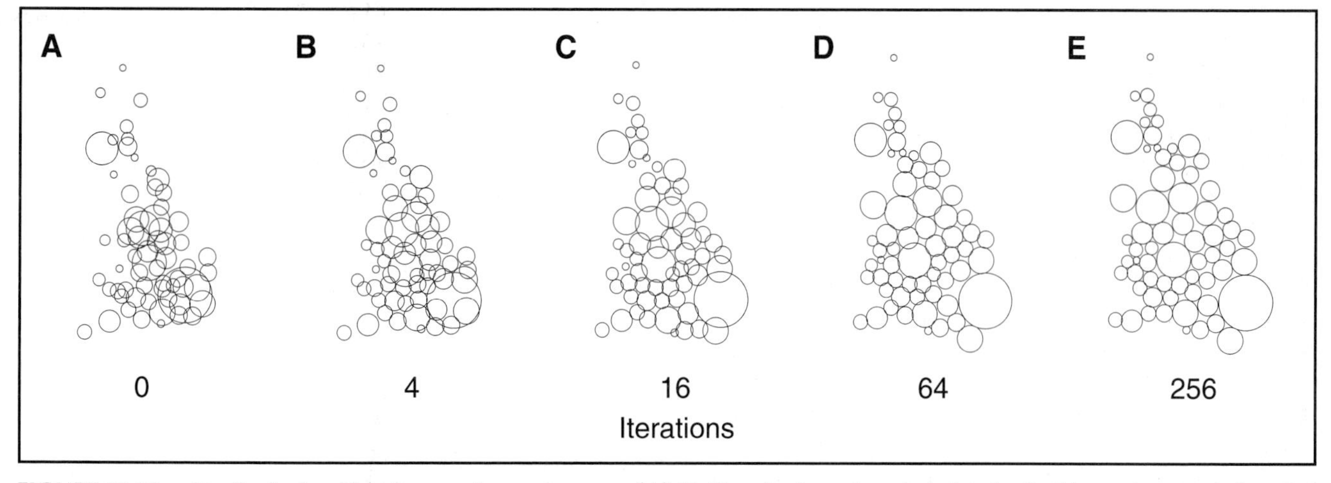

FIGURE 11.11. Dorling's algorithm for creating cartograms: (A) Uniformly shaped symbols (circles, in this case) are scaled on the basis of population, and are placed in the center of enumeration units on an equal-area projection; (B–E) the circles are gradually moved away from one another so that no two overlap. The numbers beneath each illustration represent the number of iterations in the algorithm. (After Dorling 1995b, p. 274.)

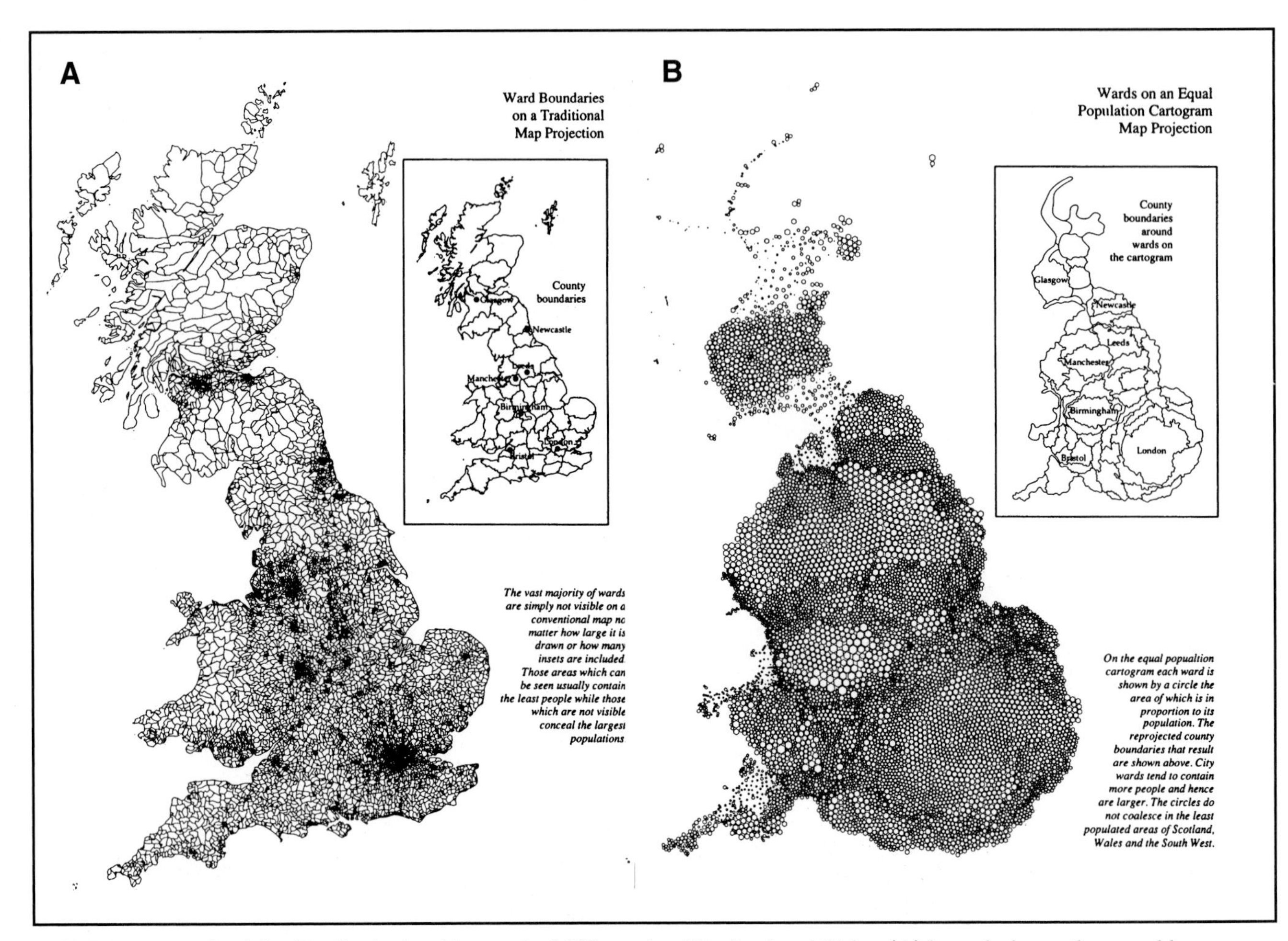

FIGURE 11.12. Applying Dorling's algorithm to the 9,289 wards of England and Wales: (A) boundaries as they would appear on a traditional equal-area projection; (B) a cartogram resulting from applying Dorling's algorithm. (From Dorling 1993, p. 171; courtesy of The British Cartographic Society.)

11.4. NOVEL METHODS FOR FLOW MAPPING

Flow maps are used to depict the movement of phenomena between geographic locations. Although a wide variety of phenomena can be portrayed on flow maps, readers are probably most familiar with those representing the flow of goods or movement of people between countries. Although general-purpose digital methods for depicting flows have not been developed, some specialized software has been devised for portraying migration and continuous vector–based flows (such as wind speed).

11.4.1 Methods for Mapping Migration

Migration data are a logical choice for digital portrayal because of the considerable number of movements that typically must be represented. For example, for the 48 contiguous U.S. states there are 2,256 possible movements that can be made between states; if we consider U.S. counties, there are more than 9 million such movements. And this does not even consider the variable of time! Clearly, a computer is essential for examining the large number of such movements.

Tobler (1987) was the first to develop a significant piece of software for displaying migration flows.* One of the simpler options in his software was the depiction of one-way migration to or from a particular state. In this case it was possible to show the migration between each state and a selected state by arrows of varying width (Figure 11.13). When one wishes to show the migration between all pairs of enumeration units simultaneously, a different approach is required because of the large number of arrows that result. The key is recognizing that many migration movements are small relative to others and that deleting such movements will allow the reader to focus on more important movements. Tobler (p. 160) indicated that by deleting migration values below the mean, he was typically able to remove 75% or more of the flow lines, while deleting less than 25% of the mi-

* A limited version of the software (known as FlowMap) is available from Professor Tobler.

FIGURE 11.13. Migration to and from California, 1965–1970. (After Tobler 1987. First published in the *American Cartographer* 14(2), p. 160. Reprinted with permission from the American Congress on Surveying and Mapping.)

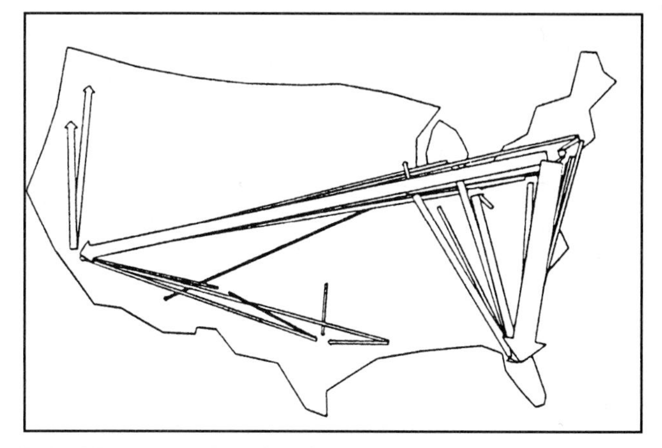

FIGURE 11.14. Net migration 1965–1970 for the 48 contiguous U.S. states, with flows below the mean net migration not shown. (Reprinted by permission from *Geographical Analysis,* Vol. 13, No. 1 (Jan. 1981). Copyright 1981 by Ohio State University Press. All rights reserved.)

grants. For example, Figure 11.14 illustrates this approach for net migration for states in the U.S. from 1965–1970.

Another interesting feature of Tobler's software was the ability to route data through enumeration units lying between the starting and ending points for migration, thus reflecting the route over which people were presumed to migrate. For example, migration data from New York to California would ideally have to be routed through Pennsylvania, Ohio, and numerous other states. The details of how Tobler achieved this are beyond the scope of this text (see Tobler 1981, 7–8 for a summary), although it should be noted that the approach used some of the same concepts implemented in his pycnophylactic method (see section 8.6). Figure 11.15 illustrates the result for the migration data used for 11.14.

One limitation of Tobler's software is that it was developed a number of years ago and so did not exhibit the kind of interaction that one might expect with modern software. For example, one might want to explore the effect of selecting different break points for deleting flow paths: for this purpose, one could imagine moving a scroll box along a scroll bar and seeing the map change dynamically as the break point is changed.

Examples of migration mapping software that have achieved some modicum of interaction include Tang's (1992) Visda and Yadav-Pauletti's (1996) MigMap. Lee et al.'s (1994) XNV, intended for mapping airline flows, also appears as though it might be useful for mapping migration data. A brief summary of each of these programs follows.

In Visda, Tang symbolizes migration data via choropleth maps. Users are able to point at an enumeration unit and receive a map of in-migration or out-migration with respect to that unit. The use of choropleth symbols avoids the overlap problem associated with Tobler's arrow symbols, but the choropleth symbols do not

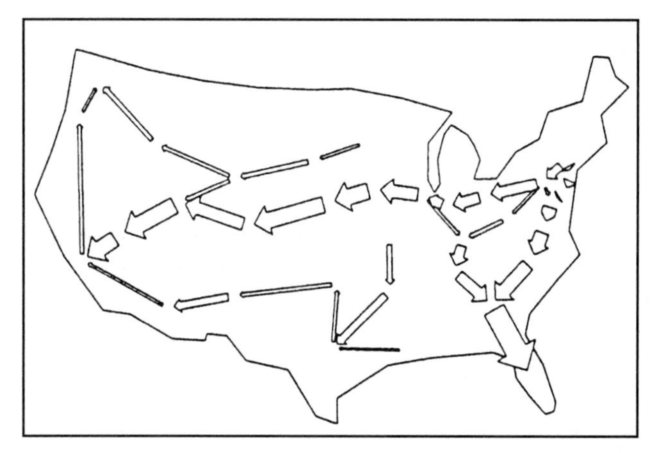

FIGURE 11.15. Migration data depicted in Figure 11.14 are rerouted to pass through states between the starting and ending points; again, flows below the mean are not shown. (Reprinted by permission from *Geographical Analysis,* Vol. 13, No. 1 (Jan. 1981). Copyright 1981 by Ohio State University Press. All rights reserved.)

connote flows as readily as the arrow symbols. Yadav-Pauletti's MigMap symbolizes migration data using both choropleth and arrow symbols, and makes heavy use of small multiples (individual maps for each time period) and animation. Although it is possible to create a number of interesting maps with her software (Color Plate 11.2), an important limitation is the inability to input one's own migration data. XNV provides exploration capability similar to what would be desirable for Tobler's software, along with the ability to save and redisplay multiple maps as separate windows; the software also permits the computation of several specialized analytical functions. Unfortunately, readers are unlikely to have access to XNV, as it must be recompiled for individual workstation environments.

11.4.2 Mapping Continuous Vector-Based Flows

Continuous vector-based flows are composed of two variables, magnitude and direction, that can change at any point. For example, wind is a continuous vector-based flow because at any point in the atmosphere we can consider the speed of the wind and the direction from which it blows. Traditionally, wind speed and direction were represented using wind arrows (as in Figure 11.16). Lavin and Cerveny (1987, 131) deemed such symbols awkward because (1) readers must visually separate them in order to create the individual distributions of wind speed and direction, and (2) the symbols representing speed (the flaglike portion) are abstract and thus time-consuming to interpret. As an alternative, Lavin and Cerveny proposed unit-vector density mapping.

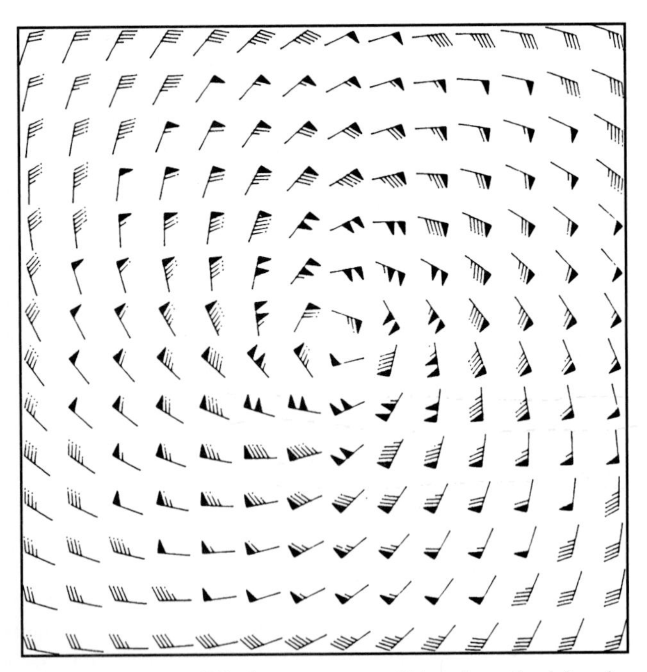

FIGURE 11.16. Wind arrows: a traditional method for depicting wind speed and direction. (From Lavin and Cerveny 1987, p. 132; courtesy of The British Cartographic Society.)

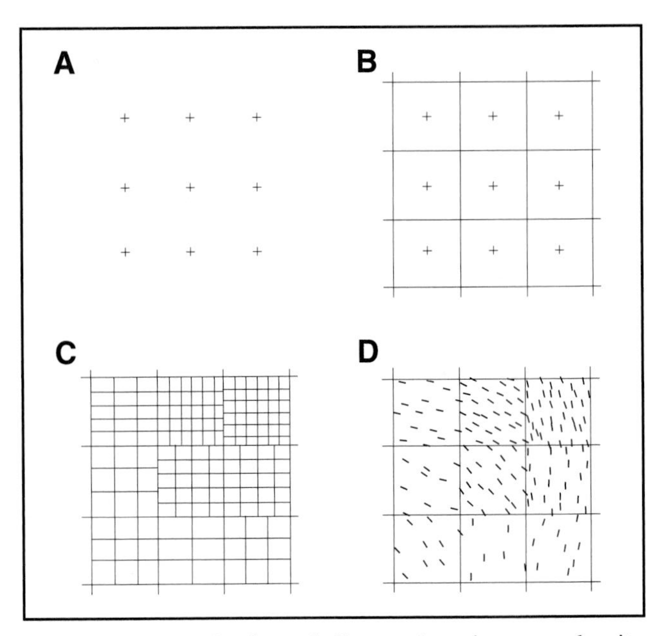

FIGURE 11.17. Lavin and Cerveny's unit-vector density method for mapping vector-based flows (data that have both magnitude and direction): (A) input is an equally spaced gridded network of magnitudes and directions; (B) a square cell is presumed around each grid point; (C) the cells in (B) are divided into subcells, the number of which is proportional to the magnitude of the data; (D) a unit-vector is placed within each subcell, with the orientation corresponding to the direction of the data. (From Lavin and Cerveny 1987, p. 134; courtesy of The British Cartographic Society.)

Unit-vector density mapping involves changing the density and orientation of short fixed-length line segments called unit vectors. The density and orientation of unit vectors are a function of the magnitude and direction of the vector-based flow being mapped. Typically, the line segments have an arrowhead to indicate movement, although they can also be used effectively without arrowheads.

Lavin and Cerveny's algorithm for unit-vector density mapping is analogous to the algorithm Lavin used in his dot-density shading technique (see section 10.3.3). The algorithm presumes that one has first created an equally spaced gridded network of vector-based flow values; thus each of the crosses in Figure 11.17A would have two data values—one for vector magnitude and one for vector direction. Creating such data requires running the gridding approach for automated contouring (section 8.2) twice, once for vector magnitude and once for vector direction. Next, it is presumed that each grid intersection is surrounded by a square cell within which the unit vectors will be placed (Figure 11.17B). Each of these square cells is then divided into subcells, with the number of subcells a function of the magnitude of the data being mapped (Figure 11.17C). Finally, individual unit vectors are placed within each subcell, with the orientation corresponding to the direction of the data being mapped (Figure 11.17D).

As an example of the unit-vector density technique, consider the map of average January surface winds for the United States shown in Figure 11.18. Lavin and Cerveny noted that this map illustrates "many of the features sought on more traditional" maps, but avoids the difficulty of determining wind speed (p. 135). In particular, they noted (1) areas where arrows appear to converge or diverge, which are indicative of low- and high-pressure areas respectively (in the extreme southwestern portion of the United States there is an apparent area of divergence); (2) the flow off the Great Lakes, which can lead to "lake-effect" precipitation; and (3) the areas of high and low wind speed defined by the differing densities of unit vectors (for example, the central high plains and portions of the Rocky Mountains, respectively).

A less obvious, but intriguing use of unit-vector density mapping, is the portrayal of topography. In this case, elevation is represented by the *density* of unit vectors, while aspect (the direction the slope faces) is depicted by the *orientation* of the unit vector; no arrow symbol is necessary because topography does not move. The density of unit vectors readily identifies the regions of highest elevation, while the orientation of vectors indicates significant characteristics of aspect. To illustrate, compare Figures 11.19 and 11.20, which contrast present-day topography with the estimated situation 18,000 years ago. In Figure 11.19, the density of unit vectors assists us in identifying major topographic features such as the Himalayas, the Greenland ice sheet, and the Andes. In Figure 11.20, these areas are again apparent, but we notice other high-elevation areas, which reflect the fact that major ice sheets once covered much of North America and Europe. Also note the orientation of the unit vectors in Figure 11.20; particularly distinctive are those radiating from source regions for glaciation in present-day Canada.

11.5. MAPPING TRUE 3-D PHENOMENA

Section 2.1.1 indicated that two kinds of volumetric phenomena are possible: 2½-D and true 3-D. 2½-D phenomena have a single z value for each X and Y position

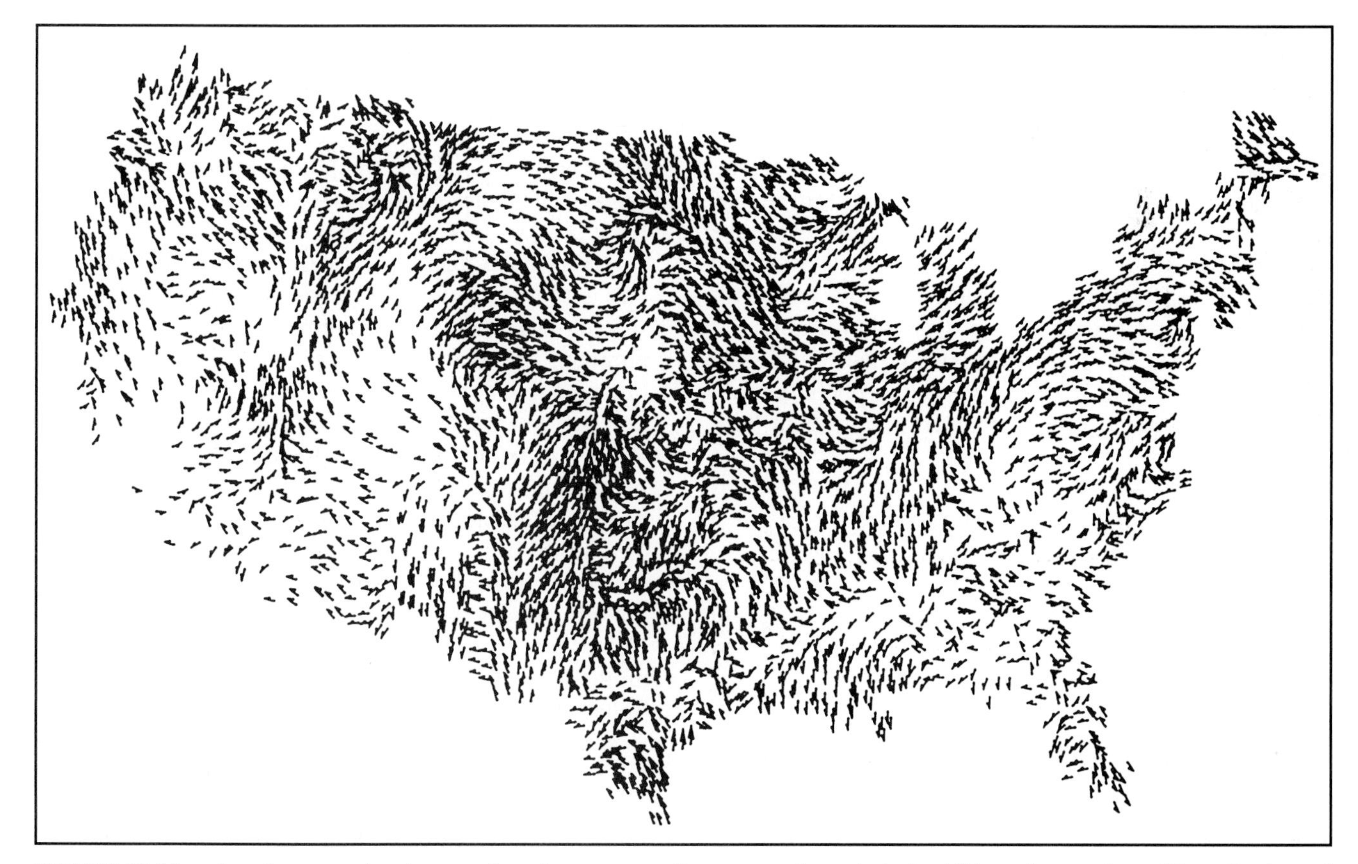

FIGURE 11.18. A unit-vector density map based on average January surface winds for 187 weather stations. (From Lavin and Cerveny 1987, p. 136; courtesy of The British Cartographic Society.)

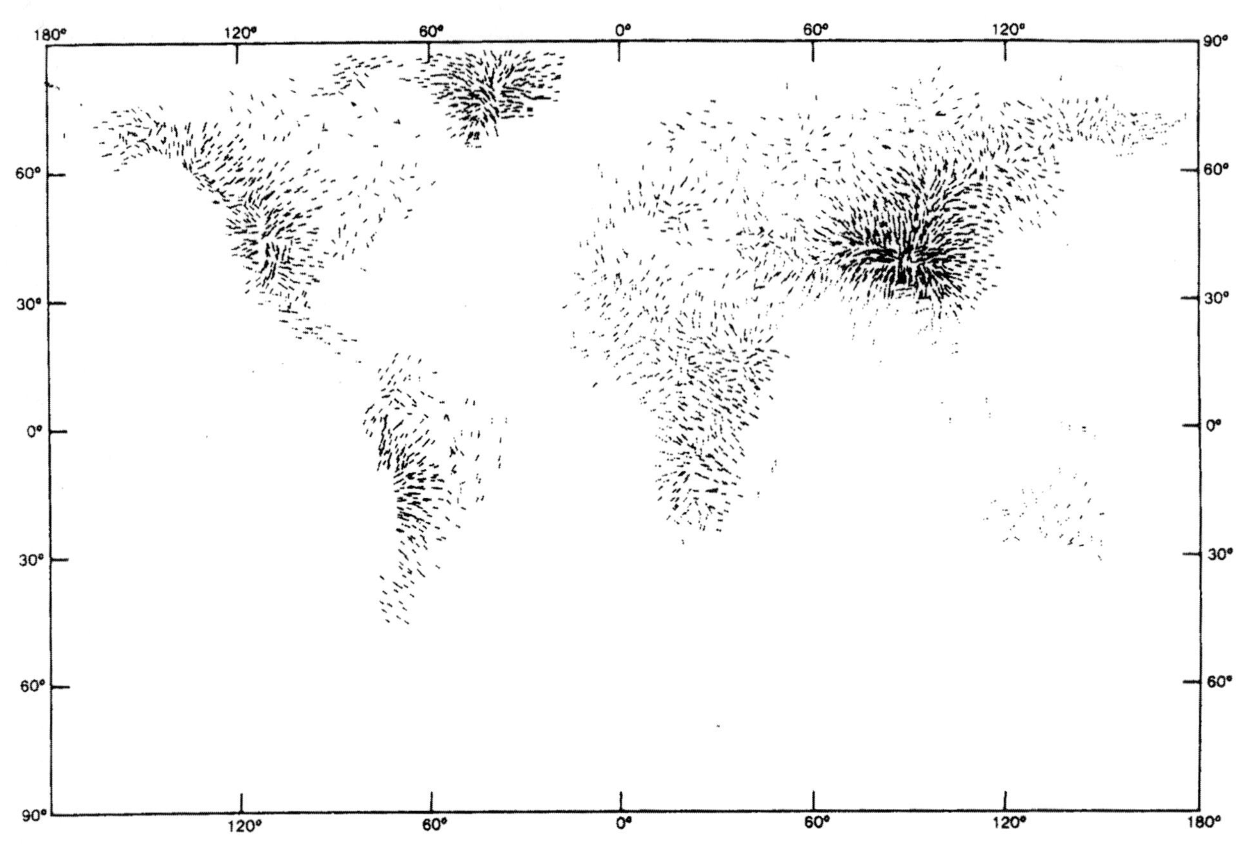

FIGURE 11.19. A unit-vector density map based on present-day topography. (From Lavin and Cerveny 1987, p. 139; courtesy of The British Cartographic Society.)

(longitude and latitude), while true 3-D phenomena have multiple z values for each X and Y. Examples of true 3-D phenomena include carbon dioxide levels in the atmosphere and water temperature in the ocean. Here we consider two approaches for symbolizing true 3-D phenomena: (1) the software package T3D, which is intended for handling a broad range of true 3-D phenomena, and (2) Waisel's (1996) use of the IBM Visualization Data Explorer (DX) to symbolize a particular phenomenon, the chemical composition of sediment deposits in New York Harbor.

11.5.1 T3D: General-purpose Software for Symbolizing True 3-D Phenomena

T3D, marketed by Fortner Software LLC, is an example of general-purpose software for symbolizing true 3-D phenomena. To illustrate some of T3D's capability, we will use a sample data set of vertical wind velocity within a simulation of a thunderstorm (the data are distributed with T3D). We will use a default "Rainbow" color scheme in which long wavelength colors (red, orange, and yellow) are assigned to upflows and short wavelength colors (blue and green) are assigned to downflows (Color Plate 11.3).*

The basic problem in symbolizing true 3-D phenomena is that only the surface of the three-dimensional structure can be seen if conventional opaque (nontransparent) shading is used, as in Color Plate 11.3A. Rather than seeing only the surface, we would like to "peer" into the data, which can be accomplished by varying the opacity of the shading. In the simplest case, this is achieved by making some portion of the data range transparent, while the remaining portion is left opaque. For example, in Color Plate 11.3B upflows are transparent, while downflows are opaque. Alternatively, one can gradually change the opacity over a selected range of data. For example, in Color Plate 11.3C the opacity for the entire range of data gradually changes from completely transparent (for the upflows) to completely opaque (for the downflows).

A second approach for peering into the data is to take a slice through the 3-D surface. By default, slices are completely opaque, although it is possible to vary their opacity too. Color Plate 11.3D illustrates horizontal and vertical opaque slices through the thunderstorm data. A third approach is to create an **isosurface,** which is the three-dimensional equivalent of a contour line.

* Unfortunately, documentation included with the data set does not indicate clearly the boundary between upflows and downflows. The use of the specified colors is presumed to approximate the boundary and is useful for illustrative purposes.

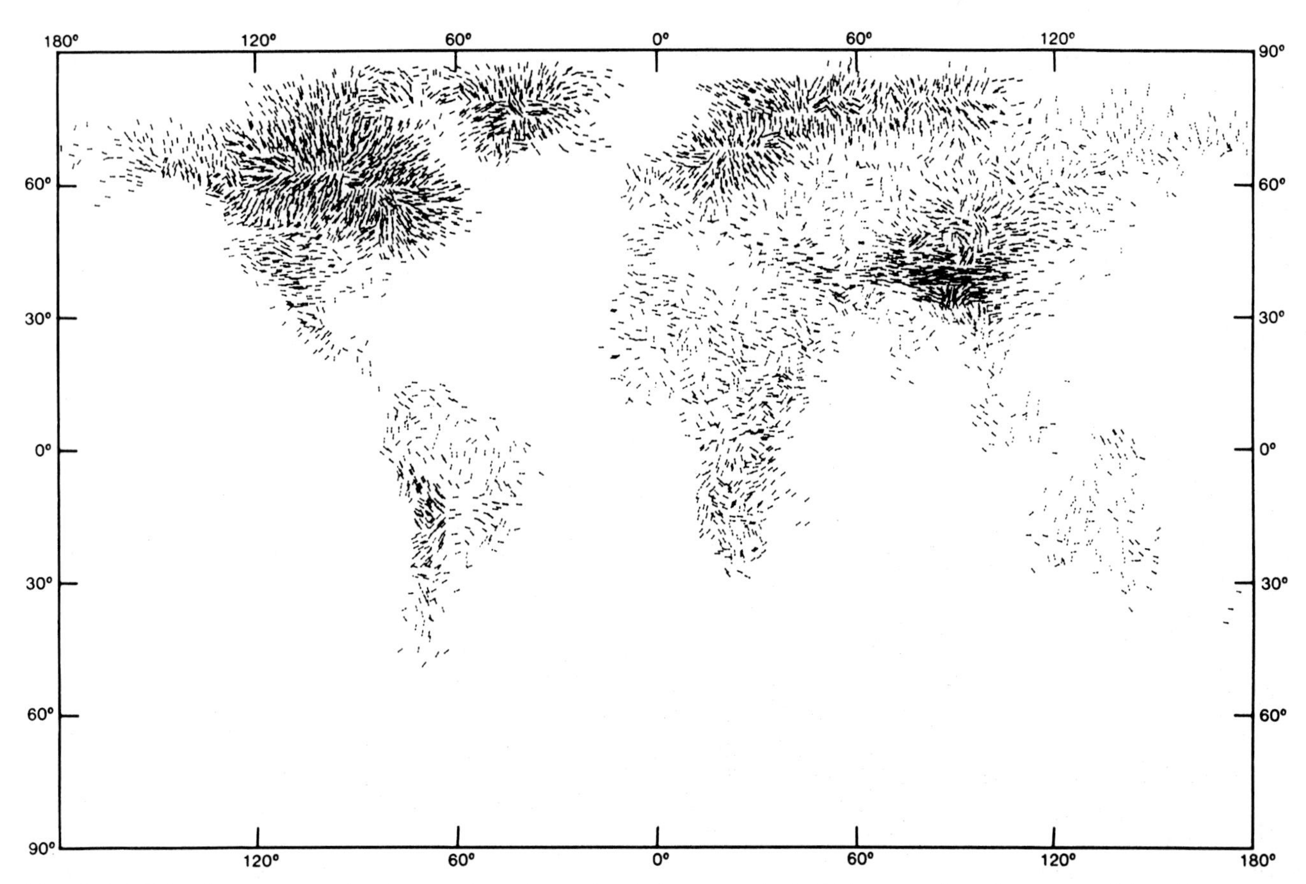

FIGURE 11.20. A unit-vector density map based on estimated topography 18,000 years before the present. (From Lavin and Cerveny 1987, p. 139; courtesy of The British Cartographic Society.)

Isosurfaces, like slices, are by default opaque (Color Plate 11.3E).

Within T3D, it is possible to combine any of the above approaches for peering into the data. For example, Color Plate 11.3F illustrates a combination of the slice and isosurface methods. Other approaches that assist in visualizing true 3-D phenomena include rotating and animating the image. One limitation of rotation in T3D (at least for the PC version) is that users cannot rotate a fully rendered image in real time. Rather, one must select a desired rotation using a solid box enclosing the 3-D data set; only when the desired rotation has been selected is the complete image rendered. In Chapter 13 we will consider another program, Vis5D, which does permit real-time rendering.*

Animation can reveal hidden three-dimensional information by either showing a sequence of movements in a rotation of a 3-D surface or sequencing a set of slices (or isosurfaces). Users specify end points in the animation sequence, and intermediate frames are interpolated automatically. Still another option with T3D is to modify the color scheme. The use of color will be described in detail when we consider another Fortner Software product, Transform, in Chapter 13.

Although T3D provides a wide range of options for symbolizing true 3-D phenomena, it is not exhaustive. One interesting possibility is the **tiny cubes method** of Nielson and Hamann (1990). In this approach a set of small cubes are regularly spaced throughout the 3-D surface; the cubes are then colored as a function of the data associated with their location (Color Plate 11.4). Keller and Keller (1993, 148) provide some suggestions on how these cubes might be used.

11.5.2 Waisel's Use of DX to Symbolize Sediment Deposits in New York Harbor

This section considers Waisel's (1996) use of the IBM Visualization Data Explorer (DX) to examine sediment chemistry in New York Harbor. DX is a form of **visualization software** that can be used to create visualizations for a wide variety of spatial data (true 3-D data being just one of them). We will consider the nature of such visualization software in greater depth in Chapter 15.

Visualizing sediment chemistry in New York Harbor is an important consideration because the harbor must be dredged periodically in order to maintain a

* Slicer Dicer, marketed by Visualogic, is another package for symbolizing true 3-D phenomena.

sufficient depth for shipping traffic. If scientists are able to visualize the level of contamination in the sediment, they may be able to decide whether costly decontamination efforts are necessary for a particular dredging operation. To visualize the sediment chemistry, Waisel used DX to map core samples taken from the bottom of the harbor. For this purpose, each core sample was partitioned into sections, and each of these sections mixed to create a homogeneous sample. These homogeneous samples were then tested for levels of potential contaminants.

Although a set of continuous 3-D data could have been interpolated from the homogeneous sample points (the result would be the sort of data used above for T3D), Waisel chose to map individual core samples and the homogeneous segments within those cores. An example of a typical map created by Waisel is shown in Color Plate 11.5. Each cylinder in the map represents one of the homogeneous samples collected within a soil core. Note that the cylinders have been color-coded based on National Oceanic and Atmospheric Administration (NOAA) guidelines for "Effects Range Low" (ERL) and "Effects Range Medium" (ERM): blue cylinders are below ERL; greenish-yellow cylinders are above ERL and below ERM; and red cylinders are above ERM.

Waisel used DX to examine sediment chemistry because of the considerable flexibility its visual programming environment provides (see Chapter 15 for a discussion of visual programming). Although Waisel's work provides a good example of the interesting kinds of maps that can be developed with such software, based on her description of the technical challenges and my own experience working with DX, it is apparent that creating such maps is more complex than using nonprogramming environments, such as the one T3D provides.

SUMMARY

In this chapter, we have examined several novel univariate mapping methods developed in recent years. First, we considered the **framed-rectangle symbol,** a point symbol created by filling a rectangular frame in proportion to the data being mapped (the approach is analogous to changing the height of a column of liquid in a thermometer). The framed-rectangle symbol has been promoted as an alternative to choropleth symbology because it eases the extraction of specific information. A disadvantage, however, is that general information is difficult to acquire because the point symbols do not match the areal extent of enumeration units.

Second, we considered a set of models that MacEachren and DiBiase have developed for representing geographic phenomena, along with a corresponding set of symbolization methods. The models form a matrix that spans the discrete-continuous and abruptness-smoothness continua introduced in Chapter 2. MacEachren and DiBiase argue that if a cartographer can determine an appropriate model for a geographic phenomenon being mapped, then an appropriate symbolization method can be chosen. For example, average farm size is arguably continuous and smooth, so an isarithmic map is appropriate. MacEachren and DiBiase also proposed a novel form of symbolization, the **chorodot map,** for a phenomenon falling in the middle of their model matrix. The chorodot map combines the features of dot and choropleth maps by shading small squares ("dots") with different intensities.

Third, we examined Dorling's novel method for creating cartograms. In this method, circles are initially constructed in proportion to the population of each enumeration unit. These circles are then gradually moved apart (to avoid overlap), while attempting to maintain the contiguity of enumeration units. Finally, the resulting circles are shaded with another variable of interest, for example the percent of the population involved in the service sector of the economy. Dorling contends that the resulting map more properly reflects the human geography of a region because of its focus on people; this is in contrast to traditional equal-area projections (and the associated choropleth map), which he argues reflect the physical extent of the region.

Fourth, we covered some of the software that has been developed for mapping migration flows and continuous vector-based flows. Tobler developed the first significant piece of software for displaying migration flows. He showed that the more important migration movements could be focused on by displaying only those movements above the mean of the data, and he developed an ingenious method for directing migration flows to reflect the route over which people were presumed to migrate. A limitation of Tobler's software was the lack of user interaction, but recently developed software such as Tang's Visda and Yadav-Pauletti's MigMap is beginning to handle this limitation.

Lavin and Cerveny's **unit-vector density mapping** provides a solution to mapping continuous vector-based flows (such as wind speed and direction). Unit-vector density mapping involves changing the density and orientation of short fixed-length line segments called unit vectors. The density of the vectors is a function of the magnitude of the vector-based flow (for example, a higher wind speed would produce a higher density), while the orientation is a function of the direction of the flow (a wind from the north produces a north-south-oriented vector). Lavin and Cerveny suggested that unit-vector density symbology provided more information than traditional wind arrow symbology because users would be able to readily interpret both wind speed and direction.

Finally, we investigated software that can assist in interpreting true 3-D phenomena (such as water temperature in the ocean). T3D is an example of software that has been developed for examining a broad range of true 3-D phenomena. T3D permits users to "peer" into true 3-D data by varying its transparency (or alternatively its opacity), slicing through the data, and viewing an **isosurface** (the three-dimensional equivalent of a contour line). T3D also permits users to rotate and animate a true 3-D data set, and to experiment with different color schemes. We will deal with the latter in Chapter 13 when we consider a companion product Transform.

The IBM Visualization Data Explorer (DX) is an example of **visualization software,** which can be used to visualize a wide variety of spatial phenomena. With respect to true 3-D phenomena, we illustrated how Waisel used DX to evaluate the chemical composition of sediment deposits in New York Harbor. We will consider the nature of visualization software and associated visual programming in greater depth in Chapter 15.

FURTHER READING

Cleveland, W. S., and McGill, R. (1984) "Graphical perception: Theory, experimentation, and application to the development of graphical methods." *Journal of the American Statistical Association* 79, no. 387:531–554.

Describes a variety of graphical and mapping techniques (one of which is the framed-rectangle symbol introduced in this chapter) and experiments to evaluate their effectiveness.

Cuff, D. J., Pawling, J. W., and Blair, E. T. (1984) "Nested value-by-area cartograms for symbolizing land use and other proportions." *Cartographica* 21, no. 4:1–8.

Introduces a novel mapping technique in which nested cartograms are centered in enumeration units and scaled proportional to the percentage of various land uses.

Dent, B. D. (1996) *Cartography: Thematic Map Design.* 4th ed. Dubuque, Iowa: William C. Brown.

Chapter 10 provides an overview of cartograms.

Dorling, D. (1993) "Map design for census mapping." *The Cartographic Journal* 30, no. 2:167–183.

A thorough discussion of Dorling's novel cartogram method. For further discussion, see Dorling (1994, 1995b, 1995c). For an interesting atlas comprised of cartograms, see Dorling (1995a).

Dougenik, J. A., Chrisman, N. R., and Niemeyer, D. R. (1985) "An algorithm to construct continuous area cartograms." *The Professional Geographer* 37, no. 1:75–81.

Describes an algorithm for a contiguous-area cartogram. In contrast to Dorling's approach, Dougenik et al.'s tends to preserve the shape of enumeration units.

Gould, P. (1993) *The Slow Plague.* Cambridge, Mass.: Blackwell.

Discusses the AIDS pandemic and methods for mapping it (see especially Chapter 9).

Gusein-Zade, S. M., and Tikunov, V. S. (1993) "A new technique for constructing continuous cartograms." *Cartography and Geographic Information Systems* 20, no. 3:167–173.

Describes an algorithm for constructing a contiguous-area cartogram that tends to preserve shape. The authors argue that in comparison to earlier algorithms (for example, Dougenik et al.), this one "obtains better resultant images, especially in cases where a sharp difference in the initial distribution of the variable exists" (p. 172).

Jackel, C. B. (1997) "Using ArcView to create contiguous and noncontiguous area cartograms." *Cartography and Geographic Information Systems* 24, no. 2:101–109.

Presents Avenue code for contiguous- and noncontiguous-area cartograms. (Avenue is a programming language used in association with ArcView.)

Lavin, S. J., and Cerveny, R. S. (1987) "Unit-vector density mapping." *The Cartographic Journal* 24, no. 2:131–141.

Provides a detailed discussion of unit-vector density mapping.

Lee, J., Chen, L., and Shaw, Shih-L. (1994) "A method for the exploratory analysis of airline networks." *The Professional Geographer* 46, no. 4:468–477.

Describes data exploration software for analyzing airline network flows.

MacEachren, A. M., and DiBiase, D. (1991) "Animated maps of aggregate data: Conceptual and practical problems." *Cartography and Geographic Information Systems* 18, no. 4:221–229.

The initial article in which MacEachren and DiBiase introduced various models of geographic phenomenon. For further discussion, see MacEachren (1994a, 57–64).

Marble, D. F., Gou, Z., Liu, L., and Saunders, J. (1997) "Recent advances in the exploratory analysis of interregional flows in space and time." In *Innovations in GIS 4*, ed. Z. Kemp, pp. 75–88. Bristol, Penn.: Taylor & Francis.

Describes prototypical data exploration software that has been developed at Ohio State University for analyzing interregional flows in both space and time. The software has the ability to map more than one variable simultaneously, which is accomplished using "projection pursuit" concepts.

Nielson, G. M., and Hamann, B. (1990) "Techniques for the interactive visualization of volumetric data." *Proceedings Visualization '90,* San Francisco, Calif., 45–50.

Describes various techniques for visualizing true 3-D phenomena.

Parks, M. J. (1987) "American flow mapping: A survey of the flow maps found in twentieth-century geography textbooks, including a classification of the various flow map designs." Unpublished MA thesis, Georgia State University, Atlanta, Georgia.

An extensive survey of various kinds of flow mapping.

Rittschof, K. A., Stock, W. A., Kulhavy, R. W., Verdi, M. P., and Johnson, J. T. (1996) "Learning from cartograms: The effects of region familiarity." *Journal of Geography* 95, no. 2:50–58.

An experimental study of the effectiveness of cartograms. Also see Griffin (1983).

Thompson, W., and Lavin, S. (1996) "Automatic generation of animated migration maps." *Cartographica* 33, no. 2:17–28.

Summarizes methods for mapping migration and develops an animated method in which either small arrows or circles appear to move from one region to another.

Tikunov, V. S. (1988) "Anamorphated cartographic images: Historical outline and construction techniques." *Cartography* 17, no. 1:1–8.

Provides a history of cartograms.

Tobler, W. R. (1987) "Experiments in migration mapping by computer." *The American Cartographer* 14, no. 2:155–163.

Describes various techniques for mapping migration.

12

Bivariate and Multivariate Mapping

OVERVIEW

Up to this point, this text has focused on the display of individual variables, or univariate mapping. Frequently, however, mapmakers need to display multiple variables. For example, a climatologist might wish to simultaneously view temperature, precipitation, pressure, and cloud cover for a geographic region. The cartographic display of such phenomena is known as ***multivariate mapping.*** *If only two variables are to be displayed, the process is termed* ***bivariate mapping.*** *Bivariate mapping is covered in section 12.1 of this chapter, while multivariate mapping is covered in section 12.2.*

A fundamental issue in both bivariate and multivariate mapping is whether individual maps should be shown for each variable (maps are compared) or whether all variables should be displayed on the same map (maps are combined). Thus, the bivariate and multivariate sections of this chapter are divided into two subsections: one for map comparison and one for map combination. The map comparison section for bivariate mapping focuses on selecting an appropriate method of classification when comparing choropleth maps. Although choropleth maps have their limitations, they are commonly used in bivariate mapping, just as in univariate mapping.

The map combination section for bivariate mapping describes how choropleth maps can be overlain to create a ***bivariate choropleth map.*** *Other symbols considered in this section include the* ***rectangular-point symbol,*** *in which* the *width and height of a rectangle are made proportional to values of two variables being mapped, and the* ***bivariate ray-glyph,*** *in which rays (straight-line segments) pointing to the left and right of a small central circle are used to represent two variables.*

The multivariate mapping section begins by considering ***small multiples****—the simultaneous comparison of three or more maps. The difficulty of synthesizing information depicted via small multiples has led to the development of numerous approaches for combining variables onto one map. Techniques discussed here include the* trivariate choropleth map *(three choropleth maps are overlaid),* the multivariate dot map *(different colored dots are used to represent multiple phenomena), and* multivariate point symbols *(a good example is the Chernoff face, in which various facial features are used to represent multiple variables).*

Although it is possible to combine a large number of variables on a single map, it is often difficult to visually interpret the resulting symbols. A solution to this problem is to explore multivariate data in an interactive graphics environment. SLCViewer, a software package developed at the Deasy Geographics Laboratory at Penn State University, provided this sort of capability by allowing users to view data as small multiples or to combine up to three variables using proportional point symbols, ***weighted isolines*** *(isolines of varying width), and area shading. Software described elsewhere in this text that also provides exploratory capability for multivariate phenomena includes Visda, Vis5D, and Project Argus.*

It should be noted that this chapter considers only methods for symbolizing multivariate data. An alternative is to manipulate the data statistically prior to symbolizing it. For example, ***cluster analysis*** *(Romesburg 1984) can be used to group observations (say, counties) based on their scores on a set of variables. A brief introduction to the use of cluster analysis and other statistical approaches for manipulating multivariate data in cartography is provided by Chang (1982).*

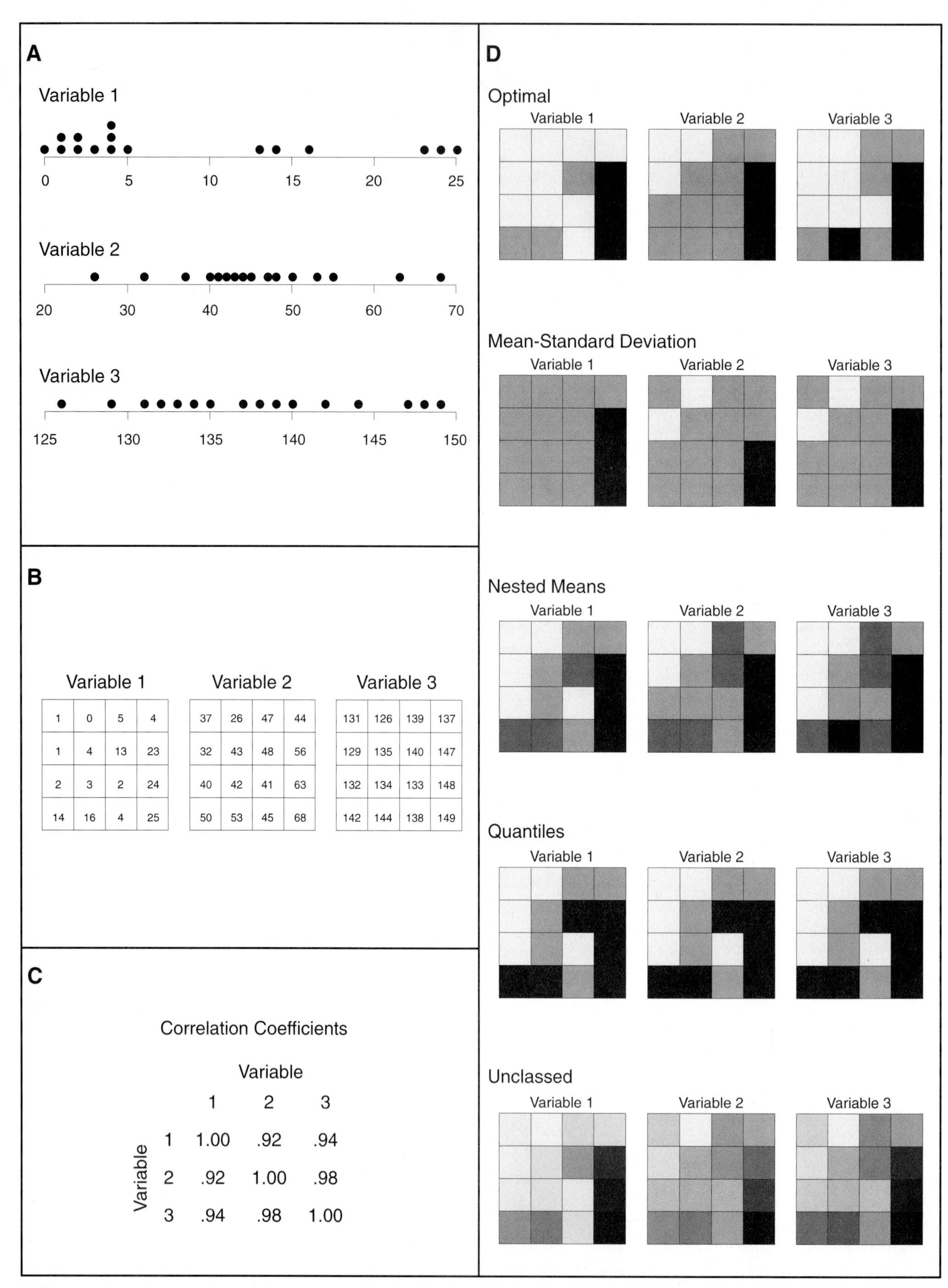

FIGURE 12.1. Using choropleth maps to compare geographic distributions: (A) three hypothetical variables (variable 1 is positively skewed, while variables 2 and 3 are normal); (B) maps of the raw data for the three variables; (C) correlation coefficients (r values) between each pair of variables; (D) maps for differing methods of classification.

12.1 BIVARIATE MAPPING

12.1.1 Comparing Maps

Choropleth Maps

Traditionally, choropleth maps have been the symbolization method commonly used to examine the relationship between two variables. As when creating individual choropleth maps, an important consideration in comparing choropleth maps is deciding whether the data should be classed, and if so which method of classification should be used. To begin, we'll assume that we wish to class the data, and thus focus on the method of classification. Initially, we will also assume that we wish to compare two variables for a single point in time (say, median income and percent of the adult population with a college education, both collected for 1980). Later, we will consider comparing data for the same variable collected over two time periods.

In selecting a method of classification, it is critical to consider the distribution of each variable along the number line. If the variables have differing distributions (for example, if one is skewed and the other is normal), certain classification methods can lead to an inappropriate visual impression of correlation between the variables. To illustrate, consider the hypothetical distributions shown in Figure 12.1A. Variable 1 is clearly positively skewed, while variables 2 and 3 appear to have normal distributions.* In Figure 12.1B, values of these variables have been assigned spatially so that extremely high correlations result in each case (the correlation coefficients, r, appear in Figure 12.1C).†

Recalling from Chapter 4 that the optimal method of classification is often recommended because it minimizes classification error, it seems natural to ask whether the optimal method might also be used for map comparison. If the optimal method is applied to all three variables shown in Figure 12.1B, we obtain the maps shown in Figure 12.1D. (Table 12.1 specifies the limits of each class.) Although these maps suggest positive associations between each pair of variables, they do not support the high correlation coefficients shown in Figure 12.1C (remember that 1.0 is the maximum possible value for r). The lack of a strong visual association between the skewed and normal distributions may not be surprising (compare the optimal maps for variables 1 and 2), but note that there is also a lack of visual association between maps of the normal distributions (variables 2 and 3). The optimal method fails to reflect the high correlations between the variables because it focuses on the precise distribution of the individual variables along the number line.

TABLE 12.1. Class Limits for Each of the Classification Methods used in Figure 12.1.

Optimal

Class	Variable 1	Variable 2	Variable 3
1	0–5	26–37	126–134
2	13–16	40–53	137–142
3	23–25	56–68	144–149

Mean–Standard Deviation

Class	Variable 1	Variable 2	Variable 3
1	none	26–32	126–129
2	0–16	37–56	131–144
3	23–25	63–68	147–149

Nested Means

Class	Variable 1	Variable 2	Variable 3
1	0–2	26–37	126–132
2	3–5	40–45	133–137
3	13–16	47–53	138–142
4	23–25	56–68	147–149

Quantiles

Class	Variable 1	Variable 2	Variable 3
1	0–2	26–41	126–133
2	3–5	42–47	134–139
3	13–25	48–68	140–149

Classification methods that are more appropriate for map comparison include mean–standard deviation, nested means, quantiles, and equal areas. As described in section 4.1.3, the mean–standard deviation method specifies class limits by repeatedly adding or subtracting the standard deviation from the mean. The results of applying the mean–standard deviation approach to the hypothetical distributions are shown in Figure 12.1D. The visual appearance of these maps supports the high correlation between the normal distributions, but not the high correlations between the skewed and normal distributions. These results are not surprising given the fact that means and standard deviations generally should be used only with normal distributions.

In the **nested-means approach,** the mean of the data is used to divide the data into two classes—values above and below the mean. The resulting classes can be further

* These visual assessments were confirmed by a Shapiro-Wilk test, an inferential test for normality (Stevens 1996, 245).

† Kendall's rank correlations (Burt and Barber 1996, 396–398), which are arguably more appropriate for skewed data, were 0.98 between variables 1 and 2 and 1 and 3 and 1.00 between variables 2 and 3.

subdivided by again computing the mean of each class, and this process can be repeated until no further subdivision is desired. The results of using nested means to create four classes are shown in Figure 12.1D. Like the mean–standard deviation approach, nested means seems to do a better job of portraying the similarity of the normal distributions. This is to be expected, as the mean used to define class breaks is an appropriate measure of central tendency for normal distributions, but not for skewed ones. Another weakness of nested means is that the number of classes can only be a power of 2 (2, 4, 8, 16, etc.); in the present case, this complicates a comparison of this approach with the other methods of classification.

The quantiles method of classification places an approximately equal number of observations in each class based on the ranks of the data (see section 4.1.2). Since the classes resulting from the quantiles method are unaffected by the magnitudes of the data, the method is arguably appropriate for comparing differently shaped distributions. For example, for the hypothetical data, the quantiles method portrays high correlations between not only the normal distributions but also the skewed and normal distributions (Figure 12.1D).

The **equal-areas method** of classification is similar in concept to quantiles, but rather than placing an equal number of observations in classes, an equal portion of the map area is assigned (the desired area in each class is simply the area of the map divided by the number of classes desired). If enumeration units are equal in size (as in the hypothetical data), the equal-area method produces a map identical to quantiles.

Up to this point, we have focused on comparing classed choropleth maps. As with univariate choropleth maps, it is natural to ask whether the issue of selecting an appropriate classification method might be obviated by simply not classing the data. To illustrate, consider the unclassed maps of the hypothetical data shown in Figure 12.1D. Here the visual impression seems similar to that for the mean–standard deviation and nested-means approaches, as variables 2 and 3 (the normal distributions) appear similar to one another, while variable 1 (the skewed distribution) does not appear as highly correlated with the other variables. This result leads one to the conclusion that unclassed maps should be used for comparative purposes only when the distributions have similar shapes.

At the beginning of this section, we assumed that we wished to compare two variables for a single point in time (say, median income and percent of the adult population with a college education for 1980). For such data, we concluded that the optimal method is inappropriate. It should be noted, however, that if we wish to compare the same variable for two points in time (say, median income for 1980 and 1990), the data could be combined into a single data set, and the optimal method could be applied to that set.

One limitation of this section is that a number of the conclusions are based on a subjective interpretation of only those variables shown in Figure 12.1. Thus, it is useful to consider some of the more formal studies done by cartographers, including those by Lloyd and Steinke, Olson, Peterson, and Muller. Lloyd and Steinke (1976, 1977) found that the visual correlation of maps is affected by the amount of blackness on each map (assuming that gray tones are used for symbolization); in other words, if maps A and B and C and D have the same statistical correlation, maps A and B will be judged more similar if their blackness levels are more similar than for maps C and D. As a result, Lloyd and Steinke argued for using equal areas (which I have indicated produces results similar to quantiles) when comparing choropleth maps.

Olson undertook two studies to determine which classification methods appeared to most accurately preserve the correlation between two variables. In her first study (Olson 1972b), she analyzed 300 pairs of variables derived from theoretically normal distributions and found that quantiles was best at reflecting the correlation between the variables. In her second study (Olson 1972a), she analyzed 300 pairs of real-world variables (which were primarily not normally distributed) and found that the mean–standard deviation and nested-means methods were most effective. These results conflict with some of my own conclusions because of Olson's stress on broad relationships (she focused on a scatterplot of the correlation coefficient and a measure of rank correlation for all pairs of variables), as opposed to looking at detailed graphs and maps of each distribution, as was done in Figure 12.1.

Studies by Peterson (1979) and Muller (1980b) dealt with people's visual comparison of both classed and unclassed choropleth maps. The studies concluded that people perceived similar correlations on pairs of classed and unclassed maps; as a result, the authors raised questions about the need to class data for choropleth mapping. Although the results of these studies contradict my recommendation to use unclassed maps only when comparing distributions having the same shape, I suspect that a detailed examination of data distributions would support my recommendation.

Miscellaneous Thematic Maps

Although it is common to hold the mapping method constant when comparing maps (for example, show two choropleth maps or two isopleth maps), useful information can often be acquired by comparing two different kinds of thematic maps. This is especially true when one map is used to show raw counts and another standardized data. For example, consider Figure 12.2, which compares a proportional symbol map of the raw number of infant mortalities in New Jersey with a choropleth map of the infant mortality rate. The proportional symbol map suggests that the "problem" of infant mortality is in the northeastern part of the state (where the largest circles are). This map by itself, however, is not very meaningful because the pattern is likely a function of population (counties with more people are apt to have more infant deaths). In contrast, the choropleth map standardizes the raw mortality data by considering the number of deaths relative to the number of live births, and suggests that the "problem" is found in three areas of the state (represented by the shades in the highest class). Unfortunately, a high rate on the choropleth map may not be meaningful if there is a low number of deaths. Only when the two maps are viewed together can the complete picture emerge: the northernmost high-rate area is most problematic because it is located where a high number of deaths also occurs.

12.1.2 Combining Two Variables on the Same Map

Bivariate Choropleth Maps

As with comparing maps, fundamental work on combining two variables on the same map has been accomplished using the choropleth method. The earliest efforts were undertaken by the U.S. Bureau of the Census in the 1970s, and involved combining two colored choropleth maps to create a **bivariate choropleth map** (see Meyer et al., 1975, for a summary of the technical aspects used at that time). Color Plate 12.1 depicts a bivariate choropleth map using a color scheme similar to one used by the Bureau of the Census, along with each of the univariate distributions used to create the bivariate map.* A listing of the factors considered in constructing the Bureau of Census color schemes is given in Table 12.2.

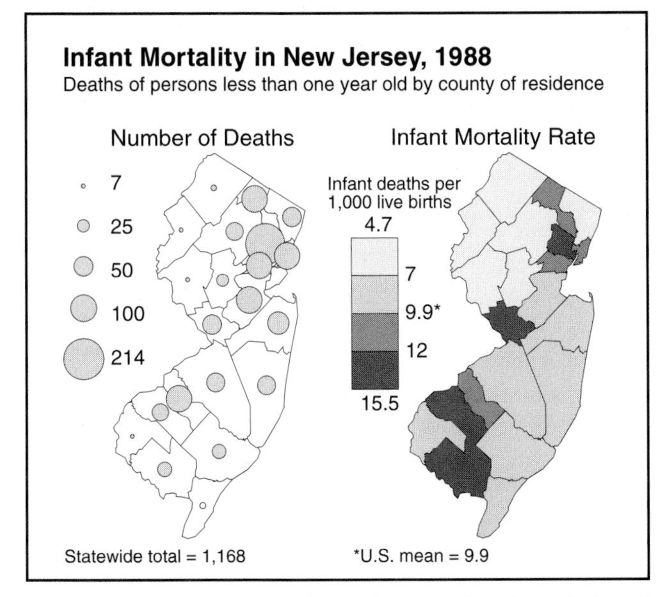

FIGURE 12.2. A comparison of proportional symbol and choropleth maps. The proportional symbol map provides information on the raw number of infant mortalities, while the choropleth map focuses on the rate of infant mortality. To understand the complete picture of infant mortality, both maps are necessary. (From Monmonier, p. 166; © 1993 by The University of Chicago Press. All rights reserved. Published 1993.)

Although bivariate choropleth maps were deemed a success from a technical standpoint, they received considerable criticism for their presumed failure to communicate either information about individual distributions or the correlation between them. In response to these criticisms, Olson (1981) conducted an experimental study using color schemes similar to those used by the Bureau of the Census. In contrast to the earlier criticism, Olson found that bivariate maps provided "information about regions of homogeneous value combinations" and that users, rather than being confused by these maps, actually had a positive attitude about them (p. 275). Olson did, however, stress that a clear legend is critical to understanding bivariate maps, both bivariate and individual maps should be shown,† and an explanatory note should be included describing the types of information that can be extracted (see p. 273 of her work for an example).

As an alternative to the Bureau of the Census colors, Eyton (1984a) developed a bivariate method based on **complementary colors,** or colors that combine to produce a shade of gray.‡ Eyton chose red and cyan as the complementary colors for his maps, presumably because they produced an attractive map. Using the CMYK system, red was generated by combining magenta (M) and yellow (Y), while cyan (C) was created directly from the cyan process color. Since overprinting colors in CMYK does not produce a true gray, Eyton used black ink (K) for areas in which a true gray was desired. A bivariate map resulting from Eyton's process is shown in Color Plate 12.2.§

Eyton argued that most of the factors listed in Table 12.2 were accounted for by his complementary method and that users appeared to understand the map more easily than one based on Bureau of the Census colors. A

* Specifications for the colors used in Color Plate 12.1 were taken from Olson (1981, 269). For a discussion of the logic behind the selection of these colors, see Olson (1975).

† The univariate maps should be shown in black-and-white so that the two distributions can be readily compared. In Color Plate 12.1 they are shown in color so that the colors composing the bivariate map are clear.

‡ Colors opposite one another on the Munsell color circle (Figure 5.18) are complementary.

§ Eyton used cross-hatching (narrowly spaced horizontal and vertical lines) for intermediate shades, while I have chosen smooth tones.

TABLE 12.2. Summary of Factors Considered in Developing Color Schemes for the U.S. Bureau of the Census Bivariate Choropleth Maps

1. All colors must be distinguishable.
2. The transition of colors should progress smoothly in a visually coherent way.
3. The individual categories of each distribution should be visually distinguishable or coherent, and the two distributions as a whole should be separable from one another.
4. The arrangement of the colors presented in the legend should correspond to the arrangement of a scatter diagram.
5. Tones should progress from lighter to darker corresponding to a change in the numerical values from low to high.
6. Extreme values (legend corners) should be represented by pure colors.
7. There should be coherence in the triangle of cells above and below the main diagonals to show positive and negative residuals.
8. To convey relationship, positive diagonals (lower left to upper right) and negative diagonals (upper left to lower right) should have visual coherence.
9. The design of the color coding scheme should take into account the difficulty in mentally sorting large numbers of colors in the legend.
10. The color scheme should relate to the data in such a way that the map relationship reflects as closely as possible the statistical relationship.
11. The crossed version of the map should be constructed as a direct combination of the specific sets of colors assigned to the two individual maps.
12. The combination of colors on the two individual maps should look like combinations of the specific colors involved.
13. The number of categories to be used should not exceed the number that can be dealt with by the reader. A 3 × 3 legend is both mechanically and visually simpler than a 4 × 4 arrangement and may actually convey more to the reader.
14. Alternatives to a rectangular arrangement to the legend should be considered. The rectangular form creates map interpretation problems and affects the message of the statistical relationship.

After Olson, 1975b, as specified by Eyton, 1984a, 480; Courtesy of Association of American Geographers.

visual comparison of Color Plates 12.1 and 12.2 suggests that Eyton was correct. The Bureau of the Census color scheme implies nominal differences and requires careful examination of the legend, while Eyton's scheme appears more logically ordered and allows patterns on the map to be discerned more easily. Note, for example, the ease with which the reddish-brown values (values above the white-gray-black diagonal) can be found using the Eyton scheme, as compared with corresponding values using the Bureau of the Census color scheme.

In addition to suggesting that complementary colors be used, Eyton also recommended that the statistical parameters of the distributions should be considered in bivariate mapping. Specifically, Eyton made use of the reduced major axis and bivariate normality. Recall from section 3.3.2 that the reduced major-axis method fits a regression line to a set of data such that the line bisects the regression lines of Y on X (Y is treated as the dependent variable) and X on Y (X is treated as the dependent variable) (Figure 12.3). The reduced major axis is thus appropriate when it is not clear which variable should be treated as the dependent variable.

A distribution is considered **bivariate-normal** if the Y values associated with a given X value are normal and the X values associated with a given Y value are also normal. Generally, this condition is met if the individual distributions of X and Y are normal. Given a bivariate-normal distribution, it is possible to construct an **equiprobability ellipse** enclosing a specified percentage of the data. The equiprobability ellipse will be centered on the mean of the data and oriented in the direction of the reduced major axis (Figure 12.3). Using this ellipse and the means of the two variables, Eyton created a bivariate map like the one shown in Color Plate 12.3.* Eyton argued that the resulting map clearly contrasted observations near the means of the data (as defined by the 50 percent equiprobability ellipse) with extreme observations (those well above the mean on both variables, those well below the mean on both variables, and those

* Given the requirement of normality, I chose to map two normal distributions: percent urban and life expectancy for females.

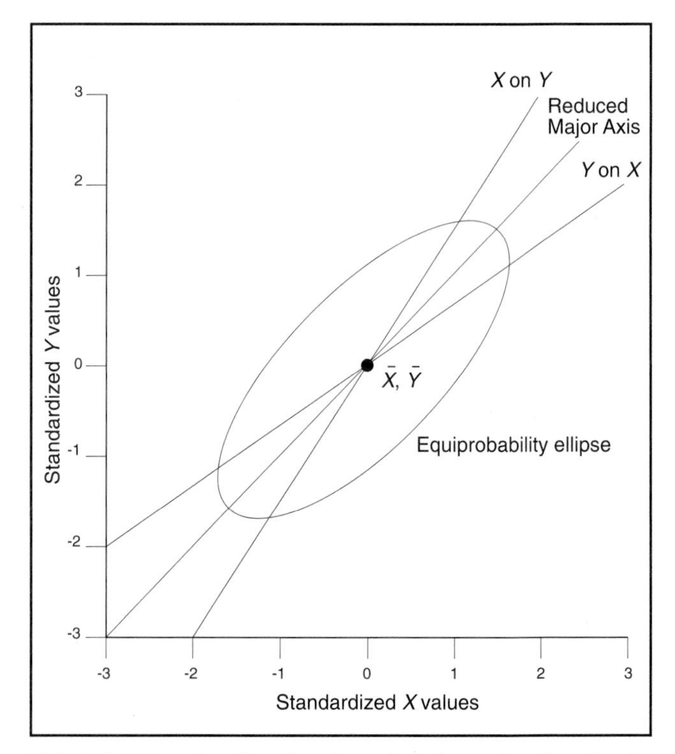

FIGURE 12.3. A reduced major axis and associated equiprobability ellipse. The reduced major axis is appropriate when one does not wish to specify a dependent variable. The equiprobability ellipse can be used to enclose a specified percentage of the data associated with a bivariate normal distribution.

high on one variable and low on the other). For example, in Color Plate 12.3 we can clearly see that the countries of Botswana, Zimbabwe, Lesotho, and Kenya (those shown in cyan) have a low percent urban and high female life expectancy.

More recently, Brewer (1994a) argued that color schemes for bivariate maps should be a function of whether the variables are unipolar or bipolar in nature. For two unipolar data sets (such as percent urban and female life expectancy), she recommended using either a complementary scheme or two subtractive primaries; the latter produces a diagonal of shades of a constant hue, as opposed to the grays resulting from complementary colors. (Brewer combined magenta and cyan to create a diagonal of purple-blues.) If one variable is unipolar and the other bipolar, she recommended using a sequential scheme for the unipolar data and a diverging scheme for the bipolar data, while if both are bipolar she recommended two diverging color schemes (Color Plate 12.4).

The approaches we have considered to this point for bivariate choropleth mapping have all been based on smooth (untextured) colors. As an alternative, both Carstensen (1982, 1986a, 1986b) and Lavin and Archer (1984) created bivariate maps using cross-hatched shading consisting of horizontal and vertical lines (Figure 12.4). Interpretation of these maps focuses on the size and shape of the boxes formed by the cross-hatched lines (low values on both variables are represented by large squares, while high values on both variables are represented by small squares). A high positive correlation is represented on the map by a predominance of squares of varying sizes within enumeration units, while a high negative correlation is depicted by rectangles.

Both Carstensen and Lavin and Archer stressed that each variable on cross-hatched bivariate maps should be depicted as unclassed (thus the legend in Figure 12.4 contains no class breaks). This suggestion contrasts with that

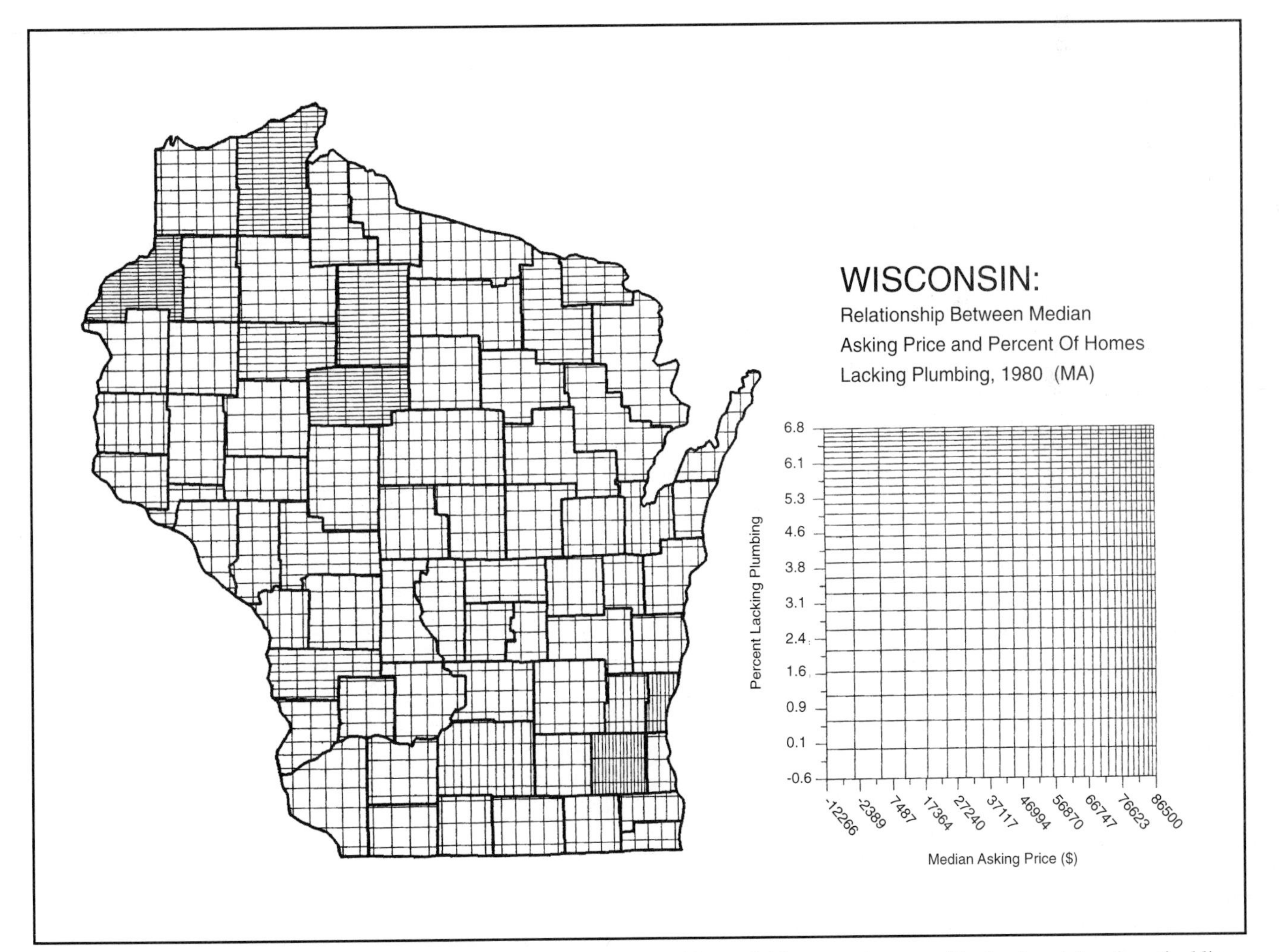

FIGURE 12.4. A bivariate choropleth map based on cross-hatching (the variables are represented by horizontal and vertical lines of varying spacing). Note that each variable is unclassed; negative values in the legend are a function of the major axis scaling used to fit the bivariate data. (From Carstensen 1986a, p. 36; courtesy of Laurence W. Carstensen.)

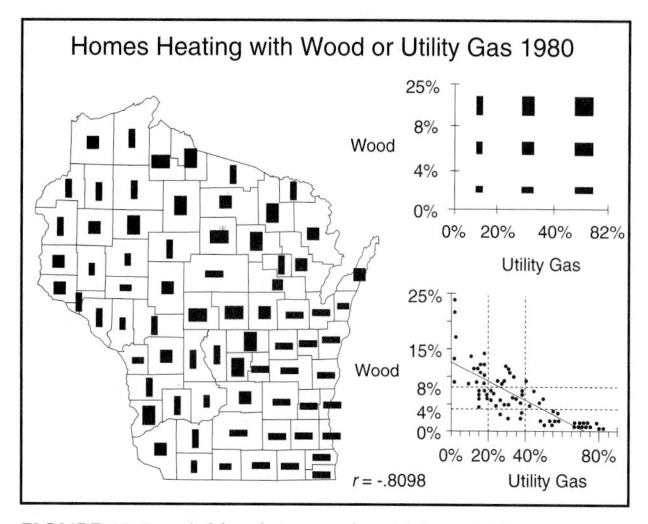

FIGURE 12.5. A bivariate map in which variables are represented by the width and height of a rectangular point symbol. (Courtesy of Sean Hartnett.)

of Eyton (1984a, 485–486), who found smooth-toned unclassed bivariate maps difficult to interpret. The reason that unclassed maps are more easily used with cross-hatching is that the line-spacing variable allows individual variables to be seen.

Carstensen (1982) found that cross-hatched bivariate maps were reasonably effective in communicating concepts about correlation, but noted two problems: the difficulty of shading small enumeration units and the unpleasant appearance of the symbology. These problems cause this method to be less frequently used than the color methods mentioned above.

Other Bivariate Maps

The preceding section focused on the use of choropleth symbology to depict two variables on the same map. An alternative is to use a **bivariate point symbol.** Figure 12.5 depicts one form of bivariate point symbol, the **rectangular point symbol,** in which the width and height of a rectangle are made proportional to each of the variables being mapped. Hartnett (1987) proposed this method as an alternative to cross-hatched symbology, arguing that examining rectangular point symbols is much easier than inspecting the small boxes formed by cross-hatched lines and that the resulting map is more aesthetically pleasing.

One problem with Hartnett's use of rectangular point symbols is that he applied them to standardized data associated with enumeration units. As indicated in Chapter 7, cartographers normally reserve point symbols for unstandardized (count) data associated with enumeration units. Rectangular point symbols, however, would be particularly appropriate for mapping true point data for two variables (say, the number of out-of-wedlock births to teenagers by city and number of hours of sex education for high school students by city).*

Another form of bivariate point symbol is the **bivariate ray-glyph,** shown in Figure 12.6. Carr et al. (1992) used this symbol to examine the relationship between trends in nitrate (NO_3) and sulfate (SO_4) concentrations at selected locations over a six-year period. Rays (the straight-line segments) pointing to the right represent sulfate concentrations, while rays pointing to the left represent nitrate concentrations. High values on both variables occur when both rays extend toward the top of the symbol, while low values on both occur when the rays extend toward the bottom of the symbol. An advantage of the ray-glyph is that the small symbols can be squeezed into a relatively restricted space. It seems likely, however, that the patterns represented by these symbols would be more difficult to interpret than those for the rectangular symbol.

A relatively common approach in bivariate mapping is to combine proportional and choropleth symbols, with the size of the proportional symbol used for count data, and a choropleth shade within the symbol used for standardized data. For example, this approach could be used

* As an alternative to using rectangles, ellipses could be used; for example, see MacEachren (1995, 95).

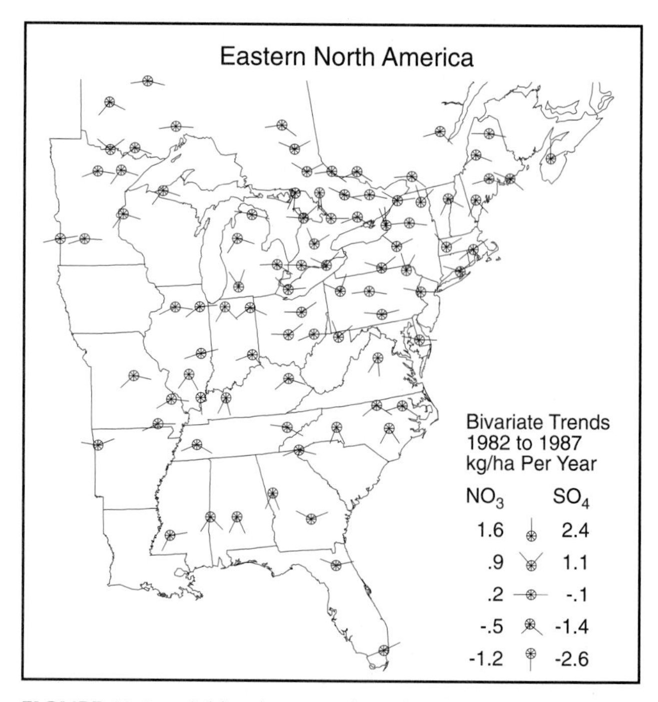

FIGURE 12.6. A bivariate map based on a ray-glyph symbol. Rays (straight-line segments) pointing to the right and left represent sulfate and nitrate concentrations, respectively. (After Carr et al. 1992. First published in *Cartography and Geograpic Information Systems* 19(4), p. 234. Reprinted with permission from the American Congress on Surveying and Mapping.)

to depict on a single map the infant mortality data presented as two maps in Figure 12.2. In section 7.6, I suggested that these same symbols could be used redundantly (both size and shading could be used to represent a single variable). Therefore, to avoid confusing readers, a legend is critical when a symbol could be interpreted in a redundant fashion.

12.2 MULTIVARIATE MAPPING

12.2.1 Comparing Maps

If more than two variables are shown, each as a separate map, the result is termed a **small multiple** (Figure 12.7). Tufte (1990, 33) argues:

> Small multiples, whether tabular or pictorial, move to the heart of visual reasoning—to see, distinguish, choose. . . . Their multiplied smallness enforces local comparisons within our eyespan, relying on the active eye to select and make contrasts.

Although much can be gleaned from small multiples, they clearly have their limitations. A general problem is that comparing two particular points or areas across the whole set of variables can be difficult. Try using Figure 12.7 to describe the nature of agriculture in the state of Michigan! In the case of choropleth maps, a particular problem is the difficulty of discerning small enumeration units (a number of countries would disappear for small multiples of choropleth maps of Africa). Problems such as these have led researchers to develop methods for combining multiple variables onto the same map.

12.2.2 Combining Three or More Variables on the Same Map

Trivariate Choropleth Maps

The notion of overlaying two choropleth maps can be extended to overlaying three choropleth maps, thus producing a **trivariate choropleth map.** This approach should be used only for three variables that add to 100 percent; examples include soil texture (expressed as percent sand, silt, and clay) and voting data for three political parties (for example, percent voting Republican, Democrat, and Independent). Colors can be assigned to the three variables using a variety of approaches: CMY (Brewer 1994a, 142), RGB (Byron 1994), and red, blue, and yellow primaries (Dorling 1993, 172); the result of using RGB is shown in Color Plate 12.5. The advantage of using variables that add to 100 percent is that the resulting colors will be restricted to a triangular two-dimensional space. Brewer (1994a) argues that if the variables do not sum to 100 percent, a three-dimensional cube-shaped legend will be required and the resulting map will be difficult to interpret.

Multivariate Dot Maps

The notion of univariate dot mapping (described in Chapter 10) can be extended to create a **multivariate dot map** if a distinct symbol or color is used for each variable to be mapped.* In the case of color, Jenks (1953) introduced the notion of pointillism for multivariate dot mapping and developed two major maps based on it (Jenks 1961, 1962). Pointillism was a technique used by 19th-century painters to create various color mixtures by having the viewer visually combine very small dots of selected colors. Jenks applied this principle by letting different-colored dots represent various crops (or farm products) (Color Plate 12.6); he argued that viewers would visually merge the separate colors to create mixtures, thus providing a more realistic view of the transitional nature of cropping practices often found in the landscape. Furthermore, Jenks noted that if the dots were made large enough, the map could provide detail regarding the location of individual crops in selected areas.

Jenks (1953) also provided some useful suggestions for color selection:

> 1. Colors should remind the map reader of the crop that they represent.
> 2. High-value, low-acreage crops such as tobacco or truck, should be of more intense hue than the more extensive and widely grown crops.
> 3. Selected minor crops, such as peanuts or soybeans, which tend to change the crop character of broader areas, should have colors of moderately high intensity. (p. 5)

Although considerable information can be obtained from colored multivariate dot maps, readers have difficulty determining the meaning of the areas of color mixture, since the legend contains only colors for individual crops. This problem might be solved if the map were viewed in an interactive data exploration environment in which individual map areas could be focused on and enlarged, and individual categories of dots could be turned on and off so that the relative contribution of each category could be determined.†

A general question that can be asked of multivariate maps is how effective they are compared to a set of individual maps of each variable (that is, how would a multivariate colored dot map compare to a set of dot maps in small multiple format). Rogers and Groop (1981) eval-

* For a discussion of the use of shape on multivariate dot maps, see Turner (1977).

† Groop (1992) has developed an automated method for creating colored dot maps, but unfortunately his method has not been published.

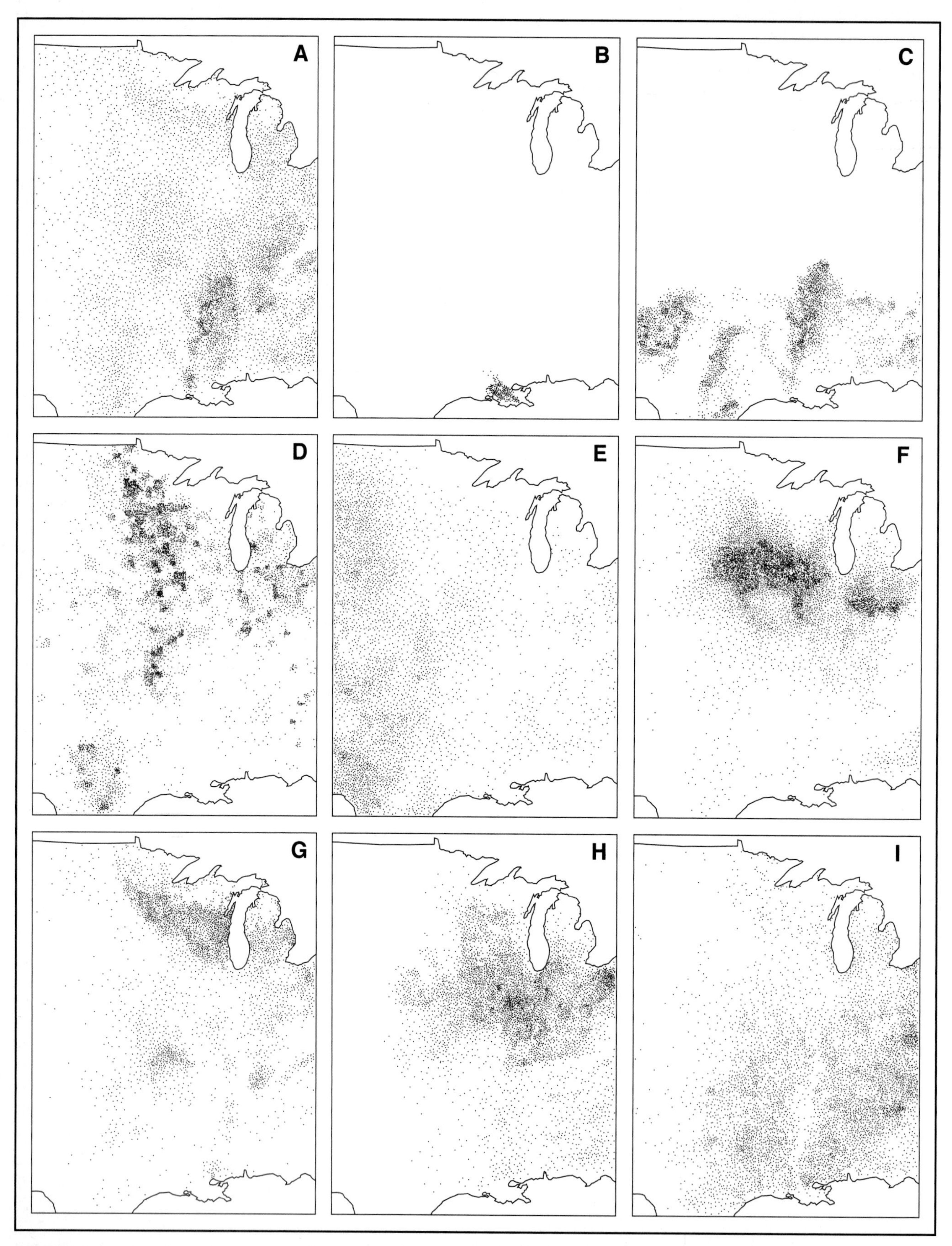

FIGURE 12.7. Small multiples: a method for multivariate mapping in which each variable is depicted as a separate map. Maps depict the following agricultural data for 1954 for the central portion of the United States: A) class V farms, B) sugar cane, C) cotton, D) turkeys, E) pasture, F) hogs and pigs, G) dairy, H) expenditures for lime, and I) residential farms. (After Lindberg 1987; courtesy of Mark B. Lindberg.)

uated this question by having readers identify regions on both a multivariate dot map (consisting of three categories) and its component univariate dot maps. Readers were asked to identify both "homogeneous regions" in which one category appeared to predominate and "mixed regions" in which there appeared to be a mixture of two or three categories. Rogers and Groop found that the multivariate dot map "was slightly more effective in communicating perceptions of both homogeneous and mixed regions" (p. 61). Their results, however, suggested that the perception of mixed regions on maps with more than three categories might not be effective.

Multivariate Point Symbol Maps

When multivariate data are depicted using a point symbol, the result is termed a **multivariate point symbol.** These symbols are obviously appropriate for point phenomena, but must also be used for areal phenomena because of the difficulty of creating multivariate areal symbols (note the limitations of the trivariate choropleth map described above when only three variables are shown). Two distinct forms of variables are commonly mapped using multivariate point symbols: related (or additive) and nonrelated (or nonadditive). *Related variables* are measured in the same units and are part of a larger whole. An example would be the percentages of various racial groups in a population: white, African-American, Asian/Pacific Islander, Native American, etc. Such variables can be depicted using the familiar **pie chart,** in which a circle is divided into sectors representing the proportion of each variable.

Nonrelated variables are measured in dissimilar units and are not part of a larger whole; variables such as percent urban and median income are examples. Multivariate point symbols used to represent such variables are commonly termed **glyphs.** Several examples of glyphs are shown in Figure 12.8. The **multivariate ray-glyph** or **star** (Figure 12.8A) is constructed by extending rays from an interior circle, with the lengths of the rays made proportional to values associated with each variable. As originally designed by Anderson (1960), rays were extended only from the top portion of the circle, but it has become common to extend the rays in all directions.* If the rays composing the multivariate ray-glyph are connected, a **polygonal glyph** or **snowflake** is created (Figure 12.8B).

Cox (1990) and her colleagues at the National Center for Supercomputing Applications (at the University of Illinois at Urbana-Champaign) have developed several novel multivariate point symbols. One of these is **three-dimensional bars,** in which the height of bars is made proportional to the magnitude of variables (Figure 12.8C); in Cox's implementation, the individual bars were shown in different colors, as opposed to using the different textures seen in Figure 12.8C. Another technique is the **data jack** (Ellson 1990), in which triangular spikes are drawn from a square central area and made proportional to the magnitude of each variable (Figure 12.8D). As with bars, the spikes of jacks can be distinguished most easily if they are

* Several terms have been used to describe this symbol. The term *star* comes from Borg and Staufenbiel (1992), while Anderson called them *metroglyphs.* I use the term *multivariate ray-glyph* because of their similarity to the bivariate ray-glyph introduced above.

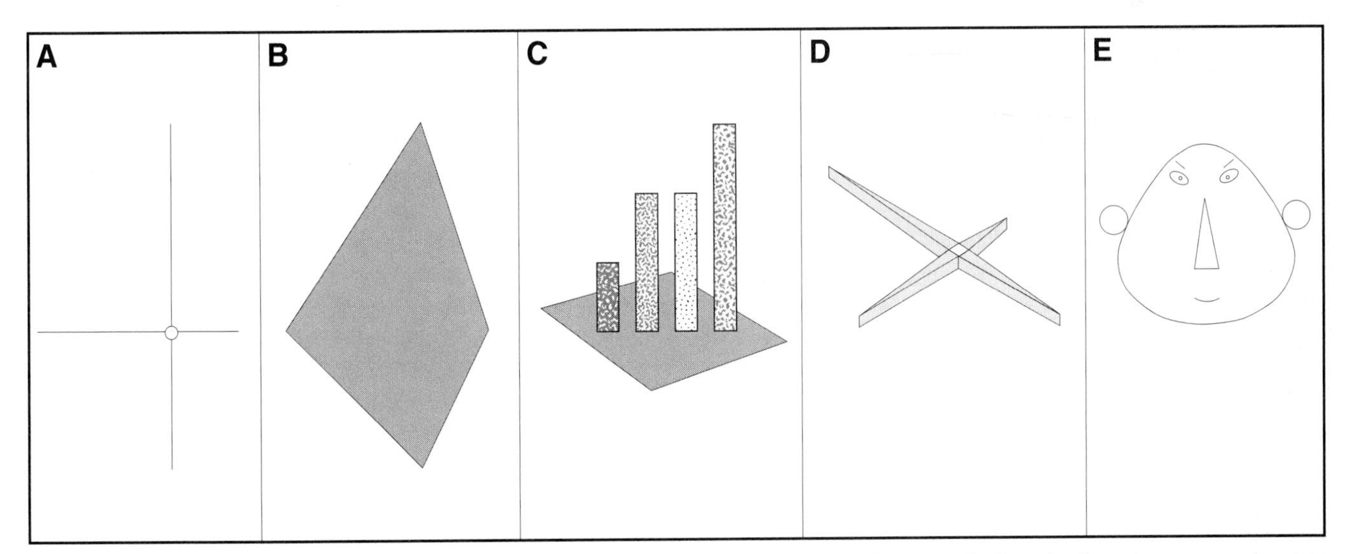

FIGURE 12.8. Examples of multivariate point symbols: (A) a multivariate ray-glyph or star: the length of rays are proportional to the values of variables; (B) a polygonal glyph or snowflake: a polygon connects the end points of the rays shown in A; (C) three-dimensional bars: the height of bars is proportional to the magnitude of variables; (D) data jacks: the spikes of the jack are proportional to the magnitude of each variable; and (E) Chernoff faces: individual facial features (such as the size of the eyes) are associated with individual variables.

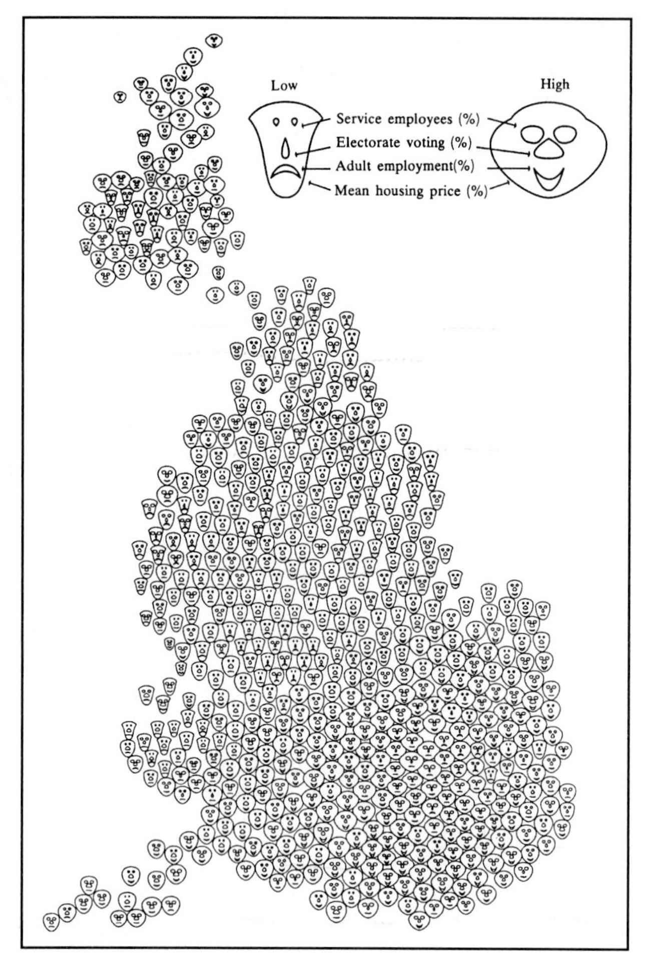

FIGURE 12.9. A cartogram in which Chernoff faces are used to display multivariate data. (Environment and Planning B: Planning and Design 1995, Vol. 22, pp. 269–290, Pion, London.)

displayed in different colors. An advantage of the three-dimensional structure of jacks is that they can be viewed from arbitrary positions in three-dimensional space.

Probably the most intriguing multivariate point symbol is the **Chernoff face** (Figure 12.8E), in which distinct facial features are associated with individual variables. For example, the fatness of the cheeks might represent one variable, while another might be represented by the size of the eyes. Figure 12.9, taken from Dorling's work with cartograms (see section 11.3) is illustrative of the use of Chernoff faces. In this case, the width of the face represents mean house price (fat cheeks indicate more expensive housing); mouth style the percent adult employment (a smile for high employment); nose size the percent voting (a larger nose indicates a higher percent); and eye size and position the percent employed in services (large eyes located near the nose indicate a high percent employed in services). The total area of the face represents the number of voters in a parliamentary constituency.

For sake of comparison, it was assumed that four variables were depicted by the symbols shown in Figure 12.8. Several of these symbols, however, can be modified to represent a considerably greater number of variables. For example, many more rays can be added to the star and up to 20 different variables can be represented by Chernoff faces (see Wang and Lake 1978, 32–35 for a variety of facial features that can be used).

From Chapter 1, you will recall that two kinds of information can be acquired from univariate thematic maps: specific and general. These same kinds of information can also be acquired from a multivariate thematic map, but readers can examine the distribution of the variables individually or holistically (in combination). In the case of multivariate point symbols, Nelson and Gilmartin (1996) summarized these kinds of information as a two-by-two matrix, with specific-general on one axis and nature of the variables (individual or holistic) on the other axis.*

To illustrate the nature of Nelson and Gilmartin's matrix, consider Figure 12.10, which uses combined star and snowflake symbols to depict "quality of life" in South Carolina counties in 1992.† An example of specific information for an individual variable would be the desirability associated with a particular variable within a county (that the southernmost county is highly desirable in terms of variable 2, median income). A holistic form of specific information would involve comparing the size of one symbol with another (noting that the southernmost symbol is larger than the one to its immediate northwest).

An example of general information for an individual variable would involve examining the distribution of a single variable across the state (for example, examining the distribution of the variable infant mortality across the state); note that for the combined star and snowflake symbol this is a difficult task. A form of holistic general information would involve examining the pattern of the size of symbols across the state (note that there appears to be a band of relatively less desirable counties running through the southern part of the state).

The apparent difficulty of visualizing the pattern of an individual variable in Figure 12.10 raises the question of whether other multivariate symbols might be more effective for one form of information than another. In one of the few experimental studies done with multivariate point symbols, Nelson and Gilmartin (1996) evaluated this question. In their experiment, readers examined four types of multivariate point symbols: a modified Chernoff face, a circle divided into quadrants, a cross, and boxed

* Although introduced in a paper dealing with multivariate point symbols, it is apparent that this notion could be applied to other forms of multivariate mapping; for example, the reader should consider applying the concept to a bivariate choropleth map.

† I make no claim that the given variables are ideal for studying quality of life in South Carolina; other individuals might wish to select their own set of variables for mapping.

letters representing variable names (Figure 12.11). Each point symbol represented four variables depicting quality of life on a map consisting of nine enumeration units.

Nelson and Gilmartin found that all symbols were processed equally well if time to examine the map was not a factor. If time was a consideration, then boxed letters were most effective (had the fastest reaction time); these were followed by crosses, divided circles, and Chernoff faces. Nelson and Gilmartin stressed that certain symbols worked better for individual variable questions, while others worked better for holistic questions. In particular, Chernoff faces and letters worked best for individual variables, while crosses and circles were more appropriate for holistic questions.

One problem with such experimental studies is that they often are unable to consider all of the factors that might affect map reader performance. In Nelson and Gilmartin's case, they mentioned aesthetics (for example, that Chernoff faces might be appropriate because of their attention-getting quality) and the number of variables mapped (they noted that other researchers found Chernoff faces effective when displaying a large number of variables) as potential factors to study. Other factors mapmakers should consider include the difficulty of discriminating variables for small point symbols (imagine interpreting the boxed letter for 20 variables), the fact that the results might differ for a larger number of enumeration units (Nelson and

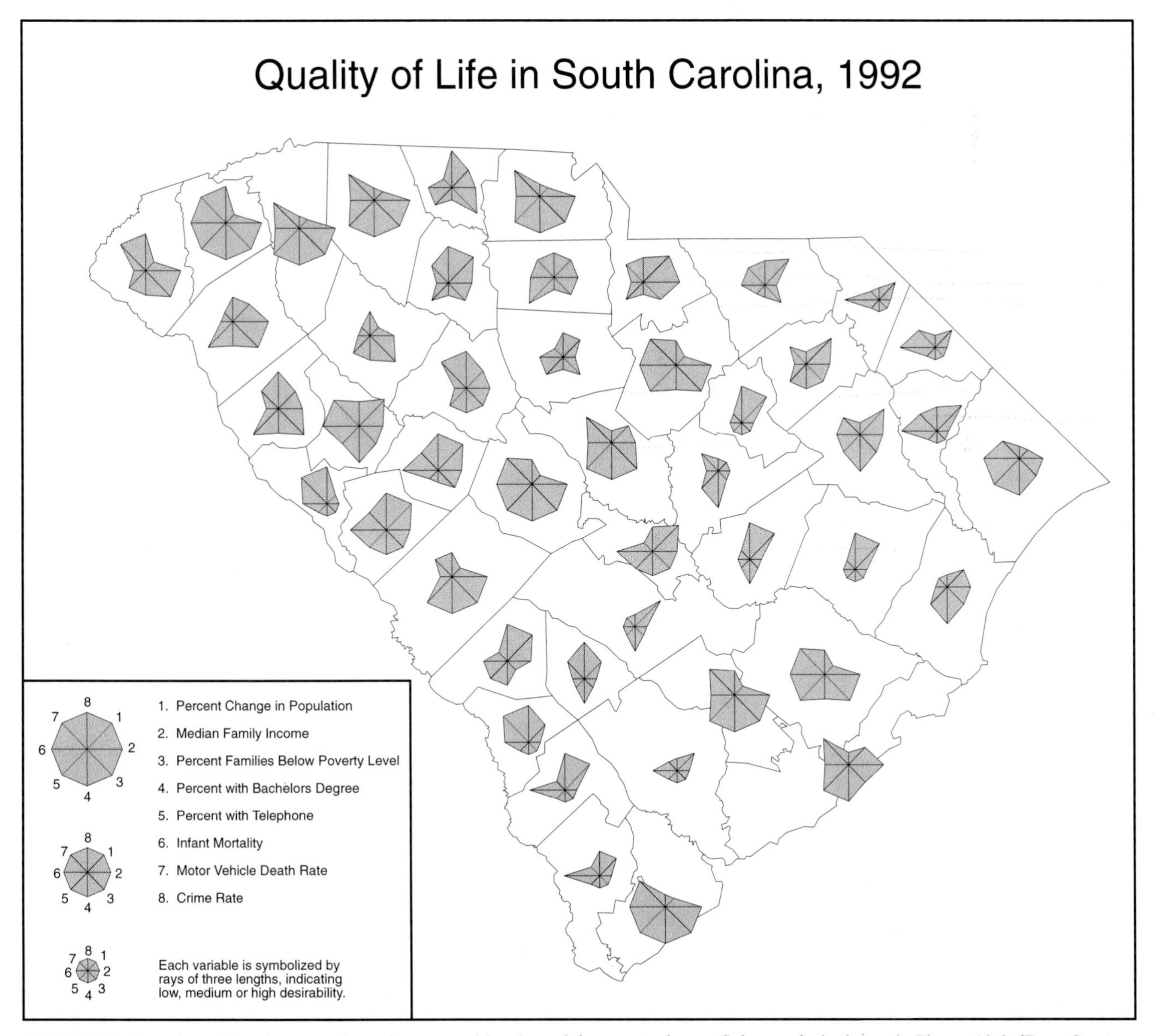

FIGURE 12.10. A multivariate map based on a combination of the star and snowflake symbols shown in Figure 12.8. (Data Source: South Carolina State Budget and Control Board 1994.)

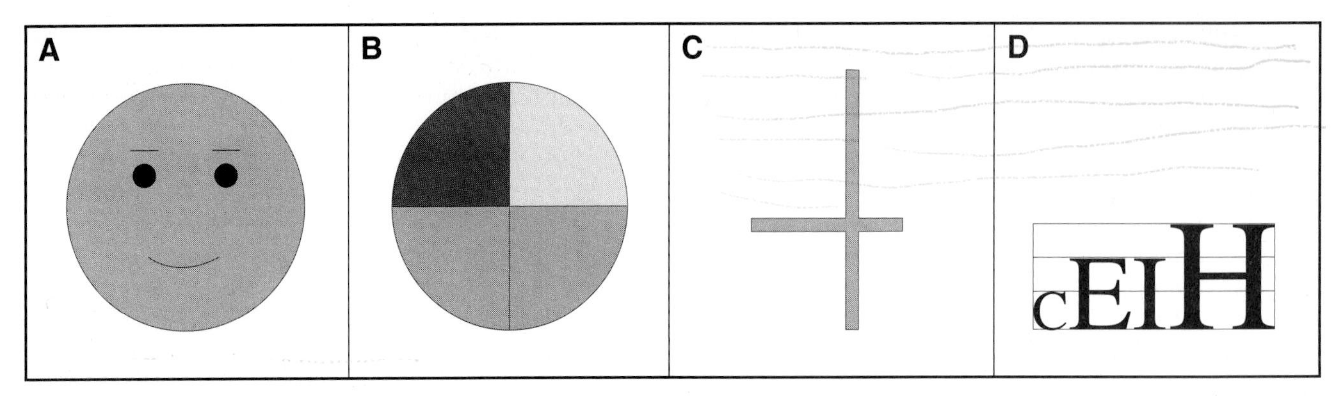

FIGURE 12.11. Multivariate symbols used in a study by Nelson and Gilmartin (1996): (A) a modified Chernoff face; (B) a circle divided into quadrants; (C) a cross; and (D) boxed letters representing variable names.

Gilmartin only considered a nine-county region), and the viewing environment (for example, in a three-dimensional environment data jacks presumably would be more effective than, say, stars).

Before leaving this section, let's consider one other multivariate point symbol method: the glyphs (or "icons") used within Exvis, a system for exploratory visualization developed at the University of Lowell. Glyphs within Exvis were selected on the basis of *preattentive processing,* or "the ability to sense differences in shapes or patterns without having to focus attention on the specific characteristics that make them different" (Grinstein et al. 1992, 638).* Examples of some of the icons used in Exvis are shown in Figure 12.12; Grinstein et al. indicate that up to 15 variables can be mapped using one of these icons by varying the angles, lengths, and intensities of the limbs composing the icon.†

Grinstein et al. note that when the icons are dense enough, "they form a surface texture" and that "structures in the data are revealed as streaks, gradients, or islands of contrasting texture" (p. 638). A geographic example is shown in Figure 12.13, where five satellite images of the eastern portion of the Great Lakes have been combined into a single image. In this case, Smith et al. (1991) argued that "We can see not only Lake Erie and Lake Ontario, but also Lake Huron, Georgian Bay, Lake Simcoe, and some of the smaller outlying lakes" (p. 197).

Although geographers have not as yet utilized icons of the form used in Exvis, Micha Pazner and his students at the University of Western Ontario have begun experimenting with icons in which small fixed symbols within a matrix of pixels are varied in lightness or color (Lafreniere et al. 1996; Pazner and Lafreniere 1997). The resulting images appear rather complex in static printed form, but Pazner et al. argue that much can be learned from these images when they are viewed in an interactive graphics environment (that is, the data are explored).

Combining Different Types of Symbols

In combining three or more variables on the same map, we have focused on using the same type of symbol for a particular application (we used an area symbol for the trivariate choropleth map and a point symbol for the multivariate point symbol map). It is also possible to combine various symbol types to display multivariate data. A good example of this is SLCViewer, a software package developed at the Deasy Geographics Laboratory at Penn State University by David DiBiase and his colleagues (DiBiase et al., 1994a, 303–309).

The purpose of SLCViewer was to explore data produced by climate models. SLCViewer permitted the analyst to view up to four climatic variables as small multiples or to overlay three variables to create a multivariate map. For the multivariate map, point, line, and area symbolization were used. For example, Figure 12.14 shows the variables mean annual evaporation, precipita-

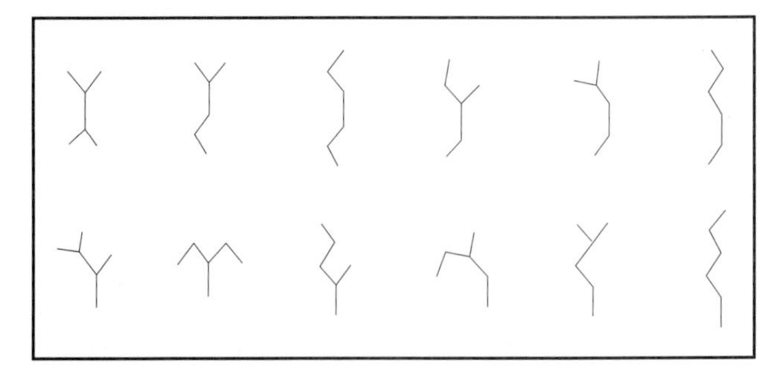

FIGURE 12.12. Some icons (a form of multivariate point symbol) used in Exvis, a system for exploratory visualization. Multiple variables can be mapped using one of these icons by varying the angles, lengths, and intensities of the limbs composing the icon. (*Visualization for Knowledge Discovery,* Grinstein, G., Sieg, J. C. J., Smith, S., Williams, M.G., International Journal of Intelligent Systems Vol. 7. Copyright © 1992, John Wiley & Sons, Inc. Reprinted by permission of John Wiley & Sons, Inc.)

* Icons in the Exvis system also have sound attributes; the role of sound for representing data will be considered in Chapter 16.

† Technically, they specify 17 variables, but two of these are accounted for by the *x* and *y* coordinate location of the icon.

FIGURE 12.13. A map created using Exvis. Five satellite images were combined to display water features in the eastern portion of the Great Lakes. (From Smith et al. 1991, p. 197; courtesy of SPIE—The International Society for Optical Engineering and Stuart Smith.)

tion, and temperature displayed as proportional symbols (circles), weighted isolines, and area (choropleth) shading respectively. **Weighted isolines** were created by making the width of contour lines proportional to the data; these isolines do not require labeling because wider lines are logically associated with more of the phenomenon being mapped.* The obvious advantage of using different symbol types is that one type does not conflict with another: each variable can be seen individually, but the variables can also be related to one another.

SUMMARY

In this chapter we have covered a variety of methods for bivariate and multivariate mapping. Remember that **bivariate mapping** refers to the display of two variables, while **multivariate mapping** refers to the display of three or more variables. Both of these techniques can be accomplished through either map comparison (a separate map is created for each variable) or by combining all variables on the same map.

With respect to bivariate map comparison, we stressed the importance of determining an appropriate method of data classification. Although optimal data classification is often recommended for univariate mapping, it is generally inappropriate for bivariate mapping because it focuses on the precise distribution of individual vari-

* In an experimental study, DiBiase et al. (1994b) found that weighted isolines were more effective than traditional labeled isolines and shadowed contours when the objective was to detect low and high areas in the data.

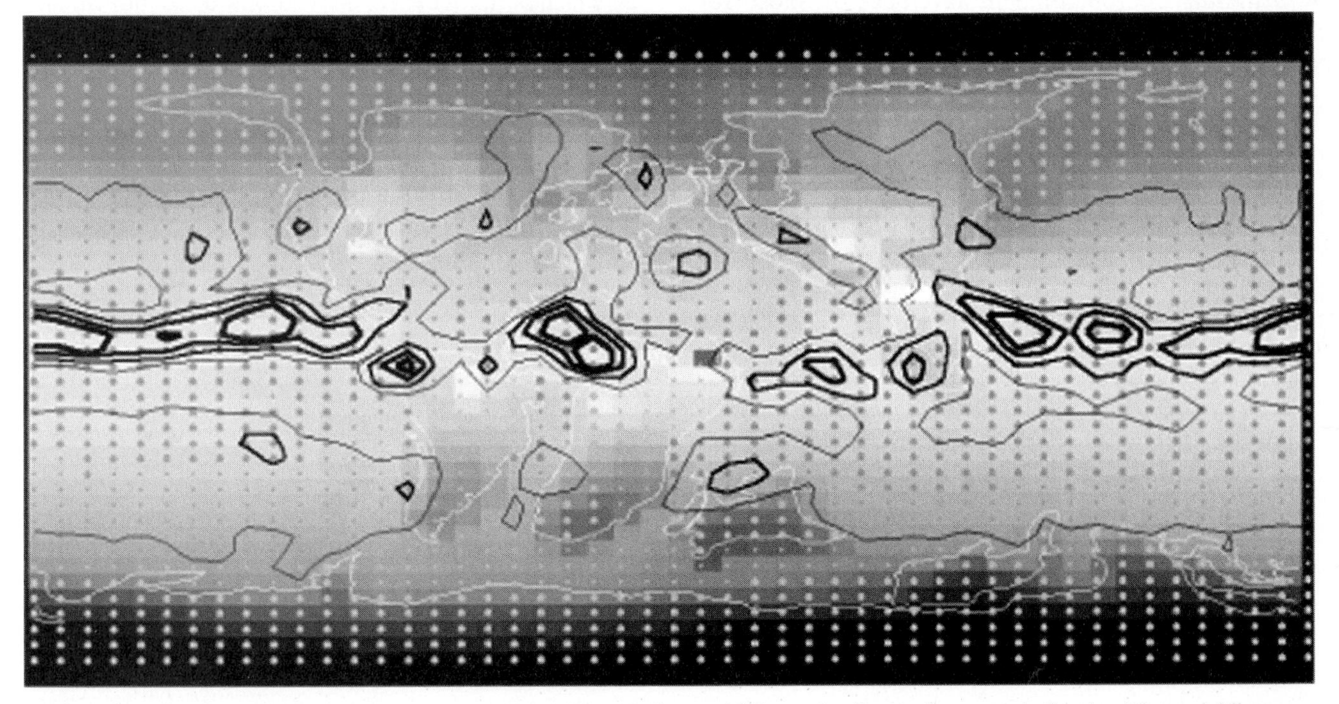

FIGURE 12.14. Using SLCViewer to create a multivariate map by combining point, line, and area symbolization. The variables mean annual evaporation, precipitation, and temperature associated with a climate model are displayed as proportional symbols (circles), weighted isolines, and area (choropleth) shading respectively. (From DiBiase et al. 1994a, p. 219; courtesy of Association of American Geographers and David DiBiase.)

ables along the number line. (An exception would be when the same variable is compared for two different time periods.) The quantiles and equal-areas methods of classification are more appropriate choices for bivariate mapping, although the mean–standard deviation and nested-means methods can be useful if the data are normally distributed. Unclassed maps can be used for map comparison, but only when the data distributions have a similar shape.

When two choropleth maps are overlain, the result is termed a **bivariate choropleth map,** a technique developed by the U.S. Bureau of the Census in the 1970s. Bivariate choropleth maps were criticized because of their presumed failure to communicate information, although Olson found they could be effective if a clear legend was used, both bivariate and individual maps were shown, and an explanatory note was included. Eyton developed a logical color scheme for the bivariate choropleth map by using **complementary colors** (colors that combine to produce a shade of gray), and he used the **reduced major axis** and **equiprobability ellipse** (an ellipse that encloses a specified percentage of the data in a scatterplot) to contrast values near the mean with more extreme observations. Brewer noted that color schemes for bivariate choropleth maps should be a function of whether the variables are unipolar or bipolar. A cross-hatched symbology for bivariate choropleth mapping has been developed, but it is little used because of its coarse appearance.

Although choropleth maps have been the most frequently used symbology when combining two variables on the same map, some interesting **bivariate point symbols** have been developed, including the **rectangular point symbol** (the width and height of a rectangle are varied), and the **bivariate ray-glyph** (straight-line segments point to either the right or left of a small central circle). The rectangular point symbol may serve as a substitute for the coarse cross-hatched symbology. An advantage of the ray-glyph is that it can be squeezed into a relatively restricted space. A bivariate point symbol can also be created by placing a choropleth shade within a proportional point symbol, but this technique should be used with caution as it may be confused with a redundant use of this symbology.

When more than two maps are compared simultaneously, the result is termed a **small multiple.** Although small multiples can be useful for overall pattern comparison, they are difficult to interpret when we wish to focus on a comparison of subregions of a map (for example, comparing small enumeration units on a choropleth map is not easy). A common alternative to the small multiple is the **multivariate point symbol** or **glyph.** Examples of glyphs include the **star** (multiple rays extend from a central circle), the **snowflake** (rays of the star are connected), **three-dimensional bars** (bars of varying height are placed alongside one another), **data jacks** (triangular spikes extend from a square central area) and **Chernoff faces** (distinct facial features are used). Although a considerable number of variables can be represented with such methods, it is questionable whether map readers can understand the resulting symbols. Multiple variables can also be combined on choropleth and dot maps to create **trivariate choropleth maps** and **multivariate dot maps.**

The difficulty of interpreting multivariate symbols has led to the development of software for exploring multivariate data. An example of such software is SLCViewer, which permitted an analyst to view up to four climatic variables as small multiples or to overlay three variables to create a multivariate map. The latter is a particularly intriguing option as proportional symbols, **weighted isolines** (the width of contour lines is proportional to the data), and choropleth shading are all included on the same map. In Chapter 13, we will consider two other packages (Vis5D and Project Argus) that can also be used to explore multivariate data.

FURTHER READING

Aspaas, H. R., and Lavin, S. J. (1989) "Legend designs for unclassed, bivariate, choropleth maps." *The American Cartographer* 16, no. 4:257–268.

Examines the effect of legend designs on the interpretation of unclassed bivariate choropleth maps.

Brewer, C. A. (1994a) "Color use guidelines for mapping and visualization." In *Visualization in Modern Cartography*, ed. A. M. MacEachren and D. R. F. Taylor, pp. 123–147. Oxford: Pergamon.

Provides guidelines for using color in both univariate and multivariate mapping.

Byron, J. R. (1994) "Spectral encoding of soil texture: A new visualization method." *GIS/LIS Proceedings,* Phoenix, Arizona, pp. 125–132.

Discusses using three colors (red, green, and blue) to represent the percentage of sand, silt, and clay, respectively, in soil.

Carstensen, L. W. J. (1982) "A continuous shading scheme for two-variable mapping." *Cartographica* 19, no. 3/4:53–70.

Introduces the use of cross-hatched shading for creating unclassed bivariate choropleth maps. Also see Carstensen (1986a, 1986b).

Chang, K. (1982) "Multi-component quantitative mapping." *The Cartographic Journal* 19, no. 2:95–103.

Considers statistical approaches (such as cluster analysis) for combining multivariate data.

Cox, D. J. (1990) "The art of scientific visualization." *Academic Computing* 4, no. 6:20–22, ff.

Describes several multivariate symbolization methods that have been used in a range of disciplines.

DiBiase, D., Reeves, C., MacEachren, A. M., Von Wyss, M., Krygier, J. B., Sloan, J. L., and Detweiler, M. C. (1994a) "Multivariate display of geographic data: Applications in earth system science." In *Visualization in Modern Cartography,* ed. A. M. MacEachren and D. R. F. Taylor, pp. 287–312. Oxford: Pergamon.

Provides an overview of various methods for displaying multivariate spatial data.

DiBiase, D., Sloan, J., II, and Paradis, T. (1994b) "Weighted isolines: An alternative method for depicting statistical surfaces." *The Professional Geographer* 46, no. 2:218–228.

An experimental study of the effectiveness of weighted isolines.

Eyton, J. R. (1984a) "Complementary-color two-variable maps." *Annals, Association of American Geographers* 74, no. 3:477–490.

Discusses the implementation of a method for bivariate choropleth mapping based on complementary colors.

Grinstein, G., Sieg, J. C. J., Smith, S., and Williams, M. G. (1992) "Visualization for knowledge discovery." *International Journal of Intelligent Systems* 7:637–648.

Describes the use of glyphs (or "icons") within Exvis.

Hancock, J. R. (1993) "Multivariate regionalization: An approach using interactive statistical visualization." *AUTO-CARTO 11 Proceedings,* Minneapolis, Minn., 218–227.

Introduces a method for multivariate regionalization based on interactive statistical visualization.

Lavin, S. and Archer, J. C. (1984) "Computer-produced unclassed bivariate choropleth maps." *The American Cartographer* 11, no. 1:49–57.

Introduces the use of cross-hatched shading for creating unclassed bivariate choropleth maps.

Lindberg, M. B. (1987) "Dot Map Similarity: Visual and Quantitative." Unpublished Ph.D. dissertation, University of Kansas, Lawrence, Kansas.

Examines the issue of comparing dot maps, both in the visual and numerical sense.

Lloyd, R. E., and Steinke, T. R. (1976) "The decisionmaking process for judging the similarity of choropleth maps." *The American Cartographer* 3, no. 2:177–184.

A study of the effect of blackness on the visual correlation of choropleth maps. Also see Lloyd and Steinke (1977).

Mersey, J. E. (1980) "An Analysis of Two-Variable Choropleth Maps." Unpublished MS thesis, University of Wisconsin-Madison, Madison, Wisconsin.

An experimental study of the effectiveness of bivariate choropleth maps.

Monmonier, M. S. (1975) "Class intervals to enhance the visual correlation of choroplethic maps." *The Canadian Cartographer* 12, no. 2:161–178.

Introduces a method for enhancing the visual correlation of choropleth maps by modifying the boundaries of class intervals. Also see Monmonier (1976).

Monmonier, M. S. (1977) "Regression-based scaling to facilitate the cross-correlation of graduated circle maps." *The Cartographic Journal* 14, no. 2:89–98.

Describes a technique for enhancing the visual correlation of proportional symbol maps.

Monmonier, M. (1993) *Mapping It Out: Expository Cartography for the Humanities and Social Sciences*. Chicago, Ill.: University of Chicago Press.

Pages 227–241 describe methods for representing geographic correlation.

Nelson, E. S., and Gilmartin, P. (1996) "An evaluation of multivariate quantitative point symbols for maps." In *Cartographic Design: Theoretical and Practical Perspectives*, ed. C. H. Wood and C. P. Keller, pp. 191–210. Chichester, England: John Wiley & Sons.

Reviews various methods for creating multivariate point symbols and describes an experimental study designed to examine the effectiveness of some of these symbols.

Olson, J. (1972a) "Class interval systems on maps of observed correlated distributions." *The Canadian Cartographer* 9, no. 2:122–131.

A study designed to determine which classification method most accurately preserves the correlation between two variables. See also Olson (1972b).

Olson, J. M. (1981) "Spectrally encoded two-variable maps." *Annals, Association of American Geographers* 71, no. 2:259–276.

An experimental study of the effectiveness of bivariate choropleth maps.

Pazner, M. I., and Lafreniere, M. J. (1997) "GIS Icon Maps." *1997 ACSM/ASPRS Annual Convention & Exposition, Volume 5* (Auto-Carto 13), Seattle, Washington, pp. 126–135.

Describes the development of icons for representing multivariate data. The icons are composed of fixed pixel patterns that vary in lightness or color.

Peterson, M. P. (1979) "An evaluation of unclassed crossed-line choropleth mapping." *The American Cartographer* 6, no. 1:21–37.

An experimental study of the effectiveness of classed and unclassed choropleth maps for comparing distributions. Also see Muller (1980b).

Rogers, J. E., and Groop, R. E. (1981) "Regional portrayal with multi-pattern color dot maps." *Cartographica* 18, no. 4:51–64.

An experimental study of the effectiveness of multivariate dot maps.

Tufte, E. R. (1990) *Envisioning Information*. Chesire, Conn.: Graphics Press.

This book provides a wealth of methods for representing both nonspatial and spatial data. Chapter 4 deals entirely with small multiples.

Turner, E. J. (1977) "The Use of Shape as a Nominal Variable on Multipattern Dot Maps." Unpublished Ph.D. dissertation, University of Washington, Seattle, Washington.

An experimental study of the use of shape on multivariate point symbol maps.

Wang, P. C. C. (ed.) (1978) *Graphical Representation of Multivariate Data*. New York: Academic Press.

A collection of chapters on the graphical representation of multivariate data. Many of the chapters focus on Chernoff faces.

13

Data Exploration

OVERVIEW

Chapters 13 to 15 focus on three topics that have received considerable attention in recent years: data exploration, animation, and electronic atlases. Data exploration is the topic of the present chapter, while animation and electronic atlases are covered in Chapters 14 and 15 respectively. Recall from Chapter 1 that data exploration involves revealing unknowns via high human-map interaction in a private setting. In this chapter, we will consider a variety of software that can assist in this process. Although the focus will be on data exploration, we will see that such software often includes a significant animation component.

We will cover the following software: ExploreMap, Project Argus, MapTime, Vis5D, Aspens, Transform, and ArcView. Additionally, we will consider Moellering's (1980) early work that preceded development of other data exploration software by approximately 10 years. The intention is not to discuss all data exploration software that has been developed, but rather to provide you with a feel for the capability of such software. An attempt has been made to select software that is available (some authors develop prototypes of software and do not distribute them), is inexpensive (all are available for free, except Transform and ArcView), and can run in a relatively low-cost hardware environment (Vis5D and Project Argus run most effectively on high-end platforms, such as Silicon Graphics workstations). A brief summary of the software to be covered follows.

Developed for exploring data commonly mapped in choropleth form, ExploreMap provides a range of options for acquiring both specific and general information. For example, a specific option enables users to point to two areas and determine the ratio of the values for those areas, while a general option enables users to toggle individual classes off and on. The purpose of ExploreMap was not to provide a complete range of exploration options for choropleth maps, but rather to serve as a prototype on which others might expand.

Project Argus, a collaborative venture of research laboratories at the University of Leicester and Birkbeck College in England, was developed to illustrate a broad range of possible data exploration functions. We'll focus on one portion of the project dealing with data collected in the form of enumeration units (the so-called enumerated software). Like ExploreMap, the enumerated software utilizes choropleth maps, but also handles proportional circle maps and Dorling's cartograms (section 11.3), and can display up to three variables at a time. Specifically, we'll consider how data can be analyzed simultaneously via maps of individual variables, bivariate overlays, point graphs (one-dimensional scatterplots), and scatterplots.

MapTime permits the exploration of temporal data associated with fixed point locations via animation, small multiples, and ***change maps*** *(maps showing the change between two moments in time). In Chapter 1, I described how the change map option for MapTime could be used to provide information about population change not seen in an animation. In this chapter, we focus on the different kinds of information provided by small multiples and animation when they are applied to hypothetical discharge values at point locations within a drainage basin.*

Vis5D was developed for handling true 3-D phenomena, or phenomena that vary continuously in three-dimensional space, such as temperature or wind speed. The "5-D" portion of Vis5D comes from its capability of handling not only the three spatial dimensions (latitude, longitude, and altitude), but also a time dimension, and a dimension for the attributes to be displayed (such as pressure, temperature, and wind speed). A number of options in Vis5D are similar to those described for T3D, a package discussed in section 11.5.1, but its multivariate and temporal capabilities and the ability to handle vector-based data (such as wind speed) make it an intriguing piece of software.

Aspens is different from the other software discussed here in the sense that it was developed solely to handle data associated with the growth in trembling aspens at Waterton Lakes National Park, Alberta, Canada. Although designed only for such data, Aspens serves as a prototype for illustrating the broader concept of ***proactive graphics,*** *or software that enables users to initiate queries using icons (or symbolic representations, as opposed to words). In addition to illustrating data exploration concepts, Aspens includes some particularly effective animations of tree growth based on pictographic leaf and tree symbols.*

Transform permits both graphs and maps to be explored. We'll focus on graphs because they provide useful illustrations of Transform's unique capabilities, and graphs are an important element of data exploration (along with maps, tables, and numerical summaries). Two unique capabilities of Transform that we'll examine are the use of specialized color schemes purported to bring out nuances in data and a "Fiddle tool," which permits shifting and compressing of color schemes.

ArcView is not simply data exploration software, but a GIS intended for both the analysis and display of spatial data. A key feature of modern GIS software such as ArcView is its ability to integrate analysis and display functions, as execution of an analytic function leads to an immediate display of the results. This is in contrast to early GIS software, which separated analysis and display functions via ***command-line interfaces.*** *Using a tutorial from ArcView, we'll see how the analysis and display functions can be integrated. Then, we'll see how that integration enhances the GIS analyst's ability to (1) explore the effect of choosing different analysis parameters, (2) view data from a variety of perspectives (for example, as a map, table, or chart), and (3) experiment with alternative symbol schemes to see if they reveal different patterns in the data.*

Although the software described in this chapter provides considerable capability for those wishing to explore spatial data, readers will undoubtedly find limitations that prevent a desired task from being accomplished. As a result, readers may wish to develop their own data exploration software. In Chapters 14 and 15, we will find that the same will be true for animations and electronic atlases. In this respect, the last section of Chapter 15 describes several tools to assist readers in developing their own software.

13.1 MOELLERING'S 3-D MAPPING SOFTWARE

Moellering's (1980) 3-D mapping software was one of the earliest attempts to demonstrate the potential of data exploration. With Moellering's software, a user could explore a three-dimensional digital elevation model (DEM) in real time, a process that Moellering termed "surface exploration."* Since Moellering's software was dependent on expensive hardware, he developed a video (Moellering 1978) to demonstrate software capabilities. In addition to surface exploration, the video illustrated two animations of spatiotemporal data: the growth of U.S. population from 1850 to 1970 and the diffusion of farm tractors in the U.S. from 1920 to 1970. These animations were created using attractive, colored three-dimensional prism maps.

Although Moellering's work suggested great potential for both data exploration and animation, there was almost no research done in these areas for approximately 10 years following his effort, in part because many cartographers lacked the necessary hardware and software. The lack of work in animation, in particular, was lamented in a review article by Campbell and Egbert (1990). Fortunately, considerable advancements have taken place since 1990, as this and the following chapter will illustrate. It is noteworthy, however, that those wishing to create animations similar to Moellering's may still find the task difficult unless they have a workstation and sophisticated visualization software, such as that described in section 15.3.4.

13.2 ExploreMap

Egbert and Slocum's (1992) ExploreMap assists users in exploring data commonly mapped with choropleth symbolization. The intention in developing ExploreMap was not to create the ultimate data exploration software, but rather to provide a prototype that others might improve upon. In contrast to Moellering's video distribution, ExploreMap and other software described in this chapter are distributed via diskette, CDROM, or over the Internet.

ExploreMap operates in two basic modes: Design and Explore. Design is necessary for creating maps, while Explore includes methods for exploring data, and thus will be our focus. Two of ExploreMap's functions, Areas and Overview, correspond to the two types of information that can be acquired from a map: specific and general (see section 1.2).

To illustrate how Areas provides specific information, consider two of its options: Single Area and Ratio of

* Technically, the data were 2½-D, but exploration was done in three-dimensional space.

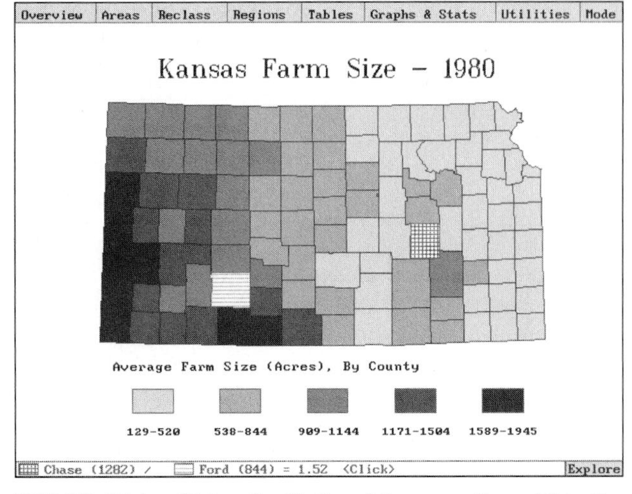

FIGURE 13.1. Using the Ratio of Areas option within ExploreMap to compare two enumeration units. The units (counties in this case) compared are highlighted on the map and the ratio of their values is shown at the bottom of the display.

Areas. The Single Area option is used to determine the exact value of an enumeration unit either by pointing at it with the mouse or typing its name. In contrast, on a traditional paper map the value for an enumeration unit is determined by visually comparing a shade on a map with shades in the legend—assuming a classed paper map, the best estimate is a range of values shown in the legend. (Data values on a paper map may be displayed within enumeration units, but these detract from the examination of map pattern and cannot be placed within small units.) Most data exploration software includes an option similar to the Single Area option.

The Ratio of Areas option is used to determine the ratio of values for two enumeration units: when a user selects two enumeration units, they are highlighted on the map, and the ratio of their values is shown at the bottom of the display (Figure 13.1). Although it remains to be seen how worthwhile such an option will be to users, it is apparent that such precise information is not available from paper maps.

The Overview function within ExploreMap includes three options: Sequenced, Classes, and Subset. Rather than presenting the entire map at once (as is done for printed maps), the Sequenced option builds the map piece by piece: the title, geographic base, and legend title and boxes appear first (in sequence), and then each class is displayed (from low to high). Two arguments can be made for **sequencing:** (1) it should enhance map understanding by providing users with "chunks" of information (Taylor 1987); and (2) it should emphasize the quantitative nature of the data (that the data are ordered from low to high values).

Experimental tests of sequencing have produced mixed results. A study by Slocum et al. (1990) found that although more than 90 percent of readers favored a sequenced approach, objective measures of map use tasks revealed no significant difference between sequenced and traditional (static) methods. As a result, Slocum et al. argued that sequenced maps may be preferred simply because of their novelty. A more recent study by Patton and Cammack (1996), however, revealed sequencing to be more effective than a static approach.

The Classes option within Overview permits users to display any combination of classes desired; for example, Figure 13.2 portrays classes 2 and 4 of a 5-class map. In an evaluation of ExploreMap, Egbert (1994, 87) found that the Classes option "was liked by all subjects without reservation"; subjects used Classes not only to identify the location of individual classes, but also to examine patterns and trends.

Although useful, the Classes option is constrained by how the data are classed (in Figure 13.2 a 5-class optimal map is used). One solution to this constraint is the Reclass function, which permits changing the method and number of classes. An option within Reclass, Compare Maps, permits showing up to four maps simultaneously (each map can have a different method of classification or number of classes). For example, Color Plate 4.1 compares four methods of classification for 5-class maps: equal intervals, quantiles, optimal based on the mean, and optimal based on the median.

An alternative to reclassing the data is the Subset option, which enables users to focus on an arbitrary range of data. To assist users in selecting data to be focused on, ExploreMap provides a dispersion graph (Color Plate 13.1A) illustrating the distribution of the data along the number line; the current class breaks, mean, median, and standard deviation can be shown on the dispersion graph (only the median is shown in Color Plate 13.1A). Users

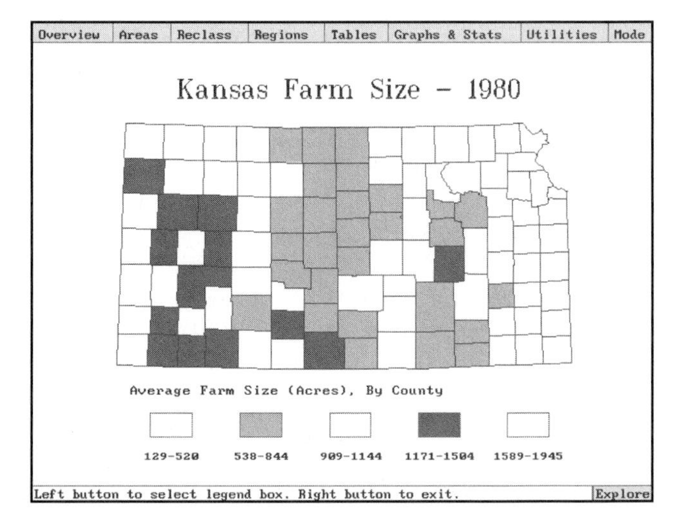

FIGURE 13.2. The Classes option for ExploreMap. Any subset of classes can be selected by clicking on the legend boxes; in this case, classes 2 and 4 are shown.

simply click on a desired range (or type a range of values), and the specified range is drawn in a highlighted color (blue is used in Color Plate 13.1B to display all values less than the median). Once an interval is examined, another interval can be selected within a few seconds.

Although Subset enables users to highlight selected portions of a data set, it is limited in that all highlighted data are shown in a single color and real-time interaction with the dispersion graph is not possible. A more flexible approach would permit a range of shades for the highlighted color (say, a range of blues instead of a constant blue). Such an approach would enable users to focus on a subset of the data, while also seeing the variation within that subset. A more flexible approach would also allow the map to change dynamically as the mouse cursor is dragged back and forth along the number line.

Interestingly, the latter capability was provided in the Xmap system developed at the Massachusetts Institute of Technology by Ferreira and Wiggins (1990). Xmap used a "density dial" to highlight portions of a data set. Ferreira and Wiggins described the use of the density dial as follows:

> What makes the density dial a truly interactive visualization tool is its ability . . . to change the group of cells that are highlighted in red . . . as fast as we can move the mouse. We get an effect that is like watching a video. We see a moving sequence of red-shaded polygons going from the least-dense tract to the most dense tract, or vice versa. This speed—and the sense of motion that comes with it—enhances our ability to remember what we have just seen of the spatial pattern. (p. 71)

Unfortunately, to my knowledge Xmap is not generally available, and so it is unlikely that readers will have access to it.

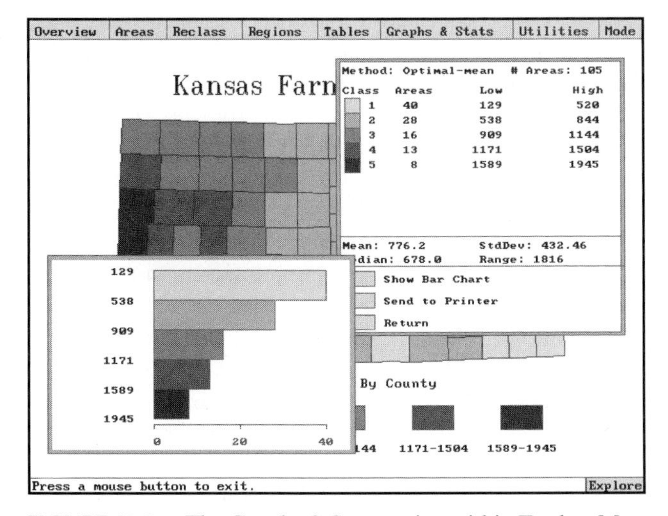

FIGURE 13.3. The Graphs & Stats option within ExploreMap. Graphical and statistical information provide an alternate view to the map of the data.

Other major functions in ExploreMap include Regions, Tables, Graphs & Stats, and Utilities. While Subset allows users to focus on a subset of data along the *number* line, Regions permits focusing on a subregion of a *mapped* display of the data. For example, a politician interested in counties comprising an economic development region could use Regions to highlight enumeration units falling within that region. The Tables and Graphs & Stats functions focus on tabular, graphical, and statistical views of the data (as opposed to spatial views of the data). For instance, Graphs and Stats can display a histogram and statistical parameters (Figure 13.3).

13.3 Project Argus

Project Argus, a collaborative venture of research laboratories at the University of Leicester and Birkbeck College in England, is intended to illustrate a broad range of possible data exploration functions. Jason Dykes has been responsible for most of the software developed in the project and has written papers (1995, 1996, 1997) summarizing its capability. Two major pieces of software are available at the web site for the project: one for data collected in the form of enumeration units (the "enumerated" software) and one for tourist data collected over time (the "time-space" software). Here we will focus on only the enumerated software.*

Like ExploreMap, the enumerated software is intended for data collected in the form of enumeration units, but it extends well beyond ExploreMap's capabilities. For example, the enumerated software can symbolize the data in more than one form (choropleth maps, proportional circle maps, and Dorling's cartograms are possible) and display up to three variables at a time. Interestingly, proportional circle maps are by default displayed using a redundant symbology of size and gray tones (see section 7.6); the gray tones can, however, also be used to display a second variable and thus create a bivariate map.

To illustrate some of the enumerated software's capability, consider Color Plate 13.2, which portrays a view created with a sample data set for Illinois included with the software. The three variables mapped are percent black, percent of population 25 years and older with any college education, and percent below the poverty level. Across the top of the screen, the variables are shown as individual choropleth maps. Choropleth maps are shown (as opposed to proportional circle maps) because the maps are based on standardized data. By default, unclassed choropleth maps are shown, although several classification methods are available within the software.

* Technically, the enumerated software is termed the "cdv for enumerated data sets." I use "enumerated" software here for simplicity.

From the discussion presented in section 12.1.1, we know that the visual correlation of the maps may be adversely affected by using unclassed maps, but unclassed maps are a useful starting point for visual analysis.

In the middle portion of the screen, we see point graphs (one-dimensional scatterplots) of two of the variables: percent black and percent of the population 25 years and older with any college education. We can visually relate these plots to the maps above and see that the overall light appearance of the percent black map is a function of a positive skew (there are a few outliers toward the positive end of the number line). To the right of the point graphs is a scatterplot illustrating the relationship between the percent black and education variables (displayed along the x and y axes, respectively). We can see that when percent black is low, a large range of education percentages are possible, but that the larger percent black values are associated with lower education values.

In the top right of the screen, we see a bivariate choropleth map (introduced in section 12.1.2) of the percent black and education variables. Higher values on these variables are represented by increasing amounts of yellow and blue respectively, while the lowest values are represented by the absence of both colors, or black in the extreme case. Note that the two highest percent black values shown in the scatterplot (they have low scores on the education variable) are depicted as a bright yellow in the extreme southern part of Illinois. Overall, the map is relatively dark because of the concentration of both variables in the lower left portion of the scatterplot. Also note that the colorings shown on the map are shown within dots on the scatterplot.

This discussion has focused on how two of the three variables shown in Color Plate 13.2 might be analyzed. When using the software, a much more complete analysis is, of course, possible. One interesting option is the use of a trivariate choropleth map (introduced in section 12.2.2), in which three choropleth maps are overlaid. It also must be recognized that the static nature of a book prevents a full understanding of the dynamic capability of the enumerated software. For example, the software permits designating an outlier on a point graph, and highlighting that point on all other displays currently in view. Alternatively, in a fashion similar to ExploreMap, it is possible to select a subrange of values on a variable, and have that subrange highlighted on the map. To illustrate these concepts, in Color Plate 13.2 a small cluster of dots is highlighted in green in the scatterplot and on other selected views. Also shown in Color Plate 13.2 is a parallel coordinate plot (see section 3.3.2) and Dorling's cartogram method (in this case illustrating the prominence of Cook County). Like much of the software in this chapter, these various display methods can best be understood by actually using the software.

13.4 MapTime

Yoder's (1996) MapTime permits users to explore temporal data associated with fixed point locations using three major approaches: animation, small multiples, and change maps. Yoder argues that animation is an obvious solution for showing changes over time because it "incorporates time itself in the presentation . . . [thus,] the cartographic presentation [is not only] a scale model of space, but of time as well" (p. 30). Animations of point data are arguably easiest to interpret when data change relatively gradually over time. For example, an animation of the change in population of U.S. cities from 1790 to 1990 (distributed with MapTime) is easy to follow because city populations gradually increase and decrease. In contrast, an animation of the change in water quality at point locations along streams and rivers would exhibit much sharper increases and decreases, and so be harder to interpret (Greenfield 1994).

Animations in MapTime are constructed using a series of **key frames** (frames associated with collected data) and **intermediate frames** (frames associated with interpolated data). A distinct advantage of MapTime is that the number of intermediate frames can be varied. Thus, if one has data for every year from 1900 to 1940, but only for every other year between 1940 and 1980, intermediate frames might be used just for the 1940–1980 data. The actual animations in MapTime clearly identify key and intermediate frames so that the user is aware which frames are based on "hard" and interpolated data.

In the context of temporal data, **small multiples** consist of a set of maps, one for each time element for which data have been collected. As an example, Figure 13.4 portrays small multiples for seven collection stations within a hypothetical drainage basin at 12-hour intervals over a four-day period (the data are distributed with MapTime). It is presumed that the streams flow from west to east, that circle size is proportional to stream discharge (in cubic feet per second), and that heavy rainfall occurred between 6 P.M. Monday and 3 A.M. Tuesday throughout the region. The small multiples reveal an initial increase in discharge for upstream stations, followed by a later increase for downstream stations; essentially, a pulse of water moves through the system.

The pulse of water is also detectable in an animation of the stream discharge data, but it is not easy to discern. Thus, small multiples are often a necessary complement to animations.* Small multiples can also assist in contrasting two arbitrary points in time. For example, one might examine the beginning and ending small multiples for stream discharge and note approximately how much change has taken place over the period at each location.

* For an experiment comparing animation and small multiples, see Koussoulakou and Kraak (1992).

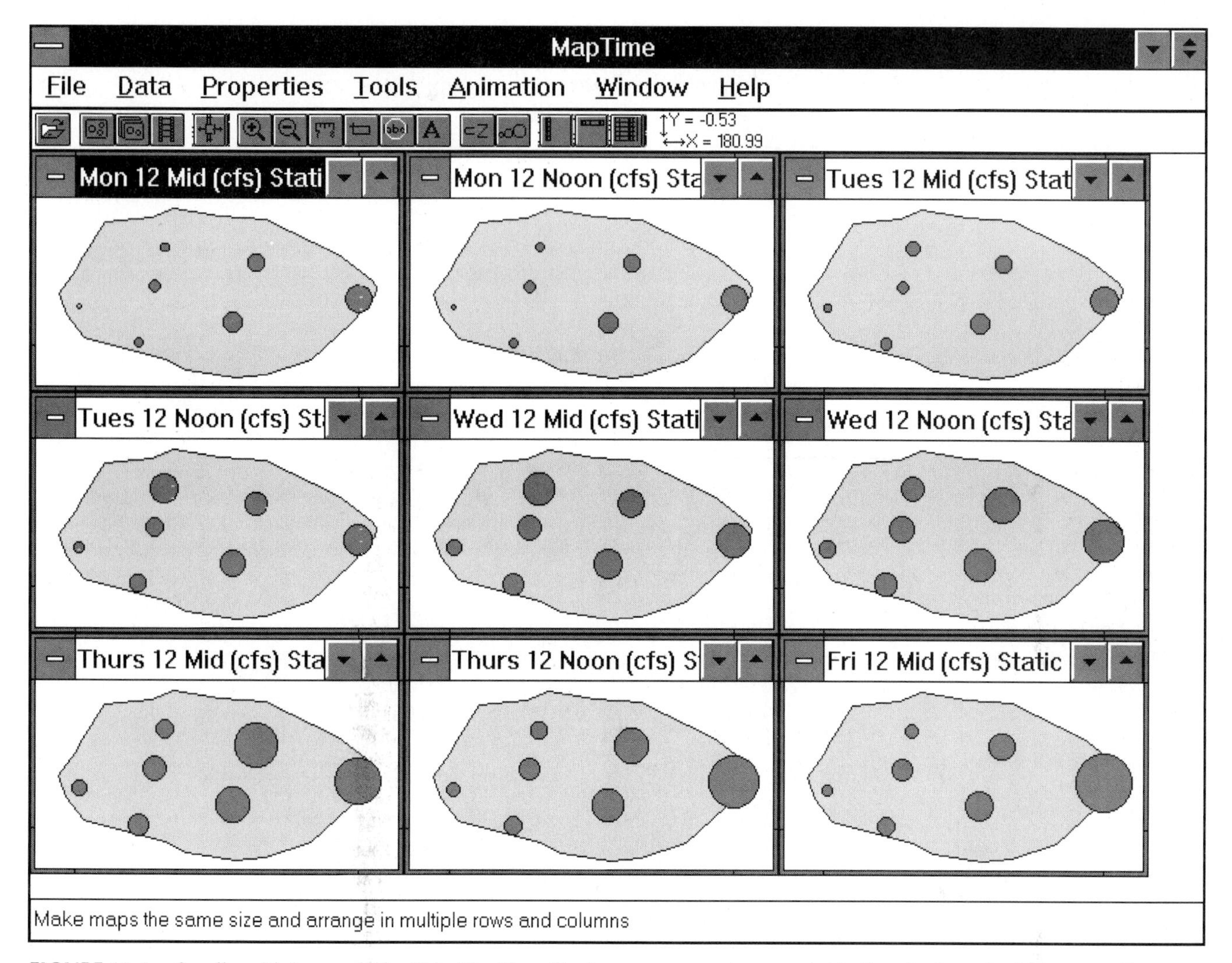

FIGURE 13.4. Small multiples created within MapTime. Each map represents one point in time for hypothetical stream discharge data at selected points. (Courtesy of Stephen C. Yoder.)

When one wishes to focus on the precise magnitude of change at selected locations or on exactly where changes have taken place, MapTime makes use of **change maps,** or maps showing the change between two moments in time. A good example of the need for a change map is the U.S. city population data mentioned above. When these data are animated, one sees a major growth in city populations in the northeast beginning in 1790, with an apparent drop in population for *some* of the largest northeastern cities from about 1950 to the present. A map showing the percent population change between 1950 and 1990, however, reveals a very distinctive pattern of population decrease *throughout* most of the northeast, as was discussed in Chapter 1 (Color Plate 1.3).

Like ExploreMap, MapTime can easily determine values associated with a particular location, or focus on a subset of the data. Additionally, a zoom feature enables users to enlarge a portion of the map. These features can be used to explore a single moment in time, as in Figure 13.5, or be implemented throughout an animation. Those who wish to experiment with different circle scaling methods (as described in Chapter 7) also will find that circle scaling exponents can easily be changed in MapTime.

13.5 Vis5D

Vis5D provides some rather intriguing possibilities for those wishing to explore true 3-D phenomena. This package was developed at the University of Wisconsin-Madison Space Science and Engineering Center, where it has been primarily used for visualizing data produced by numerical weather models. The "5-D" portion of its name comes from the notion that meteorological data can be specified in terms of three spatial dimensions (latitude, longitude, and altitude), a time dimension, and a dimension for the attributes to be displayed (such as

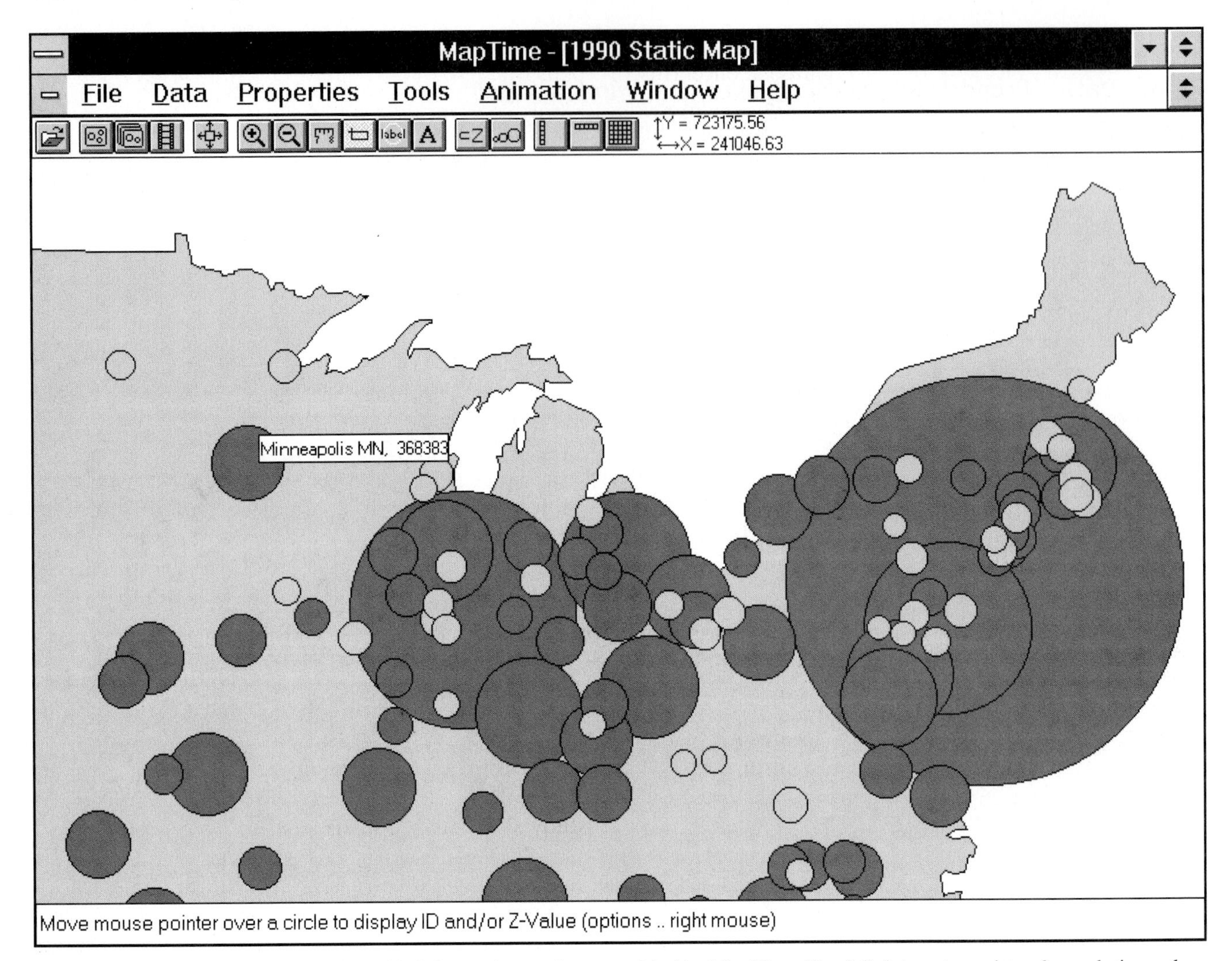

FIGURE 13.5. Zoom, subset, and specific information options provided by MapTime. The full data set consists of population values for 196 U.S. cities in 1990; here the user has zoomed in on the northeastern United States. Cities with 100,000 or more people are the subset highlighted in the dark grey shade. A specific value for Minneapolis, Minnesota, is shown. (Courtesy of Stephen C. Yoder.)

pressure, temperature, and wind speed). Although intended primarily for meteorological data, Vis5D can display any data that can be expressed in the 5-D format; for example, oceanic data have three spatial dimensions (latitude, longitude, and depth), a time dimension, and multiple attributes (such as temperature, salinity, and current speed) (Hibbard and Santek 1990).

To illustrate some of Vis5D's capability, we will use a 48-hour forecast data set for North America distributed with the software (V.GRID1.v5d). The data set consists of 29 attributes, with 49 timesteps for each attribute. Topographic base information also is provided with Vis5D (the base for North America is shown in Color Plate 13.3); the base obviously provides a necessary geographic perspective, with topography critical in meteorological applications.

One basic option in Vis5D is to construct a slice through the three-dimensional space of an attribute in either a vertical or horizontal direction. The symbology for the slice can be either isolines or hypsometric tints (to represent the magnitude of a phenomenon), or vectors (to represent the flow of a phenomenon). Color Plate 13.3 illustrates a slice of hypsometric tints for wind speed for timestep 1. The default color scheme is a traditional spectral one (red-orange-yellow-green-blue), although users can create their own schemes. Note that the altitude of the slice is 11.66 kilometers, and thus the high-valued reddish-orange area represents a portion of the jet stream.

Another basic option in Vis5D is to create an **isosurface,** a surface bounded by a particular value of an attribute. For example, Color Plate 13.4 illustrates the 45 meter per second isosurface for wind speed for timestep 33. Wind speeds inside the surface would be greater than 45 meters per second, while those outside would be less. We would expect high wind speeds to be associated with the jet stream, and the general shape of the isosurface confirms this notion.

A third option in Vis5D is "volume," in which the entire 3-D space for an attribute is symbolized at once. This might seem impossible due to symbol blockage, but it can be done if the symbology is treated as a transparent fog (if opacity is varied, as was done for Slicer; see section 11.5.1). For example, Color Plate 13.5 illustrates the volume option for wind speed for timestep 28 (a default spectral color scheme is again used). As before, the presumed location of the jet stream is indicated by its higher speed. Admittedly, the volume symbology is difficult to interpret in this static map, but Vis5D's interactive nature enables a user to "see through the fog" by varying his or her viewpoint of three-dimensional geographic space.

Potentially the most interesting feature of Vis5D is its real-time animation capability. By simply clicking on "Animate," the system will animate any currently visualized attribute. For example, after selecting the slice for wind speed shown in Color Plate 13.3, clicking on "Animate" reveals how the jet stream changes over the 48-hour period within that slice. One can of course stop the animation at any moment, and then explore a particular timestep further.

It is also important to realize that more than one variable can be explored at a time. For example, Hibbard et al. (1994) illustrate William Gray's novel idea for generating energy by creating a permanent rainstorm over a hydroelectric generator (Color Plate 13.6). They describe the process as follows:

> The white object is a balloon 7 kilometers high in the shape of a squat chimney that floats in the air above a patch of tropical ocean. The purpose of the numerical experiment is to verify that once air starts rising in the chimney, the motion will be self-sustaining and create a perpetual rainstorm. The vertical color slice shows the distribution of heat (as well as the flow of heat when model dynamics are animated); the yellow streamers show corresponding flow of air up through the chimney; and the blue-green isosurface shows the precipitated cloud ice. (p. 66)

Such capability does not come without a price, as one limitation of Vis5D is that it is most effective only on high-end workstations. (I used a Silicon Graphics Iris Onyx/2 with two 150MHz processors, 128MB RAM, and a VTX graphics subsystem.) In spite of hardware limitations, Vis5D is particularly impressive given that it is available for free via the Internet; sophisticated visualization systems offering similar capability can cost thousands of dollars.

13.6 Aspens

Buttenfield and Weber (1994) developed Aspens to assist Jelinski (1987) in exploring the growth in trembling aspen trees at Waterton Lakes National Park in Alberta, Canada. More specifically, the purpose of Aspens was to examine "the apparent contradiction that radial growth rates [in aspens] were higher where local environmental conditions (elevation, precipitation, and soils) were more harsh" (Buttenfield and Weber, p. 11). Although Aspens was designed solely to handle Jelinski's data, it was intended as a prototype for illustrating the broader concept of **proactive graphics,** a term Buttenfield and Weber coined to describe software that enables users to initiate queries using icons (or symbolic representations) as opposed to words. Ideally, proactive graphics should also avoid steep learning curves and be "responsive to commands and queries . . . not . . . anticipated by system designers" (Buttenfield and Weber, p. 10).

The philosophy of proactive graphics in Aspens is apparent in the opening screen of the study area (Color Plate 13.7), in which one finds no pull-down menus (as, for example, appear in MapTime). Rather, it is presumed that users will become aware that links to information are most often available via the color green found on various icons. For example, clicking on the green "i" provides information about using the display, while clicking on the green symbol portion of the legend leads to further information about the legend.

Interestingly, an animation begins as soon as the study area appears: six leaves (corresponding to the six major study sites) change in size and color, as each year of growth is highlighted in a bar graph; one frame from the animation appears in Color Plate 13.7. Larger leaf sizes represent larger total cumulative growth, while a darker green leaf color represents a greater incremental growth for each year. Buttenfield and Weber argued that having the animation start automatically is appropriate because, "The sooner graphical activity is recognized, the more quickly data will be explored" (p. 11). A click on the name of a study site (for example, Prairie 2 in Color Plate 13.7) leads to a screen of descriptive statistics and a histogram of the yearly incremental growth for that site. As was indicated for ExploreMap, linking statistical and graphical displays with maps is critical for exploratory analysis.

Clicking on a leaf associated with a study site leads to a more detailed view of that site: the aspen clones making up that site (Color Plate 13.8). To distinguish the clones from the generalized study site, treelike symbols are used rather than leaves. The clones are animated in a manner analogous to the general study site, and descriptive statistics and a histogram of yearly growth rates can also be obtained by clicking on a clone identifier. Within any clone, it is possible to jump to another by clicking on one of the icons in the upper right of the screen.

Buttenfield and Weber recognized that Aspens is not truly proactive because it is not possible to generate queries and commands unanticipated by system designers. They also indicated that it would be desirable to incorporate photographs of study sites, add capabilities to reverse the animation and jump to an arbitrary year, and implement functions for saving data, maps, and text information to files. There is also a need for incorporating other data that might be related to tree growth such as elevation, precipitation, and soils.

13.7 Transform

Transform is an example of commercial software for data exploration. Formerly marketed as part of Spyglass, Transform is now distributed by Fortner Software along with their Plot and T3D products (see section 11.5.1 for a discussion of some of T3D's capabilities). Although Transform contains routines for exploring data associated with both graphs and maps, we will focus on graphs because they provide useful illustrations of Transform's unique capabilities, and are an important element of data exploration (along with maps, tables, and numerical summaries).

The data shown in Figure 13.6 represent a portion of mean monthly temperatures for Springfield, Illinois (the complete data set includes mean monthly temperatures for a 40-year period). When viewed in this tabular form, it is difficult to detect trends; we can see that one column has higher values than another, but it is difficult to visualize the trend over the course of a year, and especially difficult to detect any trends over the 40-year period.

As an alternative to the tabular display, Transform permits users to view data as a graphical representation, or "image." The left-hand portion of Color Plate 13.9 rep-

	1.0	2.0	3.0	4.0	5.0	6.
1936.0	21.20	19.70	45.00	50.40	69.20	74.2
1937.0	27.30	30.40	39.00	52.60	65.40	73.4
1938.0	29.20	39.10	50.50	55.40	64.00	72.8
1939.0	35.40	30.60	44.20	50.70	68.00	75.3
1940.0	14.50	31.00	39.50	51.40	61.40	75.7
1941.0	30.90	29.10	38.00	58.30	69.40	76.1
1942.0	29.00	29.90	44.40	58.40	64.50	73.9
1943.0	27.60	35.40	38.00	51.90	62.20	76.8
1944.0	33.60	33.80	39.10	50.80	69.80	78.3
1945.0	24.40	33.30	52.30	54.70	59.40	69.5
1946.0	31.20	35.40	54.60	57.40	61.20	74.6
1947.0	33.40	25.80	35.70	52.50	61.10	71.2
1948.0	24.40	31.80	41.60	58.40	63.70	74.6
1949.0	30.40	32.90	42.20	52.70	67.60	76.8
1950.0	33.70	32.30	38.20	48.20	66.90	73.3
1951.0	29.30	32.00	37.60	50.30	67.00	70.6
1952.0	31.70	37.00	39.80	54.20	64.40	80.5
1953.0	32.70	37.00	43.90	50.20	65.70	79.2
1954.0	29.70	39.30	37.80	57.50	59.00	77.0
1955.0	27.90	31.50	40.20	59.80	65.10	69.5

FIGURE 13.6. A window from Transform showing a portion of a table of mean monthly temperatures for Springfield, Illinois. Years are shown along the y axis, while months of the year are shown along the x axis. Compare with the image of these data shown in Color Plates 13.9–13.11. (Image created with Noesys Visualization Pro, courtesy of Fortner Software LLC.)

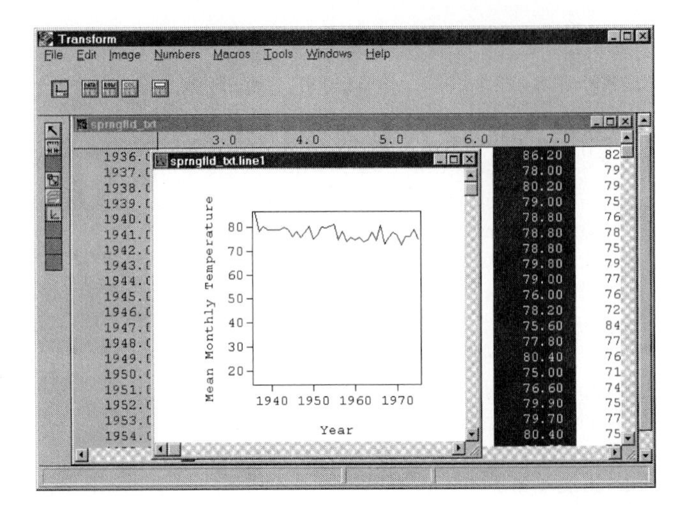

FIGURE 13.7. A graphical representation of the column for July temperatures shown in Figure 13.6. (Image created with Noesys Visualization Pro, courtesy of Fortner Software LLC.)

resents one such representation. In this case, a traditional spectral, or "Rainbow," color scheme has been used. Note that the trend over the course of a year is now more readily apparent, but detecting any trend over the 40-year period is still difficult due to variability in the data: an unusually cold or hot month appears to disrupt the overall pattern. One way to minimize such variation is to smooth the data. This is accomplished by treating cells in the left-hand portion of Color Plate 13.9 as control points for contouring, much as we did for maps in Chapter 8. The result is shown in the right-hand portion of Color Plate 13.9. Now we see the pattern over the year, but we also see a potential trend over the 40-year period, as more recent years appear to exhibit slightly cooler summers and Januarys.

In addition to representing data by images, Transform links data and images directly. For example, if one uses the mouse to point to a cell (or group of cells), that cell (or group of cells) is also highlighted in the image, as August 1949 is in Color Plate 13.9. Additionally, users can develop a graphical representation of trends along a row or column in the table, as shown in Figure 13.7.

An unusual feature of Transform is the capability to symbolize an image with what many cartographers would consider illogical color schemes. To see the effect of one of these schemes, compare Color Plates 13.9 and 13.10, which use the "Rainbow" and "Lava Waves" schemes, respectively. Based on the discussion of color in Chapter 6, you should argue that the Lava Waves scheme is illogical: the raw data increase from low to high values, but the banded appearance of the color scheme does not logically relate to these values. Although not necessarily logical, Berton (1990) argues that "such banding effects . . . can provide excellent markers for transitional areas in the data" (p. 112). It is disconcerting, however, that documentation for Transform provides no detailed discussion of either how such color schemes were developed nor how they might be used.

Another unusual feature of Transform is the "Fiddle tool," which permits shifting or compressing colors within a scheme. When using the Fiddle tool, a movement of the cursor to the left or right shifts the color scheme in that direction, while a cursor movement up or down either stretches or compresses the scheme. For example, Color Plate 13.11 shows the Rainbow scheme shifted left and compressed. Although I know of no experiments involving the Fiddle tool in a map environment, it seems feasible that it, like the various color schemes, would assist in examining the "transitional areas in the data" noted by Berton.

As indicated previously, Transform contains a range of mapping capability; for example, contour maps, block diagrams, and vector plots are possible for point data specified in two-dimensional space. Additionally, it contains an animation viewer, which can be linked to the exploration tools described here; for example, it is possible to stop an animation at an arbitrary frame and explore the data at that point.

13.8 ArcView

GISs have commonly been defined on the basis of the order in which they process data. For example, Clarke (1997, 3) defines GISs as "automated systems for the capture, storage, retrieval, analysis, and display of spatial data." Such a definition reflects, in part, the nature of early GIS software, which tended to separate the cartographic display component from the analysis of data. Typically, this separation was achieved using **command-line interfaces,** which required users to type commands specifying the processes to be performed; for example, a command to overlay two layers would need to be followed by a command to display the visual results of the overlay process. Such interfaces tended to be difficult to learn and the resulting maps were often of low quality.

Today, this situation has changed, as command-line interfaces have been replaced by **graphical user interfaces (GUIs),** which integrate the analysis and display components of GIS, are easier to learn than command-line interfaces, and provide a higher-quality cartographic output. ArcView is an example of popular GIS software that integrates analysis and display capabilities. To illustrate ArcView's capabilities, we'll use one of its tutorials (see pp. 15–29 of *Using ArcView GIS* for a full discussion of the tutorial). In the tutorial, you are asked to imagine that you wish to build a new showroom for a company in the southeastern United States. The showroom should be in a state where sales were low the previous year, in a city with more than 80,000 people, and within 300 miles of Atlanta (a presumed regional distribution center).

Determining potential locations for the showroom can be implemented as a three-step process in Arcview. First, the sales information can be displayed via a choropleth map (Color Plate 13.12A); for our purposes, we will assume that low sales are represented by the two lowest-valued classes on the map (the white and lightest red shades). Ideally, the sales information should be standardized to account for different populations of states, but here we use the raw data as provided in the tutorial. This step involves no GIS analysis (unless data classification is considered analysis), as we simply display a particular theme (sales) associated with each state.

The second step is to determine which cities have a population greater than 80,000. This is accomplished using a "query builder" function in ArcView; all cities selected using the query builder are *automatically* shown on the map as blue circles of a fixed size (Color Plate 13.12B). (The default color produced by ArcView was actually green; blue was used here because it contrasted better with the colors on the choropleth map.)

The third step is to determine which of the cities with a population greater than 80,000 are within 300 miles of Atlanta. In calculating distances of cities from Atlanta, ArcView again *automatically* displays all of the desired cities in yellow (Color Plate 13.12C). Presumably, cities in yellow that are located in those states falling in the two lowest-valued classes on the choropleth map meet the three criteria specified for locating the showroom.

Integrating analysis and display functions provides a number of advantages from a data exploration perspective. One advantage is that it enhances the efficiency of trying alternative what-if scenarios. For example, we might change the population and distance figures for locating a showroom to 150,000 and 200, respectively, and produce the map shown in Color Plate 13.12D. These sorts of changes and associated displays generally can be created in a matter of seconds.

A second advantage is that the GIS analyst can view data from a variety of perspectives. In the case of the showroom problem, ArcView permits viewing the data as a map, table, or chart (Color Plate 13.13); note that values highlighted on the map in yellow are also highlighted in the table in yellow, and that this highlighted subset is depicted within the chart.

A final advantage is that the GIS analyst can experiment with different symbol schemes to see if they reveal different patterns in the data. For example, the tutorial associated with the showroom example used a diverging color scheme (greens on one end and oranges on the other, with white in the middle). Normally, such a scheme would be restricted to bipolar data, but it could be argued that such a scheme might help the analyst locate the lower-valued sales states more quickly.

In examining the illustrations in Color Plates 13.12 and 13.13, readers unfamiliar with ArcView may have been bothered by the inability to see all of the numbers for the "Sales" portion of the "table of contents" (this is the left-hand portion of the screen and is similar in appearance to what would normally be termed a legend). The logic of not showing all of the numbers is that the GIS analyst is presumably familiar with the data (is exploring the data). Also, the analyst can easily see all of the numbers by simply changing the size of the table of contents (by using the mouse to drag the edge of the gray-shaded area). If, however, a map is to be made for presentation purposes, it is essential that GIS software provide a capability to create a map with a more conventional appearance. In ArcView, presentation is implemented using a "Layout" option; the result of using the Layout option for the showroom problem is shown in Color Plate 13.14.

13.9 OTHER EXPLORATION SOFTWARE

Other software for exploring data that readers are likely to find accessible include Tang's (1992) Visda and Yadav-Pauletti's (1996) MigMap. Visda provides capabilities for geographic brushing (section 3.3.3), interactive migration mapping (section 11.4.1), and a number of exploration tools similar to those found in ExploreMap. Yadav-Pauletti's MigMap illustrates various approaches for exploring migration data via choropleth and arrow symbology using both animation and small multiples (section 11.4.1). There has also been a fair amount of software developed that readers may find difficult or impossible to get access to. The most interesting of these include Miller's (1988) GAHM, Haslett et al.'s REGARD (1990, 1991, 1995), MacDougall's (1992) Polygon Explorer, Lee et al.'s (1994) XNV, and Scott's (1994) software for linking ArcView and STATA. Each of these is summarized briefly in the section on Further Reading.

SUMMARY

In this chapter, we have reviewed a variety of software that can assist in exploring spatial data. The purpose for exploring spatial data (or **data exploration**) is to reveal unknowns and is typically accomplished via a high degree of human-map interaction in a private setting; animation is frequently an important component of data exploration. Most of the software described here was developed after 1990, although Moellering's (1980) early work was important for developing a foundation for both data exploration and animation.

ExploreMap was one of the early data exploration software packages. Although intended solely for univariate choropleth mapping, it served as a useful prototype for other more sophisticated packages. The enumerated software within Project Argus is an example of one of these more sophisticated packages. The enumerated software can symbolize data not only in choropleth form, but as a proportional circle map, or cartogram, and it can handle up to three variables at one time. The enumerated software is also useful for illustrating some of the features we would expect to find in data exploration software. One such feature is the ability to display data in both map and graphical form: for example, single variables can be plotted as point graphs and two variables can be displayed as scatterplots. Another feature is the ability to link various displays of the data; for example, when an outlier is designated on a point graph, that point is also highlighted on all other displays currently in view.

Other data exploration software that we considered included MapTime, Vis5D, Aspens, Transform, and ArcView. MapTime permits users to explore temporal data associated with fixed point locations using three major approaches: animation, small multiples, and change maps. Vis5D enables users to explore true 3-D data that have both temporal and multivariate components. For example, we might use Vis5D to examine changes in the temperature, salinity, and current speed of some three-dimensional portion of the ocean over the course of the year. Aspens was intended for exploring a particular data set (the growth of trembling aspens), but is useful for illustrating the nature of **proactive graphics** (software that enables users to initiate queries using icons as opposed to words).

Transform provides a number of unique capabilities for exploring data. One is the use of specialized color schemes (such as a "Lava Waves" scheme) purported to bring out nuances in data. Another is the "Fiddle tool," which permits shifting and compressing of color schemes. Unfortunately, cartographers have not as yet fully experimented with such capabilities. ArcView is an example of modern GIS software that integrates analysis and display functions. Traditional GIS software separated such functions and thus was difficult to use. In contrast, modern GIS software enables users to readily try alternative what-if scenarios, to view data from a variety of perspectives (as a map, table, or chart) and experiment with different symbol schemes to see if different patterns are revealed.

In this chapter, we have seen that animation frequently is an important element of data exploration. In the following chapter, we delve into animation in greater depth, focusing primarily on stand-alone animations (those that have been developed outside of data exploration software and electronic atlases).

FURTHER READING

Buttenfield, B. P., and Weber, C. R. (1994) "Proactive graphics for exploratory visualization of biogeographical data." *Cartographic Perspectives* no. 19:8–18.

Discusses proactive graphics, which was used to create the data exploration software Aspens.

Dykes, J. A. (1997) "Exploring spatial data representation with dynamic graphics." *Computers & Geosciences* 23, no. 4:345–370.

Discusses using the programming language Tcl/Tk to create data exploration software. Tcl/Tk was used to create the software associated with Project Argus, for which Dykes was the lead programmer. Also see Dykes (1996).

Egbert, S. L., and Slocum, T. A. (1992) "ExploreMap: An exploration system for choropleth maps." *Annals, Association of American Geographers* 82, no. 2:275–288.

Discusses the development of ExploreMap and suggests a number of potential enhancements. See Egbert (1994) for information on how ExploreMap was evaluated.

Ferreira, J. J., and Wiggins, L. L. (1990) "The density dial: A visualization tool for thematic mapping." *Geo Info Systems* 1, no. 0:69–71.

Provides a detailed discussion of the data exploration software Xmap.

Haslett, J., and Power, G. M. (1995) "Interactive computer graphics for a more open exploration of stream sediment geochemical data." *Computers & Geosciences* 21, no. 1:77–87.

Describes how the data exploration software REGARD can be used to examine the geochemistry of stream sediment. For earlier developmental work, see Haslett et al. (1990, 1991).

Heywood, I., Oliver, J., and Tomlinson, S. (1995) "Building an exploratory multi-criteria modelling environment for spatial decision support." In *Innovations in GIS 2,* ed. P. Fisher, pp. 127–136. Bristol, Penn.: Taylor & Francis.

Describes the development of specialized software that can assist in exploring data commonly analyzed within a GIS.

Hibbard, W. L., Paul, B. E., Santek, D. A., Dyer, C. R., Battaiola, A. L., and Voidrot-Martinez, M.-F. (1994) "Interactive visualization of earth and space science computations." *Computer* 27, no. 7:65–72.

Describes some of the capability of Vis5D.

Lee, J., Chen, L., and Shaw, Shih-L. (1994) "A method for the exploratory analysis of airline networks." *The Professional Geographer* 46, no. 4:468–477.

Introduces the data exploration software XNV, which was developed to examine flow data associated with U.S. airlines.

MacDougall, E. B. (1992) "Exploratory analysis, dynamic statistical visualization, and geographic information systems." *Cartography and Geographic Information Systems* 19, no. 4:237–246.

Describes the data exploration software Polygon Explorer, which was developed for handling both univariate and multivariate data. Readers with some knowledge of cluster analysis will find the capability to handle multivariate data particularly intriguing.

Miller, D. W. (1988) "The great American history machine." *Academic Computing* 3, no. 3:28–29.

Summarizes the data exploration software GAHM, which was developed to encourage undergraduates to think like professional historians.

Moellering, H. (1980) "The real-time animation of three-dimensional maps." *The American Cartographer* 7, no. 1:67–75.

Describes Moellering's early efforts to explore and animate spatial data. For a related video, see Moellering (1978).

Patton, D. K., and Cammack, R. G. (1996) "An examination of the effects of task type and map complexity on sequenced and static choropleth maps." In *Cartographic Design: Theoretical and Practical Perspectives*, ed. C. H. Wood and C. P. Keller, pp. 237–252. Chichester, England: John Wiley & Sons.

An experimental investigation of sequenced versus static choropleth maps. Also see Slocum et al. (1990).

Scott, L. M. (1994) "Identification of GIS attribute error using exploratory data analysis." *The Professional Geographer* 46, no. 3:378–386.

Illustrates how data exploration software (a combination of ArcView and STATA) can be used to detect and correct the error associated with GIS coverages.

Tang, Q. (1992) "From description to analysis: An electronic atlas for spatial data exploration." *Proceedings of ASPRS/ACSM/RT 92 Convention on Mapping and Monitoring Global Change,* Vol. 3, pp. 455–463.

Describes the development of the data exploration software Visda.

Yoder, S. C. (1996) "The Development of Software for Exploring Temporal Data Associated with Point Locations." Unpublished MA thesis, University of Kansas, Lawrence, Kansas.

Provides a detailed discussion of the development of MapTime.

14

Map Animation

OVERVIEW

Animated maps **(maps characterized by continuous or dynamic change) are an important consequence of recent technological changes in cartography. As discussed in Chapters 13 and 15, animated maps can be part of data exploration software or electronic atlases; they can also exist as stand-alone entities, which is the focus of this chapter.*

Graphic scripts *(section 14.1) are dynamic maps, statistical graphics, and text blocks used to tell a story about a particular set of data. Monmonier intended graphic scripts as a form of data exploration, but implemented them as animations.*

In section 14.2, we consider fundamental work on animation by David DiBiase, Alan MacEachren, and their colleagues at Penn State University. DiBiase and colleagues have formulated a set of visual variables appropriate for animation, including **duration, rate of change, order, display date, frequency,** *and* **synchronization.** *They have also categorized animation according to whether the animation emphasizes (1) a change in a phenomenon's position or attribute, (2) the existence of a phenomenon at a particular location, or (3) an attribute of a phenomenon.*

Section 14.3 describes several examples of animations that are likely to be accessible to readers, including Wilhelmson et al.'s depiction of thunderstorm development, Treinish's portrayal of the ozone hole, Weber and Buttenfield's examination of climate change, Peterson's nontemporal approaches, and the use of animations in courseware and lectureware. The section also considers some animations that readers may not have access to, stressing the work of Dorling and Gershon.

This chapter focuses on animations that have been discussed by cartographers and other graphics specialists. Appendix E provides a more complete listing of animations, including many that are now available via the Internet. For further discussion of animation, also see sections 11.4 (flow mapping), 16.1 (data quality), and 16.5 (which summarizes a recent special electronic issue of Computers & Geosciences*). Many readers will, of course, want to develop their own animations. The last section of Chapter 15 describes a set of tools that can assist in this process.*

14.1 GRAPHIC SCRIPTS

Monmonier (1992a) defined a **graphic script** as a series of dynamic maps, statistical graphics, and text blocks used to tell a story about a particular set of data. As originally designed by Monmonier, a graphic script was displayed as an animation, without permitting the user any control over its speed or direction (whether it was played forward or backward). To illustrate, consider the portion of a graphic script shown in Figure 14.1, which Monmonier used to examine the distribution and relationship of the variables "Female Percentage of Elected Local Officials" and "Female Labor Force Participation Rate" for the 50 states. Monmonier's script is a "play in three acts":

Act I: Introduce the Variables

Act II: Variation and Covariation in Geographic Space and Attribute Space

Act III: Explore the Relationship with a Regression Model

For brevity, detail is provided only for Act I in Figure 14.1.

Figure 14.2 portrays a frame from Act I of the graphic script presented in Figure 14.1 (specifically it is associated with the line "Upward sweep by rank; downward sweep by rank"). The heights of individual bars beneath the map correspond to each of the values symbolized on the map. During an upward sweep, a bar is lit in the graph and then on the map, beginning with the lowest-ranked value and progressing to the highest; the opposite is true for a downward sweep. This approach is similar to the sequencing option used in ExploreMap. Other approaches used in Act I provide a slightly different representation of each variable.

Monmonier and Gluck (1994) evaluated the effectiveness of graphic scripts and found that users had difficulty understanding them because they had no control over them (they could not start and stop the scripts or control their speed). As a result, Monmonier and Gluck suggested that graphic scripts should automatically be stopped at key points, continuing only when the viewer pushes a "resume" key. Alternatively, they suggested that a "rerun" key would provide greater flexibility by allowing a user to repeat an immediately preceding sequence.

FIGURE 14.1 A portion of a graphic script. (After Monmonier, M. 1992a. First published in *Cartography and Geographic Information Systems* 19(4), p. 254. Reprinted with permission from the American Congress on Surveying and Mapping.)

Act I: Introduce the Variables

I-1. Introduce dependent variable "Female Percentage of Elected Local Officials, 1987," with brief title "Female Officials" and signature hue of red.
  Text block: full title, brief title, and description; pause for reading . . .
  Large map above rank-ordered bar graph, with each polygon linked to a bar.
 Upward sweep by rank [Figure 14.2]; downward sweep by rank.
 Upward sweep by value; downward sweep by value.
 Highlight highest fifth; highlight lowest fifth.
 Classify by quintiles (equal fifths); then blink-highlight each category (in sequence from highest category to lowest).
 Classify by equal thirds; then blink-highlight each category.
 Quintiles again; blink-highlight each of the nine census divisions . . .

I-2. Introduce independent variable "Female Labor Force Participation Rate, 1987," with brief title "Females Working" and signature hue of blue.
  Same layout and scenario as in I-1.

I-3. Compare geographic patterns by rapid alternation of the two variables on a single map.
  large map above hue-coded titles for both variables.
 Classify by quintiles; 13 cycles: display dependent variable on map, then display independent variable on map.

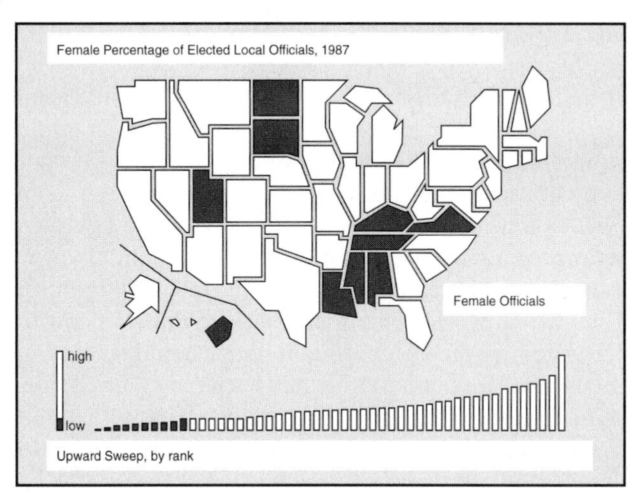

FIGURE 14.2 A frame from a graphic script illustrating "Upward sweep by rank; downward sweep by rank." Individual bars at the bottom correspond to values for each of the data shown on the map. During an upward sweep, a bar is lit in the graph and then on the map, beginning with the lowest-ranked value and progressing to the highest; the opposite is true for a downward sweep. (After Monmonier, M. 1992a. First published in *Cartography and Geographic Information Systems* 19(4), p. 248. Reprinted with permission from the American Congress on Surveying and Mapping and Mark Monmonier.)

One problem that I encountered while viewing Monmonier's graphic scripts was the lack of a verbal commentary. Monmonier and Gluck avoided this problem by having a trained moderator provide a running commentary during their evaluation of the scripts. Apparently, this was essential, as one of the participants remarked "It was all logical, but without the narrative it would have been difficult. . . . If I had to read the description and watch what was going on, I would have had a hard time" (p. 43). Monmonier (1992a) indicated that the lack of verbal commentary was a function of the large storage space required by high-quality speech; hopefully, those wishing to develop graphic scripts will find this less of a problem as technology improves.

14.2 THE FUNDAMENTAL WORK OF DIBIASE AND COLLEAGUES

In 1991, David DiBiase, Alan MacEachren, and their colleagues created a video in which they formulated visual variables unique to animation and categorized various types of animation (DiBiase et al. 1991). This video was supplemented by an article (DiBiase et al. 1992), and the visual variables were later expanded upon by MacEachren (1995). Below we consider the character of these visual variables and the various types of animation.

14.2.1 Visual Variables for Animation

In their video, DiBiase et al. described how the visual variables for static maps introduced in section 2.3 could be applied to animated maps. For example, they animated a choropleth map of New Mexico population from 1930 to 1980 to show how the visual variable lightness (as represented by gray tones) can represent changes over time, and they transformed the state of Wisconsin from its real-world geographic shape to that of a cow to illustrate how the visual variable shape could be used in animation. More importantly, however, they showed how animation requires additional visual variables beyond those used for static maps, namely duration, rate of change, and order.

In its simplest form, **duration** is defined as the length of time that a frame of an animation is displayed, with short duration resulting in a smooth animation and long duration a choppy one. More generally, a group of frames can be considered a scene (in which there is no change between frames), and we can consider the duration of that scene.* Since duration is measured in quantitative units (time), it can be logically used to represent quantitative data; for example, in their video DiBiase et al. animated electoral college votes for president, varying the duration of each scene in direct proportion to the magnitude of the victory.

Rate of change is defined as m/d, where m is the magnitude of change between frames or scenes, and d is the duration of each frame or scene. Rate of change can be defined for either geographic position or an attribute. For example, Figure 14.3 illustrates different rates of change for the position of a point feature and the attribute of circle size. The smoothness (or lack thereof) in an animation is a function of the rate of change. Either decreasing the magnitude of change between frames or decreasing the duration of each frame will decrease the rate of change and hence make the animation appear smoother.

Order is the sequence in which frames or scenes are presented. Normally, of course, animations are presented in chronological order, but DiBiase et al. argued that sometimes important knowledge can be gleaned by reordering frames or scenes. For example, they modified not only the duration of frames associated with the presidential election data, but also reordered them such that lower magnitudes of victory were presented first. I found the reordered animation difficult to interpret, but this may have been due to my inability to control the video easily (by stopping it, starting it, and changing its speed).

More recently, MacEachren (1995) has extended the visual variables for animation to include display date, frequency, and synchronization. He defines **display date** as the time some display change is initiated. For example, in an animated map of population for U.S. cities, we might note that a circle first appears for San Francisco in 1850 (if decennial census data are used).

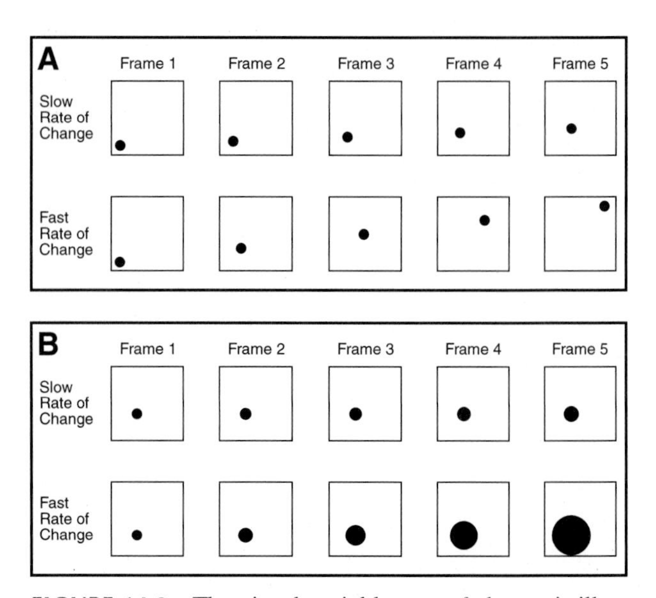

FIGURE 14.3 The visual variable rate of change is illustrated for geographic position (A) and circle size (B). Rate of change is defined as m/d, where m is the magnitude of change between frames or scenes, and d is the duration of each frame or scene. For these cases, duration is presumed constant in each frame. (After DiBiase et al. 1992, 205.)

Frequency is the number of identifiable states per unit time, or what MacEachren calls *temporal texture.* To illustrate frequency, he notes its effect on color cycling, in which various colors are "cycled" through a symbol, as is often done on weather maps to show the flow of the jet stream.

Synchronization deals with the temporal correspondence of two or more time series. MacEachren (pp. 285–286) indicates that if the "peaks and troughs of . . . [two] time series correspond, the series are said to be 'in phase,' " or synchronized.

14.2.2 Categories of Animation

In addition to introducing visual variables for animation, DiBiase et al. developed a useful categorization of animated maps, as a function of whether the animation emphasizes (1) change in a phenomenon's position or attribute, (2) the existence of a phenomenon at a particular location, or (3) an attribute of the phenomenon.

Animations Emphasizing Change

Animations emphasizing change in a phenomenon's position or attribute can be further divided into three types: time series, re-expressions, and fly-bys. A **time series,** or

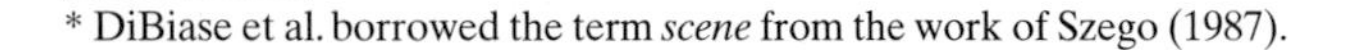

* DiBiase et al. borrowed the term *scene* from the work of Szego (1987).

an animation that emphasizes change through time, is most common. Examples would include those animations described for MapTime and Vis5D in the preceding chapter. In their video, DiBiase et al. illustrated several time series animations, including a cartogram of world population growth and the spread of AIDS in Pennsylvania.

Borrowing from the work of Tukey (1977), DiBiase et al. defined **re-expression** as an "alternative graphic representation . . . whose structure has been changed through some transformation of the original data" (p. 209). Re-expression might involve choosing subsets of a time series (DiBiase et al. termed this brushing), reordering a time series, or changing the duration of individual frames or scenes within a time series. To illustrate brushing, DiBiase et al. animated the longest sequence of victories by candidates of a single party in U.S. presidential elections (the Democratic party won all elections from 1932 to 1948).

Examples of re-expression involving reordering and changing duration were already described in the preceding section dealing with visual variables for animation. As another interesting example, DiBiase et al. animated five global-climate-model predictions for Puebla, Mexico. In reordering the temperature predictions from those that varied the least to those that varied the most, and then pacing them such that those with the greatest variation had the greatest duration, DiBiase et al. found that the "re-expression reveal[ed] a trend of maximum uncertainty during the spring planting season of April, May, and June, a pattern that had previously gone unnoticed" (p. 211).

In a **fly-by,** a user is given the feeling of flying over a three-dimensional surface. A classic example is *L.A. The Movie,* which illustrates a high-speed flight over Los Angeles, California, and vicinity (segment five of the "Best of JPL," see Appendix E). The realistic appearance of the landscape in this case was achieved by merging Landsat imagery with digital elevation data. It is important to recognize that flight in such fly-bys has usually been governed not by the user, but by the mapmaker. As a result, it is my experience that users find such fly-bys difficult to follow, although they do provide a novel look at the landscape.

Traditionally, creating fly-bys was beyond the scope of most mapmakers because of the considerable computing time required (*L.A. The Movie* required 130 hours of CPU time on a VAX 8600 mainframe). Today, this is changing as mapmakers are now able to create their own fly-bys, even within a personal computer environment. One example of this is the Truflite software, which was described in section 9.2.6. Other software with similar capability is listed in Appendix G.

In addition to empowering mapmakers to create their own fly-bys, the most sophisticated hardware and software also now enable users to control the direction of flight while the surface is being viewed, as opposed to having the path of flight governed by the mapmaker. An example is the ability of users of Vis5D (see section 13.5) to easily change their viewpoint in three-dimensional space when examining true 3-D data either for a particular time step or during an animation.

Animations Emphasizing Location

Although a static map can be used to *indicate* the location of phenomena, animation can assist in *emphasizing* location. For example, DiBiase et al. used flashing-point symbols (circles) to emphasize the location of major earthquakes (a static rendition of this concept is shown in Figure 14.4). Flashing symbols do not necessarily help users interpret a distribution, but they do attract attention, and for earthquake data serve as a sort of warning signal ("It happened once, it could happen here again").

Animations Emphasizing an Attribute

Animations can also emphasize an attribute of a phenomenon by highlighting selected portions of it. One example is the sequencing used in ExploreMap (section 13.2), in which choropleth map classes are presented in order from low to high. Another example would be Moellering's (1976) traffic-accident animation, which combined symbol duration with size to indicate severity of accidents at intersections.

14.3 EXAMPLES OF ANIMATIONS

In this section we consider a number of animations that have been discussed by cartographers and graphics specialists. Appendix E provides a more complete listing of animations, including many that are now available via the Internet.

14.3.1 Wilhelmson et al.'s Depiction of Thunderstorm Development

One of the most dramatic animations produced has been Wilhelmson et al.'s (1990) video depicting a numerical model of a developing thunderstorm (available as the first segment of *Scientific Visualizations Volume I;* see Appendix E). This animation enables the viewer to examine a thunderstorm's interior via attractive colored ribbon–like streamers and buoyant balls. Although developed primarily for research purposes, animations like this could be used to teach students basic concepts about thunderstorm development.

It is important to recognize that such dramatic animations cannot be developed without considerable effort. In this case, several researchers knowledgeable about storm development worked closely with the Vi-

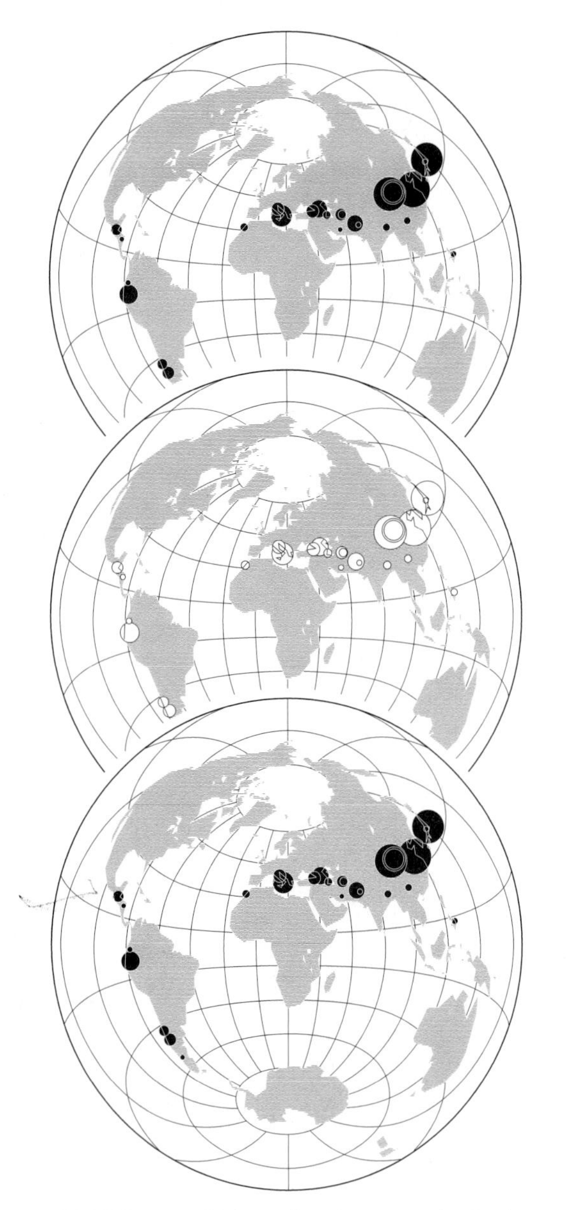

FIGURE 14.4 Using animation to emphasize location. Point symbols are flashed off and on to highlight the location of major earthquakes from 1900 to 1990. (After DiBiase et al. 1992. First published in *Cartography and Geographic Information Systems* 19(4), p. 207. Reprinted with permission from American Congress on Surveying and Mapping.)

sualization Services and Development Group at the National Center for Supercomputing Applications. Eleven months were required for development, including one person-year of effort from four scientific animators, along with input from scriptwriters, artistic consultants, and postproduction personnel. Also required was a substantial hardware component: "about 200 [hours] of computer time on a Silicon Graphics Iris 4D/240 GTX using Wavefront software and special software written by the visualization group" (Wilhelmson et al. 1990, 33).

14.3.2 Treinish's Portrayal of the Ozone Hole

Another striking animation is Treinish's (1992) *Climatology of Global Stratospheric Ozone,* which depicts the development of the ozone hole over Antarctica.* Levels of ozone (in columns of the earth's atmosphere) are symbolized using three redundant variables: height above the earth's surface (as a 2½-D surface), a traditional rainbow spectral scheme (red-orange-yellow-green-blue), and the degree of opacity (this is analogous to the transparent fog used in Vis5D). Thus, low ozone values appear in blue, are close to the earth, and are transparent, while high ozone values are red, far from the earth, and opaque. The ozone symbolization is shown wrapped around a global representation of the earth, which is colored to represent the earth's topography (Color Plate 14.1A). The complete animation consists of 4700 frames, one for each day over the period 1979–1991.

Although the ozone animation is intriguing, it is difficult to follow for three reasons. First, the entire three-dimensional structure (globe plus ozone symbolization) is frequently rotated during the animation to show interesting areas for a particular time of year; as a result, it is not possible to focus on a specific region over the entire period of observation. Second, since the animation is distributed as a video, the user is not able to easily stop and start it or focus on a selected time period, as would be possible with data exploration software. Third, the satellite that collected the data for the animation functioned on reflected sunlight, so the Antarctic winter appears as a gap in the animation; unfortunately, it is difficult to differentiate this gap from the true ozone hole.

In spite of these problems, the video claims that "the region of Antarctic ozone depletion each spring can be easily seen growing in size and severity over the last decade." I did not see this feature, but this may be due to my lack of familiarity with the ozone data, as DiBiase et al. (1992, 213) argue that those with expert knowledge about a problem are often more likely to benefit from animations. Those viewing the animation may also have had access to small multiples of the data, such as those shown in Color Plate 14.1B. These small multiples make it possible to compare one year directly with another and thus note that the ozone hole does appear larger in recent years.

* For further discussion of the ozone data on which the animation is based, see Treinish and Goettsche (1991) and Treinish (1992). A one-year animation of ozone data can be found at http://www.almaden.ibm.com/dx/samples/EarthScience.html.

14.3.3 Weber and Buttenfield's Examination of Climate Change

In recent years, there has been considerable discussion of the increased level of carbon dioxide in the earth's atmosphere and its potential impact on global warming. In considering this issue, it is natural to ask whether actual measured temperature data support the claim for global warming. To assist in answering this question, Weber and Buttenfield (1993) animated mean annual temperature data for the 48 contiguous states over a 90-year period. In contrast to the two previous animations, Weber and Buttenfield's was distributed in diskette form, thus permitting the user greater control over the speed and direction of the animation.

Raw data for the animation consisted of monthly average temperatures for 344 weather stations from 1897 to 1986. Using these raw data, yearly average temperatures were computed from October to September "to preserve yearly highs (summers) and lows (winters) in seasonal groupings" (Weber and Buttenfield, 1993, 144). Also, a three-year moving average was computed to smooth the data for purposes of animation. The actual animation is portrayed as positive or negative deviations from the mean of the 90-year period.

Individual frames of the animation illustrate a logical use of color for bipolar data (Color Plate 14.2): shades of red are used for positive deviations or temperatures warmer than normal, while shades of blue are used for negative deviations or temperatures cooler than normal. Each frame also contains a graph clearly summarizing deviations from the 90-year mean for each year, with the position of the current frame depicted by a line on the graph (see the lower right of Color Plate 14.2).

Weber and Buttenfield claim that visual patterns in the animation match those of other climatic research:

> The animation affirms a general cooling trend over the past 35 years, as well as a temporal pattern of cooling in the East and warming in the West in the last 20 years. . . . Isolines characterizing colder years are organized in a few clearly defined shapes; isoline patterns characterizing the warmest years are broken up, and these frames appear visually turbulent. (p. 147)

Although individual maps within the animation are logically designed and rather attractive, I find the overall animation difficult to follow, largely because the pattern for one year does not merge smoothly into the pattern of a preceding year, but also possibly because of my lack of experience with spatiotemporal characteristics of climate change. This again is a case in which those familiar with the phenomenon may derive greater benefit from animation; small multiples might also enable users to better understand the patterns in such data.

14.3.4 Peterson's Nontemporal Methods

The three animations described thus far all involved temporal data. In contrast, Peterson's (1993) MacChoro II software permits animating nontemporal univariate and multivariate data using choropleth symbolization. For univariate data, Peterson advocates animating both the classification method (say, equal intervals, quantiles, maximum breaks, and optimal) and the number of classes. For example, rather than displaying all four classification methods at once (as in Color Plate 4.1), MacChoro II displays them in succession as an animation.

For multivariate data, Peterson suggests that animation can be used to examine the geographic trend of related variables (those measured in the same units and part of a larger whole), and compare unrelated variables (those measured in dissimilar units and not part of a larger whole). With respect to related variables, we might animate the percentage of population in various age groups (0–4 years old, 5–9 years old, etc.) within census tracts of a city. Peterson argues that such data "will usually show a clear regionalization in a city with older populations closer to the center and younger populations nearer the periphery. An animation . . . from low to high depicts high values 'moving' from the periphery to the center" (pp. 41–42). Two unrelated variables would be median income and percent of the adult population with a high school education. Peterson argues that an animation of such variables (he recommends switching rapidly between the variables) would detect similarities or differences in the two distributions.

Although Peterson's ideas for animating nontemporal data are intriguing, his animations are difficult to follow. A small-multiples approach seems to be more effective than animation for analyzing such nontemporal data, primarily because the transitions between frames of the animation are not smooth. In fact, Peterson (p. 41) himself demonstrates his ideas in static form using small multiples.

14.3.5 Animations for Courseware and Lectureware

Numerous animations are now being used in courseware and lectureware. **Courseware** is software used to assist students in learning concepts outside the classroom, while **lectureware** is analogous software used in a more traditional lecture setting (Krygier et al. 1997). Appendix E lists several examples of geographic courseware that include animations. One of the more effective examples is *The Theory of Plate Tectonics.* An example of geographic lectureware is the *Educational Multimedia Resources* developed in the Deasy GeoGraphics Laboratory at Penn State University (Krygier et al. 1997). This lectureware contains numerous excellent examples of animation; two of the more interesting are demonstrations of the coriolis effect and how water flows deep beneath the surface of the ocean. In the latter case, one can actually "see" the water flowing; such animations should be much more effective for students than static images.

14.3.6 Other Examples of Animation

Other examples of animations include the USGS's portrayal of urban sprawl in the San Francisco–Sacramento area (Gaydos et al. 1995; Buchanan and Acevedo 1996), and the *Northern Great Plains Greenness Mapping* developed by USGS and the Eros Data Center (1989). In contrast to some of the other animations I have discussed, these are easy to follow because of their smooth transitions (Color Plate 14.3). A more complete list of available animations is given in Appendix E.

A number of animations have been developed that readers are unlikely to have access to, but which are worthy of discussion because they provide a foundation for subsequent work in animation. Perhaps most interesting is the work of Dorling (1992), who experimented with animation in three realms, which he designated "animating space," "animating time," and "three-dimensional animation." According to Dorling, animating space permits "a portion of a high-resolution picture to be shown in a window and instantaneously enlarged, reduced, and scanned as the user's hand moves a mouse or presses a button" (p. 216). To illustrate this, he described how a map of three occupational variables in the more than 100,000 enumeration units making up British census districts could be examined:

> What emerged was a picture of the way in which areas of affluence simultaneously surrounded, looped around, and appeared inside the places where the less prosperous lived. . . . One of the most interesting patterns occurs in inner London, where a thin, snakelike belt of "professionals" winds from north to south, between areas sharply differentiated as being dominated by the housing of those working in the lowest status occupations. (p. 217)

Although Dorling used the term *animation* to describe his interaction with the map, I feel it is more appropriate to describe the process as a form of data exploration. The instantaneous change in the data is a form of animation (or dynamic change), but the control exerted by the user makes *exploration* a more appropriate term.

Dorling's idea of animating time roughly equates to DiBiase et al.'s notion of time-series animation. Dorling argued that animating time is not successful when animations involve changes in color at fixed locations; he attributed this to "the brain's poor visual memory" (p. 223). His conclusion concurs with my difficulty of interpreting some of the animations we have considered (for example, Weber and Buttenfields' animation of climate change, which involved changes in the color of contour maps at fixed locations, and Peterson's animation of nontemporal data associated with choropleth maps). However, Dorling argued that animating *movement* over time can be vary effective; he claimed that his most effective animation involved moving dots within a triangle representing the proportion of votes for three major political parties (pp. 218–219).

Dorling's idea of three-dimensional animation describes the animation of data in three-dimensional space; an example would be wind speed within Vis5D. As with animations of color at fixed locations, Dorling found these confusing, arguing that although the computer can easily create three-dimensional graphics, it is more appropriate to represent data in two-dimensional form.

Dorling appeared to concur with my recommendation for using small multiples as a substitute for animation, but he indicated that this might require an animation of space (data exploration). He stated:

> Comparison requires at least one simultaneous view. The distribution of successive years' unemployment rates is often best shown on a single map, where symbols can become very small when a large number of areas are involved. Animating space can allow detailed spatial investigation of such a complex picture, which incorporates a temporal variable statically. (p. 224)

In summary, although Dorling criticized certain aspects of animation, he felt it could be especially useful for enlivening presentations.

Other animations that readers may find intriguing to read about include those by Gershon (1992), Openshaw et al. (1993), and Cox (1990). Gershon's are interesting because they contradict some of my comments about the potential difficulties of interpreting animations.

One of Gershon's approaches involved an "animation of segmented components" in which users were shown a shaded contour map in a sequenced presentation: a high narrow range of temperature appeared first, followed by a progressively wider range until the entire range to be focused on appeared; then the process was reversed with progressively narrower ranges of temperature displayed until just the highest narrow range appeared again. Gershon indicated that with this approach most of an audience of about 20 viewers "perceived the existence of structures, their locations and shapes better than in the static display of the data" (p. 270). This result is a bit surprising given the lack of success with sequencing on choropleth maps noted in section 13.2. Gershon's superior results may be a function of his particular sequencing strategy; or possibly, sequencing is more appropriate for contour than choropleth maps.

In another approach, Gershon had subjects examine the correlation between two variables as he quickly switched between maps of the variables. This approach is analogous to Peterson's suggestion for animating more than one variable. In contrast to the difficulty I had understanding Peterson's animations, Gershon argued that his audience could "visually correlate structures appearing in both [maps]" (p. 272). I suspect that Gershon's superior results were due to the similar structure of his maps: they were both of January sea surface temperatures, but for different years. As a result, users could easily focus on the differences between the maps.

SUMMARY

In this chapter, we have examined several matters pertaining to animation. First, we considered the notion of the **graphic script,** a series of dynamic maps, statistical graphics, and text blocks used to tell a story about a particular set of data. Although Monmonier initially developed graphic scripts as animations (with no potential for user interaction), it is apparent that graphic scripts will be most effective when the user has some control over the script (for example, to be able to stop and start it, and replay a portion). Graphic scripts will also be most effective if a verbal commentary is included along with maps and graphics.

Second, we considered a series of visual variables appropriate for animation, along with a categorization of various types of animation. Visual variables appropriate for animation include **duration** (the length of time that a frame of an animation is displayed), **rate of change** (m/d, where m is the magnitude of change between frames or scenes, and d is the duration of each frame or scene), **order** (the sequence in which frames or scenes are presented), **display date** (the time some display change is initiated), **frequency** (the number of identifiable states per unit time), and **synchronization** (the temporal correspondence of two or more time series). Animated maps can be categorized as a function of whether the animation emphasizes change in a phenomenon's position or attribute, the existence of a phenomenon at a particular location, or an attribute of the phenomenon. Animations emphasizing change can be further divided into three types: **time series, re-expressions,** and **fly-bys.** A time series is the most common form of animation because animating time can serve as a scale model of real world time. Some of the most interesting animations, however, do not involve time. For example, a fly-by, in which the viewer is given the impression of flying over a landscape, can be particularly dramatic.

Third, we considered several examples of stand-alone animations. Among these were Wilhelmson et al.'s depiction of thunderstorm development, Treinish's portrayal of the ozone hole, and Weber and Buttenfield's examination of climate change. Although such animations are intriguing, they are often difficult to understand, particularly for naive users. It appears that animation may be most effective when the user has some control over the pace and direction of the animation, and when the information can be displayed in a variety of forms (for example, as either an animation or a small multiple).

We have focused here on animations that have been discussed in accompanying literature by cartographers and other graphics specialists. Appendix E provides a more complete listing of animations, including many that are now available via the Internet. Our focus has been on stand-alone animations; bear in mind that animations can also be part of data exploration software (see Chapter 13) and electronic atlases (see Chapter 15). For readers interested in developing their own animations, the last section of Chapter 15 describes a set of tools that can assist in this process. Chapter 16 will consider recent uses of animation to portray data quality and in the burgeoning area of visualization.

FURTHER READING

Buchanan, J. T., and Acevedo, W. (1996) "Studying urban sprawl using a temporal database," *Geo Info Systems* 6, no. 7:42–47.

Uses animation to examine the growth of urban areas. Also see Batty and Howes (1996a, b).

Buttenfield, B. P., Weber, C., MacLennen, M., and Elliot, J. D. (1991) "Bibliography on Animation of Spatial Data." Technical Paper 91-22, National Center for Geographic Information and Analysis.

A dated but still useful bibliography on the animation of spatial data; includes a list of videotapes containing animations.

Cammack, R. G. (1991) "Cartographic animation: An exploration of a mapping technique." Unpublished M.S. thesis, University of South Carolina, Columbia, South Carolina.

An experimental study of animated choropleth maps.

Campbell, C. S., and Egbert, S. L. (1990) "Animated cartography: Thirty years of scratching the surface." *Cartographica* 27, no. 2:24–46.

A review of early animation efforts in cartography; explores the reasons for the lack of animation in cartography prior to 1990 and suggests future prospects for animation.

Cox, D. J. (1990) "The art of scientific visualization." *Academic Computing* 4, no. 6:20–22, 32–34, 36–38, 40.

Discusses a number of animations developed by Donna Cox in association with those in other disciplines at the National Center for Supercomputing Applications, where so-called Renaissance Teams were formed.

DiBiase, D., MacEachren, A. M., Krygier, J. B., and Reeves, C. (1992) "Animation and the role of map design in scientific visualization." *Cartography and Geographic Information Systems* 19, no. 4:201–214, 265–266.

Discusses visual variables for animation and categories of animation (these were introduced in section 14.2). See DiBiase et al. (1991) for a video demonstrating these concepts.

Dorling, D. (1992) "Stretching space and splicing time: From cartographic animation to interactive visualization." *Cartography and Geographic Information Systems* 19, no. 4:215–227, 267–270.

Provides a detailed discussion of Dorling's methods for animating spatial data, which were introduced in section 14.3.6.

Gersmehl, P. J. (1990) "Choosing tools: Nine metaphors of four-dimensional cartography." *Cartographic Perspectives* no. 5:3–17.

Discusses various ways in which software can be used to create animations; Gersmehl termed these "animation metaphors."

Greenfield, D. (1994) "Animating point symbols for cartographic display." Unpublished M.A. Thesis, University of Kansas, Lawrence, Kansas.

An experimental study of the effectiveness of various methods for animating data collected at point locations; one conclusion is that traditional proportional symbols are more effective than pictographic symbols.

Karl, D. (1992) "Cartographic animation: Potential and research issues." *Cartographic Perspectives* no. 13:3–9.

Introduces a number of research issues associated with animation.

Koussoulakou, A. (1994) "Spatial-temporal analysis of urban air pollution." In *Visualization in Modern Cartography,* ed. A. M. MacEachren and D. R. F. Taylor, pp. 243–267. Oxford: Pergamon Press.

Discusses and illustrates how animation can be used to display changes in air pollution over time.

Koussoulakou, A., and Kraak, M. J. (1992) "Spatio-temporal maps and cartographic communication." *The Cartographic Journal* 29, no. 2:101–108.

An experimental study comparing the effectiveness of animation and small multiples. In contrast to some of the comments I have made in this chapter, animation was found to be more effective than small multiples.

Krygier, J. B., Reeves, C., DiBiase, D., and Cupp, J. (1997) "Design, implementation and evaluation of multimedia resources for geography and earth science education." *Journal of Geography in Higher Education* 21:17–38.

Discusses the development and evaluation of multimedia resources for geography and earth science education. Many of the resources the authors have developed include map animations. The resources can be found at http://www.ems.psu.edu/Resources/Resources.html.

Monmonier, M. (1990) "Strategies for the visualization of geographic time-series data." *Cartographica* 27, no. 1:30–45.

Discusses a number of methods for visualizing spatiotemporal data; animation is considered one of the useful methods.

Monmonier, M. (1992a) "Authoring graphic scripts: Experiences and principles." *Cartography and Geographic Information Systems* 19, no. 4:247–260, 272.

A detailed discussion of graphic scripts, which were introduced in section 14.1. Also see Monmonier (1994a) and Monmonier and Gluck (1994).

Monmonier, M. (1992b) "Summary graphics for integrated visualization in dynamic cartography." *Cartography and Geographic Information Systems* 19, no. 1:23–36.

Presents some static maps and graphics that might be used in association with animated maps.

Monmonier, M. (1994b) "Minimum-change categories for dynamic temporal choropleth maps." *Journal of the Pennsylvania Academy of Science* 68, no. 1:42–47.

Discusses the problem of selecting boundaries for data classes on animated choropleth maps.

Monmonier, M. (1996) "Temporal generalization for dynamic maps." *Cartography and Geographic Information Systems* 23, no. 2:96–98.

Describes a variety of methods for smoothing map animations.

Peterson, M. P. (1995) *Interactive and Animated Cartography.* Englewood Cliffs, N.J.: Prentice Hall.

The focus of the present chapter is on the appearance of animations, as opposed to how they are created. In contrast, Peterson's text provides useful information on creating animations; see particularly Chapters 7–9.

Treinish, L. A. (1993) "Visualization techniques for correlative data analysis in the earth and space sciences." In *Animation and Scientific Visualization: Tools and Applications,* ed. R. A. Earnshaw and D. Watson, pp. 193–204. London: Academic Press.

Discusses the portrayal of the ozone hole, introduced in section 14.3.2. For other interesting work by Treinish, see http://www.research.ibm.com/people/l/lloydt/

Weber, C. R., and Buttenfield, B. P. (1993) "A cartographic animation of average yearly surface temperatures for the 48 contiguous United States: 1897–1986." *Cartography and Geographic Information Systems* 20, no. 3:141–150.

Reviews various animation efforts and discusses the creation of the animated map introduced in section 14.3.3.

Wilhelmson, R. B., Jewett, B. F., Shaw, C., Wicker, L. J., Arrott, M., Bushell, C. B., Bajuk, M., Thingvold, J., and Yost, J. B. (1990) "A study of a numerically modeled storm." *International Journal of Supercomputer Applications* 4, no. 2:20–36.

Discusses the creation of the animated map introduced in section 14.3.1.

15

Electronic Atlases and Tools for Developing Your Own Software

OVERVIEW

The first two sections of this chapter focus on electronic atlases, a topic of considerable interest to cartographers in recent years. Section 15.1 deals with the problem of defining an electronic atlas. For this purpose, we will see that it is helpful to begin by considering the nature of paper atlases. If asked to envision a ***paper atlas,*** *many of us conceive of "a bound collection of maps." Actually, modern paper atlases consist of not only maps, but also text, photographs, tables, and graphs, which when combined tell a story about a particular place or region.* ***Electronic atlases*** *can emulate such characteristics, or they can take advantage of the capabilities that computer-based systems provide, permitting data exploration, animation, and multimedia. Furthermore, electronic atlases have the potential to permit users to create their own maps and* analyze *spatial data.*

Section 15.2 presents several examples of electronic atlases. We'll begin by considering, the Domesday Project, *an early electronic atlas that was developed to commemorate the 900th anniversary of the* Domesday Book, *which was used in England primarily to determine land ownership and associated levels of taxation. The* Domesday Project *is still considered one of the premier electronic atlases ever developed, in part because of the sheer quantity of data it provided (more than 250 megabytes of digital data, 50,000 photographs, and 20 million words of text). Next, we'll consider a simple example of an Internet-based atlas, the* Digital Atlas of California. *Like a conventional paper atlas, the California atlas is composed of static maps. It is, however, available free to anyone who has Internet access, and the size of the atlas is limited only by the storage space available on a computer server.*

It is useful to contrast state and commercial atlases. State atlases generally are funded by nonprofit organizations, have a relatively small market (principally state residents and nationwide libraries), and a narrow range of use (most likely educational). In contrast, commercial atlases are funded by profit-making companies, have a very large market (potentially the entire world), and a wide range of use (educational, business, and travel). These differences are reflected in the look and feel of the resulting products (Smith and Parker 1995). To illustrate these differences, we will consider some of the capabilities of a state atlas, The Interactive Atlas of Georgia, *and two commercial atlases:* World Atlas *and the* Microsoft Encarta Virtual Globe. *The latter is representative of the degree of sophistication found in modern commercial electronic atlases.*

The next two examples we'll consider are particularly illustrative of the potential that multimedia offers. First, we'll examine Death Valley: An Animated Atlas, *which provides a series of narratives that describe the character of Death Valley via maps (many of which are animated), photographs, graphs, and spoken text. Second, we'll consider the* Grollier Multimedia Encyclopedia. ***Multimedia encyclopedias,*** *like paper encyclopedias, provide abundant information about places, people, and events, but the multimedia variant is enhanced through the inclusion of sound, video, and the ease of linking various topics. Particularly intriguing in the Grollier encyclopedia is a series of narrated map animations.*

Finally, we'll consider the TIGER Mapping Service, which illustrates the difficulty of defining what constitutes an electronic atlas. The TIGER Mapping Service permits users to create maps of base data (for example, a grid of latitude and longitude, streets, and railroads) and/or thematic data (for example, family income) for anywhere in the United States from the state level down to the block-group level. The service can be viewed as an atlas in the sense that a large variety of maps of a particular area (the United States) can be created reasonably quickly (generally on the order of seconds, although during slow Internet periods, several minutes may be required). The term atlas *may, however, be inappropriate because maps in the system are not already constructed, but must be created interactively.*

There are several cautionary notes that should be kept in mind with reference to the atlases we will consider. First, it is important to recognize that the discussions of individual atlases are not intended as reviews, but rather to illustrate the variety of characteristics found in electronic atlases today. Comprehensive reviews can be found in a variety of sources, including journals (for example, Cartography and Geographic Information Systems*) and newsletters (for example, the* Microcomputer Specialty Group of the Association of American Geographers Newsletter*). Second, bear in mind that many of these atlases are updated frequently, so the content may vary from what is presented here. Third, no attempt has been made to point out all the strengths and weaknesses of each atlas. A useful exercise would be to evaluate the correctness of symbology used in each of these atlases (for example, you might consider whether color schemes for choropleth maps are appropriately used). Finally, it must be recognized that only a sampling of atlases is provided here, and that new electronic atlases are constantly being developed. Refer to Appendix F and the home page for this book (http://www.prenhall.com/slocum) for information on other electronic atlases.*

Section 15.3 considers a range of tools for those wishing to develop their own electronic atlases, data exploration software, or map animations. The tools can be grouped into four categories: animation software, programming languages, authoring software, and visualization software. ***Animation software*** *is expressly intended for developing animations. Developing mapping software with traditional programming languages such as Basic, Fortran, Pascal, and C was tedious because of the lengthy computer code that had to be written. Today this is changing as* ***visual programming languages,*** *such as Visual Basic, ease software development by utilizing GUIs (graphical user interfaces).* ***Authoring software*** *provides a flexible set of graphical tools and multimedia/hypermedia capability that enable cartographic applications to be developed more easily than do programming languages. The full capability of authoring software cannot be achieved, however, without some knowledge of scripting languages.*

Visualization software *has been explicitly developed to create visualizations of spatial data. One form of visualization software, termed* ***data flow software*** *(or* ***modular visualization environments****) contains separate modules for importing, manipulating, and displaying data, all linked in a visual programming environment. Although visualization software holds considerable promise for creating sophisticated geographic visualizations, it is expensive, traditionally has been available only for workstations, often does not include routines that would be of interest to geographers, and generally requires some programming expertise to operate most effectively.*

15.1 DEFINING ELECTRONIC ATLASES

In attempting to define an electronic atlas, it is useful to consider the nature of paper atlases. One simple definition for a paper atlas is "a bound collection of maps." Although this definition is appropriate for many traditional atlases, it fails to consider the nature of modern paper atlases, which contain not only maps, but also text, photographs, tables, and graphs and frequently tell a story about a particular place or region (Keller 1995). For example, we might contrast the classic *Goode's World Atlas* (Espenshade 1990) with the *Historical Atlas of Massachusetts* (Wilkie and Tager 1991). The former consists of a wide variety of thematic and general-reference maps, with virtually no text or other graphic material, while the latter introduces the history of Massachusetts through a multitude of maps, but also includes ample text, graphs, and photographs. One important characteristic of both traditional and modern paper atlases is that they generally utilize a limited number of map scales and levels of generalization because of the physical constraints of the paper format; for example, a state atlas might focus on information at the county level, as opposed to showing information at both the county and township level.

In their simplest form, electronic atlases emulate the appearance of traditional paper atlases. An early example was the *Electronic Atlas of Arkansas* (Smith 1987), which consisted of alternating screens of maps and text; no interaction was permitted with either the maps or the text. A more recent example is the *Digital Atlas of California* (available at http://130.166.124.2/CApage1.html), which consists solely of thematic maps of California census tracts and block groups (see section 15.2.2 below for a more complete discussion). Like the Arkansas atlas, no interaction is possible with maps in the California atlas.

Although the digital emulation of paper atlases can be useful (for example, the *Digital Atlas of California* is available for free to anyone who has access to the Internet), there are many other reasons for creating an electronic atlas. First, users can explore data in an interactive graphics environment. This could involve pointing to an enumeration unit to determine the associated data value, zooming in on a geographic region to get more detail, or comparing any two arbitrary maps in the database. Second, animated maps are possible; as discussed in Chapter 14, animated maps are a natural choice for depicting temporal changes. Third, multimedia is readily accomplished: not only can the text, photographs, tables, and graphs found in paper atlases be used, but sound, video, and even virtual reality (see Chapter 16) are possibilities. Fourth, users have the potential to create their own maps. Although we cautioned in Chapter 1 that this may lead to improperly designed maps, such capability is essential for users who have collected their own data. Finally, an electronic atlas has the capability to analyze spatial data. As a simple example, an electronic atlas might be used to identify cities having both poor air quality and a high incidence of lung disease. A more sophisticated electronic atlas might include a broad range of analysis functions, such as are found in GIS software.

Given this range of potential functionality, the following is one possible definition for an **electronic atlas:** a collection of maps (and database) that is available in a digital environment. The more sophisticated electronic atlases enable users to take advantage of the digital environment through a variety of means, such as Internet access, data exploration, map animation, and multimedia. Electronic atlases may also permit users to create their own maps and analyze spatial data. The key point is that electronic atlases permit users to manipulate maps and associated databases in ways that were not possible with traditional printed atlases.

15.2 EXAMPLES OF ELECTRONIC ATLASES

This section considers several examples of electronic atlases. With the exception of the *Domesday Project,* they are all available at a modest cost (or for free via the Internet). Since the notion of what constitutes an electronic atlas will likely change as hardware and software evolve, readers are encouraged to examine these and other atlases (Appendix F provides a more complete list), and develop their own notion of what constitutes an electronic atlas.

15.2.1 The *Domesday Project*

In 1086, William the Conqueror enlisted a group of royal officers to conduct a survey of England, principally to determine land ownership and associated levels of taxation, but also to provide census-related and land-use information. The resulting survey was commonly referred to as the *Domesday Book* because one could not appeal a decision based on it. "As a source of geographical and historical data on a national scale, [the *Domesday Book* was] unrivaled until the nineteenth century" (Palmer 1986, 279).

In 1986, the *Domesday Project* was developed to commemorate the 900th anniversary of the *Domesday Book.* The purpose and format of the *Domesday Project* were, however, quite different. Rather than providing information for taxation, the *Domesday Project* was intended to give the general public a contemporary snapshot of the United Kingdom during the 1980s. Instead of being produced on parchment, the *Domesday Project* was created in electronic form. A key characteristic was the use of video disk technology, which permitted the inclusion of text, maps, and photographs in a user-friendly, highly interactive environment.

A particularly impressive feature of the *Domesday Project* was the large amount and type of data available. The two video disks used (known as the Community and National disks) included approximately 250 megabytes of digital data, 50,000 photographs, and 20 million words of text. The Community disk (or "people's database") contained information collected locally by more than 300,000 school children (Rhind and Mounsey 1986, 317). The National disk contained information normally collected in censuses, but also included a wealth of other data dealing with employment, disease, and the environment. In total, the National disk provided color choropleth mapping capability for "over 20,000 variables for 33 different sets of geographical areas at ten different levels of resolution" (Openshaw et al. 1986, 296).

Today, approximately 10 years after it was developed, the *Domesday Project* is recognized as one of the premier electronic atlases ever developed. Unfortunately, most of those outside of the U.K. have never been able to appreciate its capability because the system was designed for specialized BBC microcomputers used in U.K. schools. Interestingly, the system was not marketed after the late 1980s because sales were disappointing (Siekierska and Taylor 1991, 12). These low sales appear to have been the result of the "read-only" character of the video disc technology and the out-of-date information collected in 1985 (Hocking and Keller 1993).

15.2.2 Digital Atlas of California

The *Digital Atlas of California* (http://130.166.124.2/CApage1.html) consists of a large set of thematic maps depicting data at the census tract level (down to the block group level for Sacramento) for all of California. Maps in the atlas are divided into the following major categories: Population and Race, Citizenship, Income, Poverty, and Adult Educational Attainment. Within each category, numerous variables can be mapped; for example, the Population and Race category consists of 15 different variables. Maps are included both for the entire state and for major metropolitan areas. A redundant symbology is used to depict themes: proportional circles are scaled according to the magnitude of a theme in each census tract (or block group), and five colors are shown within each circle depicting a quintiles classification of the data (Color Plate 15.1).

The *Digital Atlas of California* is part of the *Electronic Map Library* being developed by Professor William Bowen at California State University, Northridge. In addition to the California atlas, the library includes atlases for the United States, New York City, Washington, D.C., and Boston, Massachusetts. Maps in these other atlases are produced in the same format as the California atlas. A distinct advantage of having these atlases available on the Internet is that they can be accessed for free by anyone with an Internet connection. Another advantage is that Professor Bowen can enlarge the *Electronic Library* indefinitely; his only limitation is the storage space available on his server. A disadvantage of the library, however, is that no interaction with the maps is possible: the maps are in GIF format, which is useful for browsing and downloading, but not for providing information about the underlying data.

15.2.3 State Atlases: *The Interactive Atlas of Georgia*

Today, a number of states have developed or are in the process of developing their own atlases; as an example, we'll consider *The Interactive Atlas of Georgia,* which is distributed by the Institute of Community and Area Development at the University of Georgia. To a certain extent, the Georgia atlas emulates traditional paper atlases, as it is possible to click on an arrow key in the menu and move through the atlas one screen (or page) at a time. In contrast to the static nature of printed atlases, however, it is also possible to interact with each of maps in the Georgia atlas.

The range of interactive tools is similar to those described for ExploreMap in Chapter 13. For example, a "Where" option permits users to obtain specific information: users select a county name from a menu, which causes the associated county to be highlighted and the name and value to appear. This option works not only for choropleth maps, which are quite common in the atlas, but also for a variety of other maps. For example, when clicking at an arbitrary location on a dot map of peanut production, the "Where" option displays the county boundary within which the location falls, along with the total peanut production for that county. Another interactive tool is the "Query" option, which enables users to specify ranges of data for multiple variables, and thus map a set of counties that meet a variety of criteria. For example, a user might want to locate counties having low crime rates and a good educational system, where the latter might be measured by the money spent on educating students.

Intended as an atlas rather than an exploration tool for a particular kind of map, the Georgia atlas not surprisingly provides a broader range of options than ExploreMap; for example, a "Cities" option enables users to locate over 3,000 places and a "County Learn" option assists users in learning county names in Georgia. The Georgia atlas is limited, however, in that it does not contain text, photographs, or graphs; nor does it contain animation or multimedia capability. Such limitations are, in part, a function of the limited funding commonly available for state atlases.

15.2.4 Commercial Atlases: *World Atlas*

The development of electronic atlases in the United States has been heavily influenced by Richard Smith, who was a leader in creating the *Electronic Atlas of Arkansas,* the first state electronic atlas. Smith was also responsible for developing *World Atlas,* one of the first widely distributed commercial electronic atlases. Two basic types of thematic maps are available in *World Atlas:* choropleth and proportional symbol (specifically, proportional squares are used). Choropleth maps are used for standardized data, while proportional symbol maps are used for raw count data (Figure 15.1).

Multimedia is incorporated into *World Atlas* as sound, video, and pictures. Audio features include national anthems, names of countries, and foreign phrases. Ideally, it would seem desirable to link these features directly with maps, but generally one must point at the name of a country within a menu (as opposed to the map) to hear an associated audio feature. Because of video's large storage requirements, video clips are shown only for each major city in the world, and only a limited number are included for each city; furthermore, the clips are small (taking up approximately one-tenth of the screen) and last only a few seconds. A brief textual phrase announces the video and background music plays during it. Examples from six clips for China include "Great Wall of China," "Streets of Beijing," and "Tiananmen Square Monument." When viewing such videos, I found myself wanting much more information, but realized that acquiring such video data would be costly and that storage space was a major limitation at the time the atlas was developed. This limitation should become less critical as CD-ROM technology and associated compression software improves, or the Internet becomes the common delivery mechanism for electronic atlases.

15.2.5 Commercial Atlases: *Microsoft Encarta Virtual Globe*

The *Microsoft® Encarta® Virtual Globe** is representative of the capability of more recently developed commercial electronic atlases. A focal point of *Encarta* is an attrac-

* Due to copyright limitations, it was not possible to provide illustrations of *Encarta* capabilities in this book.

tive perspective view of the world. Users can rotate this globe by simply clicking and dragging the mouse, or they can click on a zoom tool to move quickly through a variety of scales.

A variety of map "styles" are possible in *Encarta,* including Comprehensive (a combination of natural and human factors), Political, Satellite (how the earth looks by day or by night), Natural (for example, precipitation), Human (population and time zones), and Statistical (or thematic). All of these are interesting from a geographic standpoint, but the Statistical style is of greatest relevance to the present text. With this style, more than 150 thematic variables can be mapped at the level of individual countries (no detail is possible within a country). The legend does not appear on the screen with the mapped data, but is accessible via a separate screen. Precise values for individual countries can be obtained by simply pointing the mouse cursor at the desired country.

When the user zooms in on a particular country within *Encarta,* considerable information can be obtained for that country. Categories of information include Land and Climate (for example, a description of major rivers and lakes), Facts and Figures (such as the total population and major language spoken), Society (for example, the nature of marriage in that country and the type of government), Sights and Sounds (photographs of places and sounds of native music), Animals (pictures and sounds of common animal life), and WebLinks (links to the Internet for additional information about the country).

Although *Encarta* provides a number of useful features, it does have its limitations. One is that from a thematic mapping perspective, both the method of classification and color schemes are fixed for all Statistical maps—this is obviously a problem for expert mapmakers, but arguably less important for naive users. A second limitation is that only a perspective view of the world is possible (in other words, no projection options are available); as a result, one cannot view a distribution of an attribute for the entire world all at once (rather, panning is required). Overall, though, *Encarta* is illus-

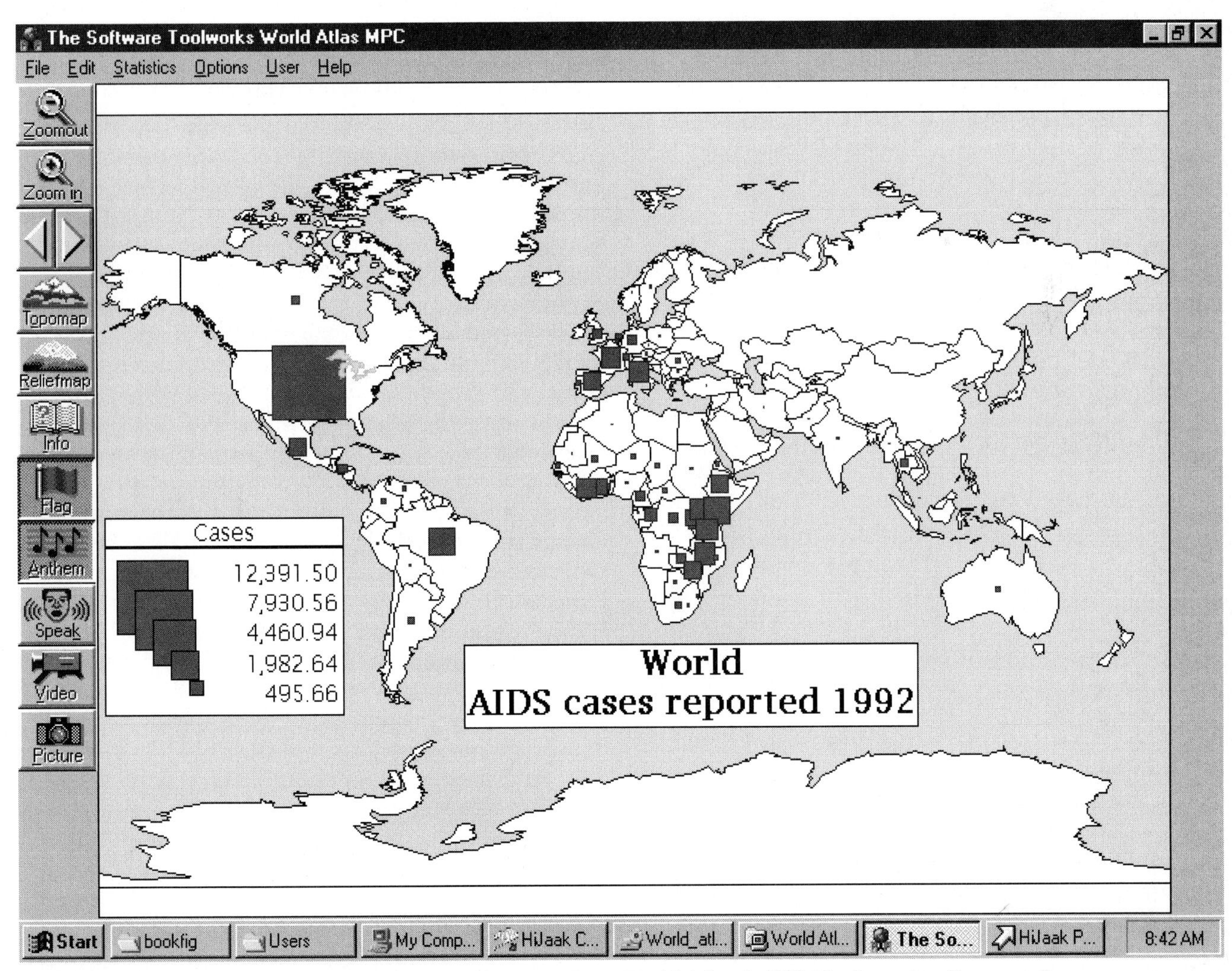

FIGURE 15.1 An example of a proportional symbol map from *World Atlas;* © 1998, The Learning Company, Inc.

trative of the vast array of geographic information that modern electronic atlases can provide.

15.2.6 *Death Valley: An Animated Atlas*

Death Valley: An Animated Atlas, developed by Hugh Howard as part of a master's thesis (Howard 1995), provides an impressive introduction to Death Valley, a national park located in southern California. Death Valley is unique in that it has the lowest elevation (282 feet below sea level) and highest summer temperatures (commonly in excess of 115°F) in the United States. It also suffers from extreme aridity (less than 2 inches of rainfall per year, with potential evaporation rates in excess of 150 inches per year), and is home to some unusual plant and animal life (for example, several species of pupfish that occur nowhere else in the world).

The atlas provides a series of narratives that describe the attributes of Death Valley via maps (many of which are animated), photographs, graphs, and spoken text. The latter is particularly useful as it permits the viewer to focus on interpreting graphics, without having to read text. The narratives include an Introduction and Quick Tour, and two major sections dealing with the physical and cultural environment.

Since static illustrations do not do the atlas justice, I'll describe how the atlas presents three factors (a rain-shadow effect, adiabatic heating, and a regional low-pressure system) involved in creating the hot, dry, and windy climate characteristic of Death Valley. The rain-shadow effect is caused by the Sierra Nevada and other mountain ranges, which prevent moist air from reaching the valley from the west. In the atlas, a shaded relief map depicts these mountain ranges, while moving arrows represent the moist air—the arrows dissolve as they pass over the mountains. Adiabatic heating occurs in the summer, as hot air rises from the valley and is replaced by cooler air from the surrounding mountains. This effect is illustrated in the atlas by moving colored arrows (red for rising hot air and blue for descending cool air). The regional low-pressure system is also a result of hot air rising from the valley; this rising air is replaced by hot, dry, tropical air from the Mexican Plateau. In the atlas, the rising hot air is illustrated by a set of red arrows moving upward. These animations (in the form of moving arrows) serve to imprint on the viewer a clearer understanding of these processes than static images could provide.

Admittedly, the Death Valley atlas is not without problems—some of the accompanying music is a bit annoying, one cannot reverse the direction of a narrative, and the text of the narrations cannot be printed. Overall, though, the atlas provides a thorough introduction to Death Valley, and a good illustration of how spoken text in a computer-based environment can be used to assist in interpreting maps and graphs.

15.2.7 Maps in Multimedia Encyclopedias: *Grollier's Multimedia Encyclopedia*

Multimedia encyclopedias are like paper encyclopedias in the sense that they provide considerable information about places, people, and events, but the multimedia variant is enhanced through the inclusion of sound, video, and the ease of linking various topics. The *Grollier Multimedia Encyclopedia* includes an "Atlas" section that provides a variety of thematic maps of the world for individual countries, and for the fifty U.S. states. These maps permit a limited degree of interactivity. For example, on a map showing numerous explorers' routes, it is possible to select one explorer's route or to show all routes simultaneously. As another example, it is possible to show a succession of colored contour maps of temperature by month for the United States by simply clicking on "Next," creating a sort of animation under control of the user. In general, however, one cannot point to locations (such as countries) to get associated attribute values (such as population density).

The most interesting feature of the Grollier encyclopedia from a cartographic standpoint is the use of multimedia in association with animated maps, which David DiBiase assisted in designing (DiBiase 1994). An example is "The American Revolution," which is particularly informative and enjoyable to watch. In this presentation, a map of the United Sttaes serves as a backdrop against which animated symbols portray the location of major battles and movement of troops. While the animation plays, an audio portion describes the battles and troop movements, along with major historical events. Like the Death Valley atlas, the combination of animated maps and audio is arguably more effective than a traditional printed atlas in which one must repeatedly move back and forth between text and maps.

Also included in "The American Revolution" are photographs of key people and events. For example, when George Washington is mentioned, a picture of him appears. Although such pictures could detract from the animation, they are entertaining and provide a useful break for the viewer. The multimedia presentations include subtitles, which are essential for the hearing impaired, and useful for older computers lacking sound cards. Overall, the multimedia presentations are particularly impressive given that DiBiase indicated the maps were created before the audio portion was available. In short, these presentations indicate that "animated cartography is not just scratching the surface anymore" (DiBiase 1994, p. 7).

15.2.8 TIGER Mapping Service

The TIGER Mapping Service (http://tiger.census.gov) provides an intriguing example of the capability of digital mapping, but also illustrates the difficulty of defining what constitutes an electronic atlas. The service permits users to create maps of base data (for example, a grid of latitude and longitude, streets, and railroads) and/or thematic data (for example, family income) for anywhere in the United States from the state level down to the block-group level. Completed maps are not part of the service, but are a function of user desires; for example, users can easily change the scale, and select from 21 base data layers and 18 different themes to create their own map. Color Plate 15.2 illustrates a sample map for the Seattle, Washington, area. In this case, base data include water bodies and interstate highways: the theme is the percent of the population 65 years and older within census tracts.

The TIGER Mapping Service can be viewed as an atlas in the sense that a large variety of maps of a particular area can be accessed reasonably quickly (creation times are generally on the order of seconds, although during slow Internet periods, several minutes may be required). The term *atlas* may be inappropriate, however, because maps in the system are not already created, but are developed interactively. The term **map server** is potentially more appropriate as such systems "serve" maps to users. An example of another such server is the Environmental Protection Agency's Maps on Demand (http://www.epa.gov/enviro/html/mod/mod.html). In the latter case, maps are not available until the next day. A number of new tools have recently been developed to assist in serving maps to the public; an example is the ArcView Internet Map Server (Limp 1997).

15.3 TOOLS FOR DEVELOPING YOUR OWN SOFTWARE

Although considerable capability now exists for those wishing to explore and animate spatial data, or examine the geography of a region via an electronic atlas, software packages frequently have weaknesses and limitations that prevent a user from accomplishing a desired task. As a result, readers may wish to develop their own data exploration software, animations, or electronic atlases. This section considers several tools that could assist in this process: animation software, programming languages, authoring software, and visualization software. The following provides an overview of each of these types of tools; particular software is listed in Appendix G.

15.3.1 Animation Software

Animation software is explicitly intended for developing animations; the most sophisticated software may also permit users to interact with the animation and thus explore data. Animation software can be categorized as either general-purpose or specialized. An example of general-purpose software is Animator Pro, which permits "virtually all of the animation techniques, including cel-based animation, tweening, optical animation (swirling, twirling, spinning, flipping, squashing . . .), and color cycling" (Peterson 1995, 149). Software for creating fly-bys, in which the viewer is given the feeling of flying over a landscape, is representative of more specialized software.

15.3.2 Programming Languages

When we think of **programming languages,** the languages Basic, Fortran, Pascal, and C may come to mind. Traditionally, developing mapping software with such languages was tedious because of the lengthy computer code that had to be written. Today this is changing as **visual programming languages** ease software development by utilizing **GUIs (graphical user interfaces).**

The programming language Visual Basic provides a good example of the capability provided by visual programming.* In Visual Basic, the programmer develops a user interface by drawing desired screen controls in a Form Window and then assigns properties to each control from a Property Window. Once a basic interface is developed, code may be written in a Code Window for events that relate to particular controls (Figure 15.2). When development is completed, the user manipulates the application in a visual fashion (by clicking on a screen icon). Other programming languages providing capability similar to Visual Basic include Delphi, Tcl/Tk, and several based on the C^{++} language (for example, Visual C^{++}). For those wishing to develop interactive graphical applications on the Internet, the programming languages Java (Halfhill 1997) and Virtual Reality Modelling Language (VRML) (Whitehouse 1996; Fairbairn and Parsley 1997) should be considered.

A distinct advantage of programming languages is that the programming logic learned may be applied when using both the authoring and visualization software described below (Slocum and Yoder 1996). One problem, however, is that extensive visualization tools generally are not included with programming languages, and thus a considerable amount of code may be required to develop sophisticated mapping software. For example, Yoder (1996) indicated that it took approximately 800 hours to develop MapTime using Visual Basic. Fortunately, mapping and GIS tools have recently become available that can be linked with programming languages, and thus reduce development time. A good example

* Source code for several Visual Basic programs of interest to cartographers is available through the Microcomputer Specialty Group (see *Using Visual Basic to Teach Programming for Geographers;* Appendix G, section 2).

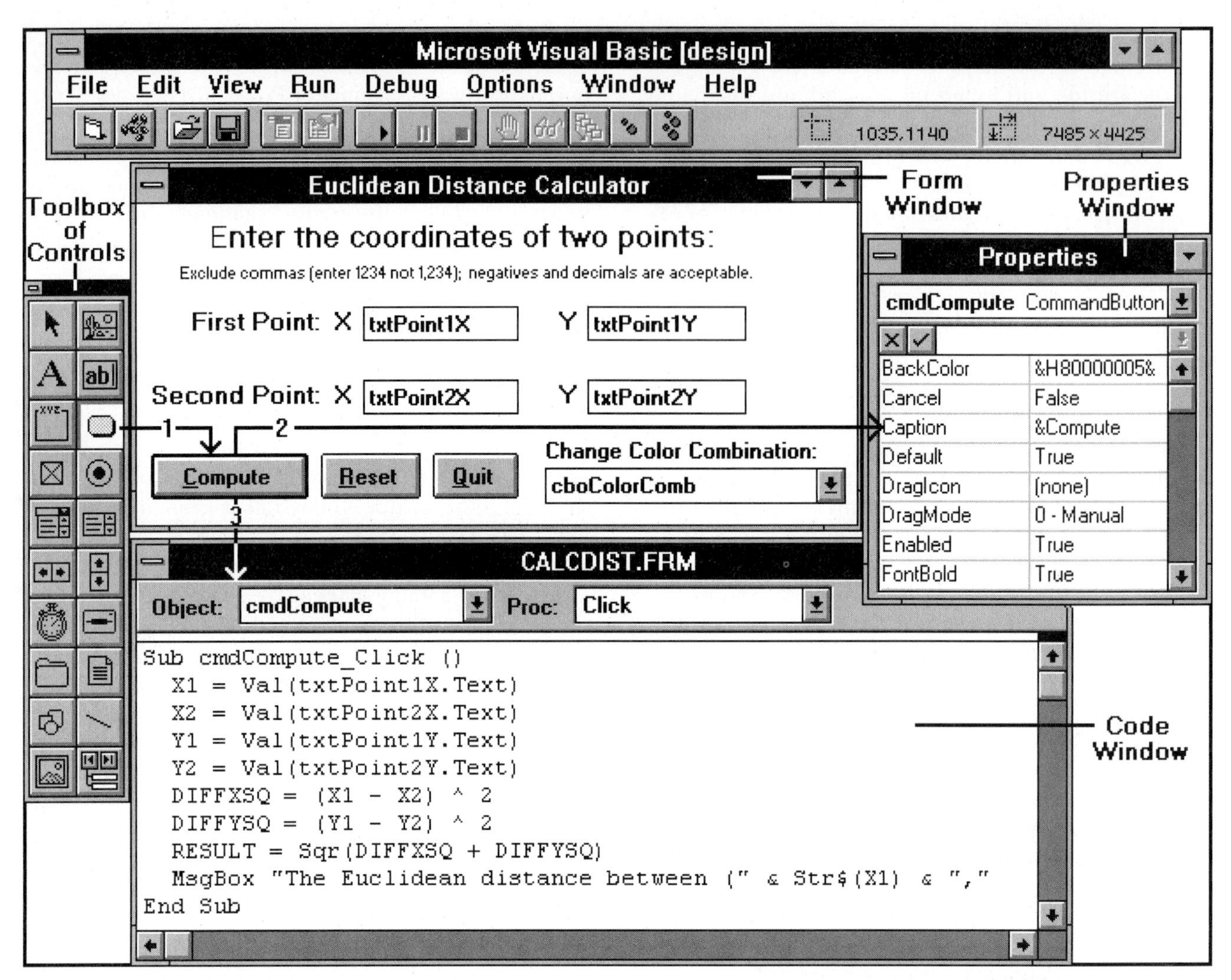

FIGURE 15.2 The Visual Basic programming environment. The programmer develops a user interface by drawing desired screen controls in the Form Window. Properties are assigned to each control from the Property Window. Once a basic interface is developed, code may be written in the Code Window for events that relate to particular controls. (From Slocum and Yoder 1996, 195; courtesy of National Council for Geographic Education.)

is MapObjects, which is available from Environmental Systems Research Institute (ESRI). Another alternative for reducing development time is to enhance already-developed mapping and GIS software by utilizing a programming language embedded in that software. For example, ArcView, also distributed by ESRI, includes the programming environment Avenue, which can be used to enhance the capability of ArcView.

15.3.3 Authoring Software

In contrast to the general-purpose nature of programming languages, authoring software is primarily used for developing business and educational presentations. For example, one might use authoring software to develop a lecture for an introductory human geography course. Macromedia's Director has been the most common authoring software used for cartographic purposes; for example, the Aspens software described in Chapter 13 was developed using Director.

The flexible set of graphical tools and multimedia/hypermedia capability provided with authoring software makes it effective for cartographic applications like those described in Chapters 13 to 15. It is important to realize, however, that the full capability of authoring software cannot be attained unless some programming is done. Generally, this is accomplished using **scripting languages,** which "rely on a more language-like syntax" than traditional programming languages (Peterson 1995, 153). Scripting languages are arguably easier to learn than traditional programming languages, but mastering a particular language can be time-consuming and is most effectively handled by those with prior programming experience.

15.3.4 Visualization Software

Although authoring software generally provides more visualization tools than programming languages, it cannot match the range of tools provided by **visualization software,** which is expressly intended for developing visualizations of spatial data. Such software is commonly broken into two types: data flow and non–data flow. **Data flow software (modular visualization environments)** contain separate modules for importing, manipulating, and displaying data. To create visualizations, the modules are linked in a visual programming environment: individual modules are moved to a desired location on the screen and then edited through a point-and-click interface. Examples of data flow software include the Application Visualization System (AVS), the IBM Visualization Data Explorer (DX), and Khoros. Non–data flow software also contain a number of modules for visualizing data, but these are not linked in a visual programming environment; examples include the Advanced Visualizer, Interactive Data Language (IDL), and Precision Visuals' Wave Command Language (PV-Wave CL).

Visualization software may seem like the obvious choice for creating sophisticated mapping packages, but the reader should be cautious in selecting this option. Such software tends to be expensive, traditionally has been available only for workstations, often does not include routines that would be of interest to a geographer, and generally requires some programming expertise to operate. Those trying to choose among programming languages, authoring software, and visualization software may find Slocum et al.'s (1994) review useful. In addition, readers should examine recent software reviews.

SUMMARY

This chapter has dealt with two topics: electronic atlases and tools for developing your own software. In the simplest sense, an electronic atlas can be viewed as a collection of maps available in digital form. The more sophisticated **electronic atlases,** however, permit users to interact with maps and/or the data underlying the maps (in other words, explore data), create map animations, create their own maps, analyze spatial data, and obtain information via multimedia (sound, video, and even virtual reality are possible).

We examined a range of electronic atlases to illustrate their varied capability. The *Digital Atlas of California* represents the simplest form of electronic atlas—a collection of maps that cannot be interacted with. Although limited in capability, this atlas is available for free via the Internet and its potential size is virtually unlimited. The *Interactive Atlas of Georgia* provides somewhat greater capability. It enables users to obtain information about particular enumeration units by pointing at the map. It also enables users to specify ranges of data for several variables, and thus map a set of counties that meet a variety of criteria.

In contrast, to these relatively simple atlases, the *Microsoft Encarta Virtual Globe* provides greater capability. *Encarta* enables users to zoom in on a particular area of the world at a variety of scales. When zoomed in on a particular country, information can be obtained about a variety of subjects (such as the physiography of a country or the characteristics of the society found there). Multimedia is an important element of *Encarta* (for example, sounds of native music and animal life are possible) and links are possible to the World Wide Web to obtain still more information. *Death Valley: An Animated Atlas* and the *Grollier Multimedia Encyclopedia* provide good examples of how multimedia can be linked effectively with map animations.

It is important to recognize that only a sampling of electronic atlases have been discussed here; Appendix F provides a more complete set. Since electronic atlases are still evolving, readers will need to consider the capability of new atlases and decide how the definition for electronic atlases given here might be modified.

The tools for developing software discussed in this chapter are potentially useful for those wishing to develop their own data exploration software, map animations, or electronic atlases. The tools can be divided into animation software, programming languages, authoring software, and visualization software. **Animation software** is expressly intended for developing map animations; examples include Animator Pro and the software used to create fly-bys discussed in Chapter 9. The trend in programming languages is toward **visual programming** that enable applications to be developed more readily by utilizing point-and-click GUIs. Another important trend is the linkage of mapping and GIS tools with programming languages; for example, MapObjects can be linked with Visual Basic.

Authoring software permits users to develop applications by writing in **scripting languages,** which have a more language-like syntax than traditional programming languages. Several pieces of cartographic software have been developed using authoring software (for example, Aspens and Weber and Buttenfield's animation of climate change discussed in Chapters 13 and 14 respectively). **Visualization software** hold considerable promise for creating geographic visualizations as it is has been expressly designed for that purpose; examples include the Application Visualization System (AVS), the IBM Visualization Data Explorer (DX), and Khoros. Such software, however, is expensive, traditionally has been available only for workstations, often does not include routines that would be of interest to geographers, and generally requires some programming expertise to operate most effectively.

FURTHER READING

Cartographic Perspectives no. 20, 1995 (entire issue).

This special issue deals entirely with electronic atlases. Several articles appearing in the issue are annotated below.

Cartwright, W. (1994) "Interactive multimedia for mapping." In *Visualization in Modern Cartography,* ed. A. M. MacEachren and D. R. F. Taylor, pp. 63–89. Oxford: Pergamon Press.

Reviews the hardware and software necessary for developing multimedia applications and describes a number of multimedia applications in cartography.

Computer Graphics 29, no. 2, 1995 (entire issue).

This special issue focuses on data flow software (modular visualization environments).

DiBiase, D. (1994) "Designing animated maps for a multimedia encyclopedia." *Cartographic Perspectives* no. 19:3–7, 19.

Describes his practical experience in developing animations for the *Grollier Multimedia Encyclopedia.*

Fairbairn, D., and Parsley, S. (1997) "The use of VRML for cartographic presentation." *Computers & Geosciences* 23, no. 4:475–481.

Discusses the use of the programming language VRML for creating cartographic applications. VRML permits the development of three-dimensional virtual worlds on the Internet. For more on virtual worlds, see section 16.4.

Halfhill, T. R. (1997) "Today the web, tomorrow the world." *Byte* 22, no. 1:68–80.

Discusses the programming language Java, which can be used to create interactive graphical applications on the Internet.

Hughes, J. R. (1996) "Technology trends mark multimedia advancements." *GIS World* 9, no. 11:40–43.

Emphasizes the technology associated with using multimedia for geographic applications.

Keller, C. P. (1995) "Visualizing digital atlas information products and the user perspective." *Cartographic Perspectives* no. 20:21–28.

Discusses the potential for electronic atlases and promotes utilizing user surveys to determine the content and format of atlases.

Limp, W. F. (1997) "Weave maps across the web." GIS World 10, no. 9:46–55.

Evaluates several pieces of software that enable users to serve maps to the public via the Internet.

Ormeling, F. (1995) "New forms, concepts, and structures for European national atlases." *Cartographic Perspectives* no. 20:12–20.

Discusses some of the capability that should be provided by electronic atlases.

Rystedt, B. (1995) "Current trends in electronic atlas production." *Cartographic Perspectives* no. 20:5–11.

Briefly summarizes a variety of electronic atlases.

Siekierska, E. M., and Taylor, D. R. F. (1991) "Electronic mapping and electronic atlases: New cartographic products for the information era—The electronic atlas of Canada." *CISM Journal ACSGC* 45, no. 1:11–21.

An early overview of electronic atlases, with emphasis on the *Electronic Atlas of Canada.*

Slocum, T. A., and Yoder, S. C. (1996) "Using Visual Basic to teach programming for geographers." *Journal of Geography* 95, no. 5:194–199.

Describes using the programming language Visual Basic to create cartographic applications.

Slocum, T. A., Armstrong, M. P., Bishop, I., Carron, J., Dungan, J., Egbert, S. L., Knapp, L., Okazaki, D., Rhyne, T. M., Rokos, Demetrius-K. D., Ruggles, A. J., and Weber, C. R. (1994) "Visualization software tools." In *Visualization in Modern Cartography*, ed. A. M. MacEachren and D. R. F. Taylor, pp. 91–122. Oxford: Pergamon Press.

Reviews a variety of tools that can be used to create software for data exploration, animation, and electronic atlases.

Smith, R. M., and Parker, T. (1995) "An electronic atlas authoring system." *Cartographic Perspectives* no. 20:35–39.

Presents the notion of developing software that is expressly designed to create electronic atlases. Smith (personal communication, 1997) indicates that the software was never actually completed.

Transactions, Institute of British Geographers (new series) 11, no. 3, 1986.

Contains numerous articles dealing with the *Domesday Project.*

16

Recent Developments

OVERVIEW

As discussed in Chapter 1, the discipline of cartography has changed considerably over the last thirty-five years, evolving from a discipline based on pen and ink to one based on computer technology. As the field continues to evolve, it is important to keep pace with recent developments. The purpose of this chapter is to examine some of these developments. Topics covered include the depiction of data quality; the use of sound to represent spatial data; research on the use of color, virtual reality and visual realism; and a special issue of Computers & Geosciences *focusing on visualization. For other recent developments, references listed in the "Further Reading" section at the end of each chapter should also be consulted.*

Data quality *refers to the notion that the data on which thematic maps are based are often subject to some form of error. For example, U.S. census data are typically based on sampling approximately one of every six housing units (U.S. Bureau of the Census 1994a, A-2), and thus are estimates of population values. There are three broad aspects of data quality that cartographers have identified as worthy of study: definitions, measures, and visualization (Morrison 1995, 6). Since a primary emphasis of this text is visualization, section 16.1 focuses on* visualization *aspects of data quality.*

Using sound to represent spatial data is referred to as ***sonification*** *or* ***acoustic visualization.*** *In examining sonification, it is useful to distinguish between realistic and abstract sounds. An example of* ***realistic sound*** *would be the narrations accompanying the map animations found in multimedia encyclopedias (as described in the preceding chapter).* ***Abstract sounds*** *are those whose meaning is not clear without using a legend. For example, clicking on counties on a choropleth map might produce sounds of differing loudness; their meaning would not be clear without a legend. Section 16.2 considers a set of* abstract sound variables *(for example,* loudness *and* pitch*) analogous to the visual variables covered in Chapter 2, and a number of mapping applications that have used sound. The discussion borrows heavily from the work of Krygier (1993, 1994).*

Although cartographers have previously developed recommendations for the appropriate use of color (see Chapter 6), these recommendations are evolving as a result of recent research. Particularly important has been the work of Cynthia Brewer at Pennsylvania State University. Section 16.3 considers two studies in which Brewer played a major role. The first (Brewer et al. 1997) involves the selection of color schemes for choropleth maps appearing in the Atlas of United States Mortality *(Pickle et al. 1996), while the second (Brewer 1996a) deals with general guidelines for diverging color schemes. Additionally, readers may wish to examine Brewer's recent work on simultaneous contrast (1996b, 1997).*

A particularly intriguing development is the attempt to create realistic representations of the earth's natural and built environments. For example, imagine that you are an urban planner in Seattle, Washington (from which beautiful Mount Ranier can be viewed), and you are considering where a new high-rise hotel should be built. In determining a suitable location, you might want to know how a potential site would impact local residents' views of Mount Ranier. In the past few years, computer-based systems have been developed that enable you to actually "see" the resulting views; this process is known as ***visual realism*** *or* ***visual simulation.*** *The notion of visual realism can be expanded to include all of the senses (sound, touch, smell, and taste), thus creating a* ***virtual reality*** *or* ***virtual world.*** *Section 16.4 considers several developments in this area.*

In 1995, the International Cartographic Association (ICA) established a Commission on Visualization, whose purpose was to:

> 1) . . . study and report on the changing and expanding role of maps in science, decision making, policy formulation, and society in general due to the advent of *intelligent* dynamic maps . . . and 2) to investigate and report on the links between *scientific visualization* and *cartographic visualization* and identify ways to facilitate exchange of ideas between cartographers and others working on problems of visualization. (MacEachren and Kraak 1997, 336)

To promote this process, a special electronic issue of *Computers & Geosciences* (1997, Vol. 23, No. 4) was developed. The special issue is available in both traditional paper form (with a companion CD-ROM) and via the Internet (http://www.elsevier.nl/locate/cgvis). Section 16.5 summarizes the contents of 12 papers developed for this special issue.

Also interesting is the work that geographers and statisticians have done to incorporate statistical concepts into geographic visualization. I have chosen not to cover this work in detail here because some of it is beyond the ability of those without a strong mathematical background. Noteworthy, however, have been the efforts of Dan Carr of George Mason University; much of his work has been published in the *Statistical Computing & Statistical Graphics Newsletter,* available at http://cm.bell-labs.com/who/cocteau/newsletter/index.html. An interesting recent effort by Carr involves splitting a standard choropleth map into "micromaps," which are linked with other information such as confidence intervals (Carr and Pierson 1996). Other work with a statistical emphasis that may be of interest includes that by Openshaw and Perrée (1996), Cressie (1992), and Pickle et al. (1996).

One of the difficulties in writing a book in a rapidly changing discipline is that it is virtually impossible to provide up-to-date information. To assist in acquiring information on recent developments, Section 16.6 provides lists of journals, conferences, and useful internet sites. The home page for this book (http://www.prenhall.com/slocum) should also be consulted for more recent information.

16.1 DEPICTING DATA QUALITY

16.1.1 What is Data Quality?

Users generally think of maps as truthful representations of reality. For example, imagine viewing a choropleth map of the United States entitled "Percent of the Population That Smokes." If a shade in the legend has class limits of 21–25 percent, then readers are apt to presume that states with that shade do, in fact, fall in the 21–25 percent range. In reality, this probably would not be correct because the map would likely be based on a sample, and thus there would be a margin of error around the sampled value (see section 3.2). As another example, imagine a dot map of bears in the United States. The naive viewer might assume that bears are found wherever dots are located on the map. Where dots are highly concentrated, there would be a high probability of bears being found, but where dots are dispersed, there would be a relatively low probability that a bear would be found at the location of a particular dot.

The notion that maps are not necessarily truthful representations of reality is commonly discussed under the headings of **data quality, data reliability,** or **data uncertainty.** Although some cartographers have suggested that one term be used in preference to others (for example, MacEachren 1992, 11 and MacEachren 1995, 458), a survey of the literature suggests that the terms tend to be used interchangeably. For our purposes, we will consider data quality to be equivalent to data reliability ("low quality" equals "low reliability") and uncertainty as an antonym for data quality and data reliability ("high uncertainty" equals "low quality").

The U.S. Federal Information Processing Standard 173, or FIPS 173 (National Institute of Standards and Technology 1994)* lists five categories for assessing data quality: lineage, positional accuracy, attribute accuracy, completeness, and logical consistency. **Lineage** refers to the historical development of the digital data. For example, if the data were acquired from a paper map, you might want to know the scale and projection of the map, whether digitizing or scanning was used to generate the digital data, and when the data were converted from analog to digital form.

Positional accuracy refers to the correctness of location of geographic features, in both horizontal and vertical form. For example, on a USGS topographic sheet, you might wish to examine the accuracy of either a stream's position or the spot height of a mountain. **Attribute accuracy** refers to the correctness of features found at particular locations. For instance, in remote sensing, you might be interested in the accuracy of a land use/land cover classification.

Logical consistency describes "the fidelity of relationships encoded in the data structure of the digital spatial data" (National Institute of Standards and Technology 1994, 23). For example, we might ask whether all polygons close correctly (in GIS terminology, this would be termed *topological correctness*). Finally, **completeness** in-

* See the January 1988 issue of *The American Cartographer* for an earlier version.

cludes "information about selection criteria, definitions used and other relevant mapping rules. For example, geometric thresholds such as minimum area or minimum width must be reported" (National Institute of Standards and Technology 1994, 24).

These five categories, and two additional ones (semantic accuracy and temporal information), are described in detail in *Elements of Data Quality,* recently published by the International Cartographic Association (ICA) Commission on Spatial Data Quality (Guptill and Morrison 1995). The categories of positional accuracy and attribute accuracy have received the most attention in the literature pertaining to visualization, so they will be focused on below.

16.1.2 Visual Variables for Depicting Data Quality

One approach for depicting data quality is to utilize the basic visual variables introduced in section 2.3. In some cases, these visual variables are suitable, while in others they can be confusing. A suitable usage would be **Tissot's indicatrix (distortion ellipse),** in which the visual variables size and shape are used to represent the ability of various map projections to maintain correct size and angular relationships at point locations (Figure 16.1). Potentially confusing would be utilizing the visual variable size to depict the reliability of stream position, as a wide line normally would be associated with a greater discharge (Figure 16.2).

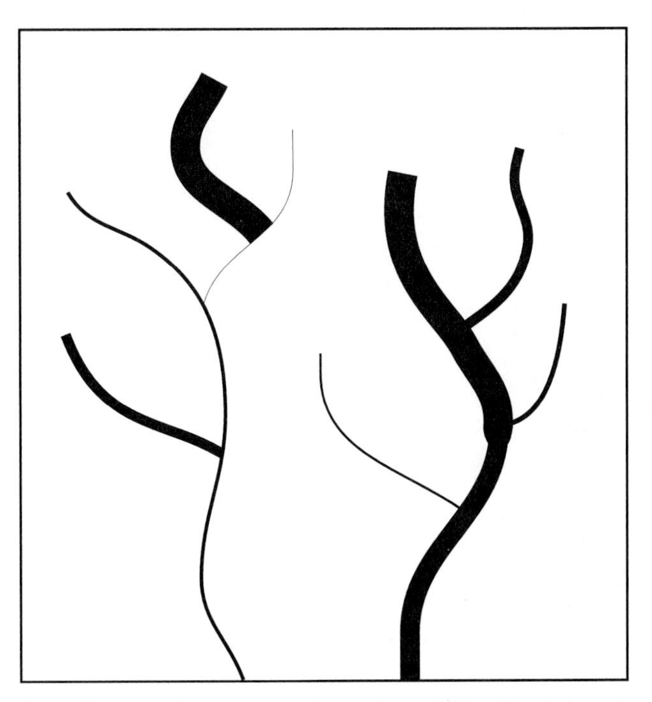

FIGURE 16.2 The visual variable size (width of line) depicts data quality associated with stream position. Here the level of data quality might be misinterpreted as stream discharge. (After McGranaghan 1993, 17.)

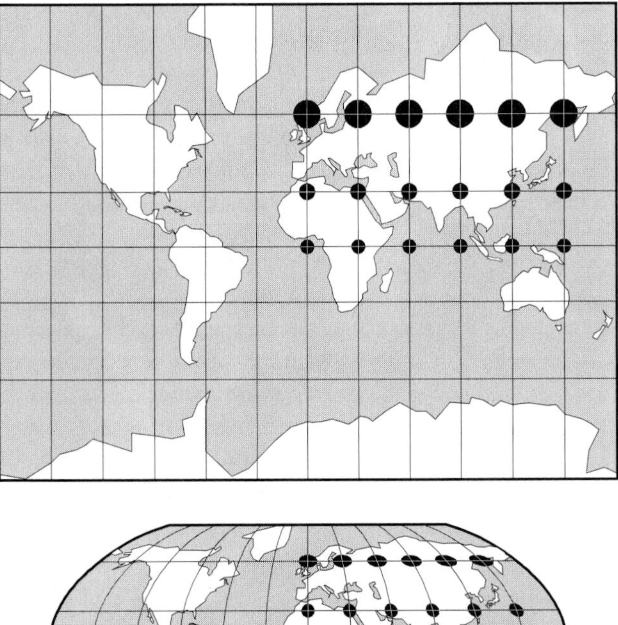

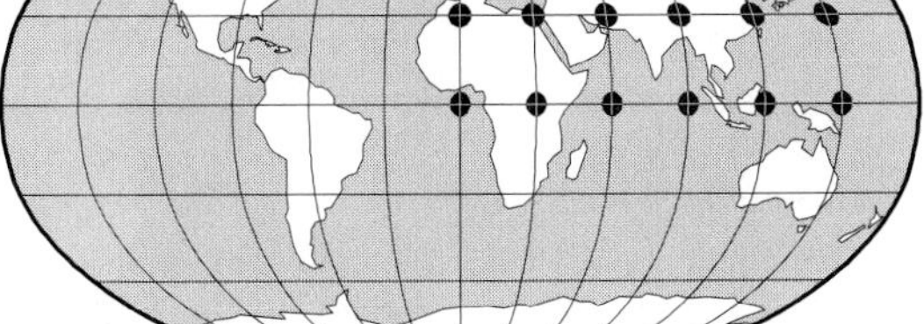

FIGURE 16.1 The visual variables size and shape represent the ability of various map projections to maintain correct size and angular relationships at point locations. Here the visual variables are easily understood. (Abler, Ronald F., Melvin G. Marcus, and Judy M. Olson, *Geography's Inner Worlds: Pervasive Themes in Contemporary American Geography.* Copyright © 1992 by Rutgers, The State University. Reprinted by permission of Rutgers University Press.)

Of those visual variables introduced in section 2.3, MacEachren (1992) argues that saturation is particularly logical for depicting data quality, with "pure hues used for very certain information [and] unsaturated hues for uncertain information" (p. 14). Referencing the work of Brown and van Elzakker (1993), MacEachren (1995, 440–441) notes that three levels of saturation can be used for up to 12 individual hues. Although portraying only three levels of data quality may sound limiting, MacEachren argues that an analyst can utilize data exploration tools to apply the three levels to only a portion of the data.

In addition to the standard visual variables introduced in section 2.3, MacEachren (1995, 275–276) proposes the visual variable **clarity,** which he argues can be subdivided into three additional visual variables: crispness, resolution, and transparency. **Crispness** refers to the sharpness of boundaries (or area fills): a crisp boundary would represent reliable data, while a fuzzy boundary would rep-

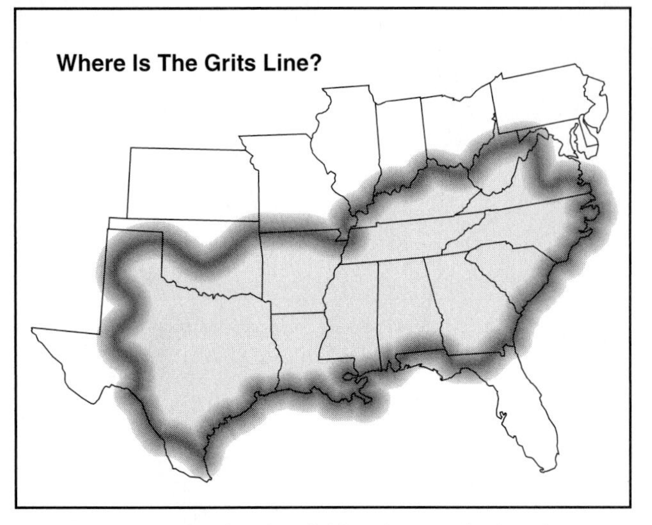

FIGURE 16.3 The visual variable crispness depicts the uncertain nature of the boundary of grits.

resent uncertain data. For example, Figure 16.3 illustrates the uncertain nature of the boundary of grits in the United States. **Resolution** is the level of precision in the spatial data underlying an attribute. For example, Figure 16.4 illustrates different resolutions for a raster database. **Transparency** is the ease with which a theme can be seen through a "fog" placed over that theme; reliable data can be easily seen through the fog, while uncertain data cannot (Figure 16.5).*

16.1.3 General Methods for Depicting Data Quality

Three general methods are used to depict data quality: individual maps can be shown both for the theme and its associated reliability (maps are compared); the theme and its reliability can be displayed on the same map (maps are combined); or data exploration tools can be utilized (MacEachren 1992). Note that the first two of these mirror the basic approaches used in bivariate and multivariate mapping (see Chapter 12), while the last approach was introduced in Chapter 13. These methods are best illustrated by considering some examples of cartographers' efforts to visualize data quality.

16.1.4 Examples of Visualizing Data Quality

Howard and MacEachren's R-VIS Software

R-VIS (for *R*eliability *VIS*ualization) was developed by Howard and MacEachren (1996) to enable environmental scientists and policy analysts to examine levels of dissolved inorganic nitrogen (DIN) in the Chesapeake Bay. Basic data input to R-VIS is a set of 49 DIN values collected at point locations by a scientific vessel.* These data are available on a quarterly basis over the year and ultimately will be available from 1984 through the year 2000. Since DIN is presumed to be a smooth continuous phenomenon, the data for each point in time can be contoured using the approaches described in Chapter 8 (triangulation, inverse distance, and kriging). Howard and MacEachren chose kriging for R-VIS because it provides

* For additional discussion on visual variables for data quality, see McGranaghan (1993). He proposes the visual variable **realism,** which refers to the "photorealism of the image" (p. 10).

* Although DIN data are collected at various depths, R-VIZ presently handles only surface data.

FIGURE 16.4 Example of differing resolutions for a raster database. (© Copyright 1995 by the Guilford Press. Reprinted with permission.)

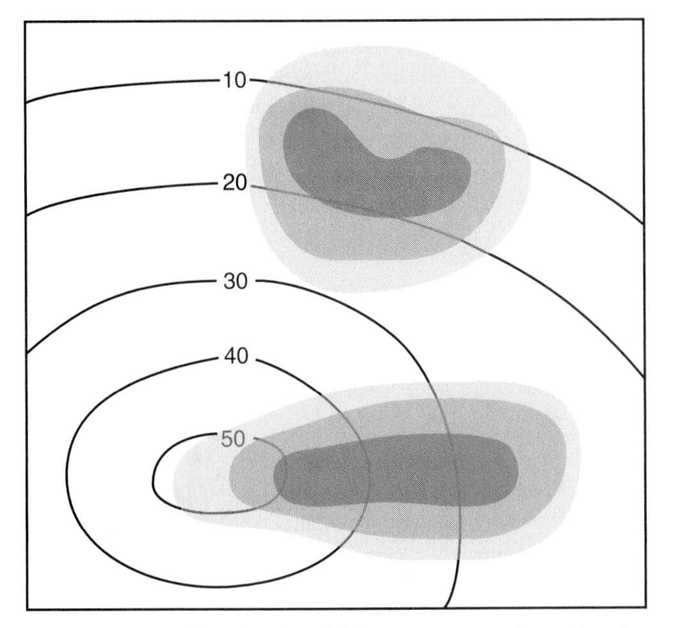

FIGURE 16.5 The visual variable transparency is used to depict the quality of the data. A light "fog" indicates reliable data, while a dark fog indicates uncertain data. (After MacEachren 1992, 15.)

a reliability map expressed as confidence intervals (see section 8.3.2), in addition to the contour map of the theme (DIN in the case of R-VIS).

The default display for R-VIS compares the contoured DIN data for a particular point in time with an associated reliability map. Howard and MacEachren initially chose a lightness scheme (light to dark red) for the contoured data, and a saturation scheme (desaturated to saturated red) for the reliability map, but the schemes "did not provide enough contrast . . . with the saturation range being particularly ineffective." As a result, they used different hues for the two maps (red for the contour map and blue for the reliability map) and varied both the lightness and saturation within each map (Color Plate 16.1).

To combine the contour map with the reliability map, two basic approaches are used in R-VIS. One is termed the "overlay method," in which area shading is the symbol for the contour map and weighted isolines (isolines of varying width; see section 12.2.2) are used for the reliability map (Figure 16.6). Howard and MacEachren indicate that the overlay method "emphasize[s] the data while allowing analysts to check the reliability in map areas that seem to be particularly good or bad in terms of meeting the EPA dissolved inorganic nitrogen targets" (p. 70).

Another approach for combining the contour and reliability maps is termed a "merged display" in which the contour and reliability maps are each displayed with a unique color (or symbol). For example, using the bivariate choropleth method described in section 12.1.2., the contour map can be shown with a lightness scheme and the reliability map with a saturation scheme. Alternatively, a merged display can be created by showing the contour map as a lightness/saturation range (or as shades of gray) and the reliability map as varying amounts of fill within a raster grid cell. Howard and MacEachren note that merged displays emphasize the *relationship* between the contour and reliability maps.

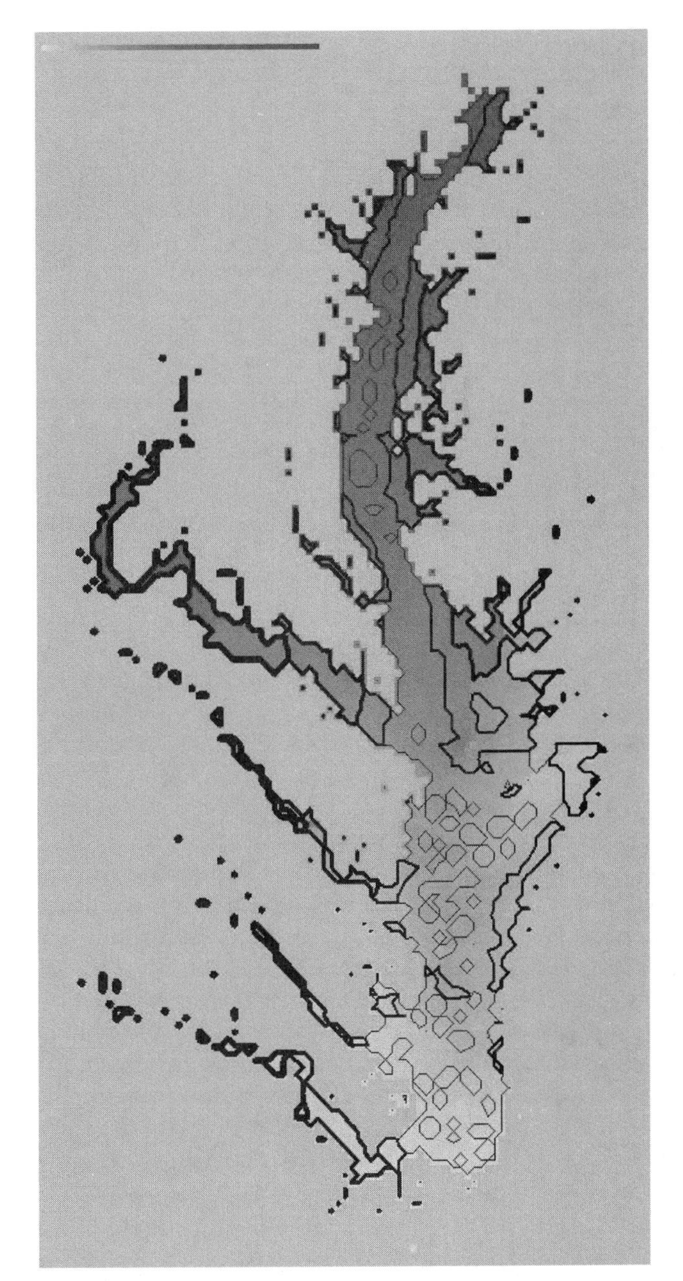

FIGURE 16.6 The "overlay method" for combining contour and reliability maps within R-VIS. The contour map is shown in shades of gray, while the reliability map is shown as weighted isolines. (Courtesy of David Howard and Alan MacEachren.)

Several exploration approaches are available within R-VIS to examine data quality. One is to alternate contour and reliability maps at the same screen location. This alternation can be under the control of the user or it can be animated. With respect to this approach, Howard and MacEachren indicate, "it is with multiple viewing that relationships between data and reliability distributions (if there are any) begin to become apparent" (p. 71). A second exploration approach is to highlight some portion of the data set by dynamically specifying the range of either contoured or reliability values to be shown. For example, users might be interested in targeting areas that exceed a particular DIN concentration. A third exploration approach is to animate the contour and reliability maps over time. Although Howard and MacEachren do not plan to release R-VIS for general use (Howard, personal communication, 1997), readers will find it useful to study their description of the software, and thus consider the capability that should be included in systems depicting data quality.

Fisher's Animation of Dot Map Uncertainty

In section 16.1.1, it was indicated that dot maps can provide misleading information when dots are dispersed because there is a low probability of finding a phenomenon at the location of a particular dot. Fisher (1996) has developed a method to depict this uncertainty via animation. In order to understand Fisher's approach, we must first review how dot maps are constructed (see section 10.3 for a more complete discussion). The first step is to delineate regions (polygons) within which the phenomenon being mapped is located; ideally, this step considers ancillary information that can assist in determining appropriate locations for dots. Second, decisions are made on dot size and unit value (the count represented by each dot). Third, dots are placed within each polygon. Fisher followed these basic steps, with the exception that he did not consider ancillary information because of the slow computer hardware he used.

Fisher used two basic algorithms to create his dot maps. In the first, an initial dot map was created by placing dots randomly within a minimum **bounding rectangle** surrounding each enumeration unit. A random dot falling within the bounding rectangle and within the enumeration unit, was plotted on the map, while a dot falling within the bounding rectangle, but outside the enumeration unit, was not plotted. In the second algorithm, the initial map was animated by selecting one dot at random, deleting that dot, and then locating a replacement dot within the enumeration unit of the deleted dot. The movement of dots on the resulting animated maps was intended to illustrate the uncertainty associated with dot placement.

Although Fisher's notion of animating dot maps is intriguing, the animations themselves are disconcerting. One problem is that relatively dense enumeration units appear just as uncertain as less dense units, while intuitively it seems that the position of dots within a dense area should appear more certain. A second problem is that the purely random placement used can lead to strange concentrations of dots (note the circular arrangement of dots in the western portion of Connecticut in the top map in Figure 16.7). These problems might be ameliorated by a revised dot placement algorithm such as the modification of Lavin's (1986) dot density method described in section 10.3.3. It is also possible that a map comparison approach might be useful in stressing areas of greater uncertainty; for example, compare the two animation frames shown in Figure 16.7 and note the quite different appearance of dot distributions in the state of Vermont (an area of low density),

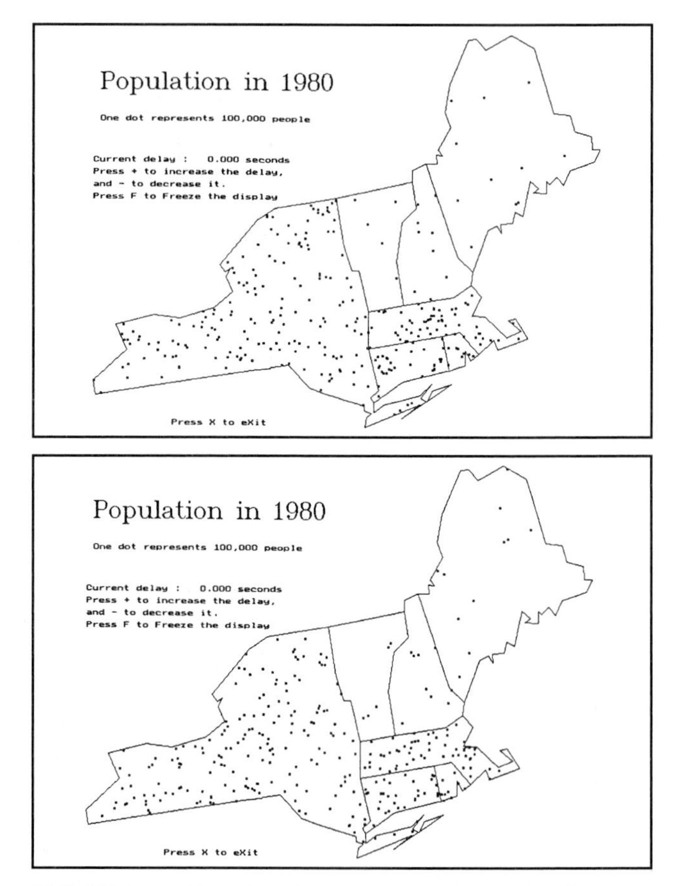

FIGURE 16.7 Two maps from an animation used to illustrate data quality. The number of dots within each data collection unit (state in this case) is held constant, but they are moved to different random locations. The randomness of location is intended to illustrate the uncertainty in the data. (After Fisher, P. F. 1996. First published in *Cartography and Geographic Information Systems* 23(4), p. 200. Reprinted with permission from the American Congress on Surveying and Mapping and Peter F. Fisher.)

and the relatively greater stability for New York State (an area of high density).

Hunter and Goodchild's Mapping of DEM Uncertainty

Topographic data are commonly available in the form of an equally spaced gridded network known as a **digital elevation model (DEM).** Such data can be readily contoured using standard mapping software such as Surfer. Hunter and Goodchild (1995) indicate one important application of the resulting contours is the specification of a particular threshold elevation: for example, in floodplain mapping, we might specify the contour above which building is permitted.

Unfortunately, users of contour maps resulting from DEMs often are unaware of the uncertainty associated with DEMs. This uncertainty can be measured using the root mean square error (RMSE), which is defined as follows:

$$\sqrt{\frac{\sum_{i=1}^{n}(Z_i - Z_i^*)^2}{n}}$$

where Z_i = elevation of the DEM at a sampled point
Z_i^* = true elevation at the sampled point
n = number of sampled points

For example, for their 7.5-minute products, the USGS analyzes 28 points in this fashion (United States Geological Survey 1990).

Hunter and Goodchild (1995) describe how the RMSE associated with the DEM can be used to specify a probability that the actual elevation value is above a particular threshold elevation. Their method is based on the assumption that the distribution of error around a particular contoured value is normal (see section 3.3.1), and that the standard deviation of this normal distribution is equal to the RMSE. This concept is illustrated in Figure 16.8. Here Z_t is the threshold elevation value of interest (for example, the elevation above which building is permitted in a floodplain). Note that if the DEM value for a raster cell (Z_{cell}) is equal to the threshold (as in Figure 16.8A), then the probability of being above the threshold is .5. If, however, the DEM value for a cell is below or above the threshold (as in Figure 16.8B and 16.8C), then the area under the normal curve above the threshold must be computed. This can be accomplished using tables of probability published in statistics books (such as Burt and Barber 1996, 194–198). Hunter and Goodchild indicate the method also has been implemented in the GIS package IDRISI.

Once the basic probabilities have been computed for each cell, contour maps of these probabilities can be constructed. Hunter and Goodchild illustrate several different ways of visualizing these probabilities. A simple black-and-white approach is shown in Figure 16.9 for a threshold contour for the DEM analyzed by Hunter and Goodchild. An obvious problem with this approach is that the neutral gray tone used to mask cells outside 2.5 to 97.5 percent probability values is identical to one of the gray tones falling within the 2.5 to 97.5 percent probability range. As a solution to this problem, Hunter and Goodchild suggest several alternative color schemes (see Plate 1 of their paper). If the 50 percent probability value is treated as a logical breakpoint in the data, then the diverging color schemes proposed by Brewer (see section 6.2) also would seem to be a logical solution to portraying such data.

Other Examples

Because of space limitations, only three examples of visualizing data quality have been covered in this section. Other noteworthy work includes Fisher's attempts to animate uncertainty on soils maps (1993), DEMs (1996), and remotely sensed images (1994b). In the following section, we will consider still another of his efforts: using sound to visualize the uncertainty in remotely sensed images (1994a). Goodchild et al. (1994) have also developed some ingenious methods for visualizing the uncertainty in remotely sensed images.

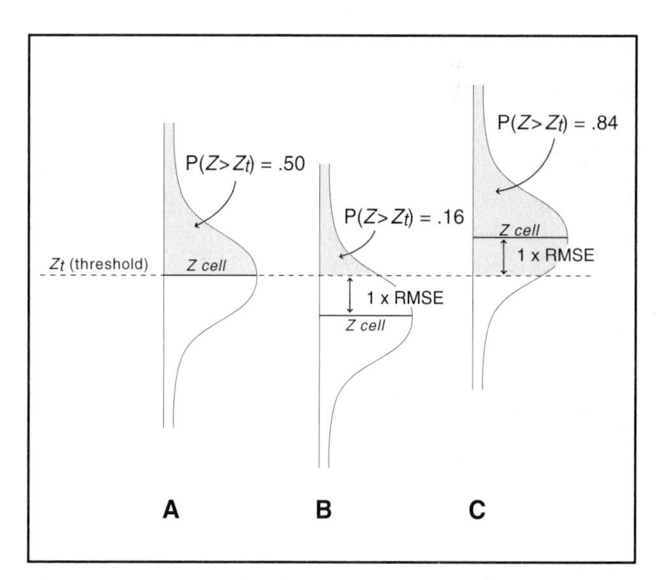

FIGURE 16.8 Determining the probability that the elevation for a raster cell in a DEM (Z_{cell}) is greater than a particular threshold (Z_t). The process involves computing the area under the normal curve above the threshold value. The standard deviation of the normal curve is a function of the root mean square error (RMSE) for the DEM. (After Hunter and Goodchild 1995, 532.)

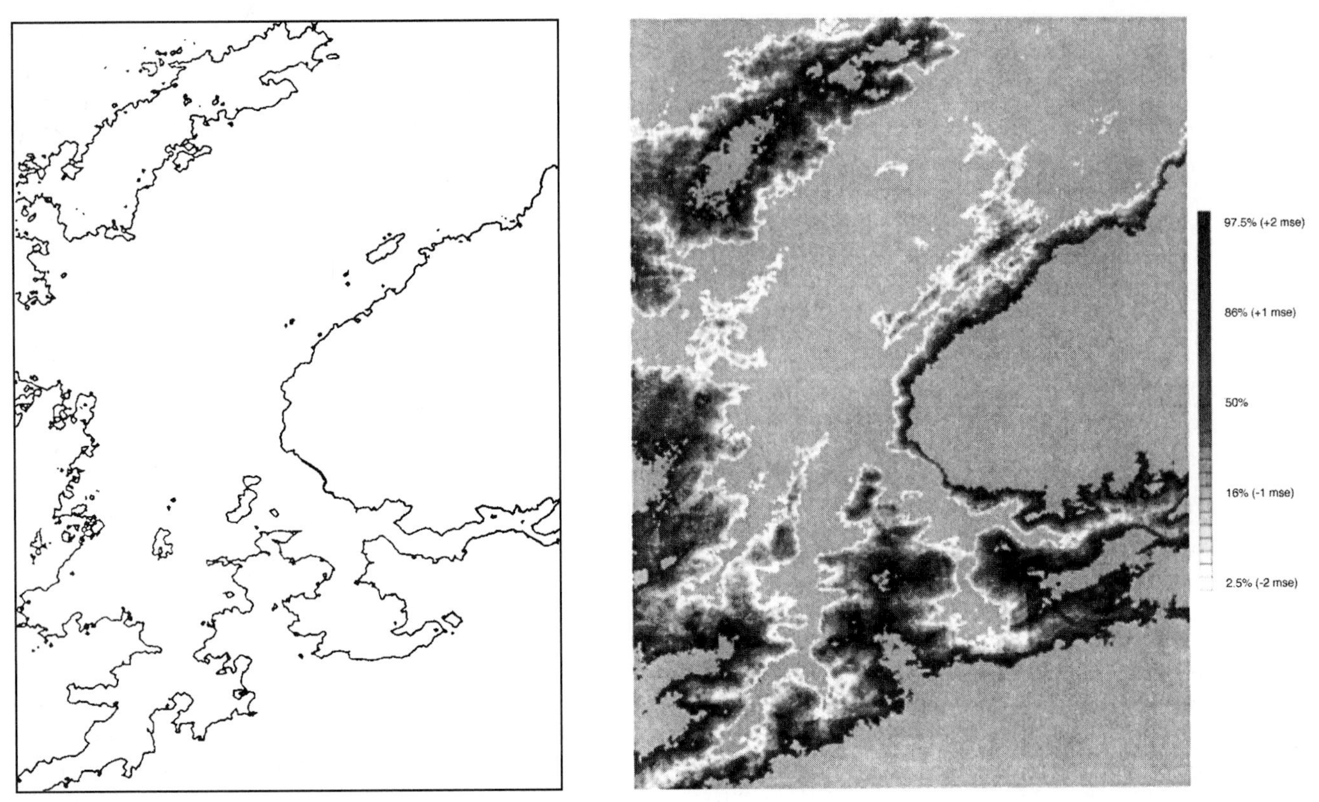

FIGURE 16.9 The contour line associated with a particular threshold (A) and the probabilities associated with exceeding that particular threshold (B). (Reproduced with permission, the American Society for Photogrammetry and Remote Sensing, Photogrammetric Engineering and Remote Sensing 61, no. 5, p. 529–537.)

16.2 USING SOUND TO REPRESENT SPATIAL DATA

16.2.1 Krygier's Work

Krygier (1994) divides the sounds used to represent spatial data into two forms: realistic sounds and abstract sounds. **Realistic sounds,** whose meaning is based on our past experience, can be subdivided into narration and mimetic sound icons (or "earcons"). An example of narration would be the audio portion of the animations included in multimedia encyclopedias such as Grollier's described in section 15.2.7 (for example, "you are now looking at major Allied troop movements during the Battle of the Bulge"). An example of a mimetic sound icon would be the sound of fire and wind in an animation of forest growth, as in a video that Krygier (1993) has developed; in this case, one actually gets the feeling of being in the forest.

Abstract sounds have no obvious meaning, and thus require a legend to explain their use. For example, imagine a map of census tracts with the title "Median Income, 1997," in which different loudness values represent different incomes (a mouse click on a tract would produce a loudness value as a function of the associated income). In order to understand the loudness values, the map reader would have to be told that a higher magnitude of loudness represents a higher median income.

Analogous to the visual variables discussed previously in this book (section 2.3), Krygier (1994, 154–146) identifies the following abstract sound variables:

location: the location of a sound in two- or three-dimensional space

loudness: the magnitude of a sound

pitch: the highness or lowness (frequency) of a sound

register: the relative location of a pitch in a given range of pitches

timbre: the general prevailing quality or characteristic of a sound

duration: the length of time a sound is (or is not) heard

rate of change: the relation between the durations of sound and silence over time

order: the sequence of sounds over time

attack/decay: the time it takes a sound to reach its maximum or minimum

As Figure 16.10 shows, Krygier argues that the bulk of these variables would be effective for ordinal-level data.

In his video, Krygier (1993) developed several applications of how these sound variables might be used. In one,

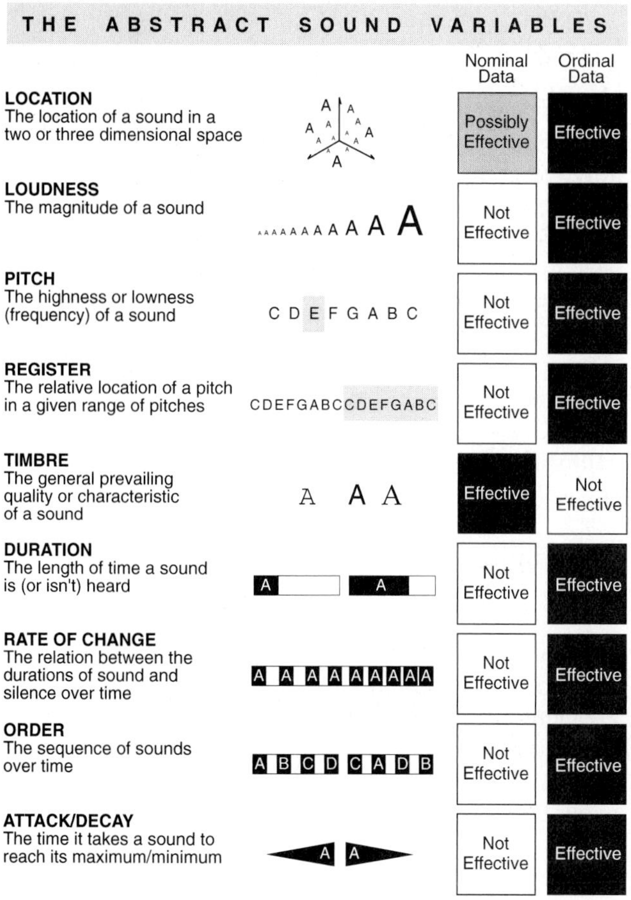

THE ABSTRACT SOUND VARIABLES

Variable	Example	Nominal Data	Ordinal Data
LOCATION The location of a sound in a two or three dimensional space	A A A A A A A A	Possibly Effective	Effective
LOUDNESS The magnitude of a sound	A A A A A A A A A A A A	Not Effective	Effective
PITCH The highness or lowness (frequency) of a sound	C D E F G A B C	Not Effective	Effective
REGISTER The relative location of a pitch in a given range of pitches	CDEFGABCCDEFGABC	Not Effective	Effective
TIMBRE The general prevailing quality or characteristic of a sound	A A A	Effective	Not Effective
DURATION The length of time a sound is (or isn't) heard	A A	Not Effective	Effective
RATE OF CHANGE The relation between the durations of sound and silence over time	A A A A A A A A A	Not Effective	Effective
ORDER The sequence of sounds over time	A B C D C A D B	Not Effective	Effective
ATTACK/DECAY The time it takes a sound to reach its maximum/minimum	A A	Not Effective	Effective

FIGURE 16.10 Abstract sound variables. (Reprinted from Krygier, J. B., *Sound and Geographic Visualization,* Copyright 1994, p. 153, with permission from Elsevier Science.)

he displayed a choropleth map animation of the rate of AIDS cases in the United States from 1980 to 1995. This animation was combined with a loudness variable representing the total number of AIDS cases in each year. When viewing the resulting animation, one gets the feeling of an impending disaster, which is exactly what Krygier was trying to create. In this same animation, Krygier also used pitch to represent the percentage increase of new AIDS cases each year (a lower pitch would represent a lower percentage increase). In this case Krygier (1994) noted: "The pitch can be heard 'settling down' as the percentage increase drops and steadies in the late 1980s. An anomaly can be heard in 1991 where the animation switches from actual AIDS cases to model predicted AIDS cases (p. 157)." It is my experience that viewers can readily hear the "settling down," but it is more difficult to detect the anomaly. This might be a case where the experience of hearing different pitches would be advantageous (a musician would have an advantage over the nonmusician).

In another application, Krygier created a multivariate display by combining traditional visual variables (lightness and size) with sound variables. For the traditional visual variables, Krygier superimposed a proportional circle map on a choropleth map. The proportional circle map depicted "median income," while the choropleth map displayed the "percentage of population not in the labor force." Values for the traditional visual variables were obtained by visually examining the map. For the sound variables, Krygier used a pitch within three different octaves (a register) to display a "drive to work index," and three levels of pitch within each of the three octaves to represent the "percentage poor." Values for the sound variables were obtained by pointing at enumeration units. For example, for an enumeration unit with a long drive to work and low rate of poverty, a high octave pitch followed by a low pitch within that octave could be heard. Krygier (1994) indicated that "After a short period of [use] . . . it becomes relatively easy to extract the four data variables" (p. 158).

16.2.2 Fisher's Work

Fisher (1994a) has developed software for utilizing sound to portray the reliability in remotely sensed images. In this software, a cursor is moved across a remotely sensed image under either user or automatic control (Color Plate 16.2). As the cursor moves, the user hears a sound representing the reliability associated with the current pixel location. For example, if duration is used, a long duration indicates a pixel with a high reliability.

In using Fisher's software, one difficulty is developing an impression of the spatial pattern of reliability. Although it is relatively easy to determine the reliability associated with specific pixels, it is not easy to determine the reliability of groups of pixels. This problem might be obviated by also using visual methods to depict reliability. For example, Goodchild et al. (1994, 163–164) describe how multiple gray-tone maps can be used to depict reliability.

16.2.3 Other Work

Other work on sound includes that by Veregin et al. (1993) and Weber (1993). Veregin et al. described the use of a "Geiger counter" to depict the uncertainty in soils maps—a greater frequency of clicks per unit time represented greater uncertainty, an approach that appears similar to Krygier's rate of change variable. Weber examined the use of musical harmonic intervals (or dyads) to represent map information (see also Weber and Yuan 1993).

16.3 RECENT RESEARCH ON COLOR

16.3.1 Brewer et al.'s Use of Color to Map Mortality

One important recent study of color is Brewer et al.'s (1997) evaluation of color schemes for choropleth maps appearing in the *Atlas of United States Mortality* (Pickle et al. 1996). Basic data depicted on maps in this atlas are death rates for various causes (such as heart disease, prostate cancer, and homicide). The data are mapped using both five- and seven-class maps. Five-class maps are based on a quintile classification in which an equal number of enumeration units (health service areas) are placed in each class, while seven-class maps are created by splitting the highest and lowest categories on the five-class map to provide greater detail on low and high death rates.

Brewer et al. evaluated sequential, diverging, and spectral color schemes (see section 6.2 for an explanation of these schemes) for possible use on the maps displayed in the atlas. A sequential scheme was considered appropriate because legend values on the maps gave the appearance of unipolar data (no dividing point was evident; see the legend for the map at the bottom of Color Plate 16.3). A diverging scheme was also considered appropriate in the sense that the midpoint of the middle class on a quantiles map is the median, and thus can serve as a useful dividing point in the data. In addition, a spectral scheme was evaluated because of its common use in visualization.

Brewer et al. took considerable care in selecting colors for each of the schemes they evaluated. For the diverging schemes, easily named colors were utilized that would unlikely be confused with one another (Boynton 1989; Sturges and Whitfield 1995). With reference to a red-blue diverging scheme they employed, Brewer et al. indicated that using easily named colors was desirable because "if you personally describe a color as red, then it is unlikely that anyone would mistake that color for a blue. [Furthermore,] novice map designers rarely have access to color-measurement instrumentation, calibrated color displays, or even color charts, when selecting colors" (p. 414). For both the diverging and sequential schemes, Brewer et al. also used colors unlikely to be confused by color-impaired readers.

The color schemes evaluated by Brewer et al. are shown in Color Plate 16.3. Note that the diverging schemes ranged from a neutral light gray (for moderate data values) to dark saturated hues (for low and high data values). The sequential schemes included two color schemes (red-yellow and purple-blue) and a gray scheme. The sequential color schemes ranged from light to dark (for example, light yellow to dark red), with saturation peaking at the median class. The latter was done "to provide visual emphasis on the critical [class] in the data" (p. 419). The spectral scheme included standard colors from the electromagnetic spectrum; these colors were carefully structured, however, so that the scheme extended from *dark* blue (for low values) to *bright* yellow (for the midpoint) to *dark* red (for high values). The spectral scheme thus provided lightness characteristics similar to the diverging schemes.

Brewer et al. evaluated the color schemes using both a set of map-reading tasks and readers' preference (based on "pleasantness" and "readability"). Since they took considerable care in designing the color schemes, it is not surprising that they found all schemes reasonably effective in the map-reading tasks (an exception was the gray scheme for an accuracy task). In terms of reader preference, the spectral and purple-green diverging schemes scored best, although a number of the other schemes were well-liked. The gray scheme was, however, decidedly less desirable. Overall, the color schemes examined constitute a useful set that readers may wish to use on their own maps. Appendix A provides CMYK specifications for these colors; CIE and Munsell specifications can be found in Brewer et al. (pp. 435–436).

16.3.2 Brewer's Guidelines for Diverging Color Schemes

In another useful study, Brewer (1996a) provides guidelines for those wishing to use diverging color schemes. Her guidelines are based on three parameters: lack of confusion in color naming, minimization of color vision impairment, and minimization of the effects of simultaneous contrast. The logic of considering color naming was discussed above, while the importance of color vision impairment and simultaneous contrast were considered in sections 6.4.3 and 5.1.4, respectively.

Table 16.1 summarizes the guidelines. The left-hand column portrays all combinations of the nine basic colors that are easily named: pink, red, orange, brown, yellow, green, blue, purple, gray. If a resulting combination is confused due to color naming, color vision impairment, or simultaneous contrast, an asterisk (*) is placed in the appropriate column; otherwise, an "OK" appears. For example, the asterisks for pink-red indicate that pink and red are confused because of color naming and color vision impairment, but are not confused by simultaneous contrast. Ideally, "OK" should appear for a particular color combination in all of the columns. Unfortunately, this occurs only for the red-blue, orange-blue, orange-purple, and yellow-purple schemes. In addition, Brewer indicated that the brown-blue and yellow-blue combinations could be used "because the simultaneous contrast confusions do not affect [isarithmic] mapping and because they can be avoided on choropleth maps by using substantial lightness differences between colors of the same hue" (p. 85).

Brewer noted, however, that the yellow-purple and yellow-blue schemes were of limited use because of the difficulty of creating different lightnesses of yellow. It is interesting that the purple-green combination used successfully in the Brewer et al. (1997) study described above was not one of those recommended by Brewer in the present work.

Although Brewer's guidelines appear to place rather severe restrictions on diverging schemes, a greater variety can be used in an interactive environment because the schemes can be selected as a function of reader characteristics (color vision impaired or not impaired) and type of map (choropleth or isarithmic). For example, if we assume a person with normal color vision, then the color vision impairment column in Table 16.1 can be ignored, and the following additional color schemes are possible: pink-yellow, pink-blue, red-yellow, red-gray, orange-green, and orange-gray. Further, if we assume an isarithmic map (in which simultaneous contrast is unimportant), pink-green, red-green, and green-purple could be added as potential schemes.

TABLE 16.1. Color Pairs Appropriate for Diverging Color Schemes

	Confusions		
Easily Named Colors	*Color Naming*	*Color Vision Impairment*	*Simultaneous Contrast*
Pink-Red	*	*	OK
Pink-Orange	*	*	OK
Pink-Brown	*	*	OK
Pink-Yellow	OK	*	OK
Pink-Green	OK	*	*
Pink-Blue	OK	*	OK
Pink-Purple	*	*	OK
Pink-Gray	*	*	OK
Red-Orange	*	*	OK
Red-Brown	*	*	OK
Red-Yellow	OK	*	OK
Red-Green	OK	*	*
Red-Blue	OK	OK	OK
Red-Purple	*	OK	OK
Red-Gray	OK	*	OK
Orange-Brown	*	*	OK
Orange-Yellow	*	*	OK
Orange-Green	OK	*	OK
Orange-Blue	OK	OK	OK
Orange-Purple	OK	OK	OK
Orange-Gray	OK	*	OK
Brown-Yellow	*	*	OK
Brown-Green	*	*	OK
Brown-Blue	OK	OK	*
Brown-Purple	*	OK	OK
Brown-Gray	*	*	OK
Yellow-Green	*	*	OK
Yellow-Blue	OK	OK	*
Yellow-Purple	OK	OK	OK
Yellow-Gray	*	OK	OK
Green-Blue	*	*	OK
Green-Purple	OK	*	*
Green-Gray	*	*	OK
Blue-Purple	*	*	OK
Blue-Gray	*	OK	OK
Purple-Gray	*	*	OK

Source: Based on Brewer 1996, p. 81.

16.4 VIRTUAL REALITY AND VISUAL REALISM

Virtual reality (or **virtual world**) refers to the use of computer-based systems for creating lifelike representations of the real world. Readers are probably most familiar with virtual reality in association with the specialized head-mounted display (HMD) and "data gloves" described in daily newspapers and news magazines. Virtual reality has, however, been used loosely to describe any lifelike representation, whether it is achieved by a static paper map, an interactive computer graphic display, or through the use of specialized apparatus such as the HMD and "data gloves." The terms **visual realism** and **visual simulation*** are often used in place of virtual reality to stress the importance of vision in creating a realistic representation. In this section, we'll consider several examples of visual realism.

Bressi (1995) provides an overview of several of the major centers where work on visual realism is taking place. One center is at the University of California at Los Angeles (UCLA), particularly within the Department of Architecture and Urban Design. Here researchers are involved in two major projects: *Virtual Los Angeles* and *Rome Reborn* (Davis 1997). The purpose of *Virtual Los Angeles* is to build a visually realistic model of the entire Los Angeles Basin, an area of more than 10,000 square miles (Color Plate 16.4). *Virtual Los Angeles* will provide not only realistic images, but also the capability for real-time fly-bys, drive-throughs, and walk-throughs. *Rome Reborn* is an attempt to recreate what Rome looked like from the ninth century B.C. to the fifth century A.D. (Figure 16.11). Rather than producing static images, *Rome Reborn* will provide "virtual actors and guides with the ability to interact with the user in touring the city or in seeing reenactments of historical events and scenes from daily life" (Favro et al. 1997, 4).

Although projects such as *Virtual Los Angeles* and *Rome Reborn* are novel and exciting, most of us do not have access to the hardware and software necessary for creating and viewing such realistic renditions. Both projects are being developed using the proprietary uSim software housed in the Department of Architecture and Urban Design at UCLA. A review of the software and associated hardware by Liggett and Jepson (1995) suggests that it will be some time before the lay mapmaker has access to such capability.

The CLRview software developed jointly by the Centre for Landscape Research (CLR) at the University of Toronto and the Swiss Federal Institute of Technology provides a second example of the use of visual realism. Hoinkes and Lange (1995), two of the leaders at these respective centers, stress that traditional GIS software

* Another term that is sometimes used is the **natural scene paradigm** (see Robertson 1991 for an introduction).

FIGURE 16.11 An example of the visual realism that can be achieved with *Rome Reborn,* a project depicting what Rome looked like from the ninth century B.C. to the fifth century A.D. (From Dean L. Abernathy's Master's thesis "A Temporal Model of the Temple of Antoninus and Faustina and Context: The Roman Forum." This project provided the graphics for Rome Reborn.)

has been largely two-dimensional in nature, while the real world is three-dimensional and should be represented as such. For example, in a traditional GIS, variation in housing density in an urban neighborhood would be represented by abstract symbols, say, differing shades of a blue tone, when the variation in density might be more realistically represented by differing concentrations of three-dimensional symbols that look like houses. In this respect, CLRview contains a variety of realistic three-dimensional symbols for depicting vegetation, buildings, and infrastructure (for example, roads, towers, and pipelines). Color Plate 16.5 is an example of the realism that can be achieved using CLRview. Unfortunately, like the uSim software mentioned above, CLRview currently requires a sophisticated hardware environment (Silicon Graphics), but a plus is that the software is available for free (see Appendix I for availability).

The Emaps package developed by Ervin (1992, 1993) provides a third intriguing example of the use of visual realism. The purpose of Emaps was to enable the planner to "make judgments . . . about dimensions and densities of view corridors, buffers, height limitations, vantage points, and special views." An example of a view created by Emaps is shown in Color Plate 16.6. Here we see that visually realistic trees and houses have been added to a traditional hue-based color scheme using a traffic light analogy: undesirable areas close to a proposed road appear in red, while more desirable areas far away appear in green, and transitional areas appear in yellow. In contrast to the sophisticated hardware required for the two previous examples, static images in Emaps were designed in a Macintosh environment using Common Lisp. Creating animated fly-bys or walk-throughs, however, required the use of additional software running on a sophisticated graphics workstation (Ervin 1993, 32). Ervin (personal communication, 1997) indicates that work on Emaps was discontinued because of changing operating systems, graphics protocols, and programming languages, but that most of the functionality is now available with a combination of the 3DAnalyst extension for ArcView and Java.

A considerable amount of work in visual realism has been done by Ian Bishop and his colleagues at the University of Melbourne in Australia. Bishop (1994) makes the argument that visually realistic images should be used whenever possible for naive users. He states:

> The non-scientific audience . . . wants abstraction minimized, information content maximized, . . . with the package digestible and non-threatening. This suggests the use of a visual realism approach which shows information consumers what will or might happen under a variety of conditions and permits them to explore the alternative environments using their natural sensory perceptions. (p. 61)

One novel recent effort by Bishop and his colleagues is the attempt to integrate software for modeling, GIS, and visualization (by utilizing linear programming routines, ARC/INFO, and IRIS Performer, respectively) into a single system to assist forest planning and management (Bishop and Karadaglis 1997; Tang et al. 1997). Although a number of researchers previously have integrated modeling and GIS software, Bishop and his colleagues are among the first to also incorporate sophisticated visualization routines. Color Plate 16.7 is an example of a visually realistic image that can be achieved using their combined software. They term this a form of **image-based visualization** because aerial photographs or remotely sensed images are used to represent different vegetation types found in a forested region. They note that **object-based visualization** (for example, using individual realistic-looking trees as in CLRview and Ervin's Emaps) is also possible for forest modeling (see *SmartForest* and *The Forestry Project* listed in Appendix I), but that suitable software "is still working towards sufficiently realistic presentation to offer clarity of interpretation to the entire range of forest users" (Tang et al. 1997).

The examples of visual realism covered thus far all involve the use of traditional computer graphics displays (such as CRTs). Researchers are also beginning to explore the potential of specialized virtual reality hardware for achieving visual realism. One example of such hardware is the CAVE, a room-size structure for experiencing virtual reality that is available at the National Center for Supercomputing Applications (NCSA) at the University of Illinois. In the CAVE, four projectors display computer-generated images onto three walls and the floor, while a "head tracker" on one of the users governs the view that all users see through stereo glasses. Interaction with the computer system is achieved via a hand-held "wand" (see http://evlweb.eecs.uic.edu/EVL/VR/systems.html#CAVE for more explanation). For another example of using specialized hardware for creating visual realism, see the paper by Neves et al. (1997) mentioned in the next section.

16.5 *COMPUTERS & GEOSCIENCES* SPECIAL ELECTRONIC ISSUE

The 1997 special electronic issue of *Computers & Geosciences* (Vol. 23, No. 4) consists of a broad range of papers dealing with visualization issues. MacEachren and Kraak, editors of the special issue, divide the papers into three broad categories: exploratory spatial data analysis, hyperlinks and the World Wide Web, and virtual reality. Within exploratory spatial data analysis, papers are further subdivided into those dealing with brushing and linking, nontemporal animation, and temporal animation. The brushing-and-linking category includes a paper by Dykes (1997) describing the use of the Tcl/Tk toolkit for exploring spatial data, and one by Cook et al. (1997) dealing with how ArcView and XGobi (a package for exploring multivariate data) can be linked to provide exploratory capability within a GIS. The paper by Dykes is closely related to the Project Argus software, which was described in Chapter 13.

Papers dealing with nontemporal animation include those by Ehlschlaeger et al. (1997), Davis and Keller (1997), and Evans (1997). The former two papers describe animating the uncertainty in terrain and soils data, respectively, while Evans analyzes the use of flickering (or blinking) to illustrate the uncertainty associated with land use/land cover maps. With respect to temporal animation, Acevedo and Masuoka (1997) discuss problems associated with developing animations of urban growth in the Baltimore–Washington, D.C., area. Their work is closely related to the animation of urban growth in the San Francisco–Sacramento region mentioned in section 14.3.6. In another paper dealing with temporal animation, Mitas et al. (1997) create numerical simulations of landscape processes (for example, erosion).

In the section dealing with hyperlinks and the World Wide Web, Cartwright (1997) describes a variety of novel approaches (such as interactive touch television and live cameras, or "Web cams," available via the World Wide Web) that have the potential for enhancing the understanding of spatial information. In this section, Kraak and van Driel (1997) also discuss the development of **hypermaps,** which they define as "georeferenced multimedia systems that can structure individual multimedia components with respect to each other and [a] map. [Hypermaps] will let users navigate multimedia data sets not only by theme but also spatially" (p. 457). Thus, a user can click on a map element and get information not only about that element (say, a Chinese restaurant on a map of a city), but also information about other Chinese restaurants within the city.

Fernandes et al. (1997) report on *Interactive Portugal,* a spatial database of digital orthophotos of Portugal that can be accessed by the general public via the World Wide Web. The database can be linked with other relevant digital data (such as topography), and various what-if scenarios can be posed (for example, what will be the impact in a particular region if an oil spill occurs?).

In the virtual reality section, Fairbairn and Parsley (1997) examine how VRML software can be utilized to provide specialized information about a college campus that could not be achieved by a traditional static paper map (such as allowing a user-defined walk-through). Finally, Neves et al. (1997) describe the use of a virtual reality environment (based on a head-mounted display) for exploring data commonly associated with GIS.

16.6 KEEPING PACE WITH RECENT DEVELOPMENTS

One problem in writing this book was keeping pace with recent developments: the field of cartography is changing so fast that any text becomes dated as soon as it is written. This problem is handled to some degree by providing updated information on the home page for the book. Those wishing detailed information on current research topics should also consider reviewing recent issues of the journals and proceedings listed in Table 16.2. Note that this table is split into *primary* and *secondary* journals and proceedings. Most of the articles within a publication in the primary section have a cartographic component, while only some of the articles in the secondary group focus on cartography.

In addition to examining the literature, you may also find it useful to attend conferences that include topics of interest to cartographers (Table 16.3). Such conferences frequently include expositions illustrating the latest technological advances in hardware and software. Being aware of such advances is important, as the field of cartography is arguably technology-driven (for example, virtual reality capability is currently not available to many because of its high cost).

It is also important to be aware of research groups and organizations that provide useful information on recent developments. The National Center for Geographic Information and Analysis (NCGIA, http://www.ncgia.ucsb.edu/) has produced numerous publications of interest to cartographers. More recently, the University Consortium for Geographic Information Science (UCGIS, http://www.ucgis.org/), a group of more than

TABLE 16.2. Journals and Conference Proceedings Containing Articles Dealing with Thematic Cartography and Visualization

Primary Journals
Cartography
Cartographic Perspectives
Cartographica
Cartography and Geographic Information Systems
The Cartographic Journal
Primary Proceedings*
ACSM/ASPRS Annual Convention (American Congress on Surveying and Mapping)
AUTO-CARTO (American Congress on Surveying and Mapping)
ICA (International Cartographic Association)
Secondary Journals
Annals, Association of American Geographers
Byte
Computer Graphics
Computers & Geosciences
Environment and Planning A
Environment and Planning B: Planning and Design
Geographical Analysis
Geo Info Systems
GIS World
IEEE Computer Graphics & Applications
IEEE Transactions on Visualization and Computer Graphics
International Journal of Geographical Information Science
Journal of Geography
Journal of the American Statistical Association
Statistical Computing & Statistical Graphics Newsletter
The American Statistician
The Professional Geographer
Secondary Proceedings*
GIS/LIS (American Congress on Surveying and Mapping)
Visualization Annual (IEEE)
Innovations in GIS (Proceedings of the U.K. National Conference on GIS Research: GISRUK)
International Symposium on Spatial Data Handling (International Geographical Union)
ACM SIGGRAPH (Association for Computing Machinists, Special Interest Group on Graphics)

*For proceedings, the name of the sponsoring organization is provided in parentheses.

30 universities, has developed a set of research priorities (UCGIS 1996), many of which are relevant to cartographers. The International Cartographic Association (ICA) has several ongoing commissions; particularly intriguing for readers of this book is the Commission on Visualization (http://www.geog.psu.edu/ica/ICAvis.html). Finally, readers may also wish to examine the Carto Project, a three-year undertaking by ACM SIGGRAPH (Association for Computing Machinery, Special Interest Group on Computer Graphics) to explore how "viewpoints and techniques from the computer graphics community can be effectively applied to cartographic and spatial data sets" (http://www.siggraph.org/~rhyne/carto/).

TABLE 16.3. Conferences Having Topics of Interest to Cartographers

Conference
ACM SIGGRAPH (Association for Computing Machinists, Special Interest Group on Graphics)
ACSM/ASPRS Annual Convention (American Congress on Surveying and Mapping)
AUTO-CARTO (American Congress on Surveying and Mapping)
GIS/LIS (American Congress on Surveying and Mapping)
International Symposium on Spatial Data Handling (International Geographical Union)
ICA (International Cartographic Association)
Visualization (IEEE)

SUMMARY

This chapter has examined a number of recent developments that have impacted the discipline of cartography. One development is a concern for depicting the quality of the data displayed on maps. In this respect, readers need to be aware that a map is not necessarily a truthful representation of reality; generally, the data underlying a map are characterized by a degree of uncertainty.

We considered several visual variables for depicting data quality, including some introduced previously in Chapter 2 (for example, saturation) and several new ones: **crispness** (the sharpness of boundaries or area fills), **resolution** (the level of precision in the spatial data underlying an attribute, represented by varying sizes of raster cells), and **transparency** (the ease with which a theme can be seen through a "fog"). We also considered general methods for depicting data quality, including showing individual maps for the theme and its associated reliability (maps are compared); displaying the theme and its reliability on the same map (maps are combined); and utilizing data exploration tools. As examples of approaches for visualizing data quality, we considered Howard and MacEachren's R-VIZ software, Fisher's animation of dot maps, and Hunter and Goodchild's mapping of the uncertainty associated with DEMs.

A second recent development is the use of sound to represent spatial data, which is referred to as **sonification** or **acoustic visualization.** Here we focused on the work of Krygier, who introduced a set of sound variables analogous to the visual variables covered in Chapter 2. Examples include *location* (the location of a sound in two- or three-dimensional space), *loudness* (the magnitude of a sound), and *pitch* (the frequency of a sound). Krygier described several mapping applications that utilized these sound variables (for example, the loudness variable was utilized to depict an impending disaster in an animation of AIDS). In addition to Krygier's work, we also considered Fisher's use of sound to depict the reliability in remotely sensed images.

A third development is increased research (particularly by Cynthia Brewer of Penn State) on the use of color on maps. We considered two studies: one involving the selection of color schemes for choropleth maps appearing in the *Atlas of United States Mortality,* and one dealing with general guidelines for diverging color schemes. Both of these studies provide a range of color schemes that should be useful to mapmakers.

A fourth development is the attempt to create realistic representations of the earth's natural and built environments. When vision is the main sense used, this process is termed **visual realism** or **visual simulation;** when all of the senses (sound, touch, smell, and taste) are used, the result is a **virtual reality** or **virtual world.** To date, the bulk of applications have focused on the use of traditional CRT displays to create realistic representations; examples include *Virtual Los Angeles,* and images created using CLRview and Emaps. Researchers are also beginning to explore the potential of specialized virtual reality hardware, such as the CAVE, to achieve greater realism.

A special electronic issue of the journal *Computers & Geosciences* contains examples of the trend in recent research efforts in cartography. Based on this issue, key areas of interest at the present time include data exploration, animation, multimedia, the World Wide Web, virtual reality, and the depiction of data quality.

In Chapter 1, it was noted that the discipline of cartography has changed considerably over the last 35 years. Although it would be comforting to say that the situation has now stabilized, such is not the case. Cartography continues to evolve, and thus if readers of this book are to keep up to date, they should read major journals and proceedings in which cartographers publish, attend conferences of interest to cartographers, and be aware of research groups and organizations that provide useful information on recent developments.

FURTHER READING

Bishop, I. (1994) "The role of visual realism in communicating and understanding spatial change and process." In *Visualization in Geographical Information Systems,* ed. H. M. Hearnshaw and D. J. Unwin, pp. 60–64. Chichester: John Wiley & Sons.

Discusses the importance of creating visually realistic images. For more recent work by Bishop, see Bishop and Karadaglis (1997).

Bressi, T. (1995) "The real thing? We're getting there." *Planning* 61, no. 7:16–20.

A concise overview of some of the work being done in virtual reality.

Brewer, C. A., MacEachren, A. M., Pickle, L. W., and Herrmann, D. (1997) "Mapping mortality: Evaluating color schemes for choropleth maps." *Annals, Association of American Geographers* 87, no. 3:411–438.

One of the experimental studies described in section 16.3.1. For work by Brewer on simultaneous contrast, see Brewer (1996b, 1997).

Carr, D. B., and Pierson, S. M. (1996) "Emphasizing statistical summaries and showing spatial context with micromaps." *Statistical Computing & Statistical Graphics Newsletter* 7, no. 3:16–23.

Introduces a novel method for choropleth mapping that involves splitting a standard choropleth map into "micromaps."

Cartographica 30, nos. 2/3, 1993.

Includes several articles dealing with the depiction of data quality.

Computers & Geosciences 23, no. 4, 1997 (entire issue).

The special issue is summarized in section 16.5.

Cressie, N. (1992) "Smoothing regional maps using empirical bayes predictors." *Geographical Analysis* 24, no. 1:75–95.

An example of sophisticated statistical techniques applied to cartography.

Davis, B. (1997) "The future of the past." *Scientific American* 277, no. 2:89–92.

A summary of some virtual reality applications.

Ervin, S. M. (1993) "Landscape visualization with Emaps." *IEEE Computer Graphics & Applications* 13, no. 2:28–33.

Describes Emaps, an early software package for creating visually realistic images.

Environment and Planning B: Planning and Design 22, no. 3, 1995.

Contains numerous articles dealing with virtual reality.

Favro, D., Frischer, B., Dorr, A., and Jepson, W. (1997) "Rome Reborn: Virtual reality model of Augustan Rome project." <http://www.aud.ucla/edu/~dabernat/rome/index.html> Accessed September 12, 1997.

Describes *Rome Reborn,* a virtual reality attempt to recreate what Rome looked like from the ninth century B.C. to the fifth century A.D.

Fisher, P. F. (1996) "Animation of reliability in computer-generated dots maps and elevation models." *Cartography and Geographic Information Systems* 23, no. 4:196–205.

An example of some of the work that Fisher has done to depict data quality. Also see Fisher (1993, 1994a, 1994b).

Goodchild, M., Chih-chang, L., and Leung, Y. (1994) "Visualizing fuzzy maps." In *Visualization in Geographical Information Systems*, ed. H. M. Hearnshaw and D. J. Unwin, pp. 158–167. Chichester: John Wiley & Sons.

Describes efforts to visualize the uncertainty in remotely sensed images.

Hoinkes, R., and Lange, E. (1995) "3-D for free: Toolkit expands visual dimensions in GIS." *GIS World* 8, no. 7:54–56.

Summarizes some of the capability found in the CLRview software mentioned in section 16.4.

Howard, D., and MacEachren, A. M. (1996) "Interface design for geographic visualization: Tools for representing reliability." *Cartography and Geographic Information Systems* 23, no. 2:59–77.

In addition to covering R-VIZ (see section 16.1.4), this paper discusses general considerations in designing interactive software for visualizing spatial data.

IEEE Computer Graphics & Applications 13, no. 2, 1993 (entire issue).

This special issue deals with visualization.

Jacobson, R. (1994) "Virtual worlds capture spatial reality." *GIS World* 7, no. 12:36–39.

Discusses the importance of creating virtual worlds and mentions the Virtual Environment Theater (VET), which provides capability similar to the CAVE discussed in section 16.4.

Krygier, J. B. (1994) "Sound and geographic visualization." In *Visualization in Modern Cartography*, ed. A. M. MacEachren and D. R. F. Taylor, pp. 149–166. Oxford: Pergamon Press.

A good introduction to the use of sound in cartography.

MacEachren, A. M. (1992) "Visualizing uncertain information." *Cartographic Perspectives* no. 13:10–19.

Covers basic principles for visualizing data quality. For more recent work, see MacEachren (1995).

MacEachren, A. M., Buttenfield, B. P., Campbell, J. B., DiBiase, D. W., and Monmonier, M. (1992) "Visualization." In *Geography's Inner Worlds: Pervasive Themes in Contemporary American Geography*, ed. R. F. Abler, M. G. Marcus, and J. M. Olson, pp. 99–137. New Brunswick, N.J.: Rutgers University Press.

A relatively early piece on visualization in geography.

Openshaw, S., and Perrée, T. (1996) "User-centered intelligent spatial analysis of point data." In *Innovations in GIS* 3, ed. D. Parker, pp. 119–134. Bristol, Penn.: Taylor & Francis.

An example of visualization work in geography with a statistical emphasis.

Pickle, L. W., and Herrmann, D. J. (eds.) (1995) "Cognitive Aspects of Statistical Mapping" *Cognitive Methods Staff,* Working Paper Series, 18. Hyattsville, Md.: Centers for Disease Control and Prevention/National Center for Health Statistics, Office of Research and Methodology.

A series of papers dealing with cognitive aspects of thematic cartography.

Pickle, L. W., Mungiole, M., Jones, G., and White, A. A. (1996) *Atlas of United States Mortality*. Hyattsville, Md.: National Center for Health Statistics.

An atlas illustrating the interdisciplinary efforts of those in cartography, statistics, and public health.

Shiffer, M. J. (1993a) "Augmenting geographic information with collaborative multimedia technologies." *AUTO-CARTO 11 Proceedings,* Minneapolis, Minn., pp. 367–376.

Describes a multimedia system useful to planners. Also see Shiffer (1995).

Appendix A

CMYK SPECIFICATIONS FOR COLOR CHOROPLETH MAPS

This appendix provides CMYK (cyan, magenta, yellow, and black) printing specifications for classes appearing on color choropleth maps in this book. For univariate maps, the CMYK values are ordered from the highest to lowest class (designated as classes 5 to 1 or 7 to 1). For bivariate maps, the values are ordered by row and column of the legend (row 1 is the topmost row and col 1 is the left-most column). Specifications for Color Plates 6.3 and 12.4 are based on CMYK values provided by Cynthia Brewer.

Color Plate 1.1

Map A

Class	C	M	Y	K
5	52	0	23	0
4	85	0	90	0
3	93	78	0	0
2	0	0	75	0
1	0	90	90	0

Map B

Class	C	M	Y	K
5	0	40	60	70
4	0	40	60	50
3	0	40	60	30
2	0	40	60	13
1	0	40	60	0

Color Plate 2.4

Map A

Class	C	M	Y	K
5	0	90	60	0
4	0	60	45	13
3	0	45	30	25
2	0	25	15	38
1	0	15	5	50

Map B

Class	C	M	Y	K
5	0	90	60	0
4	0	70	65	0
3	0	50	70	0
2	0	28	75	0
1	0	5	80	0

Color Plate 6.1

Map A

Class	C	M	Y	K
5	20	5	6	0
4	5	13	30	0
3	23	0	38	0
2	13	5	8	0
1	0	25	27	0

Map B

Class	C	M	Y	K
5	11	8	71	0
4	5	5	34	0
3	0	0	0	10
2	0	16	9	0
1	0	46	33	0

Map C

Class	C	M	Y	K
5	5	63	5	0
4	30	22	0	0
3	33	0	9	0
2	30	0	41	0
1	0	0	23	0

Map D

Class	C	M	Y	K
5	11	73	60	0
4	0	58	45	0
3	0	27	54	0
2	0	8	44	0
1	0	0	23	0

Map E

Class	C	M	Y	K
5	36	97	96	8
4	11	76	95	0
3	0	48	71	0
2	0	30	47	0
1	0	6	9	0

Map F

Class	C	M	Y	K
5	0	0	0	70
4	0	0	0	50
3	0	0	0	30
2	0	0	0	15
1	0	0	0	5

Color Plate 6.2

Map A

Class	C	M	Y	K
5	70	0	70	30
4	60	0	60	20
3	50	0	50	10
2	40	0	40	5
1	30	0	30	0

Map B

Class	C	M	Y	K
5	60	0	60	40
4	60	0	60	20
3	60	0	60	5
2	40	0	40	5
1	20	0	20	5

Color Plate 6.3

Map A

Class	C	M	Y	K
5	100	30	0	15
4	100	0	0	5
3	60	0	40	5
2	30	0	60	0
1	5	0	40	0

Map B

Class	C	M	Y	K
5	60	0	84	0
4	50	25	0	3
3	0	50	20	0
2	0	20	50	0
1	0	0	20	0

Color Plate 6.4

Class	C	M	Y	K
5	0	60	30	20
4	0	20	0	20
3	0	0	0	20
2	30	0	0	10
1	70	0	10	10

Color Plate 12.1 *(top map)*

Class	C	M	Y	K
row 1, col 1	100	0	100	0
row 1, col 2	100	30	50	0
row 1, col 3	100	100	0	0
row 2, col 1	30	0	100	0
row 2, col 2	30	30	50	0
row 2, col 3	30	100	0	0
row 3, col 1	0	0	100	0
row 3, col 2	0	30	50	0
row 3, col 3	0	100	0	0

Color Plate 12.2

Class	C	M	Y	K
row 1, col 1	0	86	82	0
row 1, col 2	50	86	82	0
row 1, col 3	0	0	0	100
row 2, col 1	0	43	41	0
row 2, col 2	0	0	0	50
row 2, col 3	100	43	41	0
row 3, col 1	0	0	0	0
row 3, col 2	50	0	0	0
row 3, col 3	100	0	0	0

Color Plate 12.3

Class	C	M	Y	K
row 1, col 1	0	100	100	0
row 1, col 2	0	0	0	100
middle	0	0	0	50
row 2, col 1	0	0	0	0
row 2, col 2	100	0	0	0

Color Plate 12.4

Left-hand legend

Class	C	M	Y	K
row 1, col 1	60	60	0	30
row 1, col 2	0	0	0	60
row 1, col 3	0	70	100	30
row 2, col 1	50	50	0	0
row 2, col 2	0	0	0	35
row 2, col 3	0	60	80	5
row 3, col 1	20	20	0	0
row 3, col 2	0	0	0	10
row 3, col 3	0	30	40	0

Right-hand legend

Class	C	M	Y	K
row 1, col 1	0	100	0	0
row 1, col 2	15	30	0	0
row 1, col 3	80	80	0	0
row 2, col 1	0	30	20	0
row 2, col 2	10	10	10	0
row 2, col 3	40	0	0	0
row 3, col 1	0	50	100	0
row 3, col 2	20	0	50	0
row 3, col 3	100	0	100	0

Color Plate 16.3 *(Seven-class maps)*

Red-Blue

Class	C	M	Y	K
7	0	75	80	40
6	0	33	42	13
5	0	10	13	1
4	0	0	0	2
3	21	2	0	0
2	60	10	0	10
1	100	45	0	15

Purple-Green

Class	C	M	Y	K
7	45	45	0	45
6	20	23	0	23
5	13	9	0	1
4	0	0	0	2
3	15	0	15	1
2	43	0	45	20
1	90	0	90	50

Pink-Yellow

Class	C	M	Y	K
7	0	70	40	50
6	0	27	17	23
5	1	7	3	1
4	0	0	0	2
3	5	0	25	4
2	25	0	75	25
1	45	0	100	57

Brown/Blue-Green

Class	C	M	Y	K
7	0	40	100	47
6	0	20	55	18
5	0	6	17	2
4	0	0	0	2
3	22	0	6	0
2	61	0	20	15
1	100	0	35	40

Spectral Scheme

Class	C	M	Y	K
7	10	80	80	0
6	0	45	100	0
5	0	15	95	1
4	0	0	100	1
3	40	0	75	0
2	77	0	32	0
1	87	28	8	0

Red-Yellow

Class	C	M	Y	K
7	0	100	70	60
6	0	78	88	30
5	0	60	100	15
4	0	35	100	5
3	0	15	95	2
2	0	3	60	4
1	0	0	35	0

Purple-Blue

Class	C	M	Y	K
7	55	65	0	75
6	50	40	0	50
5	53	35	0	15
4	63	18	0	0
3	48	7	1	0
2	20	0	1	2
1	7	0	1	0

Grays

Class	C	M	Y	K
7	0	0	0	90
6	0	0	0	75
5	0	0	0	62
4	0	0	0	47
3	0	0	0	27
2	0	0	0	12
1	0	0	0	2

Color Plate 16.3 *(Five-class maps)*

Red-Blue

Class	C	M	Y	K
5	0	65	75	15
4	0	17	25	2
3	0	0	0	2
2	37	5	0	1
1	87	28	0	8

Purple-Green

Class	C	M	Y	K
5	50	50	0	15
4	15	12	0	3
3	0	0	0	2
2	27	0	30	7
1	85	0	85	20

Pink-Yellow

Class	C	M	Y	K
5	0	63	38	25
4	0	20	15	5
3	0	0	0	2
2	21	7	65	6
1	46	0	100	30

Brown/Blue-Green

Class	C	M	Y	K
5	0	47	100	22
4	0	13	40	6
3	0	0	0	2
2	50	0	15	4
1	95	0	30	18

Spectral Scheme

Class	C	M	Y	K
5	25	100	100	0
4	0	35	100	3
3	0	0	100	1
2	65	0	85	0
1	100	45	0	15

Red-Yellow

Class	C	M	Y	K
5	0	100	100	55
4	0	62	100	25
3	0	35	100	3
2	0	12	100	2
1	0	2	60	0

Purple-Blue

Class	C	M	Y	K
5	65	60	0	60
4	65	40	0	25
3	65	20	0	0
2	37	5	0	1
1	5	0	1	0

Grays

Class	C	M	Y	K
5	0	0	0	90
4	0	0	0	65
3	0	0	0	47
2	0	0	0	15
1	0	0	0	2

Appendix B

USING THE CIE L*u*v* UNIFORM COLOR SPACE TO CREATE EQUALLY SPACED COLORS

Section 6.5.3 describes several approaches for specifying color schemes for choropleth and isarithmic maps. One of these approaches is to specify endpoints for a color scheme in RGB form, convert these to the 1976 CIE L*u*v* uniform color space, interpolate colors in the uniform color space, and then convert the interpolated colors back to RGB. This can be accomplished using the six steps listed below. (When RGB colors are referred to, it is assumed that a gamma correction has already been performed.)

Step 1. *Specify the desired endpoints of a color scheme in RGB form.* This might be accomplished directly using RGB values, or HSV might be used (since HSV values are readily converted into RGB values, as shown in Travis (1991, 82).

Step 2. *Using equations provided by Travis (pp. 93–96), convert the RGB values for the endpoints to their CIE tristimulus equivalents (X, Y, and Z).*

Step 3. *Convert the tristimulus values to the L*u*v* uniform color space using the following equations* (Wyszecki and Stiles 1982, 165):

$$L^* = 116(Y/Y_n)^{1/3} - 16 \quad \text{for } Y/Y_n > 0.008856$$
$$L^* = 903.3(Y/Y_n) \text{for } Y/Y_n \leq 0.008856$$
$$u^* = 13L^*(u' - u'_n)$$
$$v^* = 13L^*(v' - v'_n)$$

with

$$u' = \frac{4X}{X + 15Y + 3Z} \qquad v' = \frac{9Y}{X + 15Y + 3Z}$$

$$u'_n = \frac{4X_n}{X_n + 15Y_n + 3Z_n} \qquad v'_n = \frac{9Y_n}{X_n + 15Y_n + 3Z_n}$$

where X, Y, Z are the tristimulus values of a color to be converted; and X_n, Y_n, Z_n are the tristimulus values of the **white point** of the display (when all color guns fire at the maximum intensity).

Step 4. *Linearly interpolate between the $L^*u^*v^*$ values.* For example, if you computed L^* values of 50 and 60 for the endpoints, the L^* values for the intermediate classes for a five-class map would be 52.5, 55, and 57.5.

Step 5. *Convert the interpolated $L^*u^*v^*$ values back to their tristimulus equivalents using the following equations* (Tajima 1983, 313):

$$Y = \left(\frac{L^* + 16}{116}\right)^3 \cdot Y_n \text{ for } Y/Y_n > 0.008856,$$

$$Y = \frac{L^* \cdot Y_n}{903.3} \text{ for } Y/Y_n \leq 0.008856$$

$$X = \frac{9(u^* + 13v'_n L^*)}{4(v^* + 13v'_n L^*)} Y$$

$$Z = \left(\frac{39L^*}{v^* + 13v'_n L^*} - 5\right) Y - \frac{X}{3}$$

Step 6. *Using equations provided by Travis (p. 96), convert the tristimulus values to their RGB equivalents.*

Glossary

Numbers in parentheses indicate pages on which terms are introduced or discussed.

$2^1/_2$-D phenomena: a form of volumetric phenomena in which each point on the surface is defined by longitude, latitude, and a value above (or below) a zero point (e.g., the earth's topography) (p. 19).

abrupt phenomena: phenomena that change abruptly over geographic space, such as the sales tax rates for each state in the United States (p. 20).

abstract sounds: sounds that have no obvious meaning and thus require a legend for interpretation (e.g., a mouse click on a census tract produces a loudness value as a function of income) (p. 248).

accommodation: the process in which the eye automatically changes the shape of the lens to focus on an image (p. 85).

acoustic visualization: see **sonification.**

additive color: colors that are visually added (or combined) to produce other colors; for example, red, green, and blue are additive colors used in CRT displays (p. 89).

anaglyphs: two images are created, one in green or blue, and one in red; when viewed through special anaglyphic glasses, a three-dimensional view is produced (p. 154).

ancillary variables (or **ancillary information**)**:** variables used to more accurately map data associated with enumeration units (e.g., when making a dot map of wheat based on county-level data, we avoid placing dots in water bodies) (p. 168).

animated maps: maps characterized by continuous or dynamic change, such as the daily weather maps shown on television depicting changes in cloud cover (p. 8).

animation software: software that is explicitly intended for developing animations (p. 237).

anomalous trichromats: a less serious form of color vision impairment in which people use three colors to match any given color; contrast with **dichromats** (p. 88).

area cartogram: a map in which the areal relationships of enumeration units are distorted on the basis of an attribute (e.g., the size of states are made proportional to the death rate due to AIDS) (p. 181).

areal phenomena: geographic phenomena that are two-dimensional in spatial extent, having both length and width, such as a forested region; data associated with enumeration units can also be considered areal phenomena, since each unit is an enclosed area (p. 19).

arrangement: a visual variable in which the marks making up symbols are arranged in various ways, such as breaking up lines into dots and dashes to create various forms of political boundaries (p. 24).

attribute accuracy: one means of assessing data quality; refers to the correctness of features found at particular locations (e.g., the accuracy of a land use/land cover classification) (p. 242).

authoring software: used to develop business and educational application software (e.g., Director might be used to create a lecture for an introductory human geography course) (p. 238).

balanced data: two phenomena coexist in a complementary fashion, such as the percentage of English and French spoken in Canadian provinces (p. 22).

base information: information on a thematic map that is not the major theme but serves as a reference for the theme, such as county boundaries on a map of cancer death rates by county (p. 34).

bipolar and ganglion cells: cells that merge the input arriving from rods and cones in the eye (p. 85).

bipolar data: data characterized by either a natural or meaningful dividing point, such as the percentage of population change (p. 22).

bivariate choropleth map: the overlay of two univariate choropleth maps (e.g., one map could be shades of cyan and the other shades of red) (p. 197).

bivariate correlation: a numerical method for summarizing the relationship between two interval or ratio level variables (p. 48).

bivariate mapping: the cartographic display of two variables; for example, simultaneously mapping median income and murder rate for census tracts within a city (p. 193).

bivariate-normal: when *Y* values associated with a given *X* value are normal, and the *X* values associated with a given *Y* value are also normal (p. 198).

bivariate point symbol: a point symbol used to portray two variables simultaneously (e.g., representing two variables by the width and height of ellipses) (p. 200).

bivariate ray-glyph: used to map two variables by extending straight line segments to either the right or left of a small central circle (p. 200).

bivariate regression: the process of fitting a line to two interval or ratio level variables; the *dependent variable* is plotted on the *y* axis, while the *independent variable* is plotted on the *x* axis (p. 48).

boundary error: if classed data are conceived as a prism map, boundary error describes how close the resulting cliffs come to matching cliffs on an unclassed prism map of the data (p. 71).

bounding rectangle: a rectangle that just touches and completely encloses an arbitrary shape (p. 52).

box plot: a method of exploratory data analysis that illustrates the position of various numerical summaries (e.g., minimum, maximum, and median) along the number line (p. 51).

cartogram: a map that purposely distorts geographic space based on values of a theme (e.g., making the size of countries proportional to population (p. 181).

cathode ray tube (CRT): a graphics display in which images are created by firing electrons from an electron gun at phosphors, which emit light when they are struck (p. 89).

centroid: the "balancing point" for a geographic region (p. 52).

change map: a map representing the difference between two points in time (p. 9).

Chernoff face: distinct facial features that are associated with individual variables (e.g., a broader smile depicts cities with a higher per capita expenditure on public schools) (p. 204).

chiaroscuro: see **shaded relief.**

chorodot map: a combination of choropleth and dot maps; the "dot" portion derives from using small squares within enumeration units, while the "choropleth" portion derives from the shading assigned to each square (p. 180).

choropleth map: a map in which enumeration units (or data collection units) are shaded with an intensity proportional to the data values associated with those units (e.g., census tracts shaded with gray tones whose intensity is proportional to population density) (p. 25).

chroma: see **saturation.**

CIE: an international standard for color; allows cartographers to precisely specify a color so that others can duplicate that color (p. 99).

clarity: a term used to summarize visual variables for depicting data quality; see **crispness, resolution,** and **transparency** (p. 243).

classed map: a map in which data are grouped into classes of similar value, and the same symbol is assigned to all members of a class (e.g., a data set with 100 different data values might be depicted using only five shades of gray on a choropleth map) (p. 8).

clip art: pictures that are available in a digital format; such pictures can serve as the basis for creating **pictographic symbols** (p. 121).

cluster analysis: a mathematical method for grouping observations (say counties) based on their scores on a set of variables (p. 193).

CMYK: a system for specifying colors in which cyan (C), magenta (M), yellow (Y), and black (K) are used, as is commonly done for offset printing (p. 96).

cognition: the mental activity associated with map reading and interpretation, including the initial perception of the map, prior experience, and memory (p. 13).

color gamut: the range of colors produced by a graphics display device or printer (p. 95).

colorimeter: a physical device for measuring color; typically, color specifications are given in the CIE system (p. 163).

color lookup tables: used in association with a frame buffer to make rapid changes in color in a graphics display (p. 93).

color management systems: software (and associated hardware) that, in theory, enables color to be matched on different display devices (e.g., a CRT and an ink-jet printer) (p. 95).

color ramping: a method for specifying colors on a graphics display; the user selects two endpoints, and the computer automatically interpolates intermediate colors (p. 115).

color stereoscopic effect: this occurs when colors from the long wavelength portion of the electromagnetic spectrum appear nearer to a map reader than colors from the short wavelength portion (p. 108).

command-line interface: a computer interface in which the user types commands specifying processes to be performed (p. 219).

communication: the transfer of known spatial information to the public; typically, this is done via paper maps (p. 11).

compaction index: a measure of shape; the ratio of the area of the shape to the area of a circumscribing circle (p. 52).

complementary colors: colors that combine to produce a shade of gray, such as cyan and red (p. 197).

completeness: one means of assessing data quality; includes information about selection criteria and definitions used (e.g., what is the minimum area needed to define a water body as a lake?) (p. 243).

conceptual point data: data that are collected over an area (or volume) but conceived as being located at a point for the purpose of symbolization (e.g., the number of microbreweries in each state) (p. 118).

cones: a type of nerve cell within the retina that functions in relatively bright light and is responsible for color vision (p. 85).

continuous phenomena: geographic phenomena that occur everywhere, such as the distribution of snowfall for the year in Wisconsin or sales tax rates for each state in the United States (p. 20).

continuous-tone map: a map using a large number of hypsometric tints to create an unclassed isarithmic map (p. 154).

continuous vector–based flow: a flow composed of two variables, magnitude and direction, that can change at any point (e.g., at any point in the atmosphere, we can compute the speed and direction from which the wind blows) (p. 186).

contour lines: these represent the intersection of horizontal planes with the three-dimensional surface of a smooth continuous phenomenon (p. 154).

contour map: see **isarithmic map.**

control points: points that are used as a basis for interpolation on an isarithmic map, such as weather station locations (p. 136).

cornea: the protective outer covering of the eye (p. 85).

correlation coefficient: a numerical expression of the relationship between two interval-ratio level variables (p. 48).

courseware: software used to assist students in learning concepts outside the classroom (p. 227).

crispness: a visual variable for depicting data quality; a sharp (or crisp) boundary would represent reliable data, while a fuzzy one would represent uncertain data (p. 243).

cross validation: a method for evaluating the accuracy of interpolation for isarithmic mapping; involves removing a control point from the data to be interpolated and using other control points to estimate a value at the location of the removed point (p. 146).

dasymetric map: like choropleth maps, area symbols are used to represent zones of uniformity, but the bounds of zones need not match enumeration unit boundaries (p. 168).

data exploration: examining data in a variety of ways in order to develop different perspectives of the data; for example, we might view a choropleth map in both its unclassed and classed form. Data exploration can also be equated with **visualization** (p. 9).

data flow software: a form of visualization software that contains separate modules for importing, manipulating, and displaying data (p. 239).

data jack: triangular spikes are drawn from a square central area and made proportional to the magnitude of each variable being mapped (p. 203).

data quality: the notion that data for thematic maps are often subject to some form of error (p. 242).

data reliability: see **data quality;** "low data reliability" equals "low data quality" (p. 242).

data splitting: a method for evaluating the accuracy of interpolation for isarithmic mapping; control points are split into two groups, one to create the contour map, and one to evaluate its accuracy (p. 146)

data uncertainty: see **data quality;** "high uncertainty" equals "low data quality" (p. 242).

deconstruction: the process of analyzing a map (or text) to uncover its hidden agendas or meanings (p. 14).

Delaunay triangles: the type of triangles commonly used in triangulation, a method of interpolation for isarithmic maps; the longest side of any triangle is minimized and thus the distance over which interpolation takes place is minimized (p. 137).

descriptive statistics: used to describe the character of a sample or population, such as computing the mean square footage of a sample of 50 stone houses (p. 41).

dichromats: a more serious form of color vision impairment in which people use two colors to match any given color; contrast with **anomalous trichromats** (p. 88).

digital elevation model (DEM): topographic data commonly available as an equally spaced gridded network (p. 153).

digital-to-analog converter (DAC): a device within a CRT that converts a digital value to an analog voltage value that is applied to an electron gun (p. 91).

discrete phenomena: geographic phenomena that occur at isolated locations, such as water towers in a city (p. 19).

dispersion graph: data are grouped into classes, and the number of values falling in each class are represented by stacked dots along the number line (p. 44).

display date: one of the visual variables for animated maps; refers to the time that some display change in an animation is initiated (e.g., in an animation of population for U.S. cities, a circle for San Francisco appears in 1850) (p. 224).

dissemination: the distribution of maps to potential users in both paper and nonpaper forms (p. 101).

distance cartogram: a map in which real-world distances are distorted to reflect some attribute (e.g., distances on a subway map might be distorted to reflect travel time) (p. 181).

distortion ellipse: see **Tissot's indicatrix.**

dithering: colors are created by presuming that the reader will perceptually merge different colors displayed in adjacent pixels (on a graphics display) or dots (on a printer) (p. 93).

diverging scheme: a sequence of colors for a choropleth or isarithmic map in which two hues diverge from a common light hue or neutral gray, as in a dark red-light red-gray-light blue-dark blue scheme (p. 107).

dot: the individual picture elements composing an image on a raster graphics printer (p. 94).

dot-density shading: a technique for mapping smooth continuous phenomena (e.g., precipitation); closely spaced dots depict high values of the phenomenon, while widely separated dots depict low values (p. 172).

dot map: a map in which small symbols of uniform size (typically solid circles) are used to emphasize the spatial pattern of a phenomenon (e.g., one dot might represent 1000 head of cattle) (p. 28).

dot size: how large dots are on a dot map (p. 171).

duration: one of the visual variables for animated maps; refers to the length of time that a frame of an animation is displayed (p. 224).

dye-sublimation printer: a form of printing in which heat is used to move a colorant from a ribbon to paper; a special dye is used (as opposed to the wax or plastic in thermal-wax printers) and the process is implemented via sublimation (p. 95).

Ebbinghaus illusion: a circle surrounded by large circles will appear smaller than the same-size circle surrounded by small circles (p. 125).

electromagnetic energy: a wave form having both electrical and magnetic properties; refers to the way that light travels through space (p. 84).

electromagnetic spectrum: the complete range of wavelengths possible for electromagnetic energy (p. 84).

electron gun: the device used within a CRT to fire electrons at phosphors, which emit light when they are struck (p. 89).

electronic atlas: a collection of maps (and database) that is available in a digital environment; sophisticated electronic atlases enable users to take advantage of the digital environment through Internet access, data exploration, map animation, and multimedia (p. 233).

equal-area projection: a map projection that maintains correct areal relationships—an area twice another in the real world will also be twice as large on this projection (p. 181).

equal areas: a method of data classification in which an equal portion of the map area is assigned to each class (p. 196).

equal intervals: a method of data classification in which each class occupies an equal portion of the number line (p. 61).

equiprobability ellipse: if data are bivariate normal, an ellipse can be drawn that encloses a specified percentage of the data (p. 198).

expert system: a system in which a computer automatically makes decisions on symbolization by using a knowledge base provided by experienced cartographers (p. 8).

exploratory data analysis: a method for analyzing statistical data in which the data are examined graphically in a variety of ways, much as a detective investigates a crime; this is in contrast to fitting data to standard forms, such as the normal distribution (p. 50).

Eyton's illuminated contours: a raster approach for depicting topography; contours facing a light source are brightened, while those in the shadow are darkened (p. 157).

figure-ground contrast: the notion that thematic information should be seen as a figure, and base information should be seen as a ground (p. 132).

Fisher-Jenks algorithm: a method for creating an optimal classification; the optimal classification is guaranteed by *essentially* considering all possible classifications of the data (p. 71).

fishnet map: a map in which a fishnet-like structure provides a three-dimensional symbolization of a smooth continuous phenomenon (p. 154).

Flannery correction: a method of adjusting the sizes of proportional circles to account for the perceived underestimation of larger circle sizes; a symbol scaling exponent of .57 is used (p. 134).

flat-panel displays (FPDs): graphic display devices that have a small depth dimension; the liquid crystal display is one example (p. 94).

flow map: a map used to depict the movement of phenomena between geographic locations; generally, this is done using "flow lines" of varying thickness (p. 184).

fluorescent inks: specialized printing inks that produce brilliant, intense color (p. 108).

fly-by: a form of animation in which the viewer is given the feeling of flying over a landscape (p. 165).

font: the basic design of type, such as Helvetica versus Times (p. 36).

four-color process printing: a method of offset printing in which four color separations (cyan, magenta, yellow, and black) are used (p. 101).

fovea: the portion of the retina where visual acuity is the greatest (p. 85).

frame buffer: an area of memory that stores a digital representation of colors appearing on the screen (p. 90).

framed-rectangle symbol: a point symbol that consists of a "frame" of constant size, within which a solid "rectangle" is placed; the greater the data value, the greater the proportion of the frame that is filled by the rectangle (p. 178).

frequency: one of the visual variables for animated maps; refers to the number of identifiable states per unit time, as in color cycling used to portray the jet stream (p. 224).

gamma function: the relation between the voltage of a color gun and the associated luminance of the CRT display (p. 115).

general-reference map: a type of map used to emphasize the *location* of spatial phenomena (e.g., a United States Geological Survey (USGS) topographic map) (p. 2).

geographic brush: as areas of a map are selected (say using a mouse), corresponding dots are highlighted within a scatterplot matrix (p. 52).

geographic information systems (GIS): automated systems for the capture, storage, retrieval, analysis, and display of spatial data (p. 219).

geometric symbols: symbols that do not necessarily look like the phenomenon being mapped, such as using squares to depict barns of historical interest; contrast with **pictographic symbols** (p. 24).

gerrymandering: the purposeful distortion of legislative or congressional districts for partisan benefit (p. 54).

glyphs: a term applied to multivariate point symbols when the variables being mapped are in dissimilar units (not part of a larger whole) (p. 203).

goodness of absolute deviation fit (GADF): a measure of the accuracy of a classed choropleth map when the median is used as a measure of central tendency (p. 73).

goodness of variance fit (GVF): a measure of the accuracy of a classed choropleth map when the mean is used as a measure of central tendency (p. 73).

Gouraud shading: a sophisticated method for creating shaded relief; shades are *interpolated* between vertices of a polygon (or cell) boundary and thus a smoother appearance is produced than when the same shade is used throughout a polygon (p. 164).

graphical user interface (GUI): a type of computer interface in which the user specifies tasks to be performed by pointing to the desired task, typically by using a mouse (p. 219).

graphic display: used to describe the computer screen (and associated color board) on which a map is displayed in soft-copy form (p. 88).

graphic script: a series of dynamic maps, statistical graphics, and text blocks used to tell a story about a set of data (p. 222).

gray scales (or **gray curves**): a graph (or equation) expressing the relation between printed area inked and perceived blackness (p. 111).

gridding: a term applied to the inverse distance method of interpolation, in which values are estimated at equally spaced grid locations; it should be borne in mind, however, that kriging also uses a grid (p. 139).

grouped-frequency table: constructed by dividing the data range into equal intervals and tallying the number of observations falling in each interval (p. 43).

hachures: a method for depicting topography; a series of parallel lines is drawn perpendicular to the direction of contours (p. 156).

hexagon bin plot: a type of plot in which a set of hexagons is placed over a conventional scatterplot, and the hexagons are filled as a function of the number of dots falling within them (p. 48).

high-definition television (HDTV): digital television; screen resolution is higher than conventional (analog) television, and the aspect ratio is more like a movie screen (p. 102).

hill shading: see **shaded relief.**

histogram: a type of graph in which data are grouped into classes, and bars are used to depict the number of values falling in each class (p. 44).

honoring control point data: a term applied if, after an interpolation is performed, there is no difference between the original value of a control point and the value of that same point on the interpolated map (p. 146).

HSV: a system for specifying color in which hue (H), saturation (S), and value (V) are used; although the system utilizes common color terminology, colors are not equally spaced in the visual sense (p. 96).

HVC: a system for specifying color developed by Tektronix; the terms hue (H), value (V), and chroma (C) are used, and colors are equally spaced from one another in the visual sense (p. 98).

hue: along with lightness and saturation, one of three components of color; it is the dominant wavelength of light, such as red versus green (p. 24).

hypermaps: georeferenced multimedia systems; users can click items on a map and get information about that element and other related elements (e.g., a click on a Chinese restaurant can provide information about that restaurant and all other Chinese restaurants in a city) (p. 253).

hypermedia: a sophisticated form of multimedia in which various forms of media can be linked in ways not anticipated by system designers (p. 9).

hypsometric tints: the shaded areas sometimes used between contour lines on an isarithmic map (e.g., using shades of blue between contour lines to depict increasing rainfall) (p. 154).

iconic memory: refers to the initial perception of an object (in our case, a map or portion thereof) by the retina of the eye; contrast with **short-term visual store** and **long-term visual memory** (p. 13).

image-based visualization: using aerial photographs or remotely sensed images to create a visually realistic representation of the real world (p. 252).

imagesetter: a piece of hardware that converts digital files containing map information into film negatives (p. 101).

induction: see **simultaneous contrast.**

inferential statistics: used to make an inference (or guess) about a population based on a sample (p. 41).

information acquisition: the process of acquiring spatial information *while* a map is being used; contrast with **memory for mapped information** (p. 3).

ink-jet printing: a form of printing in which microscopic dots of subtractive inks are squirted onto paper (p. 94).

inset map: a supplemental map that is used to portray a congested area at an enlarged scale (p. 132).

intermediate frames: in an animation, the frames (or maps) associated with interpolated data; contrast with **key frames** (p. 214).

interquartile range: the absolute difference between the 75th and 25th percentiles of the data (p. 46).

interval: a level of measurement in which numerical values can be assigned to data, but there is an arbitrary zero point, such as for SAT scores (p. 20).

inverse distance: a method of interpolating data for isarithmic maps; a grid is overlaid on original control points, and estimates of values at grid points are an inverse function of the distance to control points (p. 139).

isarithmic map: a map in which a set of isolines (lines of equal value) are interpolated between points of known value, as on a map depicting snowfall amounts (p. 27).

isometric map: a form of isarithmic map in which control points are true point locations (e.g., snowfall measured at weather stations) (p. 136).

isopleth map: a form of isarithmic map in which control points are associated with enumeration units (e.g., assigning population density values for census tracts to the centers of those tracts and then interpolating between the centers) (p. 27).

isosurface: a surface of equal value within a true 3-D phenomenon (e.g., a surface representing a wind speed of 30 miles per hour in the jet stream) (p. 188).

jaggies: a staircase appearance to straight lines caused by displaying lines on a coarse raster graphics display (p. 90).

Jenks-Caspall algorithm: an optimal data classification that is achieved by moving observations between classes through trial-and-error processes known as *reiterative* and *forced* cycling (p. 71).

key frames: in an animation, the frames (or maps) associated with collected data; contrast with **intermediate frames** (p. 214).

kriging: a method of interpolating data for isarithmic maps that considers the spatial autocorrelation in the data (p. 141).

labeling software: software that enables type to be positioned automatically (p. 37).

large-scale map: a map depicting a relatively small portion of the earth's surface (p. 19).

laser printing: a form of printing in which a laser beam is passed over a charged belt; when the laser beam is turned on, a corresponding location on the belt is discharged, and a toner sticks to the discharged areas (p. 94).

lectureware: software used to assist students in learning material in a traditional lecture setting (p. 227).

lens: the focusing mechanism within the eye (p. 85).

level of measurement: refers to the different ways that we measure variables; we commonly consider nominal, ordinal, interval, and ratio levels (p. 20).

lightness: along with hue and saturation, one of three components of color; refers to how dark or light a color is, such as light blue versus dark blue (p. 24).

lineage: one means of assessing data quality; refers to the historical development of the data (p. 242).

linear-legend arrangement: a form of legend design for proportional symbol maps; symbols are placed adjacent to each other in either a horizontal or vertical orientation (p. 130).

linear phenomena: geographic phenomena that are one-dimensional in spatial extent, having length but essentially no width, such as a road on a small-scale map (p. 19).

liquid crystal displays (LCDs): a form of graphic display device in which liquid crystals are sandwiched between two glass plates; commonly used in laptops and overhead projection devices (p. 93).

location: a visual variable that refers to the possibility of varying the position of symbols, such as dots on a dot map (p. 25).

logical consistency: one means of assessing data quality; describes the fidelity of relationships encoded in the structure of spatial data (e.g., are polygons topologically correct?) (p. 242).

long-term visual memory: an area of memory that does not require constant attention for retainment; contrast with **iconic memory** and **short-term visual store** (p. 13).

luminance: brightness of a color as measured by a physical device (p. 163).

magnitude estimation: 1) a method for constructing gray scales in which a user estimates the lightness or darkness of one shade relative to another (p. 112); 2) a method for determining the perceived size of proportional symbols in which a value is assigned to one symbol on the basis of a value assigned to another symbol (p. 125).

map communication model: a diagram or set of steps summarizing how a cartographer imparts spatial information to a group of map readers (p. 3).

map complexity: a term used to indicate whether the set of elements composing a map pattern appear simple or intricate (complex); for example, does the pattern of gray tones on a choropleth map appear simple or intricate? (p. 57).

map server: software that enables users to create maps via the Internet (p. 237).

mathematical scaling: sizing proportional symbols in direct proportion to the data; thus, if a data value is 40 times another, the area (or volume) of a symbol will be 40 times as large (p. 121).

maximum breaks: a method of data classification in which the largest differences between ordered observations are used to define classes (p. 69).

maximum contrast shades: on a choropleth map, areal shades are selected so that they have the maximum possible contrast with one another (p. 60).

mean: the average of a data set, computed by adding all values and dividing by the number of values (p. 45).

mean center: a measure of central tendency for point data; computed by independently averaging the x and y coordinate values of each point (p. 55).

mean-standard deviation: a method of data classification in which the mean and standard deviation of the data are used to define classes (p. 69).

median: the middle value in an ordered set of data (p. 45).

memory for mapped information: remembering spatial information that was previously seen in mapped form; contrast with **information acquisition** (p. 3).

mode: the most frequently occurring value in a data set (p. 45).

modular visualization environment: see **data flow software.**

multimedia: the combined use of maps, graphics, text, pictures, video, and sound (p. 9).

multimedia encyclopedias: like paper encyclopedias, these provide information about places, people, and events, but they are enhanced through the use of sound, video, and ease of linking various topics (p. 236).

multiple regression: a statistical method for summarizing the relationship between a dependent variable and a series of independent variables (p. 50).

multivariate dot map: a dot map in which distinct symbols or colors are used for each variable to be mapped (e.g., wheat, corn, and soybeans could be represented by green, blue, and red dots) (p. 201).

multivariate mapping: the cartographic display of three or more variables; for example, we might simultaneously map ocean temperature, salinity, and current speed (p. 193).

multivariate point symbol: the depiction of three or more variables using a point symbol (p. 203).

multivariate ray-glyph: a term used when rays are extended from a small interior circle, with the lengths of the rays made proportional to values associated with each variable being mapped (p. 203).

Munsell: a system for specifying color in which colors are identified using hue, value, and chroma; colors are equally spaced from one another in the visual sense (p. 97).

Munsell curve: a gray scale commonly used for smooth (untextured) gray tones (p. 111).

natural breaks: a method of data classification in which a graphical plot of the data (such as a histogram) is examined to determine natural groupings of data (p. 70).

nested-legend arrangement: a form of legend design for proportional symbol maps in which smaller symbols are drawn within larger symbols (p. 130).

nested means: a method of data classification in which the mean is used to repeatedly divide the data set; only an even number of classes is possible (p. 195).

nominal: a level of measurement in which data are placed into unordered categories, such as classes on a land use/land cover map (p. 20).

nomograph: a graphical device that can assist in selecting dot size and unit value on a dot map (p. 171).

normal distribution: a term used to describe a bell-shaped curve formed when data are distributed along the number line (p. 44).

numerical data: a term used to describe data associated with the interval and ratio levels of measurement, as numerical values are assigned to the data (p. 22).

object-based visualization: using realistic-looking objects to represent things found in the real world (e.g., using realistic-looking trees to depict the growth of a forest) (p. 252).

offset printing: method of reproduction in which lithographic plates containing a map image are placed on drums; areas representing the image are inked, and when a paper passes over the drums, an image is transferred to paper (p. 101).

one-dimensional scatterplot: see **point graph.**

opaque symbols: proportional symbols that do not permit the reader to see base information beneath the symbols; readers also must infer the maximum extent of each symbol (p. 131).

opponent-process theory: the theory that color perception is based on a lightness-darkness channel and two opponent color channels: red-green and blue-yellow (p. 86).

optic nerve: the nerve that carries information from the retina to the brain (p. 85).

optimal: a method of classification in which like values are placed in the same class by minimizing an objective measure of classification error (e.g., by minimizing the sum of absolute deviations about class medians) (p. 70).

optimal interpolation: a term sometimes applied to kriging because, in theory, it minimizes the difference between the estimated value at a grid point and the true (or actual) value at that grid point (p. 145).

order: one of the visual variables for animated maps; refers to the sequence in which frames or scenes of an animation are presented (p. 224).

ordinal: a level of measurement in which data are ordered or ranked but no numerical values are assigned, such as ranking states in terms of where you would like to live (p. 20).

orientation: a visual variable in which the directions of symbols (or marks making up symbols) are varied, such as changing the orientation of short line segments to indicate the direction from which the wind blows (p. 24).

outliers: unusual or atypical observations (p. 43).

overview error: if classed data are conceived of as a prism map, overview error is the difference in volume between this map and an unclassed prism map of the data (p. 71).

page description language (PDL): a programming language that describes an image in terms of its vector and raster components; the most common example is PostScript (p. 94).

paper atlas: a bound collection of maps; modern paper atlases also include text, photographs, tables, and graphs, which when combined tell a story about a place or region (p. 231).

parallel computers: computers that use multiple processors (p. 38).

partitioning: a method for constructing gray scales in which a user places a set of areal shades between white and black such that the resulting shades will be visually equally spaced (p. 112).

perceptually uniform color models: see **uniform color models.**

perceptual scaling: a term used when proportional symbols are sized to account for the perceived underestimation of large symbols; thus large symbols are drawn larger than suggested by the actual data (p. 124).

perspective height: a visual variable involving a perspective three-dimensional view, such as using raised sticks (or lollipops) to represent oil production at well locations (p. 22).

physiographic method: a method for depicting topography developed by Raisz; the earth's geomorphological features are represented by standard, easily recognized symbols (p. 159).

pictographic symbols: symbols that look like the phenomenon being mapped, such as using diagrams of barns to depict the location of barns of historical interest; contrast with **geometric symbols** (p. 24).

pixels: the individual picture elements composing a raster image on a graphics display device (p. 89).

plotter: a device that uses a vector approach to create hard-copy output (p. 88).

point graph: a type of graph in which each data value is represented by a small point symbol plotted along the number line (p. 43).

point-in-polygon test: a type of test that is made to determine whether a point falls inside or outside a polygon (p. 148).

point phenomena: geographic phenomena that have no spatial extent and can thus be termed "zero-dimensional," such as the location of oil well heads (p. 19).

polygonal glyph: a symbol formed by connecting the rays of a multivariate ray-glyph (p. 203).

population: the total set of elements or things one could study, such as all stone houses within a city (p. 41).

positional accuracy: one means of assessing data quality; refers to the correctness of location of geographic features (p. 242).

power function exponent: the exponent used for the stimulus in the power function equation relating perceived size and actual size; this exponent is used to summarize the results of experiments involving perceived size of proportional symbols (p. 124).

primary visual cortex: the first place in the brain where all of the information from both eyes is handled (p. 88).

principal components analysis: a method for succinctly summarizing the interrelationships occurring within a set of variables (p. 50).

printer: a device that uses a raster approach to create hard-copy output (p. 88).

prism map: a type of map in which enumeration units are raised to a height proportional to the data associated with those units (p. 20).

proactive graphics: software that enables users to initiate queries using icons (or symbolic representations) as opposed to words (p. 217).

programming languages: users are required to write computer code in order to develop software applications (p. 237).

proportional symbol map: point symbols are scaled in proportion to the magnitude of data occurring at point locations, such as using circles of varying sizes to represent city populations (p. 26).

psychological scaling: see **perceptual scaling.**

punctual kriging: a form of kriging in which the mean of the data is assumed constant throughout geographic space (there is no trend or drift in the data) (p. 143).

pupil: the dark area in the center of the eye (p. 85).

pycnophylactic interpolation: a method of interpolating data for isopleth maps; the data associated with enumeration units are conceived as prisms, and the volume within those prisms is retained in the interpolation process (p. 148).

qualitative data: data that have a nominal level of measurement (p. 22).

quantiles: a method of data classification in which an equal number of observations is placed in each class (p. 67).

quantitative data: data that have either an ordinal, interval, or ratio level of measurement (p. 22).

quasi-continuous tone method: a method for unclassed mapping in which only a portion of the data is shown with a smooth gradation of tones; very low and very high data values are represented by the lightest and darkest tones respectively (p. 75).

random error: the notion that if we repeatedly measure a value for an enumeration unit, we will likely get a different value each time (p. 79).

range: the maximum minus the minimum of the data (p. 46).

range-graded map: a term used to describe a classed proportional symbol map (p. 123).

raster: an image composed of pixels created by scanning from left to right and from top to bottom (p. 89).

rate of change: one of the visual variables for animated maps; defined as m/d, where m is the magnitude of change between frames or scenes, and d is the duration of each frame or scene (p. 224).

ratio: a level of measurement in which numerical values are assigned to data and there is a non-arbitrary zero point, such as the percentage of forested land (p. 21).

ratio estimation: a method for determining the perceived size of proportional symbols; two symbols are compared, and the viewer indicates how much larger or smaller one is than the other (e.g., one symbol appears five times larger) (p. 125).

raw-count data: raw counts (or totals) that have not been standardized to account for the area over which the data were collected (e.g., we might map the raw number of acres of tobacco harvested in each county, disregarding the size of the counties) (p. 25).

raw table: a form of tabular display in which the actual data are listed from lowest to highest value (p. 41).

realistic sounds: sounds whose meaning is based on our past experience; contrast with **abstract sounds** (p. 248).

receptive field: a single ganglion cell corresponding to a group of rods or cones; receptive fields are circular and overlap one another (p. 85).

rectangular point symbol: the width and height of a rectangle made proportional to each of two variables being mapped (p. 200).

redistricting: the process of assigning voting precincts to legislative or congressional districts to equalize voter representation (p. 54).

reduced major-axis approach: the process of fitting a line to two variables when we do not wish to specify a dependent variable (p. 50).

redundant symbols: used to portray a single attribute with two or more visual variables, such as representing population for cities by both size of circle and gray tone within the circle (p. 133).

re-expression: an animation created by modifying the original data in some manner, such as choosing subsets of a time series or reordering a time series (p. 225).

refresh: a term used to describe how images on a CRT must be constantly redisplayed (refreshed) because phosphors making up the screen have a low persistence (p. 90).

remote sensing: the acquirement of information about the earth (or other planetary body) from high above its surface via reflected or emitted radiation; most commonly done via satellites (p. 10).

resolution: 1) for CRT displays, the number of addressable pixels; for printers, the number of dots per inch (pp. 90, 94). 2) a visual variable for depicting data quality; refers to the level of precision in the spatial data underlying the attribute (e.g., varying the size of raster cells to represent uncertainty in the data) (p. 244).

retina: the portion of the eye on which images appear after passing through the lens (p. 85).

RGB: a system for specifying color in which red (R), green (G), and blue (B) components are used, as for red, green, and blue color guns on a CRT (p. 95).

rods: a type of nerve cell within the retina that functions in dim light and plays no role in color vision (p. 85).

sample: the portion of the population that is actually examined, such as sampling only 50 stone houses out of 5000 existing within a city (p. 41).

saturation: along with hue and lightness, one of three components of color; can be thought of as a mixture of gray and a pure hue: very saturated colors have little gray, while desaturated colors have a lot of gray (p. 24).

scatterplot: a diagram in which dots are used to plot the scores of two variables on a set of *x* and *y* axes (p. 47).

scatterplot brushing: subsets of data within a scatterplot matrix are focused on by moving a rectangular "brush" around the matrix; dots falling within the brush are highlighted in all scatterplots (p. 52).

scatterplot matrix: when a matrix of scatterplots is used to examine the relationships between multiple variables (p. 48).

scientific visualization: the use of visual methods to explore either spatial or nonspatial data; the process is particularly useful for large multivariate data sets and generally requires sophisticated graphics workstations (p. 11).

scripting languages: a programming language used in authoring software; it generally has a more language-like syntax than traditional programming languages (p. 238).

semivariance: a measure of the variability of spatial data in various geographic directions (e.g., we might want to determine whether temperature values are more similar to one another along a north–south axis than an east–west one) (p. 141).

semivariogram: a graphical expression of semivariance; the distance between points is shown on the *x*-axis and the semivariance is shown on the *y*-axis (p. 143).

sequencing: displaying a map piece-by-piece (e.g., displaying the title, geographic base, legend title, and then each class in the legend) (p. 212).

sequential scheme: a sequence of colors for a choropleth or isarithmic map in which colors are characterized by a gradual change in lightness, as in varying lightnesses of a blue hue (p. 107).

serifs: short extensions at the ends of major strokes of letters (p. 36).

service bureau: a business that converts digital files of map information into film negatives that can ultimately be used in offset printing (p. 101).

shaded relief: a method for depicting topography; areas facing away from a light source are shaded, while areas directly illuminated are not; slope may also be a factor, with steeper slopes shaded darker (p. 157).

shadowed contours: a method for depicting topography; contour lines facing a light source are drawn in a normal line weight, while those in the shadow are thickened (p. 157).

shape: a visual variable in which the form of symbols (or marks making up symbols) is varied, such as using squares and triangles to represent religious and public schools (p. 24).

short-term visual store: an area of memory that requires constant attention (or activation) in order to be retained; contrast with **iconic memory** and **long-term visual memory** (p. 13).

simultaneous contrast: when the perceived color of an area is affected by the color of the surrounding area (p. 87).

size: 1) a visual variable in which the magnitudes of symbols (or marks making up symbols) are varied (e.g., using circles of different magnitudes to depict city populations) (p. 22). 2) how large a typeface is; normally expressed in points, where a point is 1/72 of an inch (p. 36).

skewed distribution: when data are placed along the number line, the distribution appears asymmetrical; most common is a *positive skew,* in which the bulk of data are concentrated toward the left and there are a few outliers on the right (p. 44).

small multiples: many small maps displayed in order to show the change in a variable over time or to compare many variables for the same time period (p. 9).

small-scale map: a map depicting a relatively large portion of the earth's surface (p. 19).

smooth phenomena: phenomena that change gradually over geographic space, such as the distribution of snowfall for the year in Wisconsin (p. 20).

snowflake: see **polygonal glyph.**

sonification: using sound to represent spatial data (p. 241).

spacing: a visual variable in which the distance between marks making up a symbol is varied, such as shading counties on a choropleth map with horizontal lines of varied spacing (p. 22).

spatial autocorrelation: the tendency for like things to occur near one another in geographic space (p. 55).

spatial dimension: a term that describes whether a phenomenon can be conceived of as points, lines, areas, or volumes (p. 19).

spectral scheme: a sequence of colors for a choropleth or isarithmic map in which colors span the visual portion of the electromagnetic spectrum (sometimes referred to as a ROYGBIV scheme) (p. 108).

splining: using a mathematical function to smooth a contour line (p. 141).

spot color: a method for offset printing in which preblended PANTONE colors are used; separate lithographic plates are required for each color and so the method is expensive (p. 101).

standard deviation: the square root of the average of the squared deviations of each data value about the mean (p. 46).

standard distance: a measure of dispersion for point data; essentially, a measure of spread about the mean center of the distribution (p. 55).

standardized data: a term used when raw-count data are adjusted to account for the area over which the data are collected; for example, we might divide the acres of tobacco harvested in each county by the areas of those counties (p. 25).

star: see **multivariate ray-glyph.**

statistical map: see **thematic map.**

stem-and-leaf plot: a method for exploratory data analysis in which individual data values are broken into "stem" and "leaf" portions; the result resembles a histogram but provides greater detail for individual data values (p. 50).

stereo pairs: a term describing when two maps of an area are viewed with a stereoscope, which enables the reader to see a three-dimensional view (p. 154).

Stevens curve: a gray scale that is sometimes used for smooth (untextured) gray tones; some cartographers have recommended that it be used for unclassed maps (p. 112).

style: whether a font is normal, bold, or italic (p. 36).

subtractive colors: the colors cyan, magenta, yellow, and black, which are used to create a multitude of other colors on printed maps (p. 94).

synchronization: one of the visual variables for animated maps; refers to the temporal correspondence of two or more time series (p. 224).

tabular error: if classed data are conceived as a prism map, tabular error is the difference in height between prisms on this map and those on an unclassed prism map of the data (p. 71).

Tanaka's method: a method for depicting topography in which the width of contour lines is varied as a function of their angular relationship with a light source; also, contour lines facing the light source are white, while those in shadow are black (p. 156).

thematic map: a map used to emphasize the *spatial distribution* (or *pattern*) of one or more geographic attributes (e.g., a map of predicted snowfall amounts for the coming winter in the United States) (p. 2).

thermal-wax printer: a form of printing in which a thermal print head is used to move colorant from a ribbon to paper; a color-impregnated wax or plastic is used (as opposed to the dye of dye-sublimation printers) (p. 95).

Thiessen polygons: polygons enclosing a set of control points such that all arbitrary points within a polygon are closer to the control point associated with that polygon than to any other polygon (p. 137).

three-dimensional bars: a term that describes how the height of three-dimensional looking bars are made proportional to each of the variables being mapped (p. 203).

three-dimensional scatterplot: a diagram that uses small point symbols in which the scores for observations on three variables are plotted on a set of x, y and z axes; viewing all points in the plot requires interactive graphics (p. 48).

time series: an animation that emphasizes change through time (p. 224).

tiny cubes method: a method in which small cubes are regularly spaced throughout a 3-D phenomenon to provide the viewer with a feel for how the phenomenon changes throughout three-dimensional space (p. 189).

Tissot's indicatrix: this concept uses the visual variables size and shape to represent the ability of map projections to maintain correct size and angular relationships at point locations (p. 243).

transparency: a visual variable for depicting data quality; refers to the ease with which a theme can be seen through a "fog" placed over that theme (p. 244).

transparent symbols: proportional symbols that enable readers to see base information beneath a symbol and to see the maximum extent of each proportional symbol (p. 131).

trend surface: a mathematical surface (e.g., a plane) fit to a set of control points so that the local trend in the data can be determined (p. 141).

triangulation: a method of interpolating data for isarithmic maps in which original control points are connected by a set of triangles (p. 137).

trichromatic theory: the theory that color perception is a function of the relative stimulation of blue, green, and red cones (p. 85).

trivariate choropleth map: the overlay of three univariate choropleth maps; this approach should be used only when variables add up to 100%, such as percent voting Republican, Democrat, and Independent (p. 201).

true 3-D phenomena: a form of volumetric phenomena in which each longitude and latitude position has multiple attribute values depending on the height above or below a zero point (e.g., the level of ozone in the earth's atmosphere varies as a function of elevation above sea level) (p. 19).

true-point data: data that can actually be measured at a point location, such as the number of calls made at a telephone booth over the course of a day (p. 118).

unclassed map: a map in which data are not grouped into classes of similar value and thus each data value can theoretically be represented by a different symbol (e.g., a data set with 100 different data values might be depicted using 100 different shades of gray on a choropleth map) (p. 8).

uniform color model: refers to a variation of the CIE color model in which colors are equally spaced in the visual sense (p. 100).

unipolar data: data that have no dividing point and do not involve two complementary phenomena, such as per capita income in African countries; contrast with **bipolar** and **balanced data** (p. 22).

unit value: the count represented by each dot on a dot map (p. 171).

unit-vector density map: a map in which the density and orientation of short fixed-length segments (unit vectors) are used to display the magnitude and direction, respectively, of a continuous vector–based flow, such as wind speed and direction (p. 186).

universal kriging: a form of kriging that accounts for a trend (or drift) in the data over geographic space (p. 143).

value: see **lightness.**

vector: the manner in which images are created as we would draw a map by hand: the hardware moves to one location and draws to the next (p. 89).

virtual reality or **virtual world:** the use of computer–based systems to create lifelike representations of the real world (p. 251).

visible light: the portion of the electromagnetic spectrum to which the human eye is sensitive (p. 84).

visual angle: the angle formed by lines projected from the top and bottom of an image through the center of the lens of the eye (p. 85).

visualization: a private activity in which previously unknown spatial information is revealed in a highly interactive computer graphics environment; sometimes called *geographic visualization* (p. 11).

visualization software: software that is expressly intended for developing visualizations of spatial data (p. 239).

visual pigments: the light-sensitive chemicals found within rods and cones (p. 85).

visual programming language: a graphical user interface is used to develop a software application, although computer code will likely also have to be written (p. 237).

visual realism or **visual simulation:** using a computer–based system to create a *visually* realistic representation of the real world (p. 251).

visual variables: the perceived differences in map symbols that are used to represent spatial phenomena (p. 22).

wavelength of light: the distance between two wave crests associated with electromagnetic energy (p. 84).

weighted isolines: the width of contour lines made proportional to the associated data; thus labeling of contour lines is not required (p. 207).

Williams curve: a gray scale that is recommended for coarse (textured) areal shades (p. 113).

Zip-a-Tone: in traditional manual cartography, areal shades could be cut from pre-printed sheets of Zip-a-Tone and stuck to a base map (p. 114)

zoom function: in an interactive graphics environment, an area to be focused on is enlarged, commonly by enclosing a desired area with a rectangular box and clicking the mouse (p. 132).

References

Acevedo, W., and Masuoka, P. (1997) "Time-series animation techniques for visualizing urban growth." *Computers & Geosciences* 23, no. 4:423–435.

Anderson, E. (1960) "A semigraphical method for the analysis of complex problems." *Technometrics* 2, no. 3:387–391.

Anderson, J. R. (1990) *Cognitive Psychology and Its Implications.* 3d. ed. New York: W. H. Freeman.

Aspaas, H. R., and Lavin, S. J.(1989) "Legend designs for unclassed, bivariate, choropleth maps." *American Cartographer* 16, no. 4:257–268.

Astroth, J. H. J., Trujillo, J., and Johnson, G. E. (1990) "A retrospective analysis of GIS performance: The Umatilla Basin revisited." *Photogrammetric Engineering and Remote Sensing* 56, no. 3:359–363.

Bachi, R. (1962) "Standard distance measures and related methods for spatial analysis." *Regional Science Association Papers* 10:83–132.

——— (1973) "Geostatistical analysis of territories." *Proceedings of the 39th Session,* Bulletin of the International Statistical Institute (Vienna), pp. 121–132.

Barnett, V., and Lewis, T. (1994) *Outliers in Statistical Data.* 3d ed. Chichester, England: John Wiley & Sons.

Barrett, R. E. (1994) *Using the 1990 U.S. Census for Research.* Thousand Oaks, Calif.: Sage Publications.

Batty, M., and Howes, D. (1996a) "Exploring urban development dynamics through visualization and animation." In *Innovations in GIS 3,* ed. D. Parker, pp. 149–161. Bristol, Penn.: Taylor & Francis.

——— (1996b) "Visualizing urban development." *Geo Info Systems 6,* no. 9:28–29, 32.

Bemis, D., and Bates, K. (1989) "Color on temperature maps." Unpublished manuscript, Department of Geography, Pennsylvania State University, University Park, Penn.

Bergman, E. F. (1995) *Human Geography: Cultures, Connections, and Landscapes.* Englewood Cliffs, N.J.: Prentice-Hall.

Bergman, L. D., Rogowitz, B. E., and Treinish, L. A. (1995) "A rule-based tool for assisting colormap selection." *Proceedings, Visualization '95,* October 29–November 3, Atlanta, Ga., pp. 118–125. (See also http://www.almaden.ibm.com/dx/vis96/proceedings/PRAVDA/index.htm.)

Bertin, J. (1981) *Graphics and Graphic Information-Processing.* Berlin: Walter de Gruyter.

——— (1983) *Semiology of Graphics: Diagrams, Networks, Maps.* Madison, Wisc.: University of Wisconsin Press.

Berton, J. A. J. (1990) "Strategies for scientific visualization: Analysis and comparison of current techniques." Extracting Meaning from Complex Data: Processing, Display, Interaction, *Proceedings, SPIE,* vol. 1259, pp. 110–121.

Birdsall, S. S., and Florin, J. W. (1992) *Regional Landscapes of the United States and Canada.* New York: John Wiley & Sons.

Birren, F. (1983) *Colour.* London: Marshall Editions.

Bishop, I. (1994) "The role of visual realism in communicating and understanding spatial change and process." In *Visualization in Geographical Information Systems,* ed. H. M. Hearnshaw and D. J. Unwin, pp. 60–64. Chichester, England: John Wiley & Sons.

———, and Karadaglis, C. (1997) "Linking modelling and visualization for natural resources management." *Environment and Planning B: Planning and Design* 24, 345–358.

Board, C. (1984) "Higher order map-using tasks: Geographical lessons in danger of being forgotten." *Cartographica* 21, no. 1:85–97.

Bockenhauer, M. H. (1994) "Culture of the Wisconsin Official State Highway Map." *Cartographic Perspectives* 18:17–27.

Bolorforoush, M., and Wegman, E. J. (1988) "On some graphical representations of multivariate data." Computing Science and Statistics, *Proceedings of the 20th Symposium on the Interface,* pp. 121–126.

Borg, I., and Staufenbiel, T. (1992) "Performance of snowflakes, suns, and factorial suns in the graphical representation of multivariate data." *Multivariate Behavioral Research* 27, no. 1:43–55.

Bowmaker, J. K., and Dartnall, H. J. A. (1980) "Visual pigments of rods and cones in a human retina." *Journal of Physiology* 298:501–511.

Boynton, R. M. (1989) "Eleven colors that are almost never confused." Human Vision, Visual Processing, and Digital Display, *Proceedings, SPIE,* vol. 1077, pp. 322–332.

Bragdon, C. R., Juppe, J. M., and Georgiopoulos, A. X. (1995) "Sensory spatial systems simulation (S^4) applied to the master planning process: East Coast and West Coast case studies." *Environment and Planning B: Planning and Design* 22, no. 3:303–314.

Brassel, K. E., and Utano, J. J. (1979) "Design strategies for continuous-tone area mapping." *American Cartographer* 6, no. 1:39–50.

Bressi, T. (1995) "The real thing? We're getting there." *Planning* 61, no. 7:16–20.

Brewer, C. A. (1989) "The development of process-printed Munsell charts for selecting map colors." *American Cartographer* 16, no. 4:269–278.

——— (1992) "Review of colour terms and simultaneous contrast research for cartography." *Cartographica* 29, no. 3/4:20–30.

——— (1994a) "Color use guidelines for mapping and visualization." In *Visualization in Modern Cartography,* ed. A. M. MacEachren and D. R. F. Taylor, pp. 123–147. Oxford: Pergamon Press.

——— (1994b) "Guidelines for use of the perceptual dimensions of color for mapping and visualization." In *Color Hard Copy and Graphic Arts III,* ed. J. Bares, pp. 54–63. Bellingham, Wash.: International Society for Optical Engineering.

——— (1996a) "Guidelines for selecting colors for diverging schemes on maps." *Cartographic Journal* 33, no. 2:79–86.

——— (1996b) "Prediction of simultaneous contrast between map colors with Hunt's model of color appearance." *Color Research and Application* 21, no. 3:221–235.

——— (1997) "Evaluation of a model for predicting simultaneous contrast on color maps." *Professional Geographer 49,* no. 3:280–294.

———, and Marlow, K. A. (1993) "Color representation of aspect and slope simultaneously." *Auto-Carto 11 Proceedings,* Minneapolis, Minn., pp. 328–337.

———, MacEachren, A. M., Pickle, L. W., and Herrmann, D. (1997) "Mapping mortality: Evaluating color schemes for choropleth maps." *Annals, Association of American Geographers* 87, no. 3:411–438.

Brown, A., and van Elzakker, C. P. J. M. (1993) "The use of colour in the cartographic representation of information quality generated by a GIS." *Proceedings, 16th Conference of the International Cartographic Association,* May 3–9, Cologne, Germany, pp. 707–720.

Buchanan, J. T., and Acevedo, W. (1996) "Studying urban sprawl using a temporal database." *Geo Info Systems* 6, no. 7:42–47.

Burt, J. E., and Barber, G. M. (1996) *Elementary Statistics for Geographers.* 2d ed. New York: Guilford Press.

Buttenfield, B. P., and Mark, D. M. (1991) "Expert systems in cartographic design." In *Geographic Information Systems: The Microcomputer and Modern Cartography,* ed. D. R. F. Taylor, pp. 129–150. Oxford: Pergamon Press.

———, and Weber, C. R. (1994) "Proactive graphics for exploratory visualization of biogeographical data." *Cartographic Perspectives* no. 19:8–18.

———, Weber, C., MacLennen, M., and Elliot, J. D. (1991) "Bibliography on animation of spatial data." Technical Paper 91–22, National Center for Geographic Information & Analysis, Buffalo, NY.

Byron, J. R. (1994) "Spectral encoding of soil texture: A new visualization method." *GIS/LIS Proceedings,* Phoenix, Ariz., pp. 125–132.

Caldwell, P. S. (1979) "Television news maps: An examination of their utilization, content, and design." Unpublished Ph.D. dissertation, University of California, Los Angeles, Calif.

——— (1981) "Television news maps: The effects of the medium on the map." *Technical Papers,* American Congress on Surveying and Mapping (ACSM)Annual Meeting, Washington, D.C., 382–392.

Cammack, R. G. (1991) "Cartographic animation: An exploration of a mapping technique." Unpublished M.S. thesis, University of South Carolina, Columbia, S.C.

Campbell, C. S., and Egbert, S. L. (1990) "Animated cartography: Thirty years of scratching the surface." *Cartographica* 27, no. 2:24–46.

Campbell, J. B. (1996) *Introduction to Remote Sensing.* 2d ed. New York: Guilford Press.

Carr, D. B. (1991) "Looking at large data sets using binned data plots." In *Computing and Graphics in Statistics,* ed. A. Buja and P. A. Tukey, pp. 7–39. New York: Springer-Verlag.

———, and Pierson, S. M. (1996) "Emphasizing statistical summaries and showing spatial context with micromaps." *Statistical Computing & Statistical Graphics Newsletter* 7, no. 3:16–23.

———, Olsen, A. R., and White, D. (1992) "Hexagon mosaic maps for display of univariate and bivariate geographical data." *Cartography and Geographic Information Systems* 19, no. 4:228–236, 271.

Carstensen, L. W. J. (1982) "A continuous shading scheme for two-variable mapping." *Cartographica* 19, no. 3/4:53–70.

——— (1986a) "Bivariate choropleth mapping: The effects of axis scaling." *American Cartographer* 13, no. 1:27–42.

——— (1986b) "Hypothesis testing using univariate and bivariate choropleth maps." *American Cartographer* 13, no. 3:231–251.

——— (1987) "A comparison of simple mathematical approaches to the placement of spot symbols." *Cartographica* 24, no. 3:46–63.

Cartwright, W. (1994) "Interactive multimedia for mapping." In *Visualization in Modern Cartography,* ed. A.M. MacEachren and D. R. F. Taylor, pp. 63–89. Oxford: Pergamon Press.

——— (1997) "New media and their application to the production of map products." *Computers & Geosciences* 23, no. 4:447–456.

Castner, H. W., and Robinson, A. H. (1969) *Dot Area Symbols in Cartography: The Influence of Pattern on Their Perception.* Washington, D.C.: American Congress on Surveying and Mapping.

Chang, K. (1977) "Visual estimation of graduated circles." *Canadian Cartographer* 14, no. 2:130–138.

——— (1980) "Circle size judgment and map design." *American Cartographer* 7, no. 2:155–162.

——— (1982) "Multi-component quantitative mapping." *Cartographic Journal* 19, no. 2:95–103.

Chrisman, N. (1997) *Exploring Geographic Information Systems.* New York: John Wiley.

Clark, W. A. V., and Hosking, P. L. (1986) *Statistical Methods for Geographers.* New York: John Wiley & Sons.

Clarke, K. C. (1995) *Analytical and Computer Cartography.* 2d ed. Englewood Cliffs, N.J.: Prentice-Hall.

——— (1997) *Getting Started with Geographic Information Systems.* Upper Saddle River, N.J.: Prentice-Hall.

Cleveland, W. S. (1993) *Visualizing Data.* Summit, N.J.: Hobart Press.

——— (1994) *The Elements of Graphing Data.* Rev. ed. Summit, N.J.: Hobart Press.

———, and McGill, M. E. (eds.)(1988) *Dynamic Graphics for Statistics.* Belmont, Calif.: Wadsworth.

———, and McGill, R. (1984) "Graphical perception: Theory, experimentation, and application to the development of graphical methods." *Journal of the American Statistical Association* 79, no. 387:531–554.

Cloke, P., Philo, C., and Sadler, D. (1991) *Approaching Human Geography: An Introduction to Contemporary Theoretical Debates.* London: Paul Chapman Publishing.

Cook, D., Symanzik, J., Majure, J. J., and Cressie, N. (1997) "Dynamic graphics in a GIS: More examples using linked software." *Computers & Geosciences* 23, no. 4:371–385.

Coulson, M. R. C. (1987) "In the matter of class intervals for choropleth maps: With particular reference to the work of George F. Jenks." *Cartographica* 24, no. 2:16–39.

Cox, C. W. (1976) "Anchor effects and estimation of graduated circles and squares." *American Cartographer* 3, no. 1:65–74.

Cox, D. J. (1990) "The art of scientific visualization." *Academic Computing* 4, no. 6:20–22, 32–34, 36–38, 40.

Crampton, J. (1995) "Cartography resources on the World Wide Web." *Cartographic Perspectives* 22:3–11.

Crawford, P. V. (1973) "The perception of graduated squares as cartographic symbols." *Cartographic Journal* 10, no. 2:85–88.

Cressie, N. (1990) "The origins of kriging." *Mathematical Geology* 22, no. 3:239–252.

——— (1992) "Smoothing regional maps using empirical bayes predictors." *Geographical Analysis* 24, no. 1:75–95.

——— (1993) *Statistics for Spatial Data.* Rev. ed. New York: John Wiley & Sons.

———, and Read, T. R. C. (1989) "Spatial data analysis of regional counts." *Biometrical Journal* 31:699–719.

Cromley, R. G. (1995) "Classed versus unclassed choropleth maps: A question of how many classes." *Cartographica* 32, no. 4:15–27.

——— (1996) "A comparison of optimal classification strategies for choroplethic displays of spatially aggregated data." *International Journal of Geographical Information Systems* 10, no. 4:405–424.

Cuff, D. J. (1972a) "The magnitude message: A study of effectiveness of color sequences on quantitative maps." Unpublished Ph.D. dissertation, Pennsylvania State University, University Park, Penn.

——— (1972b) "Value versus chroma in color schemes on quantitative maps." *Canadian Cartographer* 9, no. 2:134–140.

——— (1973) "Colour on temperature maps." *Cartographic Journal* 10, no. 1:17–21.

——— (1974a) "Impending conflict in color guidelines for maps of statistical surfaces." *Canadian Cartographer* 11, no. 1:54–58.

——— (1974b) "Perception of color sequences on maps of atmospheric pressure." *Professional Geographer* 26, no. 2:166–171.

———, and Bieri, K. R. (1979) "Ratios and absolute amounts conveyed by a stepped statistical surface." *American Cartographer* 6, no. 2:157–168.

———, Pawling, J. W., and Blair, E. T. (1984) "Nested value-by-area cartograms for symbolizing land use and other proportions." *Cartographica* 21, no. 4:1–8.

Davis, B. (1997) "The future of the past." *Scientific American* 277, no. 2:89–92.

Davis, J. C. (1975) "Contouring algorithms." *Autocarto 2 Proceedings,* pp. 352–359.

——— (1986) *Statistics and Data Analysis in Geology.* 2d ed. New York: John Wiley & Sons.

Davis, T. J., and Keller, C. P. (1997) "Modelling and visualizing multiple spatial uncertainties." *Computers & Geosciences* 23, no. 4:397–408.

DeBraal, J. P. (1992) "Foreign ownership of U.S. agricultural land through December 31, 1992." U.S. Department of Agriculture, Economic Research Service, Statistical Bulletin 853.

Declercq, F. A. N. (1995) "Choropleth map accuracy and the number of class intervals." *Proceedings of the 17th International Cartographic Conference,* vol. 1, Barcelona, Spain, pp. 918–922.

——— (1996) "Interpolation methods for scattered sample data: Accuracy, spatial patterns, processing time." *Cartography and Geographic Information Systems* 23, no. 3:128–144.

Dent, B. D. (1996) *Cartography: Thematic Map Design.* 4th ed. Dubuque, Iowa: William C. Brown.

Derrington, A., Lennie, P., and Krauskopf, J. (1983) "Chromatic response properties of parvocellular neurons in the macaque LGN." In *Colour Vision: Physiology and Psychophysics,* ed. J. D. Mollon and L. T. Sharpe, pp. 245–251. London: Academic Press.

De Valois, R. L., and Jacobs, G. H. (1984) "Neural mechanisms of color vision." In *Handbook of Physiology (Section 1: The Nervous System),* ed. J. M. Brookhart and V. B. Mountcastle, pp. 425–456. Bethesda, Md.: American Physiological Society.

DiBiase, D. (1990a) "Introduction to Macintosh graphics file formats." *Cartographic Perspectives* no. 7:15–18.

——— (1990b) "Visualization in the earth sciences." *Earth and Mineral Sciences* 59, no. 2:13–18.

——— (1994) "Designing animated maps for a multimedia encyclopedia." *Cartographic Perspectives* no. 19:3–7,19.

———, Krygier, J., Reeves, C., MacEachren, A., and Brenner, A. (1991) *Elementary Approaches to Cartographic Animation.* University Park, Penn.: Deasy GeoGraphics Laboratory, Department of Geography, Pennsylvania State University.

———, MacEachren, A. M., Krygier, J. B., and Reeves, C. (1992) "Animation and the role of map design in scientific visualization." *Cartography and Geographic Information Systems* 19, no. 4:201–214, 265–266.

———, Reeves, C., MacEachren, A. M., Von Wyss, M., Krygier, J. B., Sloan, J. L., and Detweiler, M. C. (1994a) "Multivariate display of geographic data: Applications in Earth system science." In *Visualization in Modern Cartography,* ed. A. M. MacEachren and D. R. F. Taylor, pp. 287–312. Oxford: Pergamon, Press.

———, Sloan, J. I. I., and Paradis, T. (1994b) "Weighted isolines: An alternative method for depicting statistical surfaces." *Professional Geographer* 46, no. 2:218–228.

Ding, Y., and Densham, P. J. (1994) "A loosely synchronous, parallel algorithm for hill shading digital elevation models." *Cartography and Geographic Information Systems* 21, no. 1:5–14.

Dobson, M. W. (1973) "Choropleth maps without class intervals? A comment." *Geographical Analysis* 5, no. 4:358–360.

——— (1974) "Refining legend values for proportional circle maps." *Canadian Cartographer* 11, no. 1:45–53.

——— (1980a) "Perception of continuously shaded maps." *Annals, Association of American Geographers* 70, no. 1:106–107.

——— (1980b) "Unclassed choropleth maps: A comment." *American Cartographer* 7, no. 1:78–80.

——— (1983) "Visual information processing and cartographic communication: The utility of redundant stimulus dimensions." In *Graphic Communication and Design in Contemporary Cartography,* ed. D. R. F. Taylor, pp. 149–175. Chichester, England: John Wiley & Sons.

Doerschler, J. S., and Freeman, H. (1992) "A rule-based system for dense-map name placement." *Communications of the ACM* 35:68–79.

Dorling, D. (1992) "Stretching space and splicing time: From cartographic animation to interactive visualization." *Cartography and Geographic Information Systems* 19, no. 4:215–227, 267–270.

——— (1993) "Map design for census mapping." *Cartographic Journal* 30, no. 2:167–183.

——— (1994) "Cartograms for visualizing human geography." In *Visualization in Geographical Information Systems,* ed. H. M. Hearnshaw and D. J. Unwin, p. 243. Chichester, England: John Wiley & Sons.

——— (1995a) *A New Social Atlas of Britain.* Chichester, England: John Wiley & Sons.

——— (1995b) "The visualization of local urban change across Britain." *Environment and Planning B: Planning and Design* 22, no. 3:269–290.

——— (1995c) "Visualizing changing social structure from a census." *Environment and Planning A* 27:353–378.

Dougenik, J. A., Chrisman, N. R., and Niemeyer, D. R. (1985) "An algorithm to construct continuous area cartograms." *Professional Geographer* 37, no. 1:75–81.

Dunn, R. (1988) "Framed rectangle charts or statistical maps with shading: An experiment in graphical perception." *American Statistician* 42, no. 2:123–129.

Dykes, J. A. (1994) "Visualizing spatial association in area-value data." In *Innovations in GIS,* ed. M. F. Worboys, pp. 149–159. Bristol, Penn.: Taylor & Francis.

——— (1995) "Cartographic visualization for spatial analysis." *Proceedings of the 17th International Cartographic Conference,* Vol. 1, Barcelona, Spain, pp. 1365–1370.

——— (1996) "Dynamic maps for spatial science: A unified approach to cartographic visualization." In *Innovations in GIS 3,* ed. D. Parker, pp. 177–187, Color Plates 4–5. Bristol, Penn.: Taylor & Francis.

——— (1997) "Exploring spatial data representation with dynamic graphics." *Computers & Geosciences* 23, no. 4:345–370.

Eastman, J. R. (1986) "Opponent process theory and syntax for qualitative relationships in quantitative series." *American Cartographer* 13, no. 4:324–333.

Ebinger, L. R., and Goulette, A. M. (1990) "Noninteractive automated names placement for the 1990 decennial census." *Cartography and Geographic Information Systems* 17, no. 1:69–78.

Eckhardt, D. W., Verdin, J. P., and Lyford, G. R. (1990) "Automated update of an irrigated lands GIS using SPOT HRV imagery." *Photogrammetric Engineering and Remote Sensing* 56, no. 11:1515–1522.

Egbert, S. L. (1994) "The design and evaluation of an interactive choropleth map exploration system." Unpublished Ph.D. dissertation, University of Kansas, Lawrence, Kan.

———, and Slocum, T. A. (1992) "EXPLOREMAP: An exploration system for choropleth maps." *Annals, Association of American Geographers* 82, no. 2:275–288.

———, Price, K. P., Nellis, M. D., and Lee, Re-Y. (1995) "Developing a land cover modelling protocol for the high plains using multi-seasonal thematic mapper imagery." *Technical Papers,* vol. 3, ACSM/ASPRS Annual Convention and Exposition, Charlotte, N.C., pp. 836–845.

Ehlschlaeger, C. R., Shortridge, A. M., and Goodchild, M. F. (1997) "Visualizing spatial data uncertainty using animation." *Computers & Geosciences* 23, no. 4:387–395.

Ekman, G., and Junge, K. (1961) "Psychophysical relations in visual perception of length, area and volume." *Scandinavian Journal of Psychology* 2, no. 1:1–10.

Ellson, R. (1990) "Visualization at work." *Academic Computing* 4, no. 6:26–28, 54–56.

Environmental Systems Research Institute. (1996) *Using ArcView GIS.* Redlands, Calif.

Ervin, S. M. (1992) "Integrating visual and environmental analyses in site planning and design." *GIS World* July (Special Issue):26–30.

——— (1993) "Landscape visualization with Emaps." *IEEE Computer Graphics & Applications* 13, no. 2:28–33.

Espenshade, E. B. J. (ed.)(1990) *Goode's World Atlas.* 19th ed. Chicago: Rand McNally.

Evans, B. J. (1997) "Dynamic display of spatial data-reliability: Does it benefit the map user?" *Computers & Geosciences* 23, no. 4:409–422.

Evans, I. S. (1977) "The selection of class intervals." *Transactions, Institute of British Geographers* (new series) 2, no. 1:98–124.

Eyton, J. R. (1984a) "Complementary-color two-variable maps." *Annals, Association of American Geographers* 74, no. 3:477–490.

——— (1984b) "Raster contouring." *Geo-Processing* 2:221–242.

——— (1990) "Color stereoscopic effect cartography." *Cartographica* 27, no. 1:20–29.

——— (1991) "Rate-of-change maps." *Cartography and Geographic Information Systems* 18, no. 2:87–103.

——— (1994) "Chromostereoscopic maps." *Cartouche: Newsletter of the Canadian Cartographic Association* Autumn/Winter (Special Issue): 15.

Fairbairn, D., and Parsley, S. (1997) "The use of VRML for cartographic presentation." *Computers & Geosciences* 23, no. 4:475–481.

Famighetti, R. (ed.)(1993) *The World Almanac and Book of Facts 1994.* Mahwah, N.J.: Funk & Wagnalls.

Favro, D., Frischer, B., Dorr, A., and Jepson, W. (1997) "Rome Reborn: Virtual reality model of Augustan Rome project." Reprinted from http://www.aud.ucla.edu/~dabernat/rome/index.html. Accessed September 12, 1997.

Fernandes, J. P., Fonseca, A., Pereira, L., Faria, A., Figueira, H., Henriques, I., Garcao, R., and Camara, A. (1997) "Visualization and interaction tools for aerial photograph mosaics." *Computers & Geosciences* 23, no. 4:465–474.

Ferreira, J. J., and Wiggins, L. L. (1990) "The density dial: A visualization tool for thematic mapping." *Geo Info Systems* 1, no. 0:69–71.

Fisher, P. F. (1993) "Visualizing uncertainty in soil maps by animation." *Cartographica* 30, nos. 2 and 3:20–27.

——— (1994a) "Hearing the reliability in classified remotely sensed images." *Cartography and Geographic Information Systems* 21, no. 1:31–36.

——— (1994b) "Visualization of the reliability in classified remotely sensed images." *Photogrammetric Engineering and Remote Sensing* 60, no. 7:905–910.

——— (1996) "Animation of reliability in computer-generated dots maps and elevation models." *Cartography and Geographic Information Systems* 23, no. 4:196–205.

———, and Langford, M. (1996) "Modeling sensitivity to accuracy in classified imagery: A study of areal interpolation by dasymetric mapping." *Professional Geographer* 48, no. 3:299–309.

Fisher, W. D. (1958) "On grouping for maximum homogeneity." *Journal of the American Statistical Association* 53, no. December:789–798.

Flanagan, T. J., and Maguire, K. (eds.)(1992) *Sourcebook of Criminal Justice Statistics 1991.* Washington, D.C.: U.S. Department of Justice, Bureau of Justice Statistics.

Flannery, J. J. (1971) "The relative effectiveness of some common graduated point symbols in the presentation of quantitative data." *Canadian Cartographer* 8, no. 2:96–109.

Foley, J. D., van Dam, A., Feiner, S. K., and Hughes, J. F. (1996) *Computer Graphics: Principles and Practice.* 2d ed. in C. Reading, Mass.: Addison-Wesley.

Gaydos, L. J., Acevedo, W., and Bell, C. (1995) "Using animated cartography to illustrate global change." *Proceedings of the 17th International Cartographic Conference,* vol. 1, Barcelona, Spain, pp. 1174–1178.

Gershon, N. (1992) "Visualization of fuzzy data using generalized animation." *Proceedings, Visualization '92,* October 19–23, Boston, Mass., pp. 268–273.

Gersmehl, P. J. (1990) "Choosing tools: Nine metaphors of four-dimensional cartography." *Cartographic Perspectives* 5:3–17.

Gerth, J. D. (1993) "Towards improved spatial analysis with areal units: The use of GIS to facilitate the creation of dasymetric maps." Research paper for an M.A. degree, Ohio State University, Columbus, Ohio.

Gilmartin, P. P. (1981) "Influences of map context on circle perception." *Annals, Association of American Geographers* 71, no. 2:253–258.

———, and Shelton, E. (1989) "Choropleth maps on high resolution CRTs/The effect of number of classes and hue on communication." *Cartographica* 26, no. 2:40–52.

Golden Software, (1995) *SURFER for Windows Version 6 User's Guide.* Golden, Colo.: Golden Software.

Goldstein, E. B. (1989) *Sensation and Perception.* 3d ed. Belmont, Calif.: Wadsworth.

Goodchild, M., Chih-Chang, L., and Leung, Y. (1994) "Visualizing fuzzy maps." In *Visualization in Geographical Information Systems,* ed. H. M. Hearnshaw and D. J. Unwin, pp. 158–167. Chichester, England: John Wiley & Sons.

Gould, P. (1989) "Geographic dimensions of the AIDS epidemic." *Professional Geographer* 41, no. 1:71–78.

——— (1993) *The Slow Plague.* Cambridge, Mass.: Blackwell.

Greenfield, D. (1994) "Animating point symbols for cartographic display." Unpublished M.A. thesis, University of Kansas, Lawrence, Kan.

Griffin, T. L. C. (1983) "Recognition of areal units on topological cartograms." *American Cartographer* 10, no. 1: 17–29.

——— (1985) "Group and individual variations in judgment and their relevance to the scaling of graduated circles." *Cartographica* 22, no. 1:21–37.

——— (1990) "The importance of visual contrast for graduated circles." *Cartography* 19, no. 1:21–30.

Griffith, D. A. (1993) *Spatial Regression Analysis on the PC: Spatial Statistics using SAS.* Washington, D.C.: Association of American Geographers.

———, and Amrhein, C. G. (1991) *Statistical Analysis for Geographers.* Englewood Cliffs, N.J.: Prentice-Hall.

Grinstein, G., Sieg, J. C. J., Smith, S., and Williams, M. G. (1992) "Visualization for knowledge discovery." *International Journal of Intelligent Systems* 7:637–648.

Groop, R. E. (1992) "Dot-density crop maps of the Midwest." Abstract, Association of American Geographers Annual Meeting, San Diego, California, p. 89.

———, and Cole, D. (1978) "Overlapping graduated circles: Magnitude estimation and method of portrayal." *Canadian Cartographer* 15, no. 2:114–122.

Guilford, J. P., and Smith, P. C. (1959) "A system of color-preferences." *American Journal of Psychology* 72, no. 4:487–502.

Guptill, S. C., and Morrison, J. L. (eds.)(1995) *Elements of spatial data quality.* Oxford: Elsevier Science.

Gusein-Zade, S. M., and Tikunov, V. S. (1993) "A new technique for constructing continuous cartograms." *Cartography and Geographic Information Systems* 20, no. 3:167–173.

Halfhill, T. R. (1997) "Today the web, tomorrow the world." *Byte* 22, no. 1:68–80.

Hammond, R., and McCullagh, P. (1978) *Quantitative Techniques in Geography: An Introduction.* 2d ed. Oxford: Clarendon Press.

Hancock, J. R. (1993) "Multivariate regionalization: An approach using interactive statistical visualization." *Auto-Carto 11 Proceedings,* Minneapolis, Minn., pp. 218–227.

Harley, J. B. (1989) "Deconstructing the map." *Cartographica* 26, no. 2:1–20.

Harrison, R. E. (1969) Shaded Relief. United States Department of the Interior Geological Survey. In *The National Atlas of the United States of America,* 1970, pp. 56–57. Scale 1:7,500,000.

Hartigan, J. A. (1975) *Clustering Algorithms.* New York: John Wiley & Sons.

Hartnett, S. (1987) "Employing rectangular point symbols in two-variable maps." Abstract, Association of American Geographers Annual Meeting, Portland, Oregon, p. 39.

Haslett, J., and Power, G. M. (1995) "Interactive computer graphics for a more open exploration of stream sediment geochemical data." *Computers & Geosciences* 21, no. 1:77–87.

———, Wills, G., and Unwin, A. (1990) "SPIDER —An interactive statistical tool for the analysis of spatially distributed data." *International Journal of Geographical Information Systems* 4, no. 3:285–296.

———, Bradley, R., Craig, P., and Unwin, A. (1991) "Dynamic graphics for exploring spatial data with application to locating global and local anomalies." *American Statistician* 45, no. 3:234–242.

Helmholtz, H. von (1852) "On the theory of compound colors." *Philosophical Magazine* 4:519–534.

Hering, E. (1878) *Zur Lehre vom Lichtsinne.* Vienna: Gerold.

Herzog, A. (1989) "Modeling reliability on statistical surfaces by polygon filtering." In *Accuracy of Spatial Databases,* ed. M. Goodchild and S. Gopal, pp. 209–218. London: Taylor & Francis.

Heyn, B. N. (1984) "An evaluation of map color schemes for use on CRT's." Unpublished M.S. thesis, University of South Carolina, Columbia, S.C.

Heywood, I., Oliver, J., and Tomlinson, S. (1995) "Building an exploratory multi-criteria modelling environment for spatial decision support." In *Innovations in GIS* 2, ed. P. Fisher, pp. 127–136. Bristol, Penn.: Taylor & Francis.

Hibbard, B., and Santek, D. (1990) "The VIS-5D system for easy interactive visualization." *Proceedings, Visualization '90,* San Francisco, CA., pp. 28–35.

Hibbard, W. L., Paul, B. E., Santek, D. A., Dyer, C. R., Battaiola, A. L., and Voidrot-Martinez, Marie-F. (1994) "Interactive visualization of earth and space science computations." *Computer* 27, no. 7:65–72.

Hilliard, B. (1995) "On-screen color." *Byte* 20, no. 1:101–102, 104, 106.

Hobbs, F. (1995) "The rendering of relief images from digital contour data." *Cartographic Journal* 32, no. 2:111–116.

Hocking, D., and Keller, C. P. (1993) "Alternative atlas distribution formats: A user perspective." *Cartography and Geographic Information Systems* 20, no. 3:157–166.

Hoinkes, R., and Lange, E. (1995) "3-D for free: Toolkit expands visual dimensions in GIS." *GIS World* 8, no. 7:54–56.

Horn, B. K. P. (1982) "Hill shading and the reflectance map." *Geo-Processing* 2, no. 1:65–144.

Howard, D., and MacEachren, A. M. (1996) "Interface design for geographic visualization: Tools for representing reliability." *Cartography and Geographic Information Systems* 23, no. 2:59–77.

Howard, H. (1995) "Death Valley: An animated atlas." Unpublished M.A. thesis, San Francisco State University, San Francisco, Calif.

Hsu, Mei-L. (1979) "The cartographer's conceptual process and thematic symbolization." *American Cartographer* 6, no. 2:117–127.

Hubel, D. H. (1988) *Eye, Brain, and Vision.* New York: Scientific American Library.

Hughes, J. R. (1996) "Technology trends mark multimedia advancements." *GIS World* 9, no. 11:40–43.

Hunt, R. W. G. (1987a) *Measuring Colour.* Chichester, England: Ellis Horwood.

——— (1987b) *The Reproduction of Colour in Photography, Printing and Television.* Tolworth, England: Fountain Press.

Hunter, G. J., and Goodchild, M. F. (1995) "Dealing with error in spatial databases: A simple case study." *Photogrammetric Engineering and Remote Sensing* 61, no. 5:529–537.

——— (1996) "Communicating uncertainty in spatial databases." *Transactions in GIS* 1, no. 1:13–24.

Hurvich, L. M. (1981) *Color Vision.* Sunderland, Mass.: Sinauer Associates.

———, and Jameson, D. (1957) "An opponent-process theory of color vision." *Psychological Review* 64, no. 6:384–404.

Imhof, E. (1975) "Positioning names on maps." *American Cartographer* 2, no. 2:128–144.

——— (1982) *Cartographic Relief Presentation.* Berlin: Walter de Gruyter.

Isaaks, E., and Srivastava, R. M. (1989) *Applied Geostatistics.* New York: Oxford University Press.

Jackel, C. B. (1997) "Using ArcView to create contiguous and noncontiguous area cartograms." *Cartography and Geographic Information Systems* 24, no. 2:101–109.

Jacobson, R. (1994) "Virtual worlds capture spatial reality." *GIS World* 7, no. 12:36–39.

Jelinski, D. E. (1987) "Intraspecific diversity in trembling aspen in Waterton Lakes National Park, Alberta: A biogeographical perspective." Unpublished Ph.D. dissertation, Simon Fraser University, British Columbia, Canada.

Jenks, G. F. (1953) "'Pointillism' as a cartographic technique." *Professional Geographer* 5, no. 5:4–6.

——— (1961): Crop patterns in the United States: 1959. Map, U.S. Bureau of the Census. Scale 1:5,100,000.

——— (1962): Livestock and livestock products sold in the United States: 1959. Map, U.S. Bureau of the Census. Scale 1:5,100,000.

——— (1977) "Optimal data classification for choropleth maps." Occasional Paper No. 2. Lawrence, Kan.: University of Kansas, Department of Geography.

———, and Caspall, F. C. (1971) "Error on choroplethic maps: Definition, measurement, reduction." *Annals, Association of American Geographers* 61, no. 2:217–244.

Kansas Farm Facts (1994) Topeka, Kan.: Kansas State Board of Agriculture.

Karl, D. (1992) "Cartographic animation: Potential and research issues." *Cartographic Perspectives* no. 13:3–9.

Keller, C. P. (1995) "Visualizing digital atlas information products and the user perspective." *Cartographic Perspectives* no. 20:21–28.

Keller, P. R., and Keller, M. M. (1993) *Visual Cues: Practical Data Visualization.* Los Alamitos, Calif.: IEEE Computer Society Press.

Kennedy, S. (1994) "Unclassed choropleth maps revisited/ Some guidelines for the construction of unclassed and classed choropleth maps." *Cartographica* 31, no. 1: 16–25.

Kerst, S. M., and Howard, J. H. J. (1984) "Magnitude estimates of perceived and remembered length and area." *Bulletin of the Psychonomic Society* 22, no. 6:517–520.

Kimerling, A. J. (1975) "A cartographic study of equal value gray scales for use with screened gray areas." *American Cartographer* 2, no. 2:119–127.

——— (1985) "The comparison of equal-value gray scales." *American Cartographer* 12, no. 2:132–142.

Koláčný, A. (1969) "Cartographic information: A fundamental concept and term in modern cartography." *Cartographic Journal* 6:47–49.

Kosslyn, S. M. (1994) *Elements of Graph Design.* New York: W. H. Freeman.

Koussoulakou, A. (1994) "Spatial-temporal analysis of urban air pollution." In *Visualization in Modern Cartography,* ed. A. M. MacEachren and D. R. F. Taylor, pp. 243–267. Oxford: Pergamon Press.

———, and Kraak, M. J. (1992) "Spatio-temporal maps and cartographic communication." *Cartographic Journal* 29, no. 2:101–108.

Kraak, Menno-J. (1988) *Computer-assisted Cartographical Three-dimensional Imaging Techniques.* Delft, The Netherlands: Delft University Press.

———, and van Driel, R. (1997) "Principles of hypermaps." *Computers & Geosciences* 23, no. 4:457–464.

Krygier, J. (1993) "Sound and cartographic design." Videotape. University Park, Penn.: Deasy GeoGraphics Laboratory, Pennsylvania State University.

——— (1994) "Sound and geographic visualization." In *Visualization in Modern Cartography,* ed. A.M. MacEachren and D. R. F. Taylor, pp. 149–166. Oxford: Pergamon Press.

——— Reeves, C., DiBiase, D., and Cupp, J. (1997) "Design, implementation and evaluation of multimedia resources for geography and earth science education." *Journal of Geography in Higher Education* 21:17–38.

Kumler, M. P. (1994) "An intensive comparison of triangulated irregular networks (TINs) and digital elevation models (DEMs)." *Cartographica* 31, no. 2:1–99.

———, and Groop, R. E. (1990) "Continuous-tone mapping of smooth surfaces." *Cartography and Geographic Information Systems* 17, no. 4:279–289.

Lafreniere, M., Pazner, M., and Mateo, J. (1996) "Iconizing the GIS image." *Proceedings GIS/LIS,* Denver, Colo., pp. 591–606.

Lam, N., Siu-N. (1983) "Spatial interpolation methods: A review." *Cartography and Geographic Information Systems* 10, no. 2:129–149.

Langford, M., and Unwin, D. J. (1994) "Generating and mapping population density surfaces within a geographical information system." *Cartographic Journal* 31, no. 1: 21–26.

Lavin, S. (1986) "Mapping continuous geographical distributions using dot-density shading." *American Cartographer* 13, no. 2:140–150.

———, and Archer, J. C. (1984) "Computer-produced unclassed bivariate choropleth maps." *American Cartographer* 11, no. 1:49–57.

———, and Cerveny, R. S. (1987) "Unit-vector density mapping." *Cartographic Journal* 24, no. 2:131–141.

Lee, J., Chen, L., and Shaw, Shih-L. (1994) "A method for the exploratory analysis of airline networks." *Professional Geographer* 46, no. 4:468–477.

Leonard, J. J., and Buttenfield, B. P. (1989) "An equal value gray scale for laser printer mapping." *American Cartographer* 16, no. 2:97–107.

Lewis, P. (1992) "Introducing a cartographic masterpiece: A review of the U.S. Geological Survey's digital terrain map of the United States." *Annals, Association of American Geographers* 82, no. 2:289–304.

Liggett, R. S., and Jepson, W. H. (1995) "An integrated environment for urban simulation." *Environment and Planning B: Planning and Design* 22, no. 3:291–302.

Limp, W. F. (1997) "Weave maps across the web." *GIS World* 10, no. 9:46–55.

Lindberg, M. B. (1987) "Dot map similarity: Visual and quantitative." Unpublished Ph.D. dissertation, University of Kansas, Lawrence, Kan.

——— (1990) "Fisher: A Turbo Pascal unit for optimal partitions." *Computers & Geosciences* 16, no. 5:717–732.

Lindenberg, R. E. (1986) "The effect of color on quantitative map symbol estimation." Unpublished Ph.D. dissertation, University of Kansas, Lawrence, Kan.

Lloyd, R. E., and Steinke, T. R. (1976) "The decisionmaking process for judging the similarity of choropleth maps." *American Cartographer* 3, no. 2:177–184.

——— (1977) "Visual and statistical comparison of choropleth maps." *Annals, Association of American Geographers* 67, no. 3:429–436.

MacDougall, E. B. (1992) "Exploratory analysis, dynamic statistical visualization, and geographic information systems." *Cartography and Geographic Information Systems* 19, no. 4:237–246.

MacEachren, A. M. (1982a) "Map complexity: Comparison and measurement." *American Cartographer* 9, no. 1:31–46.

——— (1982b) "The role of complexity and symbolization method in thematic map effectiveness." *Annals, Association of American Geographers* 72, no. 4:495–513.

——— (1992) "Visualizing uncertain information." *Cartographic Perspectives* 13:10–19.

——— (1994a) *Some Truth with Maps: A Primer on Symbolization and Design.* Washington, D.C.: Association of American Geographers.

——— (1994b) "Visualization in modern cartography: Setting the agenda." In *Visualization in Modern Cartography,* ed. A. M. MacEachren and D. R. F. Taylor, pp. 1–12. Oxford: Pergamon Press.

——— (1995) *How Maps Work: Representation, Visualization, and Design.* New York: Guilford Press.

———, and DiBiase, D. (1991) "Animated maps of aggregate data: Conceptual and practical problems." *Cartography and Geographic Information Systems* 18, no. 4:221–229.

———, and Kraak, Menno-J. (1997) "Exploratory cartographic visualization: Advancing the agenda." *Computers & Geosciences* 23, no. 4:335–343.

———, Buttenfield, B. P., Campbell, J. B., DiBiase, D. W., and Monmonier, M. (1992) "Visualization." In *Geography's Inner Worlds: Pervasive Themes in Contemporary American Geography,* ed. R. F. Abler, M. G. Marcus, and J. M. Olson, pp. 99–137. New Brunswick, N.J.: Rutgers University Press.

Mak, K., and Coulson, M. R. C. (1991) "Map-user response to computer-generated choropleth maps: Comparative experiments in classification and symbolization." *Cartography and Geographic Information Systems* 18, no. 2:109–124.

Malczewski, J., Pazner, M., and Zaliwska, M. (1997) "Visualization of multicriteria location analysis using raster GIS: A case study." *Cartography and Geographic Information Systems* 24, no. 2:80–90.

Marble, D. F., Gou, Z., Liu, L., and Saunders, J. (1997) "Recent advances in the exploratory analysis of interregional flows in space and time." In *Innovations in GIS 4,* ed. Z. Kemp, pp. 75–88. Bristol, Penn.: Taylor & Francis.

Martin, D. (1989) "Mapping population data from zone centroid locations." *Transactions, Institute of British Geographers (new series)* 14, no. 1:90–97.

——— (1996) "An assessment of surface and zonal models of population." *International Journal of Geographical Information Systems* 10, no. 8:973–989.

Mattson, M. (1990) "Imagesetting in desktop mapping." *Cartographic Perspectives* 6:13–22.

McCleary, G. F. J. (1969) "The dasymetric method in thematic cartography." Unpublished Ph.D. dissertation, University of Wisconsin, Madison, Wisc.

——— (1983) "An effective graphic 'vocabulary'." *IEEE Computer Graphics & Applications* 3, no. 2:46–53.

McCormick, B. H., DeFanti, T. A., and Brown, M. D. (1987) "Visualization in scientific computing." *Computer Graphics* 21, no. 6.

McCullagh, M. J. (1988) "Terrain and surface modelling systems: Theory and practice." *Photogrammetric Record* 12, no. 72:747–779.

McGranaghan, M. (1989) "Ordering choropleth maps symbols: The effect of background." *American Cartographer* 16, no. 4:279–285.

——— (1993) "A cartographic view of spatial data quality." *Cartographica* 30, no. 2/3:8–19.

——— (1996) "An experiment with choropleth maps on a monochrome LCD panel." In *Cartographic Design: Theoretical and Practical Perspectives,* ed. C. H. Wood and C. P. Keller, pp. 177–190. Chichester, England: John Wiley & Sons.

McManus, I. C., Jones, A. L., and Cottrell, J. (1981) "The aesthetics of color." *Perception* 10, no. 6:651–666.

Meihoefer, Hans-J. (1969) "The utility of the circle as an effective cartographic symbol." *The Canadian Cartographer* 6, no. 2:105–117.

Mello, J. P. J. (1993) "Printers in transition." *Byte* 18, no. 13:94–98.

Mersey, J. E. (1980) "An analysis of two-variable choropleth maps." Unpublished M.S. thesis, University of Wisconsin, Madison, Wisc.

——— (1990) "Colour and thematic map design: The role of colour scheme and map complexity in choropleth map communication." *Cartographica* 27, no. 3:1–157.

Meyer, M. A., Broome, F. R., and Schweitzer, R. H. J. (1975) "Color statistical mapping by the U.S. Bureau of the Census." *American Cartographer* 2, no. 5:100–117.

Miller, D. W. (1988) "The great American history machine." *Academic Computing* 3, no. 3:28–29, 43, 46–47, 50.

Miller, E. J. (1997) "Towards a 4D GIS: Four-dimensional interpolation utilizing kriging." In *Innovations in GIS 4,* ed. Z. Kemp, pp. 181–197. Bristol, Penn.: Taylor & Francis.

Mitas, L., Brown, W. M., and Mitasova, H. (1997) "Role of dynamic cartography in simulations of landscape processes based on multivariate fields." *Computers & Geosciences* 23, no. 4:437–446.

Moellering, H. (1976) "The potential uses of a computer animated film in the analysis of geographical patterns of traffic crashes." *Accident Analysis & Prevention* 8:215–227.

——— (1978) "A demonstration of the real time display of three dimensional cartographic objects." Videotape, Department of Geography, Ohio State University, Columbus, Ohio.

——— (1980) "The real-time animation of three-dimensional maps." *American Cartographer* 7, no. 1:67–75.

———, and Kimerling, A. J. (1990) "A new digital slope-aspect display process." *Cartography and Geographic Information Systems* 17, no. 2:151–159.

———, and Rayner, J. N. (1982) "The dual axis fourier shape analysis of closed cartographic forms." *The Cartographic Journal* 19, no. 1:53–59.

Monmonier, M. S. (1975) "Class intervals to enhance the visual correlation of choroplethic maps." *Canadian Cartographer* 12, no. 2:161–178.

——— (1976) "Modifying objective functions and constraints for maximizing visual correspondence of choroplethic maps." *Canadian Cartographer* 13, no. 1:21–34.

——— (1977) "Regression-based scaling to facilitate the cross-correlation of graduated circle maps." *Cartographic Journal* 14, no. 2:89–98.

——— (1978a) "Modifications of the choropleth technique to communicate correlation." *International Yearbook of Cartography* 18:143–157.

——— (1978b) "Viewing azimuth and map clarity." *Annals, Association of American Geographers* 68, no. 2:180–195.

——— (1979) "Modelling the effect of reproduction noise on continuous-tone area symbols." *Cartographic Journal* 16, no. 2:86–96.

——— (1980) "The hopeless pursuit of purification in cartographic communication: A comparison of graphic-arts and perceptual distortions of graytone." *Cartographica* 17, no. 1:24–39.

——— (1982) "Flat laxity, optimization, and rounding in the selection of class intervals." *Cartographica* 19, no. 1:16–27.

——— (1985) *Technological Transition in Cartography.* Madison, Wisc.: University of Wisconsin Press.

——— (1989a) "Geographic brushing: Enhancing exploratory analysis of the scatterplot matrix." *Geographical Analysis* 21, no. 1:81–84.

——— (1989b) *Maps with the News: The Development of American Journalistic Cartography.* Chicago, Ill.: University of Chicago Press.

——— (1990) "Strategies for the visualization of geographic time-series data." *Cartographica* 27, no. 1:30–45.

——— (1992a) "Authoring graphic scripts: Experiences and principles." *Cartography and Geographic Information Systems* 19, no. 4:247–260, 272.

——— (1992b) "Summary graphics for integrated visualization in dynamic cartography." *Cartography and Geographic Information Systems* 19, no. 1:23–36.

——— (1993) *Mapping It Out: Expository Cartography for the Humanities and Social Sciences.* Chicago, Ill.: University of Chicago Press.

——— (1994a) "Graphic narratives for analyzing environmental risks." In *Visualization in Modern Cartography,* ed. A. M. MacEachren and D. R. F. Taylor, pp. 201–213. Oxford: Pergamon Press.

——— (1994b) "Minimum-change categories for dynamic temporal choropleth maps." *Journal of the Pennsylvania Academy of Science* 68, no. 1:42–47.

——— (1996) "Temporal generalization for dynamic maps." *Cartography and Geographic Information Systems* 23, no. 2:96–98.

———, and Gluck, M. (1994) "Focus groups for design improvement in dynamic cartography." *Cartography and Geographic Information Systems* 21, no. 1:37–47.

———, and Schnell, G. A. (1984) "Land use and land cover data and the mapping of population density." *International Yearbook of Cartography* 24:115–121.

Moran, C. J., and Vezina, G. (1993) "Visualizing soil surfaces and crop residues." *IEEE Computer Graphics & Applications* 13, no. 2:40–47.

Morrill, R. L. (1981) *Political Redistricting and Geographic Theory.* Washington, D.C.: Association of American Geographers.

Morrison, J. L. (1984) "Applied cartographic communication: Map symbolization for atlases." *Cartographica* 21, no. 1:44–84.

——— (1995) "Spatial data quality." In *Elements of spatial data quality,* ed. S. C. Guptill and J. L. Morrison, pp. 1–12. Oxford, England: Elsevier Science.

Mower, J. E. (1993) "Automated feature and name placement on parallel computers." *Cartography and Geographic Information Systems* 20, no. 2:69–82.

Muehrcke, P. C., and Muehrcke, J. O. (1992) *Map Use: Reading, Analysis, and Interpretation.* 3d ed. Madison, Wisc.: JP Publications.

Muller, J. C. (1974) "Mathematical and statistical comparisons in choropleth mapping." Unpublished Ph.D. dissertation, University of Kansas, Lawrence, Kan.

——— (1979) "Perception of continuously shaded maps." *Annals, Association of American Geographers* 69, no. 2:240–249.

——— (1980a) "Perception of continuously shaded maps: Comment in reply." *Annals, Association of American Geographers* 70, no. 1:107–108.

——— (1980b) "Visual comparison of continuously shaded maps." *Cartographica* 17, no. 1:40–51.

Mulugeta, G. (1996) "Manual and automated interpolation of climatic and geomorphic statistical surfaces: An evaluation." *Annals, Association of American Geographers* 86, no. 2:324–342.

National Center for Health Statistics (1995) *Vital Statistics of the United States, 1992,* vol. 1: *Natality.* Washington, D.C.: Public Health Service.

National Institute of Standards and Technology (1994) "Federal Information Processing Standard 173 (Spatial Data Transfer Standard, Part 1, Version 1.1)." Washington, D.C.: U.S. Department of Commerce.

Nelson, E. S., and Gilmartin, P. (1996) "An evaluation of multivariate quantitative point symbols for maps." In *Cartographic Design: Theoretical and Practical Perspectives,* ed. C. H. Wood and C. P. Keller, pp. 191–210. Chichester, England: John Wiley & Sons.

Neves, N., Silva, J. P., Goncalves, P., Muchaxo, J., Silva, J., and Camara, A. (1997) "Cognitive spaces and metaphors: A solution for interacting with spatial data." *Computers & Geosciences* 23, no. 4:483–488.

Nielson, G. M., and Hamann, B. (1990) "Techniques for the interactive visualization of volumetric data." *Proceedings Visualization '90,* San Francisco, Calif., pp. 45–50.

Odland, J. (1988) *Spatial Autocorrelation.* Newbury Park, Calif.: Sage Publications.

Olea, R. A. (1992) "Kriging: Understanding allays intimidation." *Geobyte* 7, no. 5:12–17.

——— (1994) "Fundamentals of semivariogram estimation, modeling, and usage." In *Stochastic Modeling and Geostatistics,* ed. J. M. Yarus and R. L. Chambers, pp. 27–35, Tulsa, Okla.: American Association of Petroleum Geologists.

Oliver, M. A., and Webster, R. (1990) "Kriging: A method of interpolation for geographical information systems." *International Journal of Geographical Information Systems* 4, no. 3:313–332.

Olson, J. (1972a) "Class interval systems on maps of observed correlated distributions." *Canadian Cartographer* 9, no. 2:122–131.

——— (1972b) "The effects of class interval systems on choropleth map correlation." *Canadian Cartographer* 9, no. 1:44–49.

——— (1975a) "Experience and the improvement of cartographic communication." *Cartographic Journal* 12, no. 2:94–108.

——— (1975b) "The organization of color on two-variable maps." *Auto-Carto 2 Proceedings,* pp. 289–294, 251, 264–266.

——— (1976a) "A coordinated approach to map communication improvement." *American Cartographer* 3, no. 2:151–159.

——— (1976b) "Noncontiguous area cartograms." *Professional Geographer* 28, no. 4:371–380.

——— (1977) "Rescaling dot maps for pattern enhancement." *International Yearbook of Cartography* 17:125–136.

——— (1978) "Graduated circles." *Cartographic Journal* 15, no. 2:105.

——— (1981) "Spectrally encoded two-variable maps." *Annals, Association of American Geographers* 71, no. 2:259–276.

———, and Brewer, C. A. (1997) "An evaluation of color selections to accommodate map users with color-vision impairments." *Annals, Association of American Geographers* 87, no. 1:103–134.

Openshaw, S., and Perrée, T. (1996) "User-centered intelligent spatial analysis of point data." In *Innovations in GIS 3,* ed. D. Parker, pp. 119–134. Bristol, Penn.: Taylor & Francis.

———, Wymer, C., and Charlton, M. (1986) "A geographical information and mapping system for the BBC Domesday optical discs." *Transactions, Institute of British Geographers (new series)* 11, no. 3:296–304.

———, Waugh, D., and Cross, A. (1993) "Some ideas about the use of map animation as a spatial analysis tool." In *Visualization in Geographical Information,* ed. H. M. Hearnshaw and D. J. Unwin, pp. 131–138. Chichester, England: John Wiley & Sons.

Ormeling, F. (1995) "New forms, concepts, and structures for European national atlases." *Cartographic Perspectives* 20:12–20.

Palmer, J. J. N. (1986) "Computerizing Domesday Book." *Transactions, Institute of British Geographers (new series)* 11, no. 3:279–289.

Parks, M. J. (1987) "American flow mapping: A survey of the flow maps found in twentieth century geography textbooks, including a classification of the various flow map designs." Unpublished M.A. thesis, Georgia State University, Atlanta.

Patton, D. K., and Cammack, R. G. (1996) "An examination of the effects of task type and map complexity on sequenced and static choropleth maps." In *Cartographic Design: Theoretical and Practical Perspectives,* ed. C. H. Wood and C. P. Keller, pp. 237–252. Chichester, England: John Wiley & Sons.

Patton, J. C., and Slocum, T. A. (1985) "Spatial pattern recall: An analysis of the aesthetic use of color." *Cartographica* 22, no. 3:70–87.

Pazner, M. I., and Lafreniere, M. J. (1997) "GIS icon maps." *Auto-Carto 13 Proceedings,* Seattle, Wash., pp. 126–135.

Peddie, J. (1994) *High-resolution Graphics Display Systems.* New York: Windcrest/McGraw–Hill.

Peterson, M. P. (1979) "An evaluation of unclassed crossed-line choropleth mapping." *American Cartographer* 6, no. 1:21–37.

——— (1980) "Unclassed choropleth maps: A reply." *American Cartographer* 7, no. 1:80–81.

——— (1992) "Creating unclassed choropleth maps with PostScript." *Cartographic Perspectives* no. 12:4–6.

——— (1993) "Interactive cartographic animation." *Cartography and Geographic Information Systems* 20, no. 1:40–44.

——— (1995) *Interactive and Animated Cartography.* Englewood Cliffs, N.J.: Prentice-Hall.

——— (1997) "Cartography and the Internet: Introduction and research agenda." *Cartographic Perspectives* 26:3–12.

Pickle, L. W., and Herrmann, D. J. (eds.)(1995) *Cognitive Aspects of Statistical Mapping (Cognitive Methods Staff, working paper series 18).* Hyattsville, Md.: National Center for Health Statistics.

———, Mungiole, M., Jones, G., and White, A. A. (1996) *Atlas of United States Mortality.* Hyattsville, Md.: National Center for Health Statistics.

Powell, D. S., Faulkner, J. L., Darr, D. R., Zhu, Z., and MacCleery, D. W. (1992) "Forest resources of the United States, 1992." General Technical Report RM-234 (rev.). Washington, D.C.: U.S. Department of Agriculture.

Raisz, E. (1931) "The physiographic method of representing scenery on maps." *Geographical Review* 21, no. 2:297–304.

——— (1967) Landforms of the United States. Map, scale approximately 1:4,500,000.

"Research chases causes of crime: Chicago youths focus of study." *The State,* November 7, 1994, p. A3.

Rheingans, P. (1992) "Color, change, and control for quantitative data display." *Proceedings, Visualization '92,* Boston, Mass., pp. 252–259.

Rhind, D., and Mounsey, H. (1986) "The land and people of Britain: A Domesday record, 1986." *Transactions, Institute of British Geographers (new series)* 11, no. 3:315–325.

Rice, K. W. (1989) "The influence of verbal labels on the perception of graduated circle map regions." Unpublished Ph.D. dissertation, University of Kansas, Lawrence, Kan.

Rittschof, K. A., Stock, W. A., Kulhavy, R. W., Verdi, M. P., and Johnson, J. T. (1996) "Learning from cartograms: The effects of region familiarity." *Journal of Geography* 95, no. 2:50–58.

Robertson, P. K. (1991) "A methodology for choosing data representations." *IEEE Computer Graphics & Applications* 11, no. 3:56–67.

Robeson, S. M. (1997) "Spherical methods for spatial interpolation: Review and evaluation." *Cartography and Geographic Information Systems* 24, no. 1:3–20.

Robinson, A. H. (1956) "The necessity of weighting values in correlation analysis of areal data." *Annals, Association of American Geographers* 46, no. 2:233–236.

——— (1982) *Early Thematic Mapping in the History of Cartography.* Chicago, Ill.: University of Chicago Press.

———, Sale, R. D., Morrison, J. L., and Muehrcke, P. C. (1984) *Elements of Cartography.* 5th ed. New York: John Wiley & Sons.

———, Morrison, J. L., Muehrcke, P. C., Kimerling, A. J., and Guptill, S. C. (1995) *Elements of Cartography.* 6th ed. New York: John Wiley & Sons.

Rogers, J. E., and Groop, R. E. (1981) "Regional portrayal with multi-pattern color dot maps." *Cartographica* 18, no. 4:51–64.

Romesburg, H. C. (1984) *Cluster Analysis for Researchers.* Belmont, Calif.: Lifetime Learning Publications.

Rowe, J. (1997) "Is it real or data visualization—Windows-based packages bring terrain data to life." *GIS World* 10, no. 11:46–50, 52.

Rowles, R. A. (1991) "Regions and regional patterns on choropleth maps." Unpublished Ph.D. dissertation, University of Kentucky, Lexington, Ky.

Rundstrom, R. A. E. (1993) "Introducing Cultural and Social Cartography." *Cartographica,* 30, no. 1 (entire issue).

Ryden, K. (1987) "Environmental Systems Research Institute Mapping." *American Cartographer* 14, no. 3:261–263.

Rystedt, B. (1995) "Current trends in electronic atlas production." *Cartographic Perspectives* 20:5–11.

Sacco, W., Copes, W., Sloyer, C., and Stark, R. (1987) *Glyphs: Getting the Picture.* Providence, R.I.: Janson Publications.

Sadahiro, Y. (1995) "Size of map labels used in GIS and loss of literal information." *Cartographica* 32, no. 4:29–41.

Sampson, R. J. (1978) *SURFACE II Graphics System.* Lawrence, Kan.: Kansas Geological Survey.

Scott, L. M. (1994) "Identification of GIS attribute error

using exploratory data analysis." *Professional Geographer* 46, no. 3:378–386.

Sheppard, E., and Poiker, T. E. (1995) "GIS & Society." *Cartography and Geographic Information Systems* 22, no. 1:entire issue.

Shiffer, M. J. (1993a) "Augmenting geographic information with collaborative multimedia technologies." *Auto-Carto 11 Proceedings,* Minneapolis, Minn., pp. 367–376.

——— (1993b) "Implementing multimedia collaborative planning technologies." *URISA Proceedings,* pp. 86–97.

——— (1995) "Environmental review with hypermedia systems." *Environment and Planning B: Planning and Design* 22, no. 3:359–372.

Siekierska, E. M., and Taylor, D. R. F. (1991) "Electronic mapping and electronic atlases: New cartographic products for the information era. The electronic atlas of Canada." *CISM Journal ACSGC* 45, no. 1:11–21.

Slocum, T. A. (1983) "Predicting visual clusters on graduated circle maps." *American Cartographer* 10, no. 1:59–72.

———, and Egbert, S. L. (1993) "Knowledge acquisition from choropleth maps." *Cartography and Geographic Information Systems* 20, no. 2:83–95.

———, and McMaster, R. B. (1986) "Gray tone versus line plotter area symbols: A matching experiment." *American Cartographer* 13, no. 2:151–164.

———, and Yoder, S. C. (1996) "Using Visual Basic to teach programming for geographers." *Journal of Geography* 95, no. 5:194–199.

———, Robeson, S. H., and Egbert, S. L. (1990) "Traditional versus sequenced choropleth maps: An experimental investigation." *Cartographica* 27, no. 1:67–88.

———, Armstrong, M. P., Bishop, I., Carron, J., Dungan, J., Egbert, S. L., Knapp, L., Okazaki, D., Rhyne, T. M., Rokos, Demetrius-K. D., Ruggles, A. J., and Weber, C. R. (1994) "Visualization software tools." In *Visualization in Modern Cartography,* ed. A. M. MacEachren and D. R. F. Taylor, pp. 91–122. Oxford: Pergamon Press.

Smart, J., Mason, M., and Corrie, G. (1991) "Assessing the visual impact of development plans." In *GIS Applications in Natural Resources,* ed. M. Heit and A. Shortreid, pp. 295–303. Fort Collins, Colo.: GIS World.

Smith, Guy-H. (1928) "A population map of Ohio for 1920." *Geographical Review* 18, no. 3:422–427.

Smith, R. D. (1987) "Electronic Atlas of Arkansas: Design and operational considerations." *Proceedings of the 13th International Cartographic Conference,* vol 4, Morelia, Mexico, pp. 161–167.

Smith, R. M., and Parker, T. (1995) "An electronic atlas authoring system." *Cartographic Perspectives* 20:35–39.

Smith, S., Grinstein, G., and Pickett, R. (1991) "Global geometric, sound and color controls for iconographic displays of scientific data." Extracting Meaning from Complex Data: Processing, Display, Interaction II, *Proceedings, SPIE,* vol. 1459, pp. 192–206.

South Carolina State Budget and Control Board (1994) "South Carolina Statistical Abstract." Columbia, S.C.: Office of Research and Statistical Services.

Steenblik, R. A. (1987) "The chromostereoscopic process: A novel single image stereoscopic process." True Three-Dimensional Imaging Techniques and Display Technologies, *Proceedings, SPIE,* vol. 761, pp. 27–34.

Stevens, J. (1996) *Applied Multivariate Statistics for the Social Sciences.* 3d ed. Mahwah, N.J.: Lawrence Erlbaum Associates.

Stevens, S. S., and Galanter, E. H. (1957) "Ratio scales and category scales for a dozen perceptual continua." *Journal of Experimental Psychology* 54, no. 6:377–411.

Sturges, J., and Whitfield, T. W. A. (1995) "Locating basic colours in the Munsell space." *Color Research and Application* 20, no. 6:364–376.

Sugihara, M. (1995) "Consistent Color." *Byte* 20, no. 1:93–94, 96, 98–99.

Szego, J. (1987) *Human Cartography: Mapping the World of Man.* Stockholm, Sweden: Swedish Council for Building Research.

Tajima, J. (1983) "Uniform color scale applications to computer graphics." *Computer Vision, Graphics, and Image Processing* 21, no. 3:305–325.

Tanaka, K. (1950) "The relief contour method of representing topography on maps." *Geographical Review* 40, no. 3:444–456.

Tang, H., Bishop, I. D., and Yates, P. M. (1997) "Towards an integrated interactive visualization system for forest management." *Resource Technology '97,* Beijing, China.

Tang, Q. (1992) "From description to analysis: An electronic atlas for spatial data exploration." Proceedings of ASPRS/ACSM/RT 92 Convention on "Mapping and monitoring global change," vol. 3, pp. 455–463.

Taylor, D. R. F. (1987) "Cartographic communication on computer screens: The effect of sequential presentation of map information." *Proceedings of the 13th International Cartographic Conference,* Morelia, Mexico, vol. 1, pp. 591–611.

Taylor, J., Tabayoyon, A., and Rowell, J. (1991) "Device-independent color matching you can buy now." *Information Display* 7, no. 4/5:20–22, 49.

Thelin, G. P. and Pike, R. J. (1991) Landforms of the conterminous United States: A digital shaded-relief portrayal. Washington, D.C.: U.S. Geological Survey. Map I-2206. Scale 1:3,500,000.

Thomas, E. N., and Anderson, D. L. (1965) "Additional comments on weighting values in correlation analysis of areal data." *Annals, Association of American Geographers* 55, no. 3:492–505.

Thompson, W., and Lavin, S. (1996) "Automatic generation of animated migration maps." *Cartographica* 33, no. 2:17–28.

Thrower, N. J. W. (1959) "Animated cartography." *Professional Geographer* 11, no. 6:9–12.

——— (1961) "Animated cartography in the United States." *International Yearbook of Cartography* 20–29.

Tikunov, V. S. (1988) "Anamorphated cartographic images: Historical outline and construction techniques." *Cartography* 17, no. 1:1–8.

Tobler, W. R. (1973) "Choropleth maps without class intervals?" *Geographical Analysis* 5, no. 3:262–265.

——— (1979) "Smooth pycnophylactic interpolation for geographical regions." *Journal of the American Statistical Association* 74, no. 367:519–536.

——— (1981) "A model of geographical movement." *Geographical Analysis* 13, no. 1:1–20.

——— (1987) "Experiments in migration mapping by computer." *American Cartographer* 14, no. 2:155–163.

———, Deichmann, U., Gottsegen, J., and Maloy, K. (1997) "World population in a grid of spherical quadrilaterals." *International Journal of Population Geography* 3:203–225.

Travis, D. (1991) *Effective Color Displays: Theory and Practice.* London: Academic Press.

Treinish, L. A. (1992) "Climatology of global stratospheric ozone." Videotape, IBM Corporation.

——— (1993) "Visualization techniques for correlative data analysis in the earth and space sciences." In *Animation and Scientific Visualization: Tools and Applications,* ed. R. A. Earnshaw and D. Watson, pp. 193–204. London: Academic Press.

——— (1994) "Visualizations illuminate disparate data sets in the earth sciences." *Computers in Physics* 8, no. 6:664–671.

———, and Goettsche, C. (1991) "Correlative visualization techniques for multidimensional data." *IBM Journal of Research and Development* 35, no. 1/2:184–204.

Trifonoff, K. M. (1994) "Using thematic maps in the early elementary grades." Unpublished Ph.D. dissertation, University of Kansas, Lawrence, Kan.

——— (1995) "Going beyond location: Thematic maps in the early elementary grades." *Journal of Geography* 94, no. 2:368–374.

Tufte, E. R. (1983) *The Visual Display of Quantitative Information.* Cheshire, Conn.: Graphics Press.

——— (1990) *Envisioning Information.* Chesire, Conn.: Graphics Press.

Tukey, J. W. (1977) *Exploratory Data Analysis.* Reading, Mass.: Addison-Wesley.

Turner, E. J. (1977) "The use of shape as a nominal variable on multipattern dot maps." Unpublished Ph.D. dissertation, University of Washington, Seattle, Wash.

University Consortium for Geographic Information Science (1996) "Research priorities for Geographic Information Science." *Cartography and Geographic Information Systems* 23, no. 3:115–127.

Unwin, D. (1981) *Introductory Spatial Analysis.* London: Methuen.

U.S. Bureau of the Census (1970) "Population Distribution, Urban and Rural, in the United States: 1970." United States Maps, GE-50, No. 45. Washington, D.C.

——— (1994a) *County and City Data Book: 1994.* Washington, D.C.: U.S. Government Printing Office.

——— (1994b) *Statistical Abstract of the United States.* 114th edition. Washington, D.C.

——— (1995) *Agricultural Atlas of the United States.* Washington, D.C.: U.S. Government Printing Office.

U.S. Geological Survey (1990) "Digital Elevation Models: Data User's Guide 5." Reston, Va.

———, and EROS Data Center (1989) "Northern Great Plains Greenness Mapping." Videotape.

van Roessel, J. W. (1989) "An algorithm for locating candidate labeling boxes within a polygon." *American Cartographer* 16, no. 3:201–209.

Van Voris, P., Millard, W. D., and Thomas, J. (1993) "TERRA–Vision: The integration of scientific analysis into the decision-making process." *International Journal of Geographical Information Systems* 7, no. 2:143–164.

Veregin, H., Krause, P., Pandya, R., and Roethlisberger, R. (1993) "Design and development of an interactive "geiger counter" for exploratory analysis of spatial data quality." *GIS/LIS Proceedings,* vol. 2, Minneapolis, Minn., pp. 701–710.

Veve, T. D. (1994) "An assessment of interpolation techniques for the estimation of precipitation in a tropical mountainous region." Unpublished M.A. thesis, Pennsylvania State University, University Park, Penn.

Waisel, L. (1996) "Three-dimensional visualization of sediment chemistry in the New York harbor." *Communique (Data Explorer Newsletter)* 4, no. 1:1–3.

Waldorf, S. P. (1995) "Commercial cartography: Custom design and production." *Cartography and Geographic Information Systems* 22, no. 2:168–174.

Wang, P. C. C. (ed.) (1978) *Graphical Representation of Multivariate Data.* New York: Academic Press.

———, and Lake, L. T. G. E. (1978) "Application of graphical multivariate techniques in policy sciences." In *Graphical Representation of Multivariate Data,* ed. P. C. C. Wang, pp. 13–58. New York: Academic Press.

Wang, Z., and Ormeling, F. (1996) "The representation of quantitative and ordinal information." *Cartographic Journal* 33, no. 2:87–91.

Ware, J. M., and Jones, C. B. (1997) "A multiresolution data storage scheme for 3D GIS." In *Innovations in GIS 4,* ed. Z. Kemp, pp. 9–24. Bristol, Penn.: Taylor & Francis.

Weber, C. R. (1993) "Sonic enhancement of map information: Experiments using harmonic intervals." Unpublished Ph.D. dissertation, State University of New York, Buffalo, N.Y.

———, and Buttenfield, B. P. (1993) "A cartographic animation of average yearly surface temperatures for the 48 contiguous United States: 1897–1986." *Cartography and Geographic Information Systems* 20, no. 3:141–150.

———, and Yuan, M. (1993) "Cartographic sonification: An analysis of subject association of harmonic intervals with various cartographic adjectives." *Technical Papers,* vol. 1, ACSM/ASPRS Annual Meeting, New Orleans, La., pp. 391–400.

Weiss, M. (1995) "Final output." *Byte* 20, no. 1:109–110, 112, 114–115.

Whistler, J. L., Egbert, S. L., Jakubauskas, M. E., Martinko, E. A., Baumgartner, D. W., and Lee, Re-Y. (1995) "The Kansas State Land Cover Mapping Project: Regional scale land use/land cover mapping using Landsat Thematic Mapper data." *Technical Papers,* vol. 3, 1995 ACSM/ASPRS Annual Convention and Exposition, Charlotte, N.C., pp. 773–785.

White, D. (1985) "Relief modulated thematic mapping by computer." *American Cartographer* 12, no. 1:62–68.

White, M., and Rosenquist, E. (1995) "Following the white bronco: Where were the TV maps?" *Cartography and Geographic Information Systems* 22, no. 2:163–167.

Whitehouse, K. (1996) "VRML adds a new dimension to web browsing." *IEEE Computer Graphics & Applications* 16, no. 4:7–9.

Wilhelmson, R. B., Jewett, B. F., Shaw, C., Wicker, L. J., Arrott, M., Bushell, C. B., Bajuk, M., Thingvold, J., and Yost, J. B. (1990) "A study of a numerically modeled storm." *International Journal of Supercomputer Applications* 4, no. 2:20–36.

Wilkie, R. W., and Tager, J. (eds.)(1991) *Historical Atlas of Massachusetts.* Amherst: University of Massachusetts Press.

Wood, D., and Fels, J. (1986) "Designs on signs: Myth and meaning in maps." *Cartographica* 23, no. 3:54–103.

——— (1992) *The Power of Maps.* New York: Guilford Press.

Wright, J. K. (1936) "A method of mapping densities of population: With Cape Cod as an example." *Geographical Review* 26, no. 1:103–110.

Wyszecki, G., and Stiles, W. S. (1982) *Color Science: Concepts and Methods, Quantitative Data and Formulae.* 2d ed. New York: John Wiley & Sons.

Yadav-Pauletti, S. (1996) "MIGMAP, a data exploration application for visualizing U.S. Census migration data." Unpublished M.A. thesis, University of Kansas, Lawrence, Kan.

Yoder, S. C. (1996) "The development of software for exploring temporal data associated with point locations." Unpublished M.A. thesis, University of Kansas, Lawrence, Kan.

Yoeli, P. (1983) "Shadowed contours with computer and plotter." *American Cartographer* 10, no. 2:101–110.

——— (1985) "Topographic relief depiction by hachures with computer and plotter." *Cartographic Journal* 22, no. 2:111–124.

Young, T. (1801) "On the theory of light and colours." *Philosophical Transactions of the Royal Society of London* 92:12–48.

Zeis, M. (1993) "Color Becomes Affordable." *Byte* 18, no. 3:125 ff.

Zhan, F. B., and Buttenfield, B. P. (1995) "Object-oriented knowledge-based symbol selection for visualizing statistical information." *International Journal of Geographical Information Systems* 9, no. 3:293–315.

Index

A
Abrupt phenomena, 20
Absolute deviations about class medians (ADCM), sum of, 70, 71
Abstract sounds, 248
Accommodation, 85
Acevedo, W., 228, 253
ACM SIGGRAPH (Association for Computing Machinery, Special Interest Group on Computer Graphics), 254
ADCM. *See* Absolute deviations about class medians (ADCM), sum of
Additive color, 89–90
Adobe Systems PageMaker, 95
Advanced Visualizer, 239
Aesthetics, 109
Africa, income in, 22
Age, map users, 110
AIDS data, 33, 180–181, 249
Amacrine cells, 85, 86
Anaglyphs, 154
Animated maps, 8, 13, 179, 189
  climate change, 227
  for courseware and lectureware, 227–228
  emphasizing attribute, 225
  emphasizing change, 224
  emphasizing location, 225
  graphic scripts, 222–223
  MapTime, 214–215
  nontemporal methods, 227, 253
  of ozone hole, 226
  of segmented components, 228–229
  software, 237
  of thunderstorm development, 225–226
  of time, 228
  of urban sprawl, 228
  visual variables, 224
Animator Pro, 237
Anomalous trichromats, 88
Antarctica, ozone hole over, 9, 226
Apple ColorSync, 95
Application Visualization System (AVS), 239
Archer, J. C., 199–200
ARC/INFO, 252
ArcView, 10, 211, 219–220
Area cartograms, 181
Areal data, 19, 21, 23
  visual variables effectiveness rating, 31
Arizona, county-level map of, 58
Arkansas, electronic atlas of, 9
Arrangement, 23, 24, 31
Aspens, 211, 217–218
Association of American Geographers, 232
Atlases, electronic, 9, 13
  Death Valley: An Animated Atlas, 236
  defined, 232–233
  Digital Atlas of California, 233–234
  Domesday Project, 233
  Grolier's Multimedia Encyclopedia, 236
  Interactive Atlas of Georgia, The, 234
  Microsoft Encarta Virtual Globe, 234–236
  TIGER Mapping Service, 236–237
  vs. paper, 231
  World Atlas, 234
Atlas of United States Mortality, 241, 249–250
Attributes
  accuracy of, 242
  animations emphasizing, 225
Australia, population of, 182

B
Balanced data, 22
Base information, 34
Basic, 232, 237
Bates, K., 109
Bemis, D., 109
Bertin, J., 3, 31
Berton, J. A. J., 111, 218–219
Binary map, 174
Bipolar cells, 85, 86
Bipolar data, 22, 107
Bishop, Ian, 252
Bit-plane, 90
Bits, 90
Bivariate correlation, 48–49
Bivariate mapping, 16
  choropleth maps, 195–196, 197–200
  point symbol, 200
  ray-glyph, 200
Bivariate-normal, 198
Bivariate regression, 48, 49–50
Black-and-white maps, 110
  bit-plane considerations, 90–91
  simultaneous contrast, 87
  specifications for, 111–114
  symbol overlap on, 131
  univariate maps, 197
  visual variables for, 23, 30
Boundary error, 71
Bowen, William, 234
Box plot, 51
Boyce-Clark index, 54
Brain, 88–89
Bressi, T., 251
Brewer, Cynthia, 106–107, 109–110, 111, 114–115, 163–164, 165, 199, 241, 249–251
Brown, A., 243
Buttenfield, B. P., 217–218, 227, 228

C
Cammack, R .G., 212
Campbell, C. S., 211
Canada, languages spoken in, 22
Cancer statistics, 27, 33, 56
Carr, Dan, 200, 242
Carstensen, L. W. J., 199
Cartograms, 16, 181–183, 204
Cartography and Geographic Information Systems, 232
Carto Project, 254
Cartwright, W., 253
CAVE, 252–253
Census. *See* U.S. Bureau of the Census
Central tendency, measures of, 45–46, 55
Centre for Landscape Research (CLR), 251–252
Centroid, 52
Cerveny, R. S., 186–187, 188, 189
Chang, K., 125
Change, animation emphasizing, 224–225
Change maps, 215
Chernoff faces, 204, 206
Children, map use by, 110
Choropleth maps, 25–26, 28, 57, 119, 169, 173, 174, 229, 249
  bivariate, 195–196, 197, 198
  color schemes for, 105–117
  data quality on, 242
  data standardization, 32–33
  defined, 2
  multivariate, 201
  spatial context on, 78–81
  standardized, 6
  visual variables for, 29–32
  vs. framed-rectangle maps, 178–179
ChromaDepth lenses, 108
CI. *See* Compaction index (CI)
CIE color model, 99–100, 101, 115–116
Clarke, K. C., 154, 219
Classed maps, 8, 123
Classes option, 212
Classification methods, 61–74, 195–196
Cleveland, W. S., 43, 178
Climate change, animated map of, 227
Clip art, 121
CLRview software, 251–252
CMYK system, 96, 100, 114–115
  converting to RGB, 115
CMY system, 201
Coarse shades, 113–114
Cognition, 13–14
  vs. behaviorism, 2
Cole, D., 132
Color
  aesthetics, 109
  associations, 109
  for bivariate maps, 197–200
  board, 102
  data types, 106–107
  economic limitations, 111
  gamuts, 95
  guidelines for diverging color schemes, 250–251
  hardware to reproduce, 10, 89–95, 101–102
  hues, 31
  humans visualization of, 84–89
  lookup tables, 93
  management system, 95
  to map mortality, 249–250
  map type, 107–108
  map use and, 106, 110
  models, 95–101
  in multivariate dot maps, 201, 203
  overview of, 105–106
  presentation vs. data exploration, 110–111
  printed maps, 114–115
  ramping, 115
  recent developments in use of, 241
  specification details, 111–116
  stereoscopic effect, 108
  subtractive, 94
  to symbolize aspect and slope, 162–164
  symbol overlap on, 131
  in trivariate choropleth maps, 201
  for U.S. Bureau of the Census maps, 197–198
  vision impairment, 109–110
Colorimeter, 163
Command-line interfaces, 219
Commercial map-making, 231, 234–236
Commission on Visualization, 254
Common Lisp, 252
Communication model, 1
Compaction index (CI), 52, 53
Computers. *See also* Software
  classification methods and, 74
  for map production, 89–95
  maps on, 102
  parallel, 38
  programming languages, 232, 237–238
  virtual reality and virtual realism, 251–253

Computers & Geosciences, 241, 242, 253
Conceptual points, 26, 119–120
Cones, 85, 86
Conferences, cartography, 253–254
Confidence interval, 145
Continuous phenomena, 20, 21
Continuous-tone map, 154
Continuous vector–based flows, 186–187
Contone, 94
Contour maps, 154, 228–229, 245–246
  Hurricane Hugo precipitation, 138
  Surfer software, 139, 140
  to symbolize topography, 156–157
Contrast, 87–88
Control points, 15
  correctness of estimated data, 145–146
Cook, D., 253
Cornea, 85
Correlation, bivariate, 48–49
Cost, color use and, 111
Coulson, M. R. C., 78
Counts, 25
Courseware, animated mapping for, 227–228
Cox, C. W., 125
Cox, D. J., 203, 228
C (programming language), 232, 237
Crawford, P. V., 124
Cross-hatching, 199
Cross validation, 146
CRTS, 93–94
Cuff, D. J., 106, 109
Cycling, reiterative, 71

D
D. *See* Diameter (D)
DAC. *See* Digital-to-analog converter (DAC)
Dasymetric maps, 15, 29
  ancillary variables, 169
  benefits of, 168
  manual *vs.* automated production, 169
  for population density, 173–175
Data
  classification of, 60–82
  color selection and, 106–107
  exploration, 110–111
  proportional symbol mapping, 119–121
  raw-count, 8
  reliability, 242
  sound, 248–249
  splitting, 146
  standardization, 25, 26, 27, 32–33
  standardized, 8
  true point, 145–147
  uncertainty, 242
  weighting to account for size of enumeration units, 54–55
Data classification
  classed *vs.* unclassed mapping, 75–78
  equal areas, 196
  equal intervals, 61–67
  maximum breaks, 69–70
  mean-standard deviation, 69
  natural breaks, 70
  optimal, 70–73
  quantiles, 67–69
  selecting a method, 74
  spatial context and, 78–81
Data exploration, 9
  ArcView, 219–220
  Aspens, 217–218
  data flow software (modular visualization environments), 239
  ExploreMap, 211–213
  MapTime, 214–215
  Project Argus, 213–214
  Three 3-D mapping software, 211
  Transform, 218–219
  Vis5D, 215–217
Data jack, 203–204
Data quality
  advances in, 241
  animation of dot map uncertainty, 246–247
  defined, 242–243
  mapping digital elevation model (DEM) uncertainty, 247
  visual variables for, 243–246
Davis, J. C., 143–145
Davis, T. J., 253
Deasy GeoGraphics Laboratory, 227–228
Death Valley: An Animated Atlas, 231, 236
Declerq, F. A. N., 78, 146
Deconstruction, 14
Delaunay triangles, 137–138
DEMs, 247
Dent, B. D., 127, 128, 129
Deuteranopes, 88, 109–110
Diameter (D), 72
DiBiase, David, 179–181, 206–207, 223–226, 236
Dichromats, 88
Digital Atlas of California, 232, 233–234
Digital elevation model (DEM), 160–162, 211, 247
Digital-to-analog converter (DAC), 91
Digitizing, 169
Discontinuities, handling, 146
Discrete phenomena, 19–20
Dispersion, measures of, 46–47, 55
Display date, 224
Dissemination, map, 101–102
Distance cartograms, 181
Dithering, 93
Diverging scheme, 107
Domesday Project, 231, 233
Dorling, D., 181–183, 184, 204, 228
Dot-density shading, 172–173
Dot maps, 15, 28
  ancillary variables, 169
  animation of, 246–247
  benefits of, 168
  determining regions, 169–171
  dot-density routines, 172–173
  dot size and unit value, 171–172
  manual *vs.* automated production, 169
  multivariate, 201, 203
Dots per inch (dpi), 94
Dunn, R., 178–179
Duration, 224
DX. *See* IBM Visualization Data Explorer (DX)
Dye-sublimation printers, 95
Dykes, Jason, 213–214, 253

E
Eastman, J. R., 106
Ebbinhaus illusion, 125
Educational Multimedia Resources, 227–228
EfiColor XTension, 95
Egbert, Stephen, 109, 211–213
Ehlschlaeger, C. R., 253
Ekman, G., 124
Electromagnetic spectrum, 84
Electron gun, 89
Electronic Atlas of Arkansas, 9, 232
Electronic Map Library, 234
Elements of Cartography, 106
Elements of Data Quality, 243
Emaps, 252
England and Wales, population of, 184
Environmental Protection Agency, 237
Environmental Systems Research Institute (ESRI), 238
Equal-area projection, 181
Equal-areas method of classification, 196
Equal-energy point. *See* White point
Equal intervals, 61–67, 68, 74
Equiprobability ellipse, 198
Error, 71
  margin of, 41
  random, 79
Ervin, S. M., 252
Evans, B. J., 253
Excitation, 87
Experimental studies, 75–78
Exploratory data analysis, 50–52
ExploreMap, 211–213, 234
Exvis, 206, 207
Eye, structure of, 85
Eyton, J. R., 108, 156, 157, 197–198
Eyton's illuminated contours, 157, 159

F
Fels, J., 14
Ferreira, J. J., 213
Fiddle tool, 211, 219
Figure-ground contrast, 132
Fisher, P. F., 175, 246–247, 249
Fisher-Jenks algorithm, 71–73, 74
Fishnet maps, 20, 21, 154
Flannery, J. J., 124
Flat-panel displays (FPDs), 94
Flicker, 90
Florida population, 61, 63
  dispersion graph, 65
  maps illustrating data classification, 68
Flow maps, 2, 184–187
FlowMap software, 184–185
Fluorescent ink, 108
Fly-bys, 165, 225, 237
Fonts, 36
Forced cycling, 71
Formulas, perceptual scaling, 124–126
Fortner Software, 188, 218
Fortran, 232, 237
Four-color process printing, 101
Fovea, 85
FPDs. *See* Flat-panel displays (FPDs)
Frame buffer, 90
Framed-rectangle symbols, 178–179
Freehand software, 37, 148, 171
Frequency, 224

G
GADF. *See* Goodness of absolute derivation fit (GADF)
GAHM, 220
Galanter, E. H., 112
Gamma function, 115–116
Ganglion cells, 85, 86
Geary ratio (GR), 55
General-reference maps
  defined, 1
  design elements for, 37
Geographic brush, 52

Geographic Information System (GIS), 10, 34, 168, 169, 219, 247, 252, 253
Geologists, 20
Geometric symbols, 24, 121, 122
Gerrymandering, 54
Gershon, N., 228–229
Gilmartin, P., 78, 125, 204–206
GIS. *See* Geographic Information System (GIS)
Global topography, Treinish's depiction of, 164–165
Gluck, M., 223
Goldstein, E. B., 92
Goodchild, M., 247, 249
Goode's World Atlas, 232
Goodness of absolute derivation fit (GADF), 73
Goodness of variance fit (GVF), 73
Gould, Peter, 181
Gouraud shading, 164
GR. *See* Geary ratio (GR)
Graphical user interfaces (GUIs), 219, 232, 237
Graphic display devices, 89, 115–116
Graphic scripts, 222–223
Graphs
  dispersion, 44
  point and dispersion, 43–44
Graphs & Stats option, 213
Gray, William, 217
Gray curves, 111
Gray scales, 111
  hardware limitations, 114
Gridding, 139
Griffin, T. L. C., 125, 132, 133
Grolier's Multimedia Encyclopedia, 8, 9, 236
Groop, R. E., 132, 154, 201–202, 203
Grouped-frequency table, 43
GUI. *See* Graphical user interfaces (GUI)
Guilford, J. P., 109
GVF. *See* Goodness of variance fit (GVF)

H
Hachures, 156, 157
Hamann, B., 189
Hardware
  cathode-ray tubes (CRTs), 89–93
  color use and, 10, 111
  gray scales and, 114
  liquid crystal displays (LCDs), 93–94
  Macintosh, 252
  printers, 94–95
  Silicon Graphics Iris, 217, 226
  vector *vs.* raster graphics, 89
Harley, J. B., 14
Hartnett, S., 200
Haslett, J., 220
Helmholtz, H. von, 86
Hering, E., 86
Herzog, A., 79–81
Hexagon bin plot, 48
High-definition television (HDTV), 102
Hill shading. *See* Shaded relief
Histograms, 44, 45, 46
Historical Atlas of Massachusetts, 182–183, 232
HLS color model, 163, 164
Hoinkes, R., 251–252
Horizontal cells, 85, 86
Horn, B. K. P., 157
Howard, David, 244–246
Howard, Hugh, 236
HSV color model, 96–97, 100, 164
Hubel, David, 88
Hue, 24, 31
Hunter, G. J., 247
Hurricane Hugo precipitation maps, 138
Hurvich, L. M., 87–88
HVC color model, 98–99, 100–101
Hydrographic features, 37
Hypermaps, 253
Hypermedia, 9
Hypsometric tints, 15, 154, 155

I
IBM Visualization Data Explorer (DX), 164–165, 188, 189–190, 239
Image-based visualization, 252
Imagesetter, 101
Imhof, Eduard, 37, 157
Information, acquisition of, 3, 76–77
Inhibition, 87
Ink-jet printing, 94
Inset maps, 132
Interactive Atlas of Georgia, The, 234
Interactive Data Language (IDL), 239
Interactive Portugal, 253
Interlaced, 90
Intermediate frames, 214
International Cartographic Association (ICA), 242, 254
Internet
  atlases on, 231, 232
  cartography uses, 10
  maps on, 9–10, 83, 102, 253
  writing programs for, 237
Interpolation methods, 136–152
Interquartile range, 46
Interval measurement, 20–21
Inverse-distance interpolation methods, 138, 139–141, 146
IRIS Performer, 252
Isaaks, E. H., 145, 146
Isarithmic (contour) maps, 15, 27
  color schemes for, 105–117
Isopleth map, 27–28
Isosurface, 188, 216

J
Jameson, D., 87
Java, 237
Jelinski, D. E., 217
Jenks, George, 70, 130, 201
Jenks-Caspall algorithm, 71, 74
Jepson, W. H., 251
Journals, cartography, 253–254
Jumping Colours, 108
Junge, K., 124

K
Kansas, wheat harvest in
  choropleth mapping, 25, 26, 27–28
  data classification, 61, 62, 66
  dispersion graph, 64
  illustrating visual variables, 29–30, 32
  interpolation methods, 150
  remote sensing of, 170
  standardized mapping, 27–28
Keller, C. P., 253
Keller, M. M., 11, 189
Keller, P. R., 11, 189
Kendall's rank correlations, 195
Kenya, secondary schools in, 34, 35, 36, 37
Key frames, 214
Kimerling, A. J., 112, 162–163, 164
Kodak Precision CMS, 95
Kraak, Menno-J., 253
Krige, Daniel, 141
Kriging, 150
  Hurricane Hugo precipitation, 138
  punctual, 143
  for smooth continuous phenomena, 141–145, 146
  universal, 143
Krygier, J., 248–249
Kumler, M. P., 154

L
Labeling software, 37–38
Lambertian reflector, 161–162
Lam's algorithm, 148
Landsat Thematic Mapper, 10, 169–170, 173
Lange, E., 251–252
Langford, M., 11, 173–175
Laser printing, 94–95
Lateral geniculate nucleus (LGN), 88, 89
Lavin, S., 172–173, 186–187, 188, 189, 199–200
LCDs. *See* Liquid crystal displays (LCDs)
Lectureware, animated mapping for, 227–228
Lee, J., 185, 220
Legend, 35, 130–131
Lewis, P., 159
Liggett, R. S., 251
Light, visible, 84
Lightness, 23, 24, 30, 31, 32
Limiting variables, 169
Lineage, 242
Linear data, 23
Linear-legend arrangement, 130
Linear phenomena, 19
Liquid crystal displays (LCDs), 93–94
Lloyd, R. E., 196
Location, 25
  analysis of, 52–54
  animation emphasizing, 225–226
Logical consistency, 242–243
Louisiana, cancer in, 56–57
Luminance, 163

M
MacChoro II, 227
MacDougall, E. B., 220
MacEachren, Alan, 3, 11, 57, 58, 77–78, 88, 130, 179–181, 223–226, 243–246, 253
Macromedia's Director software, 238
Magnitude estimation, 112, 125
Maine, agricultural map of, 75
Major axis approach, 50
Mak, K., 78
Map communication models, 3–7
Map complexity, 57, 58
MapObjects software, 238
Maps on Demand, 237
MapTime software, 9, 15, 132, 210, 214–215, 216, 237
MapViewer software, 123
Marlow, K. A., 163–164, 165
Massachusetts Institute of Technology, 213
Masuoka, P., 253
Mathematical scaling, 121–123, 126, 127
Mattson, M., 101–102
Maximum breaks, 64, 65, 66, 68, 69–70, 74
MC. *See* Moran coefficient (MC)
McCullagh, M. J., 146–147
McGill, R., 178

McManus, I. C., 109
Mean, 45–46
Mean center, 55
Mean-standard deviation, 64, 65, 66, 68, 69, 74, 195
Measurement, level of, 20–22
Median, 45
Meihoefer, Hans-J., 126–127, 128
Memory, 3, 13–14
Mersey, J. E., 77–78, 106
Microsoft Encarta Virtual Globe, 234–236
Microsoft ICM, 95
MigMap software, 185, 220
Migration, 184–186
Miller, D. W., 220
Mimetic sound icons, 248
Mining, 141
Minnesota, small towns in, 55
Mitas, L., 253
Mode, 45
Modeling the semivariogram, 143
Moellering, H., 162–163, 164, 211, 225
Monmonier, M. S., 102, 179, 222–223, 223
Moran coefficient (MC), 55–57
Muller, J. C., 196
Multimedia encyclopedias, 231
Multimedia presentation, 9
Multiple regression, 50
Multivariate mapping, 16
  choropleth maps, 201
  comparing, 201, 202
  dot maps, 201, 203
  symbols, 203–207
Mulugeta, G., 147–148
Munsell system, 97–99, 100–101, 109, 111–115
Murder rates, 42, 44, 45, 49, 50, 51
  histogram of, 46
Music, 249

N
National Center for Geographic Information and Analysis (NCGIA), 253
National Center for Supercomputing Applications (NCSA), 203, 252–253
Natural breaks, 64, 65, 66, 68, 70, 74
Nelson, E. S., 204–206
Nested-legend arrangement, 130
Nested means, 195–196
New Jersey, infant mortality in, 197
News maps, 102
Newspapers, 15
New York harbor, sediment in, 188, 189–190
Nielson, G. M., 189
Nominal data, visual variables effectiveness rating, 31
Noninterlaced, 90
Nontopographic phenomena, symbolizing, 154–156
Normal distribution, 44
Numerical data, 22
  visual variables effectiveness rating, 31
Numerical summaries, 44–47, 48–50

O
Object-based visualization, 252
Offset printing, 101
Ohio, population of, 122
Oil production, 22
Olson, J., 109–110, 126, 196, 197
Openshaw, S., 228
Opponent-process theory, 86–87
Optic chiasm, 88, 89
Optic nerve, 85, 88–89
Optimal interpolation, 145
Optimal (median) classification, 64, 65, 66, 68, 70–73, 74, 195
  advantages and disadvantages of, 73
  with spatial constraint, 78
Order, 224
Ordinal data, 31
Ordinal measurement, 20
Orientation, 23, 24, 30, 31
Outliers, 43
Overlap, 131–133
Overview error, 71
Ozone hole, 9, 226

P
Page description language (PDL), 94
PANTONE Personal Color Calibrator, 95, 101
Paper dissemination, 101–102
Partitioning, 112
Pascal, 232, 237
Pattern comparison, 4
Patton, D. K., 212
Pazner, Micha, 206
Peddie, J., 93–94
Perceptual scaling, 122, 123–126
Persistance, 90
Perspective block diagram, 155
Perspective height, 22–24, 30, 31
Peterson, M. P., 196, 227, 228
Phenomena
  categories of, 19–20
  vs. data, 20
Phosphors, 89–90
Photocopying, 101
Physiographic method, 159–160
Pictographic symbols, 24, 121
Pike, R. J., 11–12
Pixels, addressable, 90
Plotters, 89
Point data, 23, 55
Pointillism, 201
Point-in-polygon test, 148
Point phenomena, 19
Point symbol maps, 203–206
Political boundaries, 34
Polygonal glyph, 203
Polygonal snowflake, 203
Polygon Explorer, 220
Population density, 41
  Dasymetric mapping of, 173–175
  remote sensing to assess, 11
Positional accuracy, 242
Postmodernism, 14–15
PostScript software, 94, 101–102
Power function exponent, 124
Preattentive processing, 206
Precision Visuals' Wave Command Language (PV-Wave CL), 239
Presentation map, 75, 110–111
Primary visual cortex, 89
Principal components analysis, 50
Printers, 89, 94–95, 101
Prism maps, 20, 21, 29–30, 154
Proactive graphics, 217
Project Argus, 213–214
Proportional symbol maps, 6, 26–27
  data selection, 119–121
  Kansas wheat harvest, 28
  legend design, 130–131
  overlapping symbols, 131–133
  redundant symbols, 133–134
  scaling symbols, 121–129
  uses of, 118
Protanopes, 88, 109–110
Psychological scaling. *See* Perceptual scaling
PYCNO, 149–150
Pycnophylactic approach, 15, 148–150

Q
Quadrant approach, 138
Qualitative data, 22
Quantiles, 64, 65, 66, 67–69, 74, 195, 196
Quantitative data, 22
Quark XPress, 95
Quartiles, 67
Quasi-continuous tone method, 75
Quintiles, 67

R
Raisz, E., 159–160
Random error, 79
Range, 46
Range grading, 123, 126–130
Ratio, 21–22
Ratio estimation, 125
Ratio of Areas, 212
Raw table, 41–43
Ray-glyph maps, 203
Realistic sounds, 248
Rectangular point symbol, 200
Redistricting, 54
Redundant symbols, 133–134
Re-expression, 225
Refreshing, 90
REGARD, 220
Regression
  bivariate, 48, 49–50
  multiple, 50
Religion, 20
Remote sensing, 10–11, 168, 169–171, 173–174, 247
Reproduction, map. *See* Dissemination, map
Residual, 146
Resolution, 90, 244
Retina, 85, 86
Reviews, of atlases, 232
RGB color model, 95–96, 100, 201
  from CMYK system, 115
Road networks, 131
Robertson, P. K., 3
Robinson, A. H., 106, 131
Rods, 85, 86
Rogers, J. E., 201–202, 203
Rome Reborn, 251, 252
Rotation, 189
ROYGBIV scheme. *See* Spectral scheme
R-VIS software, 244–246

S
Sample, 41
SAT scores, 21
Saturation, 24, 31
Scaling
  proportional symbols, 121–123
  small *vs.* large-scale maps, 19
Scanning, 169
Scatterplots, 47
  brushing, 52, 53
  matrix, 47
  three-dimensional, 47

Scott, L.M., 220
Scripting languages, 238–239
SD. *See* Standard distance, SD
Semivariance, 141–143
Semivariogram, 143
Sequential scheme, 107
Serifs, 36
Service bureaus, 101–102
Shaded relief, 108, 157–162
  computation of, 160–162
Shadowed contours, 157
Shape, 23, 24, 31
Shapiro-Wilk test, 195
Shelton, E., 78
Simultaneous contrast, 87–88
Size
  map, 22, 23, 30, 31
  type, 36
Skewed distributions, 44
SLCViewer, 206–207
Slocum, T. A., 109, 211–213, 239
Slope, depicting, 157
Small multiples, 201, 202, 214, 215
Smith, P. C., 109
Smith, Richard, 9, 234
Smooth continuous phenomena, 15, 19, 20, 21, 27, 187, 226
  interpolation methods for, 136–152
  inverse distance, 139–141
  kriging, 141–145
  symbols for, 153–176
  Tobler's pycnophylactic approach, 148–150
  triangulation of, 137–139
  visual varibles and, 22–24
Smooth shades, 111–114
Snowfall, 15
Soft-copy maps, 115–116
Software
  animated mapping, 226, 227, 228, 237
  authoring, 238–239
  availability of, 210
  data exploration, 52, 210–221
  data symbolizing, 15
  dot-mapping, 172
  drawing, 148
  for electronic atlases, 234
  fly-bys, 165
  GIS, 34
  graphic design, 171
  for handling overlap, 132
  interpolation methods, 139, 140
  labelling, 37–38
  map scaling, 123
  migration mapping, 184–186
  multimedia, 231–232
  for multivariate mapping, 206–207
  programming languages, 237–238
  pycnophylactic interpolation, 149–150
  for 3-D phenomena, 188–190
  virtual reality, 251–253
  for visual data quality, 244–246
  visualization, 239
Soil maps, 32, 247
Sound
  advances in use of, 241
  to represent spatial data, 248–249
Source, 35
South Carolina
  forest cover in, 20, 21, 22
  quality of life in, 204, 205
Spacing, 22, 23, 30, 31
Spatial context, 78–81
Spatial data
  analysis of, 41–52
  autocorrelation, 55–57
  choropleth mapping, 25–26
  data standardization, 32–33
  definition of, 18
  dot mapping, 28
  geographic phenomena, 19–20
  isopleth map, 27–28
  map design elements, 33–38
  measurement levels, 20–22
  proportional symbol maps, 26–27
  visual variables, 22–25, 29–32
Spatial location
  centroid, 52
  indices for measuring shape, 52–54
Spatial pattern, 55–57
Spectral scheme, 108
Splining, 141
Spot color, 101
Springfield, Illinois, temperatures for, 218–219
Spyglass, 218
Srivastava, R. M., 145, 146
Standard deviation, 46
Standard distance (SD), 55
Standard error of the estimate, 145
Standardization, data, 119–121
Standard observer, 100
STATA, 220
States, map-making by, 231, 232, 233–234
Static maps, 177–192
Statistics, descriptive *vs.* inferential, 41
Steinke, T. R., 196
Stem-and-leaf plot, 50–52
Stereo pairs, 154
Stevens, S. S., 112
Stevens curve, 112–114, 114
Style, type, 36
Surfer software, 139, 140
Swiss Federal Institute of Technology, 251
Switzerland, students in, 79–80
Symbols
  combining different types of, 206–207
  proportional symbol mapping, 15
  redundant, 15
  for smooth continuous phenomena, 153–176
Synchronization, 224

T
Tables, 41–43, 47
Tabular error, 71
Tanaka, K., 156–158
Tanaka's method, 156–157
Tang, Q., 52, 185, 220
Taylor, J., 99
T3D software, 188
Tektronix Corporation, 98–99
Television, maps on, 83, 102
Tcl/Tk toolkit, 253
Temperature maps, 15, 21, 218, 225
Terrain, 165
Texture, 22, 25, 224
Thelin, G. P., 11–12
Thematic map, 1, 2–3
Thermal-dye printers. *See* Dye-sublimation printers
Thermal-wax transfer printers, 95
Thiessen polygons, 137–138
3-D phenomena (true), 19
  bars, 203
  difficulties with, 177
  mapping, 187–190
  mapping software, 211
  symbol, 122
  true, 16
  visual varibles and, 22–24
Thunderstorm development, animated map of, 226
TIGER mapping service, 231–232, 236–237
Time series, 224–225
Tiny cubes method, 189
Tissot's indicatrix (distortion ellipse), 243
Title, 35
Tobler, W. R., 8, 75, 79, 114, 148–150, 184–185
Topography, 2, 15
  color use, 162–164
  contour-based methods, 156–157
  hachures, 156
  methods for symbolizing, 154–156
  shaded relief, 157–162
Totals. *See* Counts
Transform, 218–219
Transparency, 244
Transparent symbols, 131–133
Treinish, L. A., 9, 164–165, 226
Trend surface, 141
Triangulation, 137–139, 146
Trichromatic theory, 85–87
Trifonoff, K. M., 110
Tristimulus values, 100
True point data, 26, 119–120
TruFlite software, 165
Tufte, E. R., 201
Tukey, John, 50–52, 225
2½ d. *See* Smooth continous phenomena
Two-dimensional symbol, 122
Type positioning, automated, 37–38
Typography, 36

U
Unclassed map, 8
Understanding, ease of, 147
Uniform color models, perceptually, 100
Unipolar data, 22
  color selection and, 107
United States, maps of
  corn/wheat growth patterns, 4
  foreign-owned agricultural land, 76, 77
  high school education, 76, 77, 113
  Landforms of the Conterminous United States, 12
  microbreweries and brewpubs, 24, 119, 121, 126, 129, 130, 132
  murder rates in, 42, 44, 45, 46, 47, 48, 49, 50, 51, 52
  out-of-wedlock births to teenagers, 120, 127, 129, 131, 134
  population density in 1990, 6
United States Geological Survey (USGS), 2, 159, 160
  Landforms of the Conterminous United States, 12
Unit value, 171–172
Unit-vector density mapping, 186, 188, 189
Univariate mapping
  continuous vector–based flows, 186–187
  Dorling's cartograms, 181–183
  flow maps, 183–187
  framed-rectangle symbols, 178–179
  geographic phenomena and chorodot map, 179–181
  true 3-D phenomena, 187–190

University Corsortium for Geographic Information Science (UCGIS), 253–254
University of California at Los Angeles (UCLA), 251
University of Illinois, 252–253
University of Melbourne (Australia), 252
University of Toronto, 251
University of Wisconsin–Madison Space Science and Engineering Center, 215–217
Unwin, D. J., 11, 173–175
Urban sprawl map, 228
U.S. Bureau of the Census, 47, 55, 197, 198, 241
U.S. Federal Information Processing Standard 173, 242

V
Valiant Vision, 108
van Driel, R., 253
van Elzakker, C. P. J. M., 243
Van Voris, P., 10
Variables
  dependent and independent, 47–48
  related, 169
Veregin, H., 249
Videocassettes, 102
Virtual Los Angeles, 251
Virtual reality, 237, 251–253
Virtual Reality Modeling Language (VRML), 237
Visda software, 52, 185, 220
Vis5D software, 215–217, 226
Vision, 109–110
  beyond the eye, 88–89
  color perception theories, 85–87
  color vision impairment, 88
  eye structure, 85
  simultaneous contrast, 87–88
  visible light, 84
Visual angle, 85
Visual Basic, 232, 237, 238
Visual hierarchy, 132
Visualization, 11–13, 13, 239
Visual pigments, 85
Visual programming languages, 237
Visual variables, 22–25
  for depicting data quality, 243–244
Volumetric phenomena, 19

W
Waisel, L., 188, 189–190
Waldorf, S. P., 101
Wavefront software, 226
Wavelength of light, 84
Weather maps, 102, 186–187, 188
  climate change, 227
  ozone hole, 226
  thunderstorm development, 226
Weather stations, 15
Weber, C. R., 217–218, 227, 228, 249
Weighted isolines, 207
Weighted mean center, 55
Weighted standard distance, 55
Whistler, J. L., 169, 170
White point, 99
Wiesel, Torsten, 89
Wiggins, L. L., 213
Wilhelmson, R. B., 226
Williams curve, 113–114
William the Conqueror, 233
Wind, 186–187
Wood, D., 14
World Atlas, 234, 235
World map, slave trade, 2
World Wide Web, 253

X
Xerography, 101
Xmap system, 213
XNV, 185–186, 220

Y
Yadav-Pauletti, S., 185, 220
Yoder, S. C., 214–215, 237
Young, T., 85–87

Z
Zero points, 21–22
Zip-a-Tone shades, 114
Zoom function, 132–133